LONDON MATHEMATICAL SOCIETY LECTURE NOTE SERIES

Managing Editor: Professor J.W.S. Cassels, Department of Pure Mathematics and Mathematical Statistics, University of Cambridge, 16 Mill Lane, Cambridge CB2 1SB, England

The books in the series listed below are available from booksellers, or, in case of difficulty, from Cambridge University Press.

34	Representation theory of Lie groups, M.F. ATIYAH *et al*
36	Homological group theory, C.T.C. WALL (ed)
39	Affine sets and affine groups, D.G. NORTHCOTT
46	p-adic analysis: a short course on recent work, N. KOBLITZ
49	Finite geometries and designs, P. CAMERON, J.W.P. HIRSCHFELD & D.R. HUGHES (eds)
50	Commutator calculus and groups of homotopy classes, H.J. BAUES
57	Techniques of geometric topology, R.A. FENN
59	Applicable differential geometry, M. CRAMPIN & F.A.E. PIRANI
66	Several complex variables and complex manifolds II, M.J. FIELD
69	Representation theory, I.M. GELFAND *et al*
74	Symmetric designs: an algebraic approach, E.S. LANDER
76	Spectral theory of linear differential operators and comparison algebras, H.O. CORDES
77	Isolated singular points on complete intersections, E.J.N. LOOIJENGA
79	Probability, statistics and analysis, J.F.C. KINGMAN & G.E.H. REUTER (eds)
80	Introduction to the representation theory of compact and locally compact groups, A. ROBERT
81	Skew fields, P.K. DRAXL
82	Surveys in combinatorics, E.K. LLOYD (ed)
83	Homogeneous structures on Riemannian manifolds, F. TRICERRI & L. VANHECKE
86	Topological topics, I.M. JAMES (ed)
87	Surveys in set theory, A.R.D. MATHIAS (ed)
88	FPF ring theory, C. FAITH & S. PAGE
89	An F-space sampler, N.J. KALTON, N.T. PECK & J.W. ROBERTS
90	Polytopes and symmetry, S.A. ROBERTSON
91	Classgroups of group rings, M.J. TAYLOR
92	Representation of rings over skew fields, A.H. SCHOFIELD
93	Aspects of topology, I.M. JAMES & E.H. KRONHEIMER (eds)
94	Representations of general linear groups, G.D. JAMES
95	Low-dimensional topology 1982, R.A. FENN (ed)
96	Diophantine equations over function fields, R.C. MASON
97	Varieties of constructive mathematics, D.S. BRIDGES & F. RICHMAN
98	Localization in Noetherian rings, A.V. JATEGAONKAR
99	Methods of differential geometry in algebraic topology, M. KAROUBI & C. LERUSTE
100	Stopping time techniques for analysts and probabilists, L. EGGHE
101	Groups and geometry, ROGER C. LYNDON
103	Surveys in combinatorics 1985, I. ANDERSON (ed)
104	Elliptic structures on 3-manifolds, C.B. THOMAS
105	A local spectral theory for closed operators, I. ERDELYI & WANG SHENGWANG
106	Syzygies, E.G. EVANS & P. GRIFFITH
107	Compactification of Siegel moduli schemes, C-L. CHAI
108	Some topics in graph theory, H.P. YAP
109	Diophantine analysis, J. LOXTON & A. VAN DER POORTEN (eds)
110	An introduction to surreal numbers, H. GONSHOR
111	Analytical and geometric aspects of hyperbolic space, D.B.A. EPSTEIN (ed)
113	Lectures on the asymptotic theory of ideals, D. REES
114	Lectures on Bochner-Riesz means, K.M. DAVIS & Y-C. CHANG
115	An introduction to independence for analysts, H.G. DALES & W.H. WOODIN
116	Representations of algebras, P.J. WEBB (ed)
117	Homotopy theory, E. REES & J.D.S. JONES (eds)
118	Skew linear groups, M. SHIRVANI & B. WEHRFRITZ
119	Triangulated categories in the representation theory of finite-dimensional algebras, D. HAPPEL
121	Proceedings of *Groups - St Andrews 1985*, E. ROBERTSON & C. CAMPBELL (eds)
122	Non-classical continuum mechanics, R.J. KNOPS & A.A. LACEY (eds)
124	Lie groupoids and Lie algebroids in differential geometry, K. MACKENZIE
125	Commutator theory for congruence modular varieties, R. FREESE & R. MCKENZIE
126	Van der Corput's method of exponential sums, S.W. GRAHAM & G. KOLESNIK
127	New directions in dynamical systems, T.J. BEDFORD & J.W. SWIFT (eds)
128	Descriptive set theory and the structure of sets of uniqueness, A.S. KECHRIS & A. LOUVEAU
129	The subgroup structure of the finite classical groups, P.B. KLEIDMAN & M.W.LIEBECK

130   Model theory and modules,   M. PREST
131   Algebraic, extremal & metric combinatorics,   M-M. DEZA, P. FRANKL & I.G. ROSENBERG (eds)
132   Whitehead groups of finite groups,   ROBERT OLIVER
133   Linear algebraic monoids,   MOHAN S. PUTCHA
134   Number theory and dynamical systems,   M. DODSON & J. VICKERS (eds)
135   Operator algebras and applications, 1,   D. EVANS & M. TAKESAKI (eds)
136   Operator algebras and applications, 2,   D. EVANS & M. TAKESAKI (eds)
137   Analysis at Urbana, I,   E. BERKSON,   T. PECK, & J. UHL (eds)
138   Analysis at Urbana, II,   E. BERKSON,   T. PECK, & J. UHL (eds)
139   Advances in homotopy theory,   S. SALAMON, B. STEER & W. SUTHERLAND (eds)
140   Geometric aspects of Banach spaces,   E.M. PEINADOR and A. RODES (eds)
141   Surveys in combinatorics 1989,   J. SIEMONS (ed)
142   The geometry of jet bundles,   D.J. SAUNDERS
143   The ergodic theory of discrete groups,   PETER J. NICHOLLS
144   Introduction to uniform spaces,   I.M. JAMES
145   Homological questions in local algebra,   JAN R. STROOKER
146   Cohen-Macaulay modules over Cohen-Macaulay rings,   Y. YOSHINO
147   Continuous and discrete modules,   S.H. MOHAMED & B.J. MÜLLER
148   Helices and vector bundles,   A.N. RUDAKOV et al
149   Solitons nonlinear evolution equations & inverse scattering, M. ABLOWITZ & P. CLARKSON
150   Geometry of low-dimensional manifolds 1,   S. DONALDSON & C.B. THOMAS (eds)
151   Geometry of low-dimensional manifolds 2,   S. DONALDSON & C.B. THOMAS (eds)
152   Oligomorphic permutation groups,   P. CAMERON
153   L-functions and arithmetic,   J. COATES & M.J. TAYLOR (eds)
154   Number theory and cryptography,   J. LOXTON (ed)
155   Classification theories of polarized varieties,   TAKAO FUJITA
156   Twistors in mathematics and physics,   T.N. BAILEY & R.J. BASTON (eds)
157   Analytic pro-p groups,   J.D. DIXON, M.P.F. DU SAUTOY, A. MANN & D. SEGAL
158   Geometry of Banach spaces,   P.F.X. MÜLLER & W. SCHACHERMAYER (eds)
159   Groups St Andrews 1989 volume 1,   C.M. CAMPBELL & E.F. ROBERTSON (eds)
160   Groups St Andrews 1989 volume 2,   C.M. CAMPBELL & E.F. ROBERTSON (eds)
161   Lectures on block theory,   BURKHARD KÜLSHAMMER
162   Harmonic analysis and representation theory for groups acting on homogeneous trees,
       A. FIGA-TALAMANCA & C. NEBBIA
163   Topics in varieties of group representations,   S.M. VOVSI
164   Quasi-symmetric designs,   M.S. SHRIKANDE & S.S. SANE
165   Groups, combinatorics & geometry,   M.W. LIEBECK & J. SAXL (eds)
166   Surveys in combinatorics, 1991,   A.D. KEEDWELL (ed)
167   Stochastic analysis,   M.T. BARLOW & N.H. BINGHAM (eds)
168   Representations of algebras,   H. TACHIKAWA & S. BRENNER (eds)
169   Boolean function complexity,   M.S. PATERSON (ed)
170   Manifolds with singularities and the Adams-Novikov spectral sequence,   B. BOTVINNIK
171   Squares,   A.R. RAJWADE
172   Algebraic varieties,   GEORGE R. KEMPF
173   Discrete groups and geometry,   W.J. HARVEY & C. MACLACHLAN (eds)
174   Lectures on mechanics,   J.E. MARSDEN
175   Adams memorial symposium on algebraic topology 1,   N. RAY & G. WALKER (eds)
176   Adams memorial symposium on algebraic topology 2,   N. RAY & G. WALKER (eds)
177   Applications of categories in computer science,   M.P. FOURMAN, P.T. JOHNSTONE,
       & A.M. PITTS (eds)
178   Lower K- and L-theory,   A. RANICKI
179   Complex projective geometry,   G. ELLINGSRUD, C. PESKINE, G. SACCHIERO
       & S.A. STRØMME (eds)
180   Lectures on ergodic theory and Pesin theory on compact manifolds,   M. POLLICOTT
181   Geometric group theory I,   G.A. NIBLO & M.A. ROLLER (eds)
182   Geometric group theory II,   G.A. NIBLO & M.A. ROLLER (eds)
183   Shintani zeta functions,   A. YUKIE
184   Arithmetical functions,   W. SCHWARZ & J. SPILKER
185   Representations of solvable groups,   O. MANZ & T.R. WOLF
186   Complexity: knots, colourings and counting,   D.J.A. WELSH
187   Surveys in combinatorics, 1993,   K. WALKER (ed)
190   Polynomial invariants of finite groups,   D.J. BENSON
191   Finite geometry and combinatorics,   F. DE CLERCK et al
197   Two-dimensional homotopy and combinatorial group theory,   C. HOG-ANGELONI
       W. METZLER & A.J. SIERADSKI (eds)

London Mathematical Society Lecture Note Series. 197

Two-Dimensional Homotopy and Combinatorial Group Theory

Edited by

Cynthia Hog-Angeloni
University of Frankfurt

Wolfgang Metzler
University of Frankfurt

and

Allan J. Sieradski
University of Oregon

CAMBRIDGE
UNIVERSITY PRESS

Published by the Press Syndicate of the University of Cambridge
The Pitt Building, Trumpington Street, Cambridge CB2 1RP
40 West 20th Street, New York, NY 10011–4211, USA
10 Stamford Road, Oakleigh, Melbourne 3166, Australia

© Cambridge University Press 1993

First published 1993

Library of Congress cataloguing in publication data available

British Library cataloguing in publication data available

ISBN 0 521 44700 3 paperback

Transferred to digital printing 2004

Contents

Editors' Preface   v

Addresses of Authors   vii

I   **Geometric Aspects of Two-Dimensional Complexes**   1

Cynthia Hog-Angeloni and Wolfgang Metzler

1   Complexes of Low Dimensions and Group Presentations . . .   1

2   Simple–Homotopy and Low Dimensions   11

3   P.L. Embeddings of 2-Complexes into Manifolds   29

4   Three Conjectures and Further Problems   44

II   **Algebraic Topology for Two Dimensional Complexes**   51

Allan J. Sieradski

1   Techniques in Homotopy   51

2   Homotopy Groups for 2-Complexes   62

3   Equivariant World for 2-Complexes   75

4   Mac Lane-Whitehead Algebraic Types   88

III   **Homotopy and Homology Classification of 2-Complexes**   97

M. Paul Latiolais

1   Bias Invariant & Homology Classification   97

2   Classifications for Finite Abelian π_1   111

3   Classifications for Non-Finite π_1 (with Cynthia Hog-Angeloni)   117

IV   Crossed Modules and Π_2 Homotopy Modules   **125**

Micheal N. Dyer

1   Introduction 125

2   Crossed and Precrossed Modules 126

3   On the Second Homotopy Module of a 2-Complex 140

4   Identity Properties 148

V   Calculating Generators of Π_2   **157**

William A. Bogley and Steve J. Pride

1   The Theory of Pictures 157

2   Generation of Π_2 167

3   Applications and Results 176

VI   Applications of Diagrams to Decision Problems   **189**

Günther Huck and Stephan Rosebrock

1   Introduction 189

2   Decidability and Dehn's Algorithm 190

3   Cayley Graph and van Kampen Diagrams 192

4   Word Hyperbolic Groups and Combings 197

5   Curvature Tests 203

**VII Fox Ideals, $\mathcal{N}$-Torsion and Applications to Groups and
3-Manifolds**   **219**

Martin Lustig

1   Fox ideals 220

2   Applications of Fox ideals: Tests for the rank, the deficiency
     and the homological dimension of a group 225

3    $\mathcal{N}$-torsion: Basic theory 230

4   $\mathcal{N}^1(G)$, Nielsen equivalence of generating systems and Hee-
    gaard splittings 233

5   $\mathcal{N}$-torsion as generalization of the bias and (simple)-homotopy
    of (G, m)-complexes 245

VIII (Singular) 3-Manifolds   **251**

Cynthia Hog-Angeloni and Allan J. Sieradski

1   3–Manifolds 251

2   Singular 3–Manifolds 274

**IX   Cancellation Results for 2-Complexes and 4-Manifolds
      and Some Applications**   **281**

Ian Hambleton and Matthias Kreck

1   A Cancellation Theorem for 2-Complexes 281

2   Stable Classification of 4-Manifolds 286

3   A Cancellation Theorem for Topological 4-Manifolds 290

4   A Homotopy Non-Cancellation Theorem for Smooth
    4-Manifolds 296

5   A Non-Cancellation Example for Simple-Homotopy Equivalent
    Topological 4-Manifolds 299

6   Application of Cancellation to Exotic Structures on
    4-Manifolds 302

7   Topological Embeddings of 2-Spheres into 1-Connected
    4-Manifolds and Pseudo-free Group Actions 305

X   J. H. C. Whitehead's Asphericity Question   **309**

William A. Bogley

1   Introduction 309

2   The Context of Whitehead's Question 310

3   Structural Results 312

4   Reductions, Evidence and Test Cases 314

5   On the π_1-Kernel 320

6   Acyclic Coverings 322

7   Finitely Generated Perfect Subgroups 326

8   Kaplansky's Theorem 328

9   Framed Links 330

10   Open Questions (with J. Howie) 333

XI   Zeeman's Collapsing Conjecture   335

Sergei Matveev and Dale Rolfsen

1   Introduction 335

2   Collapsing 338

3   Some Special Ways of Collapsing $P^2 \times I$ 339

4   1-Collapsibility Modulo 2-Expansions 348

5   Zeeman Conjecture for Special Polyhedra 349

6   Generalizing (Z) to Higher Dimensions 361

7   Open Problems 363

XII The Andrews-Curtis Conjecture and its Generalizations   365

Cynthia Hog-Angeloni and Wolfgang Metzler

1   Introduction 365

2   Strategies and Characterizations 366

3   Q^{**}-Transformations and Presentations of Free Products . . . 373

4   Some Further Results 380

Bibliography   381

Index   408

Editors' Preface

It is well known that techniques developed for manifolds of higher dimensions don't suffice to treat open problems in dimensions three and four. The latter are inextricably tied to questions about the (simple-) homotopy type of 2-skeleta or -spines of these low-dimensional manifolds and, hence, to presentations of groups.

Basic work on two-dimensional homotopy dates back to K. Reidemeister and J.H.C. Whitehead. For instance, Whitehead gave an algebraic description of the homotopy type of 2-complexes. But, until the early 70's, one didn't have examples of 2-complexes with different homotopy type but equal fundamental groups and Euler characteristic. Since then considerable advances have been made, yielding, in particular, remarkable partial results on famous open problems like Whitehead's question, whether subcomplexes of aspherical 2-complexes are always aspherical themselves. The authors of this book have contributed to this development.

Because of its relations to decision problems in combinatorial group theory, two-dimensional homotopy probably will never take the shape of a complete theory. However, the occurrence of certain notions (e.g., the Reidemeister-Peiffer identities of presentations) in different questions is far from being accidental. The time has come to collect the present knowledge in order to stimulate further research.

This book contains the elements of both a textbook and a research monograph, and, hence, addresses students as well as specialists. Parts of the book have already been used to substantiate courses with concrete geometric and/or algebraic material. A student reader should know already some (algebraic) topology and algebra. We start with two introductory chapters on low-dimensional complexes. These are followed by chapters on prominent techniques including their applications to manifolds. Concluding chapters treat the present status of three famous conjectures (Whitehead, Zeeman, Andrews-Curtis). The coherent organization of the book includes cross references as well as a common index and an ample bibliography. But the chapters can also be read independently; they range from an introduction to the specific topic(s) to a survey of latest results, with guidelines to the current literature. Particular emphasis is placed on covering open problems.

This book project was initiated by Wolfgang Metzler and presented to the majority of the authors at a workshop on Geometric Topology and Combinatorial Group Theory held in Luttach/Southern Tyrol (Italy) in August 1991. During this meeting, an approximate table of contents was developed.

The book demonstrates and documents mathematical cooperation. We mention, in particular, long-term activities of Micheal Dyer, Allan Sieradski and their former students, of Wolfgang Metzler and his former students and of Paul Latiolais' Fall Foliage Seminars, which have grown together in recent years. In addition to personal contact, modern electronic communication of (drafts of) sections or whole chapters was a basic ingredient in the production of manuscripts. This includes the illustrations that were based on sketches of the authors and drawn by Allan Sieradski using the Postscript drawing application Adobe Illustrator. The final layout was done by the editors.

All chapters were refereed twice, according to the usual textbook/journal standards, once by one or more member(s) of the team of authors, second by an external referee who had the option of remaining anonymous. As nobody chose to do so, we express our gratitude for their valuable service to: Juan Alonso (Stockholm), Stefan Bauer (Göttingen), Gerhard Burde (Frankfurt/Main), David Gillman (Los Angeles), Mauricio Gutierrez (Medford), Jens Harlander (Frankfurt/Main), Wolfgang Heil (Tallahassee), James Howie (Edinburgh), John Ratcliffe (Nashville), Nancy Waller (Portland) and Perrin Wright (Tallahassee). With their help, we have, in particular, tried to avoid mathematical and typographical errors. Our editorial efforts included eliminating conflicting use of terminology in different chapters; but we did not insist on standardized notations.

We invite all readers to communicate any remaining errors to us; we are also eager to learn about further progress on the mathematical substance of this book, which might be integrated into revisions of the text.

Last, but not least, we want to thank Roger Astley and David Tranah of Cambridge University Press for their continuous encouragement and advice throughout the project.

Cynthia Hog-Angeloni, Wolfgang Metzler, Allan J. Sieradski

Brombach (Taunus)
August 1993

Addresses of Authors

William A. Bogley
Department of Mathematics
Oregon State University
Corvallis, OR 97331-4605 USA

Micheal N. Dyer
Department of Mathematics
University of Oregon
Eugene, OR 97403 USA

Ian Hambleton
Dept. of Mathematical Sciences
McMaster University
Hamilton, Ontario
Canada L8S 4K1

Cynthia Hog–Angeloni
Fachbereich Mathematik
der Universität Frankfurt
Postfach 11 19 32
60054 Frankfurt am Main/Germany

Günther Huck
Department of Mathematics
Northern Arizona University
Flagstaff, AZ 86011 USA

Matthias Kreck
Fachbereich Mathematik
der Universität Mainz
Saarstrasse 21
55122 Mainz/Germany

M. Paul Latiolais
Dept. of Mathematical Sciences
Portland State University
Portland, OR 97207-0751 USA

Martin Lustig
Institut für Mathematik
der Ruhr–Universität
Postfach 10 21 48
44780 Bochum/Germany

Sergei Matveev
Chelyabinsk State University
129 Kashirin Brothers St.
454136 Chelyabinsk/Russia

Wolfgang Metzler
Fachbereich Mathematik
der Universität Frankfurt
Postfach 11 19 32
60054 Frankfurt am Main/Germany

Steve J. Pride
Department of Mathematics
Glasgow University
University Gardens
Glasgow G12 8QW U.K.

Dale Rolfsen
Department of Mathematics
University of British Columbia
Vancouver B.C. Canada V6T 1Y4

Stephan Rosebrock
Fachbereich Mathematik
der Universität Frankfurt
Postfach 11 19 32
60054 Frankfurt am Main/Germany

Allan J. Sieradski
Department of Mathematics
University of Oregon
Eugene, OR 97403 USA

Chapter I

Geometric Aspects of Two-Dimensional Complexes

Cynthia Hog-Angeloni and Wolfgang Metzler

The aim of this introductory chapter is to provide a geometric background for the algebra and homotopy theory to follow. We also focus on geometric questions of intrinsic interest. The algebraic topology of later chapters is meant to contribute to their understanding and treatment. What we present is an extract of courses the authors have given on Low Dimensional Topology. These courses were enriched by selected topics from further chapters of this book and/or some of the material which this article only summarizes, together with references to other sources.

1 Complexes of Low Dimensions and Group Presentations

A crucial tool for dealing with geometric and homotopy theoretic problems in topology is the decomposition of certain spaces into disjoint unions of *cells* e_i^n of various dimensions, each (open) cell e_i^n as a subspace being homeomorphic to an open unit disc $\overset{\circ}{D}{}^n = \{x \in \mathbb{R}^n \mid |x| < 1\}$. Thus, we may obtain the structure of a *cell complex* K for the underlying topological space $|K|$. We denote by K^n (the *n-skeleton*) the subspace comprised of the cells of K of dimensions $\leq n$ with the induced cell structure. It is also standard terminology to indicate by the superscript n of K^n that K is a complex with cells of at most dimension n (sometimes it is required that there *are* n-cells in an n-dimensional complex K).

The various notions of complexes differ by the conditions that are imposed on the closure $\bar{e}_i \subseteq |K|$ of a cell $e_i \in K$ and/or the *boundary* $\partial e_i = \bar{e}_i - e_i$. These conditions also regulate how the cells are "glued" together to yield the topology of $|K|$.

We assume that the reader is familiar with *simplicial complexes* (finite or infinite). Here each closed cell $\bar{e}_i$ is a union of cells and is homeomorphic to a simplex with all its faces by a homeomorphism which maps open cells to open cells. (A homeomorphism between cell-complexes which maps open cells to open cells is called *cellular* or an *isomorphism*.) As a "gluing condition" one mostly confines to the *weak topology* with respect to the $\bar{e}_i^n$, i.e.,

(1) a subset of $|K|$ is closed iff its intersection with each $\bar{e}_i^n$ is closed.

Between 1939 and 1950, J.H.C. Whitehead published several papers which, on one hand, contain masterpieces of simplicial techniques. Some of these are basic for our geometric questions and will be cited in several sections of this article. On the other hand, Whitehead gradually was led to the insight that many of his results in homotopy theory actually hold for a generalization of simplicial complexes, where proofs and constructions can avoid a lot of hard work (checking the strong conditions of simplicial complexes and maps). In [Wh49$_1$], he introduced the notion of "CW-complexes," for which the assumptions on the ∂e_i are far less restrictive than in the simplicial case. We give an inductive definition of CW-complexes which is particularly convenient in low dimensions. For the equivalence to Whitehead's original definition, see Schubert ([Schu64], III. 3, Exercise 1 and Sieradski ([Si92], Chapter 15).

1.1 Inductive construction of CW-complexes

Definition: A *CW-complex* K is a space $|K|$ with a cell decomposition, whose skeleta are inductively constructed as follows:

(a) K^0 is a discrete space, each point being a 0-cell.

(b) K^n is obtained by attaching to K^{n-1} a disjoint family D_i^n of closed n-discs via continuous functions $\varphi_i : \partial D_i^{n-1} \to K^{n-1}$, i.e.: take the topological sum $K^{n-1} + \sum D_i^n$ and pass to the quotient space given by the identifications $x \sim \varphi_i(x)$, $x \in \partial D_i^n$. Each $\overset{\circ}{D}_i^n$ then projects homeomorphically to an n-cell e_i^n. φ_i is called an *attaching map* for e_i^n.

(c) $|K| = \bigcup_{n=0}^{\infty} |K^n|$ is assigned the weak topology with respect to the $\bar{e}_i^n$ (as in (1)).

More generally, a cell complex is called a CW-complex, if it is isomorphic to one obtained by the preceding construction[1].

Remarks: (c) can be verified inductively for the skeleta (Exercise), hence it holds automatically if K is of finite dimension. The "W" in CW is motivated by <u>W</u>eak topology, the "C" (<u>C</u>losure finite) by the fact that each $\bar{e}_i^n$ is contained in a finite union of cells (Exercise).

In contrast to the situation of simplicial complexes, $\bar{e}_i^n$ itself is not necessarily a union of cells; see Figure 1, where ∂e^2 is a point of e^1, but not a 0-cell:

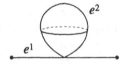

Figure I.1.

The following facts on CW-complexes can also be found in [Schu64] and [Si92]:

(2) A CW-complex K is finite (i.e., consists of finitely many cells) iff $|K|$ is compact.

(3) A covering space of a CW-complex K can be (uniquely) decomposed as a CW-complex $\tilde{K}$ such that the projection map $\tilde{K} \to K$ sends each cell $\tilde{e} \in \tilde{K}$ homeomorphically to a cell $e \in K$. (Note that the corresponding statement for closed cells is false in general, as the universal covering space of $S^1 = e^0 \cup e^1$ shows.) An attaching map for $\tilde{e}^n$ is obtained by appropriately lifting an attaching map $\varphi : \partial D^n \to K^{n-1}$ of e^n to $\tilde{K}^{n-1} = p^{-1}(K^{n-1})$.

We will deal mainly with finite CW-complexes and infinite ones that arise as covering complexes of (finite) complexes.

---

[1] Attaching maps in this general case are those of an isomorphic model according to (a), (b), (c), composed with a cellular homeomorphism. We shall tacitly assume similar extensions of definitions to be made later on. Note that specific attaching maps are not considered data of the complex; compare § 2.1.

1.2　Questions of subdivision and triangulation

A 1-dimensional CW-complex is a graph. It may contain loops and more than one edge between two vertices; but by introducing new vertices, it can be subdivided[2] to become simplicial (details left as an exercise):

Figure I.2.

However, already in dimension 2 – due to the generality of attaching maps – there exist nontriangulable CW-complexes: Consider the finite(!) 2-complex K^2 of Figure 3, an *infinitely crumpled curtain* with three 0-cells, three 1-cells and one 2-cell, the attaching map of which oscillates on e_1^1:

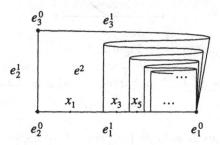

Figure I.3.

Note that all other marked points (e.g., the x_i) and lines of the drawing don't indicate further cells.

Theorem 1.1　*The CW-complex K^2 of Figure 3 is not homeomorphic to any simplicial complex.*

Proof: Triangulating appropriate neighbourhoods of the points x_i, $i = 1, 3, 5,$ $7, \ldots,$ one obtains the local homology groups $H_2(|K|, |K| - \{x_i\}) = \underbrace{\mathbb{Z} \times \ldots \times \mathbb{Z}}_{(i-1) \text{ factors}}$. The number of "sheets" of the curtain coming together at x_i

---

[2]The general notion of subdivision for CW-complexes will be introduced below; see also footnote 5.

is thus seen to be an invariant of the local homeomorphism type of $|K|$ at x_i : $|K|$ has *infinitely many* distinct local homeomorphism types. Any triangulation of $|K|$ would have to be finite, as $|K|$ is compact (see (2)). But a finite simplicial complex has only *finitely many* local homeomorphism types, given by the stars of the simplices. Thus no such triangulation exists. ☐

In the beginning of this century, the "Hauptvermutung" of combinatorial topology was raised, the question, whether homeomorphic simplicial complexes are combinatorially equivalent (i.e., if they become isomorphic after simplicial subdivisions). The terminology naturally generalizes to CW-complexes: A *subdivision* K' of a CW-complex K is a CW-complex K' with $|K'| = |K|$ and the property that each cell $e \in K$ is the union of certain cells $e'_i \in K'$. K and L are *combinatorially equivalent*, if they admit subdivisions K' (of K) and L' (of L) which are isomorphic. But the answers to the Hauptvermutung question are different in the simplicial and the CW-case:

After contributions of prominent mathematicians (e.g., Papakyriakopoulos, Moise, Bing, Milnor, Stallings, Kirby, Siebenmann), the answer to the simplicial Hauptvermutung is known to be "Yes" for (locally finite[3] simplicial complexes of) dimensions ≤ 3; see [Br69], and "No" in dimensions ≥ 4. (Closed 4-manifolds with exotic differentiable structures were first exhibited by M. Kreck [Kr84$_2$]; that such examples yield combinatorially distinct triangulations follows from [Mu60] together with [Ce68]).

But the Hauptvermutung for CW-complexes fails already in dimension 2, as the following example(s) will show:

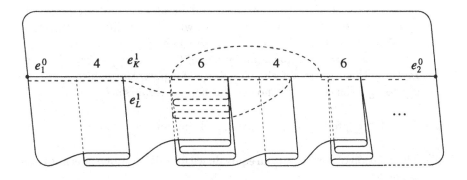

Figure I.4.

---

[3] A CW-complex is *locally finite*, if each point has a neighbourhood which meets only finitely many cells. (For simplicial complexes, this is equivalent to saying that the star of every simplex is finite (Exercise).) K is locally finite iff $|K|$ is locally compact; see [Schu64].

In Figure 4, K and L are finite CW-complexes with two 0-cells, three 1-cells and two 2-cells. They differ in their dissection of $|K| = |L|$ by the middle 1-cell e_K^1 resp. e_L^1. The cell e_K^1 contains a sequence of (open) subintervals, where four or six sheets of 2-dimensional material meet, the sequence converging towards e_2^0 alternatingly: $4, 6, 4, 6, \ldots$; e_L^1 is defined by the broken line and its periodic and shrinking continuation towards e_2^0 with the "rhythm" $4, 4, 6, 6, 4, 4, 6, 6, \ldots$ of subintervals. It requires a little visualization to make sure that for the resulting 2-cells of L – above and below e_L^1 – none of the CW-conditions are violated.

Theorem 1.2 *K^2 and L^2 of Figure 4 are homeomorphic but combinatorially inequivalent.*

Proof: We use some easy facts on the sheets, which result from local homology considerations as in the proof of Theorem 1.2.

a) Any subdivision K' of K, which is finite by (2), contains a 1-cell $e_{K'}^1$ (the one adjacent to e_2^0) characterized by the fact that almost all of the 4- and 6- sheeted subintervals are carried by it; L' analogously contains an $e_{L'}^1$.

b) Any homeomorphism of $|K|$ to $|L|(= |K|)$ must map a 4- resp. 6-sheeted subinterval to a 4- resp. 6-sheeted subinterval. Remembering a), a cellular homeomorphism of K' to L' thus in particular would have to map $e_{K'}^1$ to $e_{L'}^1$.

c) In e_K^1 – hence also in $e_{K'}^1$ – any two subintervals of type 4 are separated by a type 6 subinterval (and vice versa), whereas the different "rhythm" $4, 4, 6, 6, \ldots$ in e_L^1 implies that $e_{L'}^1$ contains adjacent subintervals of one type without such a separation by one of the other type. The different rhythms thus contradict the monotony of a potential homeomorphism $e_{K'}^1$ to $e_{L'}^1$. Hence there is no cellular homeomorphism $K' \to L'$. □

We refer to [Me67] for further subdivision phenomena, for instance:

(4) The relation of combinatorial equivalence for CW-complexes is not transitive.

Geometric pathologies arising from general attaching maps don't deserve too much interest of their own. But their occurrence either suggests the restriction to piecewise linear CW-complexes (see 1.4 and § 3 below) or it forces specific care in the proof of certain statements (see § 2.3 below).

1.3 Reading off presentations for π_1 of a CW-complex

The following material may be found in many textbooks. Proofs are based either on cellular/simplicial approximation (e.g., [Schu64]) or on the Seifert-van Kampen theorem (e.g. [Ma67], [Si92]); see also Chapter II, § 1 and § 2.

(5) If K^1 is connected[4], select a vertex e^0 as a basepoint and a *spanning tree*, i.e., a tree that consists of some edges and all vertices. Each remaining edge e_i^1, together with an orientation, determines a closed path from e^0 as in Figure 5 (by connecting the initial and terminal vertex of e_i^1 with e^0 on the tree). The elements a_i of $\pi_1(|K^1|, e^0)$ given by these paths constitute a free basis of $\pi_1(|K^1|, e^0)$, i.e., $\pi_1(|K^1|, e^0)$ is the free group $F(a_i)$.

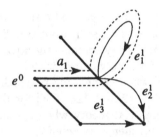

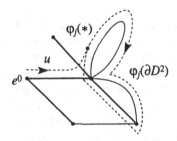

Figure I.5. Generating loops   Figure I.6. Defining relations

(6) If K^2 is connected, select a basepoint, a spanning tree and paths a_i for the 1-skeleton as in (5). For each 2-cell e_j^2, an attaching map $\varphi_j : \partial D^2 \to K^1$, together with a base point $*$, an orientation of ∂D^2 and a connecting arc u from e^0 to $\varphi_j(*)$ (as in Figure 6) defines an element $R_j \in \pi_1(|K^1|, e_0)$ – a word R_j in the $a_i^{\pm 1}$ – which is trivial in $\pi_1(|K^2|, e^0)$. Moreover, $\pi_1(|K^1|, e^0) \to \pi_1(|K^2|, e^0)$ is surjective with the normal closure $N(R_j)$ (j ranging over all 2-cells) as the kernel.

Thus $\pi_1(|K^2|, e^0)$ has the *presentation* $\langle a_1, a_2, \dots | R_1, R_2, \dots \rangle$, i.e., the fundamental group $\pi_1(|K^2|, e^0)$ is given by *generators* a_i and *defining relations* R_j as the quotient $F(a_i)/N(R_j)$.

(7) If K has K^2 as its 2-skeleton, then the natural map $\pi_1(|K^2|, e^0) \to \pi_1(|K|, e^0)$ is an isomorphism.

---

[4]A CW-complex is connected iff its 1-skeleton is pathwise connected (Exercise).

A crucial point of this chapter (and book) is that a finite, connected 2-complex not only determines a fundamental group, but – via (6) – a certain class of presentations, which doesn't contain all presentations of π_1. Many homotopy invariants can be derived from any member of this presentation class. The first one to be mentioned is the Euler characteristic $\chi(|K|) = 1 - \#(a_i) + \#(R_j)$; see § 2.3.

As an example, note that for a compact, connected and simply connected K^2 the following statements are equivalent: (a) $|K|$ is *contractible* (i.e., $|K| \simeq *$); (b) $\chi(|K|) = 1$; (c) $H_2(|K|) = \{0\}$; (d) the presentations read off from $|K|$ according to (6) are *balanced*, i.e., the number of defining relations equals the number of generators (Exercise).

1.4 *PLCW*-complexes

In his papers [Wh39] and [Wh41$_2$], Whitehead didn't use general CW–complexes, but he confined his attention to specific ones which still admit simplicial[5] subdivisions, so–called "membrane complexes". They combine the advantage of a) providing complexes with "few cells" for a polyhedron with b) the piecewise linear (= p.l.) geometry, a main tool when embedding polyhedra into manifolds; see § 3. We refer to [Hu69], [RoSa72] and [Ze63-66] for the p.l. category, which is the proper home for simplicial theory when specific triangulations become irrelevant. In this category, membrane complexes are just what is given by the following definition:

Definition: A *PL CW-complex* is a locally finite CW-complex K together with a p.l. structure for $|K|$ such that

(a) all closed cells and all skeleta are subpolyhedra;

(b) K^n is obtained from K^{n-1} by a family of p.l. attaching maps $\varphi_i :$ $\partial D_i^n \to K^{n-1}$ such that each subpolyhedron $e_i \cup K^{n-1}$ is p.l. homeomorphic, rel. K^{n-1}, to $D_i^n \underset{\partial D_i^n}{\cup} C(\varphi_i)$, where $C(\varphi_i)$ is the p.l. mapping cylinder of φ_i.

Here D^n is meant to be equipped with the p.l. structure given by a fixed homeomorphism of D^n to the n-cube I^n. b) involves results on p.l. mapping cylinders (existence, triangulations) which date back to [Wh39]; see the

---

[5]A *simplicial subdivision* K' of a CW-complex K is a subdivision of K by a simplicial complex K'.

discussion and the references in ([CoMeSa85], § 2). They yield, in particular, that by attaching finitely many n-cells to a $PL\,CW$-complex K^{n-1} via p.l. attaching maps, the result naturally becomes a $PL\,CW$-complex. Hence there is an inductive construction yielding finite $PL\,CW$-complexes. Exercise: Derive concrete triangulations in the 2-dimensional cases (8), (9) which follow.

Some classes of finite 2-dimensional PL CW-complexes:

(8) *Reidemeister complexes* (see [Re32]): For the construction of K^2 from K^1, let each edge of K^1 be equipped with a linear structure; each ∂D_j^2 is subdivided as a polygon; $\varphi_j : \partial D_j^2 \to K^1$ maps each edge linearly onto an edge of K^1 or onto a vertex. φ_j thus defines a closed *edge path* in K^1. The notion of a Reidemeister complex agrees with the one of a combinatorial CW-complex of dimension 2; see Chapter II, § 1.2.

After subdividing K^1, any (finite) $PL\,CW$-complex becomes a Reidemeister complex (Exercise). Simplicial 2-complexes are Reidemeister complexes in which each triangle of a simplicial K^1 may be filled in at most once.

(9) *Standard complexes of (finite) presentations:* As an "inverse" process to 1.3, we may associate to a finite presentation $\mathcal{P} = \langle a_1, \ldots, a_g | R_1, \ldots, R_h \rangle$ a Reidemeister-complex $K_{\mathcal{P}}$ with one vertex. Its 1-skeleton is a bouquet of circles with an oriented e_i^1 for each a_i:

Figure I.7.

The 2-cells e_j^2 correspond bijectively to the R_j, which determine closed edge paths φ_j as the corresponding attaching maps.

Note that the sequence $R_1, \ldots, R_h$ may contain repetitions and/or the trivial word $R_j = 1 \in F(a_i)$. Those relators nevertheless must *not* be suppressed when constructing the standard complex $K_{\mathcal{P}}$, which hence may contain different 2-cells with the same boundary and/or closed 2-cells that are 2-spheres.

There is no general rule as to whether the R_j are assumed to be (cyclically) reduced or whether R_j may contain adjacent inverse letters a_i, a_i^{-1}. Reducing such "spurs" in the φ_j does not affect the (simple-)homotopy type of $K_\mathcal{P}$, see Lemma 2.1 of § 2.1 below; but it may drastically change the embedding behavior of $K_\mathcal{P}$ into 4-space, see § 3.2 below.

Amongst the standard complexes are those for canonical dissections of closed 2-manifolds, ([Schu64], III 5.8; [Si92], Chapter 13), but in general more than two local sheets may meet at an e^1. It is nevertheless possible to apply certain moves to finite 2-dimensional CW-complexes which reduce the local complexity by achieving general position; see Ikeda [Ik71], Wright [Wr73], [Wr77] and Remark 1 after the proof of Theorem 3.1 in § 3.1. The result is a compact 2-dimensional polyhedron in which each point has a star of one of the following p.l. types:

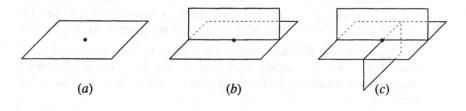

(a) $\qquad\qquad\qquad$ (b) $\qquad\qquad\qquad$ (c)

Figure I.8.

Such a polyhedron is called a *closed fake surface*; its *intrinsic 2-skeleton* consists of points of type (a); type (b) resp. (c) defines the *intrinsic 1-resp. 0-skeleton*[6]. As closed 2-manifolds show, these intrinsic skeleta in general don't dissect a closed fake surface into a cell complex. This additional requirement gives rise to the following definition:

(10) A *special* (or: *standard* polyhedron) is a closed fake surface $|K|$ with the property that the components of the intrinsic skeleta are the cells of a $PL\,CW$-decomposition K (for the given p.l. structure) of $|K|$.

Standard polyhedra are also Reidemeister complexes (Exercise).

In the literature, one sometimes has to deduce the precise meaning of "2-complex" from the context; and if specific names are given, their meaning is not always the same.

---

[6]Casler [Ca65] and Wright [Wr77] define the intrinsic 1-skeleton to consist of the points of type (b) *or* (c); consequently, these authors don't introduce the separate notion of the intrinsic 2-skeleton. Compare also the notions of Chapter XI, § 5.1.

2   Simple–Homotopy and Low Dimensions

2.1   A survey on geometric simple–homotopy

Throughout this section, all spaces are assumed to be compact.

In the definition of a CW-complex in order to attach an n-cell e^n, we had to form the quotient map $p : K^{n-1} + D^n \twoheadrightarrow K^{n-1} \cup_\varphi D^n$ that identifies each $x \in \partial D^n$ with $\varphi(x)$, where $\varphi : \partial D^n \to K^{n-1}$ is the attaching map for e^n. The restriction of p to D^n yields a characteristic map for e^n:

Definition: A map $\Phi : D^n \to K$ is *characteristic* for an n-cell e^n of K if Φ maps the interior $\overset{\circ}{D}{}^n$ homeomorphically onto e^n and $\Phi|\partial D^n$ is an attaching map.

Note that attaching maps and characteristic maps are by no means unique. In the case of characteristic maps, an easy example is obtained by a self-homeomorphism of I^n on the one hand and shrinking $I^{n-1} \times \{1\}$ in I^n to a point (while mapping the interior homeomorphically) on the other hand. Nevertheless, it is true that

(11a) Exercise: Two attaching maps for an n-cell e^n are homotopic up to composition with an orientation reversing map of ∂D^n to itself.

(11b) Exercise: Two characteristic maps for e^n determine homotopic maps $(D^n, \partial D^n) \to (K^n, K^{n-1})$ up to composition with an orientation reversing homeomorphism of D^n to itself.

We now want to describe particularly simple operations which do not change the homotopy type of a given complex L, namely the attaching of an n-ball along an $(n-1)$-ball in its boundary: (D^n, D^{n-1}) are assumed to form a *standard ball pair*, i.e., a pair homeomorphic to $(I^{n-1} \times I, I^{n-1} \times \{1\})$. Let $\varphi : D^{n-1} \to L^{n-1}$ and define $|K|$ to be $|L| \cup_\varphi D^n$. Then $|K|$ has a natural cell decomposition consisting of cells of L together with new e^n and e^{n-1}, and the associated characteristic map $\Phi : D^n \to K$ fulfills

(12) $\Phi|D^{n-1}$ maps to the subcomplex L^{n-1} and[7]

(13) $\Phi|\overline{\partial D^n - D^{n-1}}$ is characteristic for e^{n-1}.

---

[7]Equivalently, the roles of D^{n-1} and $\overline{\partial D^n - D^{n-1}}$ could be interchanged simultaneously in (12) and (13) which seems more adequate in some applications; see also Figure 9.

This gives rise to the following

Definition: An *elementary collapse* (of dimension n) $K \overset{e}{\searrow} L$ is the transition from K to L if L is a subcomplex[8] of K, $K = L \cup e^n \cup e^{n-1}$ where $e^n, e^{n-1} \notin L$ such that there exists a standard ball pair (D^n, D^{n-1}) and a characteristic map $\Phi : D^n \to K$ for e^n fulfilling (12) and (13)[9]. The inverse operation $L \overset{e}{\nearrow} K$ is called an *elementary expansion* (of dimension n):

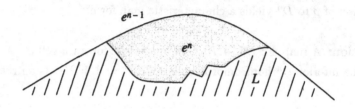

Figure I.9.

Note that for $K \overset{e}{\searrow} L$ (resp. $K \overset{e}{\nearrow} L$) there exists a (strong) deformation retraction (resp. the inclusion) from $|K|$ to $|L|$. A sequence $K_0 \overset{e}{\searrow} \ldots \overset{e}{\searrow} K_m$ is called a *collapse* $K_0 \searrow K_m$ resp. an *expansion* $K_m \nearrow K_0$. K_m is called a *spine* of K_0. A sequence where both, collapses and expansions, are allowed, is abbreviated by $K_0 \nearrow\!\!\searrow K_m$. $|K_0|$ and $|K_m|$ are then related by a *deformation* defined to be a composition of retractions and inclusions as above. Sometimes the maximal dimension of involved cells will be added in the notation and we shall speak of an n-(dimensional) *deformation* $K_0 \overset{n}{\nearrow\!\!\searrow} K_m$. Finally, we define a *simple-homotopy equivalence* (in short: *sh-equivalence*) to be a map which is homotopic to a deformation. This concept is due to J. H. C. Whitehead [Wh50]; we also refer to [Co73].

The definition of an elementary collapse makes sense not only in the *CW*-setting. For *polyhedra* K and L, D^n is attached along a p.l. embedding φ; hence $\overline{|K| - |L|}$ is a p.l. ball D^n, which has $|L| \cap D^n$ as a face; compare ([RoSa72], Chapter III). If K and L are *PLCW-complexes*, a *CW*-collapse as in the above definition is called a *PLCW*-collapse and induces a p.l. collapse $|K| \searrow |L|$; see ([CoMeSa85], Remark 4). Finally, if K and L are *simplicial*

---

[8]A subcomplex of a *CW*-complex K is a subset L of cells which constitutes a *CW*-structure for the subspace $|L| = \cup e_i$, $e_i \in L$. This holds iff $\bar{e}_i \subseteq |L|$ for all $e_i \in L$. The skeleta of a *CW*-complex are examples of subcomplexes.

[9]e^{n-1} is then called a *free face* of K. If e^n, $e^{n-1} \in K$ and $K^{n-1} \cup e^n \overset{e}{\searrow} K^{n-1} - e^{n-1}$, then e^{n-1} is a *free face* of e^n. A free face of K is always a free face of the corresponding e^n.

complexes, e^n and e^{n-1} are supposed to be (open) simplices; see [Hu69]. In all these settings, the basic terms can be defined analogously and lead to a simple-homotopy theory. The latter though, roughly speaking, turns out always to be the same; see Proposition 2.3 and (18) below for a precise meaning when comparing the CW with the $PLCW$ setting; for CW versus polyhedra, we refer to ([RoSa72], Appendix B5); polyhedra versus simplicial complexes see ([Hu69], Chapter II). To avoid confusion, we shall sometimes put a superscript $\overset{CW}{\searrow}$, $\overset{p.l.}{\searrow}$, $\overset{PLCW}{\searrow}$, $\overset{s}{\searrow}$ on the arrow to indicate the appropriate setting.

Returning to CW-complexes, we note that

(14) (Exercise:) the elementary steps in a deformation can be reordered (up to isomorphism[10] of the resulting complex) so that first all expansions are done in increasing dimensions and then all collapses in decreasing dimensions.

An important application of deformations arises when deforming the attaching map of an n-cell to some n-complex:

Lemma 2.1 *Let K be an n-dimensional complex and let $\varphi \simeq \psi : S^{n-1} \to K$ with $\varphi(S^{n-1})$, $\psi(S^{n-1}) \subseteq K^{n-1}$ be attaching maps for n-cells e_φ^n resp. e_ψ^n. Set $K_\varphi := K \cup_\varphi D^n$ and $K_\psi := K \cup_\psi D^n$. Then $K_\varphi \overset{n+1}{\diagdown} K_\psi$.*

Proof: Attach $S^{n-1} \times I$ in the boundary of an $(n+1)$-ball $D^n \times I$ via the given homotopy $H : S^{n-1} \times I \to K$

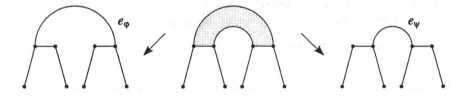

Figure I.10.

and let $K^{n+1} = K \cup_H (D^n \times I)$ be the resulting space with its naturally induced cell decomposition. It has a single $(n+1)$-cell with two free faces e_φ

---

[10]The original deformation may remove and reintroduce certain material which, in the resulting one, has to be formally separated.

and e_ψ. Then $K_\varphi \overset{e}{\nearrow} K^{n+1} \overset{e}{\searrow} K_\psi$ by a collapse through the top e_φ resp. the bottom e_ψ of the $(n+1)$-cell. $\qquad\square$

Here, only one $(n+1)$-cell is expanded to a CW-complex and collapsed immediately afterwards. Such a deformation is called an $(n+1)$-dimensional *exchange* or *transient move*.

If we wish to do homotopies on attaching maps for cells of dimensions lower than the one of the complex, the following lemma (on simple-homotopy extension) is useful:

Lemma 2.2 *Let K^n be a finite CW-complex with m-skeleton K^m, $m < n$. Then for a given deformation $K^m \overset{m+1}{\diagdown} L^m$ there exists a finite CW-complex L^n with m-skeleton L^m such that $K^n \overset{n+1}{\diagdown} L^n$ holds.*

Proof: $K^m \overset{m+1}{\diagdown} L^m$ yields a deformation $K^m \nearrow P^{m+1} \searrow L'^m$ such that α) L'^m is isomorphic to L^m and β) the $(m+1)$-cells of P^{m+1} can also be expanded to K^{m+1} to result in Q^{m+1}; compare (14). But in general the collapse corresponding to $P^{m+1} \searrow L'^m$ cannot be performed in Q^{m+1}. Instead, using the strong deformation retraction induced by the collapse $P^{m+1} \searrow L'^m$, reattach the $(m+1)$-cells of K^{m+1} differently to P^{m+1} so that they avoid the free m-faces of that collapse[11]. As the preceding lemma shows, the resulting complex Q'^{m+1} can be obtained from Q^{m+1} by an $(m+2)$-deformation; and afterwards the collapse $P^{m+1} \searrow L'^m$ can be performed in Q'^{m+1}. In total, we have constructed the sequence $K^{m+1} \nearrow Q^{m+1} \overset{m+2}{\diagdown} Q'^{m+1} \searrow L'^{m+1}$, where L'^{m+1} is the union of L'^m and the re-attached $(m+1)$-cells of K^{m+1}. A further $(m+2)$-deformation of L'^{m+1} to an L^{m+1} with m-skeleton L^m then follows from disjointness considerations; compare [Co73], 5.2.b). The proof is completed by an inductive application of this step. $\qquad\square$

The two lemmas together yield the

Proposition 2.3 *Every n-dimensional finite CW-complex $(n+1)$-deforms to a $PLCW$-complex.*

Proof: Assume that the m-skeleton of a given finite CW-complex K is already $PLCW$, which it is for $m = 0$. As long as $m < n$, deform the attaching maps of the $(m+1)$-cells to become p.l.. Let L^{m+1} be the $PLCW$-complex

---

[11] Attaching maps for the new $(m+1)$-cells are thus given by those of K^{m+1} composed with the deformation.

obtained by attaching the $(m+1)$-cells via these p.l. maps to K^m. By Lemma 2.1, K^{m+1} $(m+2)$-deforms to L^{m+1}; by Lemma 2.2 this induces a deformation $K \overset{n+1}{\nearrow\!\!\!\searrow} L$, where the $(m+1)$-skeleton L^{m+1} of L now is a $PLCW$-complex. $\square$

(15) Similarly, we can apply an $(n + 1)$-deformation to standardize the 1-skeleton of any finite connected CW-complex K^n to a bouquet of circles (see § 1.4, Figure 7):

Attaching maps for 1-cells that are not in the spanning tree are always homotopic to the constant map to the basepoint, and when they are all deformed away from the tree (except for the basepoint), the tree can be collapsed. The two lemmas again convert this procedure into an $(n + 1)$-dimensional deformation. $\square$

Each choice of p.l. approximations for attaching maps during the inductive construction applied to K in the proof of Proposition 2.3, gives rise to a $PLCW$-complex which will be called a $PLCW$-approximation of K.

(16) Exercise: Any two $PLCW$-approximations P, Q of a finite n-dimensional CW-complex K fulfill $P \overset{n+1,PLCW}{\nearrow\!\!\!\searrow} Q$.

(17) Exercise: Let $K \overset{e}{\searrow} L$ where K and L are finite CW-complexes. Then there are $PLCW$-approximations P resp. Q for K resp. L for which $P \overset{e,PLCW}{\searrow} Q$ holds.

These two facts together imply that

(18) finite $PLCW$-complexes P, Q which are simple-homotopy equivalent in the CW-sense, are related by a deformation where all intermediate steps are $PLCW$:

In a deformation from P to Q that uses intermediate stages K_i, choose $PLCW$-approximations P_i for each K_i. By (17), there are $PLCW$-approximations P_i', P_{i+1}' for K_i and K_{i+1} such that P_i' and P_{i+1}' are related by an elementary collapse in the $PLCW$-sense. But then by (16), also P_i and P_{i+1} are related by a deformation where all intermediate steps are $PLCW$. $\square$

Note that this procedure will in general raise the dimension of a deformation by 1.

In § 2.3 (Remark 2 after the proof of Theorem 2.4), we shall see that 2-dimensional *PLCW*-complexes which 3-deform into each other in the *CW*-sense, actually are related by a 3-dimensional *PLCW*-deformation.

The fundamental questions of simple-homotopy theory are, whether every homotopy equivalence is simple or (if not) whether homotopy equivalent complexes are always simple-homotopy equivalent. The answer to these questions turns out to be different according to the dimension of the complexes:

(19) In dimension 1, using Nielsen's theorem [Ni19] on the decomposition of automorphisms of free groups, we shall see in § 4.1 that in fact every continuous map between finite, connected 1-complexes which induces an isomorphism of fundamental groups is homotopic to a deformation of dimension 2.

(20) For all dimensions ≥ 3, there exist finite, connected complexes which are homotopy equivalent but simple-homotopy distinct; see [Co73], 24.4.

(20) needs the algebraic theory of simple-homotopy, in particular the *Whitehead group* Wh(π) of a group π; see [Wh50] and [Co73]. The first examples arose in the p.l. classification of certain closed 3-manifolds, the lens spaces (see § 3.1) and their higher dimensional analogues (Reidemeister [Re32] and Franz [Fr35]).

In dimension 2, the one of our main interest, an additional difficulty is the realization of the algebraic topological construction by different presentations of the same group; compare § 4.4. Nevertheless

(21) there exist finite, connected 2-complexes which are homotopy equivalent but simple-homotopy distinct (Metzler [Me90] and Lustig [Lu91$_1$]); see Chapter VII, Theorem 3.2.

A related question is, whether finite, homeomorphic *CW*-complexes are of the same simple-homotopy type. In the p.l. case for dimensions ≤ 3, this is a consequence of the Hauptvermutung (compare § 1.2) and the fact that

(22) a subdivision of a *CW*-complex can be realized by a deformation (Theorem 12 of [Wh50] and [Co73], 25.1).

But the topological invariance of simple-homotopy type holds for arbitrary dimensions and *CW*-complexes[12], which is a consequence of a famous *Theorem of Th. Chapman* [Ch74] *and J. West* [We71]; see also [BrCo74]:

---

[12]For dimension 2, this follows also from § 2.3, Theorem 2.7.

(23) *If K and L are finite CW-complexes (resp. polyhedra) then $f : K \to L$ is a simple-homotopy equivalence if and only if $f \times Id_Q : K \times Q \to L \times Q$ is homotopic to a homeomorphism.*

In (23), Q denotes the Hilbert cube $\prod_{j=1}^{\infty} I_j$.

On the other hand, *collapsibility*[13] (even of $PLCW$-complexes) is not a topological invariant as long as subdividing is not allowed: Goodrick [Go68] has given examples of triangulations of the topological n-ball for $n \geq 3$ which are not simplicially collapsible. However, by Zeeman ([Ze63-66], Chapters I and III), it follows that, for any triangulation of a topological n-ball, there is an integer k such that the k-th barycentric subdivision of the triangulation is simplicially collapsible. A result of Chillingworth ([Ch67], [Ch80]) shows that simplicial collapsibility of complexes of dimension of at most 3 carries over to subdivisions. Together with the Hauptvermutung, this implies that a $PLCW$-complex that has dimension at most 3 and is topologically homeomorphic to a collapsible $PLCW$-complex becomes (simplicially) collapsible after subdivision.

2.2   Some examples

Every compact, connected n-dimensional p.l. manifold M with nonempty boundary collapses to a spine of dimension $\leq n - 1$: Each (open) n-simplex is accessible from ∂M by a path which misses the $(n - 2)$-skeleton. Thus all n-simplices can be removed by collapses:

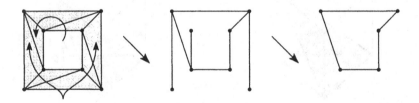

Figure I.11.

In some cases it may be necessary to do some expansions before getting down to a "small" complex. In particular, contractible complexes K do not necessarily collapse to a point.

---

[13] K is called *collapsible*, i.e., $K \searrow *$, if K has a spine which is a point.

A famous example for this is the *dunce hat* H introduced by E. C. Zeeman [Ze64$_1$]. It is the standard complex of the presentation $\langle a | aaa^{-1} \rangle$:

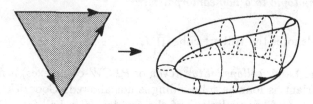

Figure I.12. Dunce hat construction

The dunce hat doesn't allow any collapse, as each 1-cell locally bounds at least two sheets. By use of local homology as in § 1, this "can't start"–argument generalizes to any CW-decomposition of $|H|$ and to the p.l. setting. Nevertheless we will establish p.l. collapses

$$(24) \qquad |H| \nearrow |H| \times I \searrow *$$

via suitable $PLCW$-decompositions and -collapses[14] of $|H| \times I$:

The collapse $(|H| \times I \searrow |H|)$ is obvious: $|H| \times I$ inherits a natural "cylindrical" structure $H \times I$ from H which can be collapsed with decreasing dimensions through the "tops" $e^n \times \{1\}$ down to the "bottom" $H \times \{0\}$. For the collapse $(|H| \times I \searrow *)$ we subdivide $H \times I$ by a "slanted sheet" as in Figure 13 and collapse the 3-dimensional material from above and below; see Figure 14:

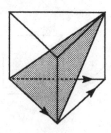
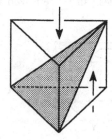

Figure I.13.   Figure I.14

What is left, is the edge a of H, crossed with I, together with the sheet, the slanted edges of which are identified:

---

[14]Later on we will be less strict with the notation, when p.l. collapses are applied to a $PLCW$-complex.

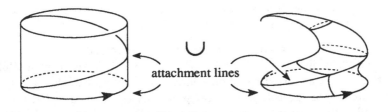

Figure I.15.

If in a next step we begin by collapsing through $a \times \{1\}$

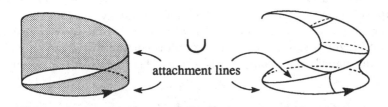

Figure I.16.

and then through $(\{\text{base point}\} \times I)$, we get rid of all 2-dimensional material of $a \times I$. At this point, only the sheet is left over and can easily be collapsed to the basepoint of $H \times \{0\}$.   □

Even though the dunce hat embeds into 3-space, it is not quite easy to visualize. *Bing's house*, which is another example of a contractible but non-collapsible complex, naturally occurs as a spine of the 3-ball:

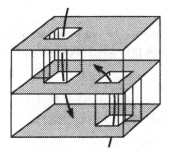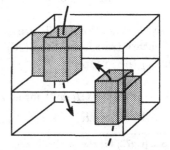

Figure I.17. Floors and panels of Bing's house; enter the bottom room from the roof and the top room from below.

We have given a modified version[15] of Bing's house which has the further advantage of being a special polyhedron.

The same "can't start"-argument as for the dunce hat implies the non-collapsibility, while {Bing's house} $\searorrow I^3 \searrow *$ yields the contractibility. Note that we have started with something obviously collapsible, namely the 3-ball, but one of the collapsing strategies "got stuck": we ended up with a 2-complex that doesn't have any free faces. One dimension lower though *in order to check the collapsibility of a complex K^2, we can just do collapses "helter-skelter" (R. Lickorish) and see whether eventually we come down to a point.* To see this, observe first that $K^n \searrow *$ implies that no collapsing strategy gets stuck prior to the collapse of all of the highest dimensional material. (A strategy that hasn't removed all n-cells yet, could be extended by the first collapse of the remaining n-cells in the original strategy.) In particular, when dealing with a collapsible 2-complex, any sequence of collapses can be continued until a graph is obtained. But a graph is contractible if and only if it is collapsible namely if and only if it is a tree. This completes the argument. □

2.3   3-deformation types and Q^{**}-transformations

The process in § 1.3 to read off a presentation $\mathcal{P} = \langle a_1, \ldots, a_g | R_1, \ldots, R_h \rangle$ for the fundamental group of a finite, connected CW-complex K^2 involves the choice of a spanning tree, base points, orientations, connecting arcs and attaching maps. But different choices only amount to

(25)   a) replacing an R_j by $wR_j^{\pm 1}w^{-1}$ for some $w \in F(a_i)$ (compare Exercise (11a) above on different attaching maps for a 2-cell e_j^2), and

b) replacing the a_i (in the R_j) by the result of a change of basis[16] or of an isomorphism applied to $F(a_i)$.

These modifications of $\mathcal{P}$ are among those that can be achieved using the elementary transformations (26), (27), (28), which preserve (the isomorphism type of) the group $\pi = F(a_i)/N(R_j)$:

(26)   a) $R_j \to wR_jw^{-1}$ for some j and $w \in F(a_i)$ *(conjugation)*,

b) $R_j \to R_j^{-1}$ for some j *(inversion)*, and

---

[15] See [Co73] for the original design.

[16] If T and T' are two different spanning trees for K^1, then one can add to T an edge e^1 in $T'-T$ and remove one of the $T-T'$ edges of the path which connects the distinct endpoints of e^1 in T. Finitely many such changes convert T into T'. They yield concrete elementary Nielsen transformations (compare (27)) which change the basis a_i of $\pi_1(|K^1|, e^0)$ with respect to T into a basis with respect to T'; see [Re32].

c) $R_j \to R_j \cdot R_k$ or $R_k \cdot R_j$ for some $k \neq j$ *(multiplication on right or left).*

In each elementary step, the relators with an index $\neq j$ remain unchanged.

Definition: According to [Ra68$_2$], a finite sequence of moves of type (26) – where j may vary – applied to a finite presentation $\mathcal{P} = \langle a_1, \ldots, a_g | R_1, \ldots, R_h \rangle$ is called a *Q-transformation.* We speak of *Q*-transformations* if, in addition, the following elementary operations are admitted:

(27) in the relators one replaces a_i throughout by

    a) a_i^{-1},   or   b) $a_i a_k$ or   c) $a_k a_i$, $k \neq i$.

    (An elementary operation consists of exactly one of the cases a) b) c). The generators with index $\neq i$ remain unchanged.)

The operations of type (27) generate all *free (= Nielsen) transformations*[17]; see [Ni19].

Q**-transformations[18] are obtained if, moreover, we allow

(28) to enlarge $\mathcal{P}$ by a new generator a and a new relator $R = a$ *(prolongation)*, or to perform the inverse operation if the generator a appears in no other relator than $R = a$.

The *Q**-class* of $\mathcal{P}$ is also called the *presentation class* of $\mathcal{P}$ and will be denoted by $\Phi(\mathcal{P})$.

Because of the remark leading to (25),

(29) a finite, connected CW-complex K^2 determines a well defined *presentation class* $\Phi(K^2)$ of $\pi = \pi_1(|K^2|)$,

given by $\Phi(K^2) = \Phi(\mathcal{P})$ for any $\mathcal{P}$ which is read off from K^2 according to § 1.3.

But there is an even stronger connection between Q^{**}-classes and geometry. It is motivated by the (generalized) Andrews-Curtis problem which asks if a simple-homotopy equivalence between finite 2-dimensional CW-complexes K^2, L^2 can always be turned into a 3-deformation, i.e.,

(30) whether $K^2 \searrow L^2$ implies $K^2 \overset{3}{\searrow} L^2$; see § 4.1 below.

---

[17]In fact, c) and *permutations* of the a_i can be effected by a) and b) (Exercise); inversion and multiplication on one side likewise generate all transformations of the relators in (26) c).

[18]Instead of Q^{**}-transformations ([Me76] and [Me79$_1$]), [AnCu66] and [Cr79$_1$] use the terminology *extended Nielsen transformations*.

The central result of this subsection is that the *3-deformation types* of compact, connected 2-complexes bijectively correspond to presentation classes:

Theorem 2.4 *The assignment $K^2 \to \Phi(K^2)$ induces a bijection between 3-deformation types of compact, connected 2-dimensional CW-complexes and Q^{**}-classes of finite presentations, the inverse bijection being induced by associating to a finite presentation $\mathcal{P}$ its standard complex $K_{\mathcal{P}}$.*

Remark: For the p.l. version of this result, see P. Wright [Wr75]; compare also [Yo76]. In the CW-case we will have to deal with technical complications arising from general attaching maps (see the proof of Lemma 2.6 below).

Proof of Theorem 2.4: A) We first consider the transition $\mathcal{P} \to K_{\mathcal{P}}$. Note that the operations (27) and (28) together allow

(31) to enlarge $\mathcal{P}$ by a new generator a and a new relator $R = w^{-1}a$ (*generalized prolongation*), where w is a word in the original $a_i^{\pm 1}$, or to perform the inverse operation (if possible).

Moreover, free transformations of the generators (27) can be replaced by α) generalized prolongations and their inverses, and β) Q-transformations (Exercise). Thus we must only check the effect of α) and β) on the associated standard complexes $K_{\mathcal{P}}, K_{\mathcal{P}'}$:

α) A generalized prolongation induces a 2-expansion from $K_{\mathcal{P}}$ to $K_{\mathcal{P}'}$, its inverse a 2-collapse.

β) An elementary Q-transformation replaces the attaching map of e_j^2 (associated to R_j in $K_{\mathcal{P}}$) by one for the 2-cell $e_j'^2$ corresponding to the result of (26) a) b) or c). In each case, the attaching maps of e_j^2 and $e_j'^2$ are homotopic maps from S^1 to $K_{\mathcal{P}} - \{e_j^2\} = K_{\mathcal{P}'} - \{e_j'^2\}$. Hence, by § 2.1., Lemma 2.1, there is a transient 3-deformation $K_{\mathcal{P}} \overset{3}{\wedge} K_{\mathcal{P}'}$.

With respect to later considerations, we note that in this transition from presentation classes to 3-deformation types,

(32) all complexes, including the 3-dimensional intermediate stages of β), can be assumed to be *PLCW*.

B) The direction $K^2 \to \Phi(K^2)$ will use two lemmas, the proofs of which we postpone till after (34). The first one and its proof is a generalization of

([Wr75], Lemma 2) to CW-complexes and arbitrary dimensions[19]:

Lemma 2.5 *If a CW-complex K is transformed into L by expanding finitely many $(n+1)$-cells e_i^{n+1} (and their free faces) and afterwards collapsing these e_i^{n+1} again (from potentially different free faces), then there is a transformation from K to L by finitely many transient $(n+1)$-moves.*

This result will help to avoid an accumulation of 3-cells; see A) β).

Lemma 2.6 *If a connected CW-complex contains a 3-cell e^3 with two different free faces e^2, e'^2, then the process of reading off a presentation for π_1 yields relators R for e^2, R' for e'^2 which satisfy* [20]

(33) $R = wR'^{\pm 1}w^{-1} \cdot S$,

   where S is a consequence of those relators which correspond to 2-cells in $\partial e^3 - (e^2 \cup e'^2)$.

Continuing the proof of Theorem 2.4, let K^2 and L^2 be related by a 3-deformation. Because of (14), we may assume $K^2 \nearrow K_1^2$, $L_1^2 \searrow L'^2$ where K_1^2 and L_1^2 fulfill the hypothesis of Lemma 2.5 with $n = 2$, and L'^2 is isomorphic to L^2. Hence, there is a deformation $K^2 \nearrow K_1^2 \overset{3}{\rightsquigarrow} L_1^2 \searrow L'^2$, the middle symbol indicating a sequence of transient 3-moves. By (25), we may choose any spanning tree etc. when reading off presentation classes. Thus in $K^2 \nearrow K_1^2$, an elementary 1-expansion may be taken to solely enlarge the tree; for a 2-expansion the tree may be kept and the presentation may be enlarged by a generalized prolongation.

Hence, we get $\Phi(K^2) = \Phi(K_1^2)$ and likewise $\Phi(L_1^2) = \Phi(L'^2) = \Phi(L^2)$, the last equality trivially holding because of the isomorphism between L'^2 and L^2. We finish the argument by showing $\Phi(K_1^2) = \Phi(L_1^2)$: If, in the chain of transient 3-deformations from K_1^2 to L_1^2, a 3-cell e_j^3 is collapsed from the same free face that was used for its expansion, then the 2-complex hasn't changed during this deformation; if the two free faces are different, then (33) ensures that the corresponding relators before and afterwards differ by a Q-transformation.

---

[19] An n-dimensional and simplicial version of Lemma 2.5 and Theorem 2.4 is given in [KrMe83]. In our case $n = 2$, Lemma 2.5 can be replaced also by an algebraic argument using Lemma 2.6; see [DeMe88].

[20] By (25) above, an identity of type (33) is independent of the choice of geometric data in § 1.3.

C) If $\mathcal{P}$ is read off from K^2, then the composed transitions $K^2 \to \mathcal{P} \to K_\mathcal{P}$ give rise to a 3-deformation: The contraction of a spanning tree of K^2 can be achieved by a 3-deformation (see (15) above) without changing the R_j to be read off; the 2-cell corresponding to R_j can then be normalized to the one of $K_\mathcal{P}$ by an application of Lemma 2.1.

Once more we observe that

(34) if K^2 is a $PLCW$-complex, then $K^2 \overset{3}{\searrow} K_\mathcal{P}$ can be achieved by a $PLCW$-3-deformation.

For the opposite direction of compositions, it is obvious that the data in reading off a presentation from $K_\mathcal{P}$ can be chosen such that $\mathcal{P} \to K_\mathcal{P} \to \mathcal{P}$.

Hence the assignments $K^2 \to \Phi(K^2)$ and $\mathcal{P} \to K_\mathcal{P}$ induce inverse maps between 3-deformation types of compact connected 2-dimensional CW-complexes and Q^{**}-classes of finite presentations. □

Remarks: Using (32) and (34), it follows 1) that Theorem 2.4 is also true in the $PLCW$-case; 2) (via the transition to presentation classes:) $PLCW$-complexes K^2 and L^2 which fulfill $K^2 \overset{3}{\searrow} L^2$ as CW-complexes are already related by a $PLCW$-3-deformation (compare (18) above). The latter holds iff there exists a p.l. 3-deformation from $|K^2|$ to $|L^2|$; see [Wr75].

Proof of Lemma 2.5: Let D be the given deformation from K to L. Enumerate the $(n+1)$-cells $e_1^{n+1}, \ldots, e_k^{n+1}$ in the order in which they appear in D, and let e_i^n denote the free face through which e_i^{n+1} is eventually collapsed. Construct a transient deformation D' from K to L in the following manner: When e_1^{n+1} is expanded in D, let it also be expanded in D' but immediately collapsed via e_1^n. Instead of expanding e_2^{n+1} in D, we expand $e_2'^{n+1}$ (in D'), which is obtained from e_2^{n+1} by identifying[21] each point of $\partial e_2^{n+1} \cap e_1^n$ with its image under a retraction map $\partial e_1^{n+1} \to (\partial e_1^{n+1}) - (e_1^n)$. e_2^n must be a free face of $e_2'^{n+1}$: this could fail only if, in D both $e_1^n \cap \partial e_2^{n+1}$ and $e_2^n \cap \partial e_1^{n+1}$ were nonempty, which is impossible since it would block the collapse of e_1^{n+1} *and* e_2^{n+1} in D. Collapse $e_2'^{n+1}$ via e_2^n.

Similarly, replace each subsequent e_i^{n+1} of D by $e_i'^{n+1}$ in D', which is obtained from e_i^{n+1} by dragging each point of ∂e_i^{n+1} along previous retraction maps of D'. $e_i'^{n+1}$ can be collapsed immediately via e_i^n, for otherwise there would exist some "cycle" of nonempty intersections: $e_i^n \cap \partial e_{i_1}^{n+1} \neq \emptyset$, $e_{i_1}^n \cap \partial e_{i_2}^n \neq \emptyset, \ldots, e_{i_m}^n \cap \partial e_i^{n+1} \neq \emptyset$ with disjoint indices; and D could never have entered the cells $e_i^{n+1}, e_{i_1}^{n+1}, \ldots, e_{i_m}^{n+1}$. □

---

[21] Similar quotients turn attaching/characteristic maps for e_2^{n+1} into analogous maps for (a well defined) $e_2'^{n+1}$, although different ones might be used for the expansion and the collapse; see footnote 22.

Proof of Lemma 2.6: By the freeness of e^2 and e'^2, we have characteristic maps $\Phi, \Phi' : I^3 \to \bar{e}^3$ such that

(35) $\Phi|I^2 \times \{0\}$ is characteristic for e^2, $\Phi^{-1}(\bar{e}^2) = I^2 \times \{0\}$;
     $\Phi'|I^2 \times \{1\}$ is characteristic for e'^2, $\Phi'^{-1}(\bar{e}'^2) = I^2 \times \{1\}$.

It would be easy to prove (33) if we could assume in addition that Φ equals Φ'. But in general $\Phi|I^2 \times \{1\}$ and $\Phi'|I^2 \times \{0\}$ will not be injective. We use homotopy considerations[22] to bypass the problem of finding a characteristic map for e^3 which simultaneously establishes e^2 and e'^2 as free:

Select a point p of e'^2. Although $\Phi^{-1}(p)$ in general will be different from a single point, it is not hard to see that

(36) $\Phi^{-1}(p)$ is a compact, *connected* subset of $\partial I^3 - (I^2 \times \{0\})$.
     (Compare Φ and Φ', details left as an exercise.)

$\Phi^{-1}(p)$ then is contained in the interior $\overset{\circ}{M}$ of a compact, connected(!) p.l. submanifold M^2 of $\partial I^3 - (I^2 \times \{0\})$, which fulfills $\Phi(M^2) \subseteq e'^2$.

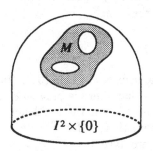

Figure I.18.

As $\Phi|(\partial I^3 - \overset{\circ}{M}) : (\partial I^3 - \overset{\circ}{M}) \to \partial e^3$ does not meet p, we can deform this map rel. $I^2 \times \{0\}$ to miss e^2 entirely. This homotopy extends to a homotopy of $\Phi|\partial I^3$, which maps $M^2 \times I$ to $\bar{e}'^2$. Applying cellular approximation rel. $I^2 \times \{0\}$ ([Schu64], III. 3.5), we get an $f : \partial I^3 \to \partial e^3$ such that

(37)   a) $f \simeq \Phi|\partial I^3$ rel. $I^2 \times \{0\}$;

---

[22]The proof given here has been communicated privately since 1976; see ([Me79₁], footnote 2). R.A.Brown [Br92] established the existence of simultaneous characteristic maps for free faces of CW-3-cells. The analogous question for faces of n-cells, $n > 3$, seems to be open.

b) $f^{-1}(e^2) = \overset{\circ}{I}{}^2 \times \{0\}$ (this property can be preserved throughout the deformation);

c) $f^{-1}(e'^2) \subseteq M^2$ and $f(M^2) \subseteq \bar{e}'^2$;

d) f maps the 1-skeleton (according to Figure 19) of a CW-decomposition of ∂I^3 to the portion of ∂e^3 that lies in the 1-skeleton.

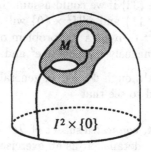

Figure I.19.

As all of ∂M is mapped to $\partial e'^2$ by f, this cellular dissection of ∂I^3 and (37) give rise to an identity of the form[23]

$$(38) \quad R = wR'^\gamma w^{-1} \cdot S, \quad \gamma \in \mathbb{Z},$$

with R, R', S having the same meaning as in (33). (In short: The connectivity of $\Phi^{-1}(p)$ and of M yields that there is only *one* "R'-patch" in (38).) But f is homotopic to $\Phi|\partial I^3$; hence, γ is the incidence number $[e^3, e'^2]$, which is independent of the characteristic map for e^3; compare Chapter II, § 3. Using Φ', i.e., the freeness of e'^2, we have $\gamma = [e^3, e'^2] = \pm 1$. $\qquad\square$

Theorem 2.7 *The presentation class* $\Phi(K^2)$ *is an invariant of the homeomorphism type*[24] *of* $|K^2|$.

Proof: Let K^2, L^2 be finite, connected CW-complexes which are homeomorphic. Without loss of generality, we may assume $|K| = |L|$. Select a finite number of points x_i such that each 2-cell of K^2 and each 2-cell of L^2 contains at least one x_i, and none of the x_i lies in $|K^1| \cup |L^1|$. There also exist pairwise disjoint 2-disc-neighbourhoods $U(x_i)$ which still don't meet the 1-skeleta. Define $\mathcal{J}$ to be $|K| - \underset{i}{\cup} \overset{\circ}{U}(x_i) = |L| - \underset{i}{\cup} \overset{\circ}{U}(x_i)$.

---

[23] Compare footnote 20.

[24] Compare (23) above. In the p.l. case, a proof of Theorem 2.7 could be based on the Hauptvermutung; see § 1.2. But with respect to Theorem 1.2 of § 1.2, the CW-case also has to take care of combinatorially distinct 2-complexes.

Every 2-cell is punctured at least once; hence $\pi_1(\mathcal{J})$ is free (see below for a concrete basis). We will show that

(39) by expressing the boundaries $\partial U(x_i)$ ($\overset{\wedge}{=}$ defining relators of a presentation) with respect to any[25] basis ($\overset{\wedge}{=}$ generators of the presentation) of $\pi_1(\mathcal{J})$, we obtain a member of $\Phi(K^2)$.

As (39) symmetrically holds for $\Phi(L^2)$ as well as for $\Phi(K^2)$, this yields $\Phi(K^2) = \Phi(L^2)$. Thus it suffices to prove (39):

Reindex the $U(x_i)$ such that e_j^2 contains $U(x_{j,1}, \ldots U(x_{j,k_j}))$, $k_j \geq 1$. Subdivide K^2 by connecting the $\partial U(x_{j,\nu})$ with ∂e_j^2 according to Figure 20. This figure also indicates further obvious notations.

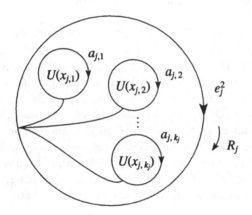

Figure I.20.

Note that $\pi_1(\mathcal{J})$ is freely generated by a basis of $\pi_1(K^1)$ together with the $a_{j,\nu}$, $\nu \geq 2$. (To see this, collapse the $e_j^2 - \overset{k_j}{\underset{\nu=1}{\cup}} U(x_{j,\nu})$ from $\partial U(x_{j,1})$). With respect to this basis the $\partial U(x_{j,\nu})$ give rise to the relators $a_{j,1} = R_j \cdot a_{j,k_j}^{-1} \cdot \ldots \cdot a_{j,2}^{-1}, a_{j,2}, \ldots, a_{j,k_j}$. Hence we have to show that the presentation $\mathcal{P} = \langle \pi_1(K^1) - \text{basis}, a_{j,\nu}(\nu \geq 2) | R_j \cdot a_{j,k_1}^{-1} \cdot \ldots \cdot a_{j,2}^{-1}, a_{j,2}, \ldots, a_{j,k_j} \rangle$ (with j ranging over the 2-cells of K^2) is Q^{**}-equivalent to (the) one read off from K^2. But this follows from the chain of Q^{**}-transformations

$$\mathcal{P} \rightarrow \langle \pi_1(K^1) - \text{basis}, a_{j,\nu}(\nu \geq 1) | R_j \cdot a_{j,k_1}^{-1} \cdot \ldots \cdot a_{j,2}^{-1}, a_{j,1}, \ldots, a_{j,k_j} \rangle$$
$$\rightarrow \langle \pi_1(K^1) - \text{basis}, a_{j,\nu}(\nu \geq 1) | R_j, a_{j,1}, \ldots, a_{j,k_j} \rangle \rightarrow \langle \pi_1(K^1) - \text{basis} | R_j \rangle. \quad \square$$

---

[25] A different one would result in a Q^{**}-transformation of type (27).

Discussion and Additions:

If one enlarges Q^{**}-transformations by the elementary operation which allows to pass from $\mathcal{P} = \langle a_1, \ldots, a_g | R_1, \ldots, R_h \rangle$ to $\mathcal{P} = \langle a_1, \ldots, a_g | R_1, \ldots, R_h, 1 \rangle$, or to perform the inverse operation (if possible), then all[26] *Tietze transformations* are obtained. By these it is possible to convert any two (finite) presentations of a given group π into each other (Exercise, *Theorem of Tietze* [Ti08]). Q^{**}-transformations, however, preserve the *deficiency*[27] $h - g$ of $\mathcal{P}$, or, equivalently, 3-deformations preserve the Euler characteristic $\chi(|K^2|) = 1 - g + h$. The theorem of Tietze yields that

(40) any two finite, connected CW-complexes K^2 and L^2 with isomorphic fundamental groups fulfill[28] $K^2 \vee S_1^2 \vee \ldots \vee S_k^2 \overset{3}{\wedge} L^2 \vee S_1^2 \vee \ldots \vee S_\ell^2$ for suitably chosen finite numbers k, ℓ of 2-spheres.

But until the mid seventies it was unknown, whether (a finitely presentable group) π and the deficiency together are sufficient to determine a presentation class or not: Although an algebraic homotopy classification of finite, connected 2-complexes had been obtained (see Chapter II, § 4), one didn't have examples with $K^2 \not\simeq L^2$ but equal π_1 and χ. Such examples were first given by Dunwoody [Du76] and Metzler [Me76]; see also Sieradski [Si77]. A systematic treatment of this phenomenon is a main topic of Chapter III. Presentation classes even contain genuine simple-homotopy information, as simple-homotopy type and homotopy type differ already in dimension 2, see the references given in (21) above. At the time of writing, only (30) contributes a question mark in the following chain of implications:

(41)

$$K^2 \overset{3}{\wedge} L^2 \underset{?}{\overset{?}{\rightleftarrows}} K^2 \wedge L^2 \rightleftarrows K^2 \simeq L^2 \rightrightarrows$$

$$\pi_1(|K^2|) \approx \pi_1(|L^2|) \text{ and } \chi(|K^2|) = \chi(|L^2|),$$

where K^2, L^2 are finite, connected CW-complexes.

Because of (40), all distinguishing examples necessarily have a "small" Euler characteristic; in many cases the minimal value of χ is required; see Chapters III, VII and IX.

---

[26] The usual list of Tietze transformations (generalized prolongation, adding an arbitrary consequence of the R_j, and the inverses of these operations) is easily seen to be equivalent to this larger one; see the exercise after (31).

[27] Sometimes the deficiency is defined to be $g - h$.

[28] From this it can be deduced that there is no algorithm which decides whether two arbitrary finite presentations are Q^{**}-equivalent or not; see [Me85] and Chapter XII, § 2.1.

Every finite, connected CW-complex K^3 with $K^3 \searrow K^2$ determines a presentation class by defining $\Phi(K^3) = \Phi(K^2)$. It is well defined for, if L^2 is another 2-spine of K^3, then $K^2 \nearrow K^3 \searrow L^2$ implies $\Phi(K^2) = \Phi(L^2)$ (Theorem 2.4). The analogous definition can be made in the p.l. case (see the remarks following the proof of Theorem 2.4).

In particular, *every compact, connected (p.l.) 3-manifold M^3* with nonempty boundary *determines a presentation class* $\Phi(M^3)$. If $\partial M = \emptyset$, we first remove the interior $\overset{\circ}{D}{}^3$ of a p.l. 3-ball D^3 in M^3 and define $\Phi(M^3)$ to be the presentation class of the complement. By connectedness this doesn't depend on the choice of D^3 ([He76], Th. 1.5) and, because of the Hauptvermutung for 3-manifolds, $\Phi(M^3)$ *is a topological invariant.* It may be considered as the "algebraic essence" which remains when one "forgets" that relators live on a surface (compare § 3.1 as well as Reidemeister [Re33] and Singer [Si33]).

Clearly

(42) $\Phi(S^3) = \Phi(D^3) = \Phi_0,$

where Φ_0 is the class of the empty presentation or of a *trivial presentation* $\langle a_1, \ldots, a_g | a_1, \ldots, a_g \rangle$.

Presentation classes Φ and Ψ may be added by taking "disjoint" representatives $\langle a_i | R_j \rangle$, $\langle b_k | S_\ell \rangle$ and passing to the class of $\langle a_i, b_k | R_j, S_\ell \rangle$. The sum corresponds to the one-point union of 2-complexes and the free product of the groups which are presented; see Chapter XII, § 3. Thus one gets an *abelian semigroup $\mathcal{H}$ of presentation classes* with the neutral element Φ_0. But several interesting algebraic questions on this semigroup are still unsolved; see § 4.4.

3 P.L. Embeddings of 2-Complexes into Manifolds

If K is a subcomplex in the interior of a triangulated p.l. manifold M, then a sufficiently small "regular" p.l. neighbourhood $N(|K|)$ of K in M may be obtained as follows: Thicken the 0-simplices to p.l. balls, and connect those by "tubes", for which there is a connecting 1-simplex in K; see Figure 21.

The result is a "handlebody around" K^1. Continue to thicken, if K contains simplices of dimensions ≥ 2; see again Figure 21: they likewise give rise to handles of corresponding *indices* ≥ 2 which are attached inductively to the boundary of the previous thickening.

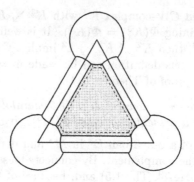

Figure I.21.

This intuitive idea simultaneously leads to the theory of *handlebodies* (see [Hu69], [Ki89] and [RoSa72]) and to Whitehead's theory of regular neighbourhoods [Wh39]. Both will be needed in this section.

A formal definition of regular neighbourhoods may be given as follows:

Suppose X is a closed subpolyhedron of some polyhedron Y with triangulations (L, K) for (Y, X). Let L'' be a second derived subdivision of L inducing the subdivision K'' of K, and let $N(L'', K'')$ be the simplicial neighbourhood of K'' in L'', i.e., the union of all simplices of L'' which have at least one vertex in K'', together with their faces. Then $|N(L'', K'')|$ is called a *regular neigbourhood $N_Y(X)$ of X in Y*.

We list some basic facts, for which we refer to Cohen [Co69] and [RoSa72]; see also [Ze63-66]. Rourke and Sanderson present a treatment of regular neighbourhoods in (general) polyhedra based on [Co69]:

(43) Let N_1, N_2 be two regular neighbourhoods of X in Y. Then there exists a p.l. homeomorphism $h : Y \to Y$ with $h(N_1) = N_2$ and $h|X = \mathrm{id}_X$. If X is compact, h has compact support.

(44) If X is a compact subpolyhedron of Y, then any regular neighbourhood $N_Y(X)$ fulfills $N_Y(X) \searrow X$.

Regular neighbourhoods of polyhedra in the interior of a p.l. manifold M are themselves p.l. manifolds. An essential tool in this manifold case (compare the handle idea above) is that

(45) if Y is a compact subpolyhedron of $\overset{\circ}{M}$ and if $Y \searrow X$, then every regular neighbourhood of Y is also a regular neighbourhood of X.

In the case of $X = *$, (45) and (43) together yield *Whitehead's characterization of p.l. balls*:

(46) A p.l. manifold collapses to a point if and only if it is a p.l. ball.

In fact, all regular neighbourhoods in the interior of a p.l. manifold M can be characterized by a *collapsing criterion*:

(47) Let X be a compact subpolyhedron of $\overset{\circ}{M}$ and let $N \subseteq \overset{\circ}{M}$ be a polyhedral neighbourhood of X. Then N is a regular neighbourhood of X iff

    (i) N is a p.l. submanifold of M, and

    (ii) $N \searrow X$.

A submanifold $F^k \subset M^n$ is *properly* embedded if $F \cap \partial M = \partial F$. *Cutting along* a properly embedded $(n-1)$-dimensional submanifold is the same as removing a regular neighbourhood of F and passing to $\overline{M - N(F)}$.

Throughout this section we confine ourselves to p.l. spaces and p.l. embeddings. All complexes are assumed to be compact, connected and $PLCW$.

3.1 3-dimensional thickenings

At the beginning of § 2.2 we saw that every (compact, connected, see above) 3-dimensional manifold M^3 with boundary has a 2-dimensional spine K^2.

The other way round, i.e., starting with a complex, several questions arise:

(48) Does K^2 embed into the interior of some (orientable)(closed) 3-manifold?

(49) If so, are regular neighbourhoods ("thickenings") uniquely determined by K^2?

(50) If not, does there exist an L^2 with $L^2 \overset{3}{\wedge} K^2$ (or $L^2 \wedge K^2$, or $L^2 \simeq K^2$, or with $\pi_1(K^2) \approx \pi_1(L^2)$) such that L^2 embeds in some 3-manifold?

The discussion of these questions is the main topic of this subsection[29]. For an approach, we wish to visualize a manifold M^3 (compact, connected, nonempty

---

[29]Throughout this subsection a) any spine of M^3 will be assumed to be 2-dimensional and to be contained in $\overset{\circ}{M}$, compare § 2.2 and [Ze64₁] (first collapse away an open collar from the boundary); b) a handlebody always consists of one 0-handle and some 1-handles (orientable or nonorientable); see [He76].

boundary) by means of some spine K^2. Let $\mathcal{P} = \langle a_1, \ldots, a_g | R_1, \ldots, R_h \rangle$ be a presentation read off from K^2. A regular neighbourhood of the 1-skeleton of K^2 in M^3 is just a handlebody V, each 1-handle corresponding to an a_i. The boundary ∂V intersects K^2 in a collection of pairwise disjoint simple closed curves k_j, namely the boundaries of the discs $\overline{K^2 - N_{K^2}(K^1)}$ which run around the handles according to R_j.

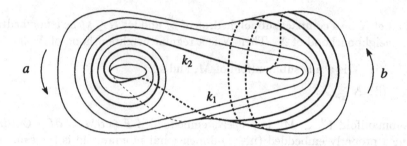

Figure I.22. $(V; k_1, \ldots, k_h)$; $\mathcal{P} = \langle a, b | a^3 bab, a^2 b^3 \rangle$

These discs thicken to $D_j^2 \times I$, where $\partial D_j^2 \times I$ must be identified with a neighbourhood of k_j in ∂V. The resulting manifold is a regular neighbourhood $N(K^2)$ of K^2 in M^3. But the p.l. type of M is also a regular neighbourhood of K^2 in M. This can be shown by addition of a collar to ∂M and application of (47) (Exercise); compare footnote 32 below. Hence (43) implies that $N(K^2)$ and M^3 are p.l. homeomorphic: the tuple $(V; k_1, \ldots, k_h)$ carries all of the information about the manifold we started with[30]. A curve k in a surface is *2-sided* if its regular neighbourhood is an annulus. We summarize:

(51) K^2 with presentation $\langle a_1, \ldots, a_g | R_1, \ldots, R_h \rangle$ has a 3-dimensional thickening if and only if it is possible to draw 2-sided curves k_j on the boundary of a handlebody V with g handles, (one handle for each a_i,) such that k_j reads R_j on V.   □

The reader who tries to use (51) to answer (48) for concrete examples, will soon establish this question negatively. In fact, any spine can be altered by a little 3-deformation to become non-thickenable; see Remark 2 after the proof of Theorem 3.1.

Perhaps even more obvious is the observation that

---

[30]Each $D_i^2 \times I$ cuts an annulus out of ∂V and fills the resulting holes with discs. An easy computation of the Euler-characteristic shows that if $g = h$, ∂M is connected, then $\partial M = S^2$, in which situation we can fill in a 3-ball to obtain a *closed* 3-manifold $\hat{M}^3$. The tuple $(V; k_1, \ldots, k_h)$ is then called a *Heegaard-diagram* of $\hat{M}^3$.

(52) if K^2 (p.l.) embeds into some M^3, then the link (i.e., the boundary of a regular neighbourhood) of each point must be planar.

In particular this test can be applied to the single vertex of a standard complex $K_{\mathcal{P}}$. Its link is called the *link graph* (or *Whitehead graph*) graph of $K_{\mathcal{P}}$ and of $\mathcal{P}$.

For a standard complex K^2, L. Neuwirth [Ne68] exhibited *an algorithm to decide whether K^2 is a spine of some orientable manifold*; see also Chapter VIII, § 1.3. Neuwirth's algorithm, which we are going to describe, is a refinement of (52). We may assume that every 1-cell of K is incident with a 2-cell and that every 2-cell is incident with some 1-cell because wedging on spheres doesn't change the outcome of (48) (as opposed to that of (50)!).

Starting with a spine K^2 of an orientable manifold M^3 as above, using (51) we obtain a tuple $(V; k_1, \ldots, k_h)$. Now cut each handle of V. The k_j give rise to arcs running on a sphere with $2g$ holes:

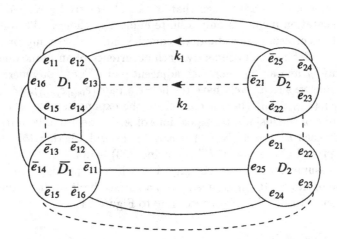

Figure I.23.   ($\mathcal{P}$ as in Figure 22)

Here the e_{ik} (resp. $\bar{e}_{ik}$) for $k = 1, \ldots, r(i)$ denote the intersection points of the boundary of the thickened i^{th} cutting disc with the curves k_j. For each i, they are assumed to be ordered clockwise (resp. counterclockwise) according to an orientation of S^2. The reader should verify that identifying D_i with $\bar{D}_i$ such that e_{ik} falls onto $\bar{e}_{ik}$, converts Figure 23 into Figure 22. He should further observe that prolonging the arcs (which have remained of the k_j) to the centers of the D_i resp. $\bar{D}_i$, yields the link graph $\Gamma = K^2 \cap \partial N_M$ (vertex of K^2).

Following Neuwirth, let E denote the set consisting of all points e_{ik} and $\bar{e}_{ik}$ and define permutations A and C by $C := \prod_{i=1}^{g} (e_{i1} \ldots e_{ir(i)})(\bar{e}_{ir(i)} \ldots \bar{e}_{i1})$ and A a product of disjoint transpositions interchanging the endpoints of the arcs. If X is a permutation of E, denote by $|X|$ the number of its cycles (orbits in E). Note that $|C|$ equals the number of vertices of Γ while $|A|$ equals the number of its edges. In the case of a connected Γ the number of components of $S^2 - \Gamma$ equals $|AC|$. Then the Euler formula yields

(53a) $|C| - |A| + |AC| = 2$.

If Γ has L components,

(53b) $|C| - |A| + |AC| = 2L$

holds by summation over the components. Equation (53) assures that each component of Γ is embedded on a 2-sphere. This description in terms of permutations has the advantage that it can abstractly be read off from a given presentation involving only a finite number of choices. Thus it tells us whether the corresponding standard complex has a thickening or not: the set E can be defined independently, each occurrence of an a_i in some relator defining an e_{ik} and an $\bar{e}_{ik}$. For each adjacent pair $a_i^{\pm 1}, a_j^{\pm 1}$ of generators in a cyclic relator we successively have to choose a pair $(\bar{e}_{ik(i)} e_{jk(j)})$, $(\bar{e}_{ik(i)} \bar{e}_{jk(j)})$, $(e_{ik(i)} e_{jk(j)})$, $(e_{ik(i)} \bar{e}_{jk(j)})$ (bars according to the exponents of a_i and a_j) constituting one of the disjoint transpositions of A. There are only finitely many choices for each k. For a given A it can be checked whether (53) holds. It has been proved above that if K^2 is a spine, (53) is fulfilled. For proving the converse, assume first $L = 1$. For any A which corresponds to the relations of K^2, the orbits of AC in E give rise to a surface built up from the D_i and $\overline{D}_i$ together with labelled discs according to Figure 24.

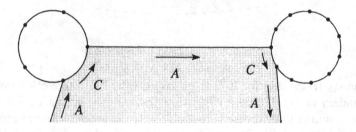

Figure I.24.

If (53) is fulfilled, the surface is a 2-sphere in which Γ embeds in a way that allows the coherent identification of D_i and $\overline{D}_i$. In the case $L > 1$ the same procedure applies to each component of Γ, yielding a collection of surfaces. (53b) guarantees that each of these is a 2-sphere. This implies that again Γ embeds in S^2 in a way that permits the thickening of K^2 to a 3-manifold. $\square$

Remarks: 1) By similar techniques, with the use of an additional permutation B, Neuwirth's algorithm is able to detect whether there is a thickening of K^2 whose boundary is a 2-sphere, thus answering (48) for being a spine of a *closed* orientable 3-manifold; see [Ne68].
2) The interested reader may extend the algorithm to the nonorientable case.

We now turn our attention to question (49), to which the answer is also "no". Counterexamples are provided by punctured *lens spaces*. These are closed 3-manifolds $L(m,n)$ which have been introduced by Tietze 1908. They are quotients of a lens shaped D^3 with its hemispheres divided into m equal sectors,

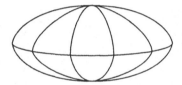

Figure I.25.

where the upper hemisphere is identified with the lower hemisphere after a twist of $2\pi \cdot \frac{n}{m}$; $(m,n) = 1$. By Alexander [Al19] (compare Stillwell [St80], 8.3.5), $L(5,1)$ and $L(5,2)$ are not homeomorphic, a fact that is not changed after removing the interior of a p.l. 3-ball. But a little ball missing in the center of the lens can be enlarged to hollow out all of the 3-material. Now each point below the equator is identified to some point above the equator, hence our spine is the quotient of the upper hemisphere modulo the identifications on the equator. To see what these are, consider the oriented edges on the equators. Since m and n are relatively prime, the $2\pi\frac{n}{m}$ - twist acts transitively on them. Thus we see that the punctured lens space $\overline{L(m,n) - D^3}$ collapses to the standard complex of $\langle a | a^m \rangle$ independently of n even though the homeomorphism type in general differs for different choices of n.

For examples of non-homeomorphic knot spaces with a common spine compare Mitchel, Przyticki and Repovs [MiPrRe86].

On the other hand, question (49) has a positive answer for the restricted class of special polyhedra. The following theorem is essentially due to B.G. Casler [Ca65]; compare also Wright [Wr77] for (a) and Quinn [Qu81] for (b).

Theorem 3.1

(a) *Every compact, connected 3-manifold with nonempty boundary has a spine which is a special polyhedron.*

(b) *A special polyhedron thickens to at most one 3-manifold.*

Proof of (a): The proof works with the "banana and pineapple trick" (compare [Ze63-66], Chapter III, Remark 2): Let $K^2 \swarrow M^3$ be a spine. In M^3, thicken edges of K^2 to sufficiently small, disjoint "bananas"; see Figure 26 (a):

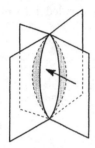
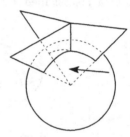

Figure I.26. (a) "banana"   Figure I.26 (b)   "pineapple"

As $K^2 \cup \{$ bananas $\}$ collapses to the M^3-spine K^2, Corollary 4 of [Ze63-66], Chapter III yields:

(54) $M^3 \searrow K^2 \cup \{$ bananas $\}$.

This may also be obtained by stopping certain original collapses earlier (i.e., at the bananas); compare [Co77], 3.1. By collapsing each banana through a longitudinal piece of the peel, we get $M^3 \searrow K_1^2$, where K_1^2 has the property that at most 3 sheets meet at every edge. Next thicken vertices of K_1^2 to sufficiently small, disjoint "pineapples", and apply the same argument(s) as for (54) to this pineapple situation; then collapse each pineapple through a piece of the pineapple peel; see Figure 26 (b). Continue to collapse until no further collapses are possible. The resulting K_2^2 is a spine of M^3 which consists of the union of a closed fake surface L^2 and a graph G. By further modifications as above, we may assume that G intersects L^2 in points of the intrinsic 2-skeleton of L^2.

If $L^2 \neq \emptyset$, thicken each component of G to a handlebody as in Figure 27. Collapse the interior of each handlebody minus a meridian disc per 1-handle through a disc of intersection with L^2; see Figure 27:

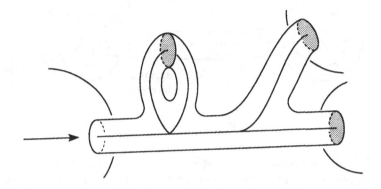

Figure I.27.

We end up with a closed fake surface K_3^2. Again $M^3 \searrow K_3^2$ holds.

The case $L^2 = \emptyset$ can be treated by first creating a Bing's house according to Figure 17 near a vertex of G (Exercise).

By similar modifications, K_3^2 now can be changed into a special polyhedron which is a spine of M^3:

If K_3^2 is not a 2-manifold, choose a collection of disjoint arcs whose endpoints lie in the intrinsic 1-skeleton of K_3^2 while their interior runs in the intrinsic 2-skeleton of K_3^2. Each circle component of the intrinsic 1-skeleton should be hit by an endpoint of some arc and the components of the intrinsic 2-skeleton of K_3^2 should be dissected into discs. Now thicken all these arcs to disjoint cylinders and collapse through one end.

In the case of a 2-manifold, again create a Bing's house as shown in Figure 17 on it, and then proceed as above.

Proof of (b): We will show that K uniquely determines its regular neighbourhood in M^3; by the argument preceding (51) this is p.l. homeomorphic to M^3.

Let K be a special polyhedron and let N_0 be a regular neighbourhood of the intrinsic 0-skeleton in K. Components of N_0 are homeomorphic to the cone on the 1-skeleton of a tetrahedron which thickens uniquely by including the cone on the entire tetrahedron. Denote this thickening of the 0-skeleton by T_0. Next, let N_1 be a regular neighbourhood in K of the intrinsic 1-skeleton.

$T_0 \cup N_1$ can be described as T_0 with copies of $Y \times I$ (where Y denotes the cone on 3 points) attached by embeddings $Y \times \{0,1\} \subset \partial T_0$:

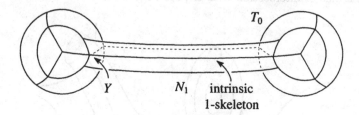

Figure I.28.

Copies of $Y \times I$ uniquely thicken to copies of $D^2 \times I$ that are attached to T_0 by the unique (up to isotopy) extension of the above embeddings to $D^2 \times \{0,1\}$. Denote the resulting thickening of the intrinsic 1- and 0-skeleton of K by T_1. Now $T_1 \cup K$ is just the handlebody T_1 with a collection of discs D_i^2 attached along their boundaries. These discs uniquely thicken to balls $D_i^2 \times I$ which have to be attached along the annulus $\partial D_i^2 \times I$. This is precisely where an attempt to thicken K may run into an obstacle: some ∂D_i^2 could have a Moebius strip as a regular neighbourhood on ∂T_1, in which case it is impossible to thicken D_i^2. The $D_i^2 \times I$ can be expanded if and only if all ∂D_i^2 are two-sided on ∂T_1. But then there are only two isotopy classes for the extension of the embeddings $\partial D_i \subset \partial T_1$ to $\partial D_i^2 \times I$. They differ by multiplication with (-1) in the I-factor, which extends to all of $D_i^2 \times I$ thus yielding homeomorphic thickenings. □

Remarks: 1) *Every compact K^2 3-deforms to a special polyhedron* [Wr73]: The steps in the above proof of a) can be mimicked to work in the absence of a surrounding 3-manifold. Note that however, for the application of the "pineapple-trick", the regular neighbourhoods of the vertices must first be prepared to have a planar boundary graph (Exercise).
2) The above methods of proof show how a *special spine* (i.e., a spine which is a special polyhedron) K^2 of M^3 can be altered by a local modification to become non-thickenable: in the neighbourhood of some point of the intrinsic 1-skeleton deform one of the 3 sheets by adding a little loop:

Figure I.29.

This can be achieved by a 3-deformation of K (Exercise, use § 2, Lemma 2.1). Following the permutation of the sheets along the loop shows that the corresponding handle is a solid Klein bottle. Hence the boundary curve of the enclosed disc D has a Moebius strip as a regular neighbourhood on the boundary of the Klein bottle whence it is impossible to thicken D.

For *all* special polyhedra, a 3-dimensional thickening can be achieved which is only slightly more general than a 3-manifold. This notion is due to F. Quinn [Qu81]:

Definition: A *singular 3-manifold* $\check{M}^3$ is a compact connected polyhedron in which the link of each point is D^2 (*boundary point*), S^2 (*inner point*) or the projective plane P^2 (*singular point*). The set of boundary points is assumed to be nonempty.

Now turn back to the "bad" situation of part (b) in the proof of Theorem 3.1, when we had a Moebius strip on ∂T_1 with some ∂D_i^2 as its center line. In this case we glue a further disc into the boundary curve of the Moebius strip and then cone the resulting projective plane P_i^2 to some centerpoint $c_i \in \overset{\circ}{D_i^2}$. The subcone over ∂D_i^2 is (identified with) the disc D_i^2. Note that all points different from the c_i have manifold-neighbourhoods. It is possible to collapse $c_i \cdot P_i^2$ to the Moebius strip plus the subcone over its center line. The other way round, the $c_i \cdot P_i^2$ can actually be obtained by expansions on $T_1 \cup K$. Having done this for all those $\partial D_i^2 \subset \partial T_1$ for which D_i^2 could not be thickened to $D_i^2 \times I$, we get a well defined thickening $\check{M}^3 = \check{M}^3(K^2)$ (compare [Qu81]):

Theorem 3.2 *Every special polyhedron K^2 determines a singular 3-manifold $\check{M}^3(K^2) \supset K^2$ such that $\check{M}^3(K^2) \searrow K^2$.*   □

A corresponding modification of the proof of Theorem 3.1 (a) yields that special spines also exist for singular 3-manifolds (Exercise, consider in addition "pineapples" which are cones on projective planes.). In Chapter VIII, § 2.1 it will be shown that these transitions from singular 3-manifolds to special polyhedra and vice versa induce inverse bijections between presentation classes and equivalence classes of singular 3-manifolds under certain elementary surgery moves.

Finally we consider questions (50). Again, in general the answer is already "no" for the weakest condition. This leads to the question

(55) what (finitely presented) groups can occur as fundamental groups of 3-manifolds.

For instance, in Hempel's book ([He76], Chapter 9) there is a complete list of the abelian groups which arise as fundamental groups of 3-manifolds; among (nontrivial) finitely generated abelian groups there are only $\mathbb{Z}, \mathbb{Z} \oplus \mathbb{Z}, \mathbb{Z} \oplus \mathbb{Z} \oplus \mathbb{Z}, \mathbb{Z}_n$ and $\mathbb{Z} \oplus \mathbb{Z}_2$. The subcase of closed, orientable 3-manifolds (cyclic groups, $\mathbb{Z} \oplus \mathbb{Z} \oplus \mathbb{Z}$) is due to Reidemeister [Re36].

Stallings has shown [St62] that there is no algorithm to decide whether a given presentation determines a 3-manifold group.

Thus for the remainder of questions (50), we need at least to add the assumption that the underlying group is a 3-manifold group. Still then there are homotopy types which nevertheless don't contain an embeddable 2-complex; see [HoLuMe85], where it is shown that Dunwoody's exotic presentation of the trefoil group [Du76] defines a $K_{\mathcal{P}}$ which is not homotopy equivalent to any 3-manifold spine.

It is not known to the authors whether there are different simple homotopy types which coincide as homotopy types one of which embeds whereas the other doesn't.

For further material on (singular) 3-manifolds, we refer to Chapter VIII and XII.

3.2   4- and 5-dimensional thickenings

We present some relevant examples and basic facts. The arguments which are given use handlebody and regular neighbourhood theory.

In § 2.2 it was already mentioned that the dunce hat H embeds in $\mathbb{R}^3$. Taking a regular neighbourhood $N^3(H)$ with respect to such an embedding, we may form $N^3(H) \times I$ which fulfills $N^3(H) \times I \searrow H \times \{\frac{1}{2}\}$. By (47) this implies that $N^4(H) = N^3(H) \times I$ is a 4-dimensional regular neighbourhood of H in $\mathbb{R}^4$, where H is embedded as $H \times \{\frac{1}{2}\} \subset N^3 \times I \subset \mathbb{R}^4$. But we also have $N^3(H) \times I \searrow H \times I \searrow *$, the last collapse given by (24). Hence, because of (47),

(56) H has a 4-dimensional regular neighbourhood $N^4(H)$ which is a p.l. 4-ball.

(Alternatively, one can deduce (56) by establishing $N^3(H)$ to be a 3-ball.)

As H itself is contractible but not collapsible, one may ask whether there exist regular neighbourhoods of H in the interior of p.l. 4-manifolds which are different from a 4-ball. Indeed, such an example was given by B. Mazur [Ma61]; see also [Ze64₁]. It is constructed as follows:

Attach a 1-handle to D^4. The boundary is the double[31] of a solid 3-dimensional torus T. Attach a 2-handle along the curve $k \subset T$ as in Figure 30:

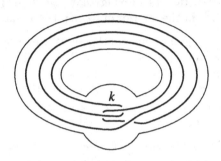

Figure I.30.

In order to determine the 2-handle, one has to choose in addition a curve k' "following" k in the boundary of a regular neighbourhood of k in T, the difference between two choices (up to isotopy) being their twisting around k; see Figure 31.

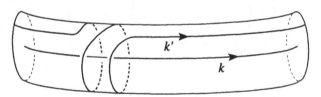

Figure I.31.

This *framing curve* k' bounds a 2-disc in the portion of the 2-handle which lies in the boundary of the resulting 4-manifold $M^4(H)$.

As k approximates the attaching map of the 2-cell of H, it follows (see [Ze64₁]) that

(57) the dunce hat H is a spine[32] of $M^4(H)$.

---

[31]The *double* $2M$ of a manifold M with nonempty boundary is obtained by taking two copies of M and identifying corresponding points of their boundaries.

[32]That $M^4(H)$ occurs as a regular neighbourhood of H in the interior of some p.l. 4-

But, by a particular choice of k' (namely the one "parallel" to k according to the projection of k in Figure 30), Mazur obtained[33] $\pi_1(\partial M^4) \neq \{1\}$. Hence

(58) this *Mazur-manifold* is different from D^4.

The following considerations reveal that linking phenomena as in Figure 30 are the essential obstruction to proving that, for $\Phi(K^2) = \Phi_0$, a regular neighbourhood $N^4(K^2)$ in the interior of some p.l. 4-manifold is a p.l. 4-ball:

Let $N^4(K^2)$ be such a regular neighbourhood. Without loss of generality we may assume K^2 to be simplicial, hence $N^4(K^2)$ has a handle structure[34] given by thickening the simplices of K^2. Choose a spanning tree for K^1 to read off a presentation $\mathcal{P}$. As $\Phi(\mathcal{P}) = \Phi_0$ holds, a presentation $\mathcal{Q}$ which is obtained from $\mathcal{P}$ by generalized prolongations, Q-transforms to a trivial presentation $\mathcal{R} = \langle a_1, \ldots, a_g | a_1, \ldots, a_g \rangle$; compare the exercise after (31).

Unite the handles along the spanning tree to one 0-handle and introduce handles according to the (generalized) prolongations to form $\mathcal{Q}$, see [RoSa72], p. 79; this yields a new handle structure on $N^4(K^2)$. The boundary of the union of the 0-handle and the 1-handles is the double of a solid 3-dimensional handlebody; see Figure 32.

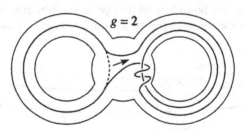

Figure I.32.

The 2-handle attaching curves of the handle structure of $N^4(K^2)$ according to $\mathcal{Q}$ approximate the relator words of $\mathcal{Q}$ by running through the 1-handle parts of Figure 32. We now apply the elementary Q-transformations from $\mathcal{Q}$ to $\mathcal{R}$, leading to further handle structures of $N^4(K^2)$: conjugation of a relator corresponds to an isotopy of an attaching curve; inversion doesn't change the curves at all; multiplication of relators corresponds to 2-handle addition

---

manifold can then be seen by addition of a collar to ∂M^4 and application of (47); compare the exercise preceding (51).

[33] The presentation for $\pi_1(\partial M^4)$ of [Ma61] can be verified by use of the Wirtinger method and the Seifert-van Kampen Theorem (Exercise).

[34] All handlebodies arising in the following considerations will be orientable because of $\pi_1(K^2) = \{1\}$.

([RoSa72], p.50), which means the changing of the attaching curve of a 2-handle by connecting it with a narrow band to the framing curve of another 2-handle (for details see [Ki89], pp. 10-11). One of the attaching curves in Figure 32 is obtained in such a way (the arrow indicating the change). In general these moves produce linking and knotting of the curves. The linking and knotting may be concentrated in the 0-handle part according to Figure 32 such that the relator words run "parallel" through the 1-handle parts; but it may block isotopies for free reductions to obtain the relator words of $\mathcal{R}$ (and already for intermediate stages between $\mathcal{Q}$ and $\mathcal{R}$) as in the case of the dunce hat; see Figure 30.

This potential obstruction to $N^4(K^2)$ being a 4-ball may be turned positively: Define a 2-dimensional CW-complex to be a *generalized dunce hat*, if it is the standard complex of a presentation $\langle a_1, \ldots, a_g | R_1, \ldots, R_g \rangle$, the relators of which may be non-reduced, but such that each R_j freely reduces to a_j. Then the above considerations yield (compare [Me85] and [Hu90]):

Theorem 3.3 *If* $\Phi(|K^2|) = \Phi_0$, *then a regular neighbourhood* $N^4(|K^2|)$ *of a compact, connected subpolyhedron* $|K^2|$ *in a p.l. 4-manifold* [35] *M already occurs as a regular neighbourhood of a generalized dunce hat in M.*   □

The arguments for proving Theorem 3.3 can be "copied" one dimension higher; in addition, the linking obstructions against free reductions of attaching curves vanish: $N^5(K^2)$ thus turns out to be a 5-dimensional regular neighbourhood of the collapsible $PLCW$-complex $K_{\mathcal{R}}$. Hence, (44) and (46) together imply that $N^5(K^2)$ is a p.l. 5-ball (Andrews and Curtis [AnCu65]; see also [Yo76]):

Theorem 3.4 *If* $|K^2|$ *is a compact, connected subpolyhedron of a p.l. 5-manifold M, and if* $\Phi(|K^2|) = \Phi_0$ *holds, then any* [36] *regular neighbourhood* $N^5(|K^2|)$ *of* $|K^2|$ *in M is a p.l. 5-ball.*   □

Theorem 3.4 has the immediate consequence:

(59) If $\Phi(|K^2|) = \Phi_0$, then the double $2N^4$ of an $N^4(K^2)$ is p.l. homeomorphic to S^4,

---

[35]The general case can be reduced to the case $|K^2| \subset \overset{\circ}{M}$ by addition of a collar to ∂M; compare footnote 32.

[36]Compare the preceding footnote.

as $N^4 \times I$ may serve as an $N^5(K^2)$ and $\partial(N^4 \times I)$ is p.l. homeomorphic to $2N^4$. In particular, although (58) holds, we have:

(60) The double of the Mazur-manifold is p.l. homeomorphic to S^4.

In this case, the conclusion of Theorem 3.4 for $M^4 \times I$ alternatively could be deduced from the collapses $M^4 \times I \searrow H \times I \searrow *$; compare (56) above.

We remark that for an arbitrary compact contractible polyhedron $|K^2|$ (i.e., without the assumption $\Phi(|K^2|) = \Phi_0$) Freedman's work [Fr82] yields the conclusions of Theorem 3.4 and of (59) in the topological category (see [Me85], footnote p.36).

By thickening the simplices of a triangulation (remember the construction of the Mazur-manifold), *every compact, connected polyhedron $|K^2|$ can be p.l. embedded into the interior of some p.l. 4-manifold.* As opposed to the 3-dimensional situation, the multitude (not the existence) of regular manifold neighbourhoods is the central topic in dimension 4. But embeddings[37] into $\mathbb{R}^4$ resp. S^4 do not always exist; the first examples (finite, connected 2-complexes which generalize the nonplanar Kuratowski graphs) are due to van Kampen [vK32]. Standard complexes of finite presentations embed into S^4; in fact, G. Huck [Hu90] has constructed embeddings with "nice" complements; his method easily generalizes to an embedding of every finite, 2-dimensional $PLCW$-complex K^2 into S^4, if K^2 has a planar 1-skeleton.

In contrast, there may occur rather "complicated" complements for embeddings of a given 2-complex into S^4: by J.P. Neuzil [Ne73] and B.M. Freed [Fr76], *the dunce hat H embeds into S^4 in infinitely many distinct ways,* distinguished by the fundamental groups of the complements.

We also recommend the study of Akbulut-Kirby [AkKi85], Gompf [Go91] and of Chapter IX of this book as important completions to this section.

4 Three Conjectures and Further Problems

Low dimensional topology carries many unsolved problems. We first introduce three prominent ones, each of which will be the topic of a separate chapter in this book.

---

[37]By [We67], topological embeddability of a compact polyhedron $|K^2|$ into $\mathbb{R}^4$ implies p.l. embeddability.

4.1 (Generalized) Andrews-Curtis conjecture

Theorem 3.4 of the preceding section was the reason for J.J. Andrews and M.L. Curtis to conjecture that every finite, contractible CW-complex K^2 fulfills $\Phi(K^2) = \Phi_0$; see [AnCu65]. This *Andrews-Curtis conjecture*, henceforth abbreviated by (AC), generalizes to question (30) of § 2.3 above, whether a simple-homotopy equivalence between finite 2-dimensional CW-complexes can always be replaced by a 3-deformation. The expectation "yes" as an answer to (30) is the *generalized Andrews-Curtis conjecture* (AC'). But there exist several notorious examples, for which the implication $K^2 \simeq * \Rightarrow K^2 \overset{3}{\diagdown} *$ seems debatable; moreover, we are convinced that counterexamples to (AC') for nontrivial fundamental groups are even more likely to exist; see Chapter XII.

The following relations are of interest:

(61) A 3-deformation class of finite, contractible 2-complexes that does not contain any 3-manifold spine would disprove (AC),

as the "trivial" class $(\overset{\Delta}{=} \Phi_0)$ for instance contains D^2. On the other hand,

(62) a counterexample to (AC) which contains a 3-manifold spine K^2 would disprove the 3-dimensional Poincaré conjecture,

which states that every closed, connected, simply connected 3-manifold is homeomorphic to S^3: Any 3-manifold thickening of K^2 would constitute a fake[38] 3-ball, as $\Phi(D^3) = \Phi_0$ (see (42) above). $\qquad\qquad\square$

The remark after (60) suggests that counterexamples to (AC) may also be related to exotic p.l. structures on D^5 resp. S^4; compare [Go91].

Another motivation for the study of the generalized Andrews-Curtis problem is C.T.C. Wall's result [Wa66] on the dimensions of deformations. He proved:

(63) Let $f : K_0 \to K_1$ be a simple-homotopy equivalence of connected, finite CW-complexes, inducing the identity on the common subcomplex L. If $n = \max(\dim(K_0 - L), \dim(K_1 - L))$, then f is homotopic rel. L to a deformation[39] $K_0 \overset{n+1}{\diagdown} K_1$ which leaves L fixed throughout, provided $n \geq 3$.

---

[38] A *fake 3-ball* is a compact, contractible 3-manifold which is not homeomorphic to D^3. The 3-dimensional Poincaré conjecture is equivalent to claiming that fake 3-balls don't exist (Exercise).

[39] Note that Wall's notation of the dimension of a deformation differs from ours for transient moves.

What can be said about the corresponding result for $n \leq 2$? For $n = 0$ it is trivially true. The case[40] $n = 1$ is easy if L is empty: By (15) we may assume that K_0 and K_1 are a bouquet of circles. Up to homotopy, f is determined by the images of these circles according to the isomorphism $f_* : \pi_1(K_0) \to \pi_1(K_1)$ of free groups. But Nielsen's theorem [Ni19] then yields that f is homotopic to a composition of a finite sequence of cellular homeomorphisms and of maps arising from 2-dimensional transient moves:

Figure I.33.

These cases were already mentioned in [Wa66]. If L is nonempty but connected, the case $n = 1$ was achieved by a "Nielsen theorem for free groups with operators" in [Me79₁]; see also [Br76₁] and [DeMe88]. Wall [Wa80] subsequently removed the connectivity hypothesis on L using a direct geometric argument patterned on Stallings' proof of Grushko's theorem [St65₁].

Hence 2 is the only value of n, where (63) is an open question[41]. For $L = \emptyset$ a positive solution would imply the generalized Andrews-Curtis conjecture. Thus we propose to name the case $n = 2$ of (63) the *relative generalized Andrews-Curtis conjecture*/problem (rel. AC')[42]

4.2   Zeeman collapsing conjecture

In this section we work within the p.l. category. As in § 2.2 (24), $|K| \searrow |L|$ is often proved by performing CW-collapses on $PLCW$-decompositions of $|K|$ and $|L|$ within this p.l. structure.

---

[40] In the case $n = 1$ it is sufficient to solely assume f to be a homotopy equivalence. That f is *simple* then is part of the conclusion.

[41] In this case (63) still implies $K_0 \overset{4}{\nearrow\searrow} K_1$ rel. L.

[42] As for (AC'), one could in this case also drop the requirement that the resulting deformation is homotopic (rel. L) to the given map f.

The collapse of { dunce hat} $\times I$ in § 2.2 (24) shows that H fulfills the

Zeeman Conjecture (Z): A compact contractible polyhedron crossed with the unit interval collapses to a point: $K^2 \simeq * \Rightarrow |K^2| \times I \searrow *$.

Zeeman also established the "Zeeman-property" $|K| \times I \searrow *$ for the standard complexes of $\langle a, b | ab, a^n b^{n+1} \rangle$ and $\langle a, b | ab^n, ab^{n+1} \rangle$. But so far it is not even known whether all presentations of the form $\langle a, b | a^m b^n, a^r b^s \rangle$ with $ms - nr = \pm 1$ fulfill $|K| \times I \searrow *$; see Chapter XI, §3, for a detailed discussion.

(Z) is of particular interest because of the following:

(64) Zeeman's Conjecture implies the 3-dimensional Poincaré-Conjecture.

According to Zeeman [Ze64$_1$] the following type of argument to prove (64) is probably due to M.L. Curtis:

Let $\hat{M}^3$ be a closed connected and simply connected 3-manifold. Remove the interior of a p.l. 3-ball to form $M = \hat{M} - \overset{\circ}{D}^3$. M is connected and simply connected and we have $\chi(M) = \chi(\hat{M}) + 1 = 1$. Collapse M to some spine K^2. By the exercise in the end of §1.3, K^2 is contractible. Hence our assumption yields $M^3 \times I \searrow K^2 \times I \searrow *$. Thus (46) implies that $M^3 \times I$ is a p.l. 4-ball with $M^3 \times \{0\}$ being contained in the p.l. 3-sphere $\partial(M^3 \times I)$, compare (56) above. But ∂M^3 is the boundary of the open ball which was removed, so ∂M is a 2-sphere. Hence, by the theorem of Alexander-Schoenflies [Mo77], see also [Ha89], M^3 is homeomorphic to D^3. Therefore $\hat{M}^3$ is an ordinary 3-sphere after all, which proves (64). $\qquad\square$

Note that it suffices that *some* spine of M^3 fulfills $|K^2| \times I \searrow *$.

As $|K^2| \nearrow |K^2| \times I \searrow *$ would be a 3-deformation,

(65) Zeeman's Conjecture implies the Andrews-Curtis Conjecture.

But it is (even) unknown whether (Z) can be established under the hypothesis of (AC); compare the examples preceding (64), which are contained in the trivial presentation class.

The general case of deformations between compact, connected 2-dimensional polyhedra can be related to collapses of $K \times I$ as follows:

$$K^2 \times I \searrow L'^2 \quad \text{or} \quad L^2 \times I \searrow K'^2 \underset{?}{\rightleftharpoons} K^2 \nwarrow L^2,$$

where K'^2 resp. L'^2 is p.l. homeomorphic to K^2 resp. L^2.

The *generalized Zeeman-Conjecture* (Z') is the positive expectation to the questioned implication. $(Z') \Rightarrow (Z)$ and $(Z') \Rightarrow (AC')$ are obvious.

4.3 Whitehead asphericity conjecture as a special problem of dimension 2

A connected 2-dimensional CW-complex K is called *aspherical* if $\pi_2(K^2) = \{0\}$ (compare Chapter X, § 1). In [Wh41₁], J.H.C. Whitehead raised the following question:

(66) Is any (connected) subcomplex of an aspherical 2-complex itself aspherical?

Or, reversely: is it possible to create nontrivial π_2-elements by omitting material from an aspherical 2-complex? The *Whitehead asphericity conjecture* claims (66) to be true.

Notice that, in general, the inclusion of a connected subcomplex K^2 into L^2 does not induce an injective homomorphism $\pi_2(K^2) \to \pi_2(L^2)$:

(67) Example: $K^2 = S^1 \vee S^2$, $L = D^2 \vee S^2$.

Here $\pi_1(K^2)$ is infinite cyclic with generator a; $\pi_2(K^2)$ is isomorphic to $\mathbb{Z}\pi_1$ (i.e., the free module[43] of rank 1 over the group ring $\mathbb{Z}\pi_1$) generated by the homotopy class of a homeomorphism f of S^2 onto the 2-sphere of K. Consider $(a - 1) \cdot [f] \in \pi_2(K^2)$. A realization of this element as a combinatorial map g (see Chapter II, § 1.2) can be sketched as in Figure 34.

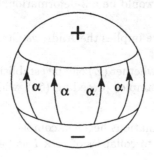

Figure I.34.

---

[43]See Chapter II, § 2.1 for an explanation of $\pi_2(K)$ as a module over $\pi_1(K)$.

As a map to L^2, g can be homotoped to become the constant map, i.e., $[g] = 0 \in \pi_2(L^2)$. (The map $\pi_2(K^2) \to \pi_2(L^2)$ is the augmentation homomorphism $\mathbb{Z}\pi_1 \to \mathbb{Z}$.) $\square$

Other variations of Whitehead's question likewise have a definitive answer:

If K is allowed to be 3-dimensional, then by addition of two-dimensional material one can annihilate $\pi_2(K) \neq \{0\}$. This is shown by the

Example (Adams [Ad55]): Let the 2-skeleton of K^3 be K^2 of (67) and – with notations as above – attach a 3-ball via any map in the homotopy class $(2a - 1) \cdot [f] \in \pi_2(K^2)$. Let $L^3 \supset K^3$ be obtained by enlarging the 2-skeleton of K^3 as for $L^2 \supset K^2$ above; $L^3 - K^3$ is an open 2-cell. Similarly to the argument after (67), the attaching map of the 3-cell of L is homotopic to f, hence by § 2.1, Lemma 2.1, $L^3 \searrow D^2 \vee D^3 \searrow *$, whence $\pi_2(L^3) = \{0\}$. $\square$

The straightforward generalization of (66):

$$K^n \subset L^n \quad \text{with} \quad \pi_n(L^n) = \{0\} \overset{?}{\Rightarrow} \pi_n(K^n) = \{0\}$$

is true in dimension 1. In contrast to the example (67), the homomorphism $\pi_1(K^1) \to \pi_1(L^1)$ induced by the inclusion of 1-complexes is always injective. This can easily be deduced from § 1.3. On the other hand, the generalization of (66) is false for dimensions ≥ 3. As an example, take $K^n = S^{n-1} \subset D^n = L^n$. For $n = 3$, the Hopf-map (see [Ho31]) gives a nontrivial element of $\pi_3(S^2)$; for $n \geq 4$, we have $\pi_n(S^{n-1}) \approx \mathbb{Z}_2$ (see [Hu59], Theorem 15.1).

So, as in § 4.1, (63), only the 2-dimensional case remains unsolved.

4.4 Further open questions

Wall's proof of (63) for $n \geq 3$ turns algebraic simple-homotopy operations on chain complexes of universal covering spaces (see Chapter II) into geometric deformations. The analogous construction for $n = 2$ is impossible in general: it cannot be performed preserving the fundamental group throughout, as defining relations have to take care of "more non-commutativity" than the algebra of chain complexes. The same obstacle is the origin of the next two open questions:

(68) Which torsion values lie in $\text{Wh}^*(\pi)$ ($\subseteq \text{Wh}(\pi)$), that is, which ones occur as $\tau(K, L)$ for a finite CW-pair, such that L is a deformation retract of K, $\dim(K - L) \leq 2$ and $\pi = \pi_1(L)$? Is $\text{Wh}^*(\pi)$ always trivial; is it a subgroup?

See [Me76], [Co77], [Me79$_1$], [Me79$_2$] [Me85]. A deep result of O.S. Rothaus is that there exist examples $\tau \in$ Wh(π) for dihedral groups π with $\tau \notin$ Wh$^*(\pi)$; see [Rot77].

(69) Let π be finitely generated, $\pi = F(a_1, \ldots, a_g)/N$. Then $N/[N, N]$ is a $\mathbb{Z}(\pi)$-module, the *relation module*. Suppose the relation module is finitely generated by h elements. Does it follow that N is normally generated by h elements $R_1, \ldots, R_h$, i.e., that π has the presentation $\langle a_1, \ldots, a_g | R_1, \ldots, R_h \rangle$ with (not more than) h defining relators? (Weaker, but also unsolved: is π finitely presented at all?)

See Gruenberg [Gr79], Harlander [Ha92]. Here is an idea for constructing counterexamples; compare [HoLuMe85] and Chapter XII, § 3.1: Let $\tilde{U}, \tilde{V}, \tilde{W}$ be finitely presented groups with elements $u \in \tilde{U}, v \in \tilde{V}, w \in \tilde{W}$ of finite, pairwise relatively prime order. Define U, V, W to be the quotients of $\tilde{U}$ resp. $\tilde{V}$ resp. $\tilde{W}$ by adding the relations $u = 1$ resp. $v = 1$ resp. $w = 1$. Then $\pi = U * V * W$ can be presented by forming the "disjoint" union of presentations of minimal deficiency for $\tilde{U}, \tilde{V}, \tilde{W}$, enlarged by the relations $uvw = 1$ and $[v, w] = 1$. But the last relator becomes superfluous in the relation module, as it projects to $[N, N]$. Although in certain (nontrivial) cases (with U, V, W finite abelian $\neq$ cyclic) there exists also a presentation for π which dispenses with one defining relation (see [Ho-An88]), we can't prove and don't think that this holds in general.

Turning to the semigroup $\mathcal{H}$ of presentation classes at the end of § 2.3, there are the questions

(70) whether Q^{**}-classes of balanced presentations $\langle a_1, \ldots, a_g | R_1, \ldots, R_g \rangle$ for $\pi = \{1\}$

  a) are invertible,

  b) are of finite resp. infinite order.

Craggs and Howie [CrHo87] have given a characterization of invertible classes. [Me79$_2$] contains an argument showing that the truth of (AC) would imply that a factorization of any element $\Phi \in \mathcal{H}$, $\Phi \neq \Phi_0$ by factors $\neq \Phi_0$ must terminate after finitely many steps. But further analogies to the existence of the prime factorization of 3-manifolds (compare [He76]) are unknown so far.

In this section we have not listed again questions which were already mentioned earlier in this chapter. We also recommend the lists of problems Kirby [Ki78] and Wall [Wa79]. They have stimulated research since they appeared, and although progress has been made in the meantime, they are still worth consideration.

Chapter II

Algebraic Topology for Two Dimensional Complexes

Allan J. Sieradski

This chapter presents homotopy classifications of two dimensional CW complexes and maps between them. Cases of these abstract classifications are detailed in Chapter III. Simplicial techniques are invoked in Section 1 to analyze maps of balls and spheres into 2-complexes. This analysis is applied in Section 2 to study the long exact sequence of homotopy groups for a 2-complex and to derive J. H. C. Whitehead's equivalence of the homotopy theory of 2-complexes with the purely algebraic theory of free crossed modules. Cellular chain complexes of universal coverings of 2-complexes are developed in Section 3. This equivariant world provides the foundation for the treatment in Section 4 of an abelianized version of Whitehead's equivalence, namely, the theory of algebraic 2-type of 2-complexes due to S. Mac Lane and Whitehead.

1 Techniques in Homotopy

In this section, we use simplicial approximations of maps between simplicial complexes to construct combinatorial approximations of maps between CW complexes, at least in dimensions one and two.

1.1 Simplicial Techniques

We view real m-space $\mathbb{R}^m$ as a real vector space and we assume that the reader is familiar with the concepts of finite simplicial complexes K in $\mathbb{R}^m$ and simplicial maps $\phi : K \to M$ between such complexes. We don't distinguish notationally between a simplicial complex and the associated topological subspace of $\mathbb{R}^m$, and let context convey the object under consideration.

Simplicial Approximations   Not every (topological) map $f : K \to M$ between simplicial complexes K and M respects their simplicial structure. But we shall show that each such map is approximated by a simplicial map, in the following sense: A simplicial map $\phi : K \to M$ is a *simplicial approximation* to a map $f : K \to M$ provided, for each point $x \in K$, the image point $\phi(x)$ belongs to a closed simplex $\bar{t}$ in M whenever the image point $f(x)$ belongs to the corresponding open simplex t in M.

A simplicial approximation ϕ serves as a controlled representative of the homotopy class of the map f. By definition, the line segments $[\phi(x), f(x)]$ for all points $x \in K$ lie in simplexes in M and so provide a homotopy $\phi \simeq f : K \to M$ relative to any subspace of K on which the map f and the simplicial map ϕ agree.

Simplicial approximations are constructed using the following lemma. For each vertex $v \in K^0$ of a simplicial complex K, let $Star_K(v)$ be the union of all open simplexes of K having v as a vertex. Let $Star K$ denote the cover of K by its open vertex stars $Star_K(v)$ $(v \in K^0)$. The open coverings $Star K$ and $Star M$ of two simplicial complexes K and M provide this *simplicial approximation test*, whose proof is left as an exercise:

Lemma 1.1 *A map* $f : K \to M$ *admits a simplicial approximation if and only if the open covering* $Star K$ *refines* $f^{-1}(Star M)$; *moreover, any vertex assignment* $\phi : K^0 \to M^0$ *such that* $Star_K(v) \subseteq f^{-1}(Star_M(\phi(v)))$ *for all* $v \in K^0$ *determines a simplicial approximation* $\phi : K \to M$ *to* f. $\square$

Any map f passes this test for a sufficiently fine barycentric subdivision of the domain complex K. Here is a sketch of this fundamental result.

The *barycenter* of a k-simplex $s = v_0 \ldots v_k$ is its point with equal barycentric coordinates: $b(s) = \sum_{i=0}^{k} \frac{1}{k+1} v_i$. The *barycentric subdivision* of a simplicial complex K is the simplicial complex $Sd\, K$ whose vertices are the barycenters $b(s)$ of the simplexes s of K and whose simplexes $b(s_0)b(s_1) \ldots b(s_k)$ are determined by the chains of proper faces $s_0 < s_1 < \ldots < s_k$ of simplexes of K. The simplicial complex $Sd\, K$ has the same underlying space as K but its *mesh*, or maximal simplex diameter, is decreased by the factor $d/(d+1)$, where $d = dim\, K$. This fact and Lebesgue's covering lemma show that for any given open covering $\mathcal{U}$ of K, there is some iterated barycentric subdivision K^* of K whose star covering $Star K^*$ refines $\mathcal{U}$.

Coupled with the simplicial approximation test (1.1), these observations yield the following *simplicial approximation theorem*:

Theorem 1.2 *Let $f : K \to M$ be any map of finite simplicial complexes. For some iterated barycentric subdivision K^* of K, there is a simplicial approximation $\phi : K^* \to M$ of $f : K \to M$.* $\square$

The *relative barycentric subdivision $Sd_L \, K$* of K modulo a simplicial subcomplex L consists of all simplexes $t = v_0 v_1 \ldots v_j b(s_1) \ldots b(s_k)$ associated with sequences $v_0 v_1 \ldots v_j = s_0 < s_1 < \ldots < s_k$ of proper faces of simplexes s_0 of L and simplexes $s_1, \ldots, s_k$ of $K - L$. The case $j = -1, s_0 = \varnothing$, and $t = b(s_1) \ldots b(s_k)$ is allowed.

Clearly, $Sd_L \, K$ retains L as a subcomplex, and so Sd_L only creates shrinkage in the size of simplexes of K away from L. Nevertheless, there is this relative version of Theorem 1.2 due to Zeeman [Ze64$_2$]:

Theorem 1.3 *Let $f : K \to M$ be a map of finite simplicial complexes that is simplicial on a subcomplex L of K. For some barycentric subdivision K^* modulo L of K, there is a simplicial map $\phi : K^* \to M$ and a homotopy $f \simeq \phi$ relative L.* $\square$

We use these approximation techniques to analyze maps of such spaces as the unit ball-sphere pair (B^{n+1}, S^n) of $\mathbb{R}^{n+1}$. We first analyze maps of an $(n+1)$-dimensional complex onto an n-dimensional complex. The generic situation is the projection of $\mathbb{R}^{n+1} = \mathbb{R}^n \times \mathbb{R}^1$ onto $\mathbb{R}^n$. The image of the $n+1$-ball B^{n+1} under this projection is the n-ball B^n and the pre-image of a smaller ball concentric with B^n consists of two small n-dimensional hemispherical boundary caps about the poles of $S^n = \partial B^{n+1}$ as well as an $n+1$-dimensional cylinder in the interior of B^{n+1} spanning these caps. Moreover, these hemispherical caps acquire opposite orientations in $S^n = \partial B^{n+1}$ from an orientation of their image in B^n. We use simplicial linkages to show that an analogous situation always holds for simplicial maps.

Simplicial Linkages   Let (K, L) be any triangulation of any manifold pair $(N^{n+1}, \partial N^{n+1})$ in $\mathbb{R}^{n+1}$, such as (B^{n+1}, S^n). Let $\phi : K \to M$ be a simplicial map to an n-dimensional simplicial complex M. If ϕ maps an open $(n+1)$-simplex $t = v_0 \ldots v_{n+1}$ of K onto an open n-simplex $w = u_0 \ldots u_n$ of M, the surjective assignment of the $n+2$ vertices of t to the $n+1$ vertices of w makes exactly one duplicate assignment $\phi(v_i) = \phi(v_j)$, where $0 \le i < j \le n$. So ϕ maps two n-faces $s = v_0 \ldots \hat{v}_i \ldots v_{n+1}$ and $s' = v_0 \ldots \hat{v}_j \ldots v_{n+1}$ of t onto w, folding t along their shared $(n-1)$-face $v_0 \ldots \hat{v}_i \ldots \hat{v}_j \ldots v_{n+1}$ and identifying their two unshared vertices v_i and v_j ($\hat{v}$ denotes that the vertex v is deleted).

Since each n-simplex s of the manifold interior $K - L$ is a face of exactly two $(n + 1)$-simplexes t and t' of the triangulation K, it follows that each component of the pre-image $\phi^{-1}(w)$ of an open n-simplex w of M can be expressed as a sequence

$$s_1 < t_1 > s_2 < \ldots > s_j < t_j > s_{j+1} < \ldots > s_{r-1} < t_{r-1} > s_r$$

of distinct open n-simplexes s_j and open $(n+1)$-simplexes t_j of K such that, for all $1 \leq j < r$, s_j and s_{j+1} are distinct faces of t_j (see Figure 1). The single vertex b_{j+1} of s_j that s_{j+1} lacks is called the vertex *behind* s_{j+1} and the single vertex f_j of s_{j+1} that s_j lacks is called the vertex *in front of* s_j. Any such sequence of simplexes in K is called a *simplicial linkage* in K.

By maximality of components, a simplicial linkage that arises as a component of the pre-image $\phi^{-1}(w)$ of an open n-simplex w of M either begins and ends at the same n-simplex $s_1 = s_r$ in $K - L$ or joins distinct n-simplexes $s_1 \neq s_r$ of the boundary subcomplex L. It is called a *toroidal* simplicial linkage or a a *cylindrical* simplicial linkage, accordingly.

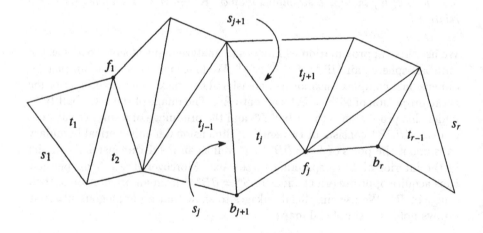

Figure II.1. Simplicial linkage

We now show that *the ordered end n-simplexes of an $n + 1$-dimensional cylindrical simplicial linkage in $\mathbb{R}^{n+1}$ are oppositely oriented simplexes of the boundary subcomplex L*. The following definitions make this claim precise.

Let $H\{v_0, \ldots, v_n\}$ be the *hyperspace* in $\mathbb{R}^{n+1}$ of affine combinations $\sum_{i=0}^{n} \omega_i v_i$ (i.e., $\sum_{i=0}^{n} \omega_i = 1$) of the vertices $v_0, \ldots, v_n$ of an n-simplex. A point $p \in \mathbb{R}^{n+1}$

lies in the hyperspace $H\{v_0, \ldots, v_n\}$ if and only if the $(n+2) \times (n+2)$ matrix

$$(v_0 \mid \ldots \mid v_n \mid p) = \begin{pmatrix} v_{0,1} & \cdot & \cdot & \cdot & v_{0,n+1} & 1 \\ \vdots & \vdots & \vdots & \vdots & \vdots & \vdots \\ v_{n,1} & \cdot & \cdot & \cdot & v_{n,n+1} & 1 \\ p_1 & \cdot & \cdot & \cdot & p_{n+1} & 1 \end{pmatrix},$$

whose rows involve the Euclidean coordinates of the $n+1$ vectors $v_0, \ldots, v_n$, and p, has trivial determinant. Thus, the complement in $\mathbb{R}^{n+1}$ of the hyperspace $H\{v_0, \ldots, v_n\}$ has two components, called the *sides* of H in $\mathbb{R}^{n+1}$, whose points p are characterized by the two possible signs of the non-zero determinant $Det(v_0 \mid \ldots \mid v_n \mid p)$. This sign is called the *sign of the ordered n-simplex* $s = v_0 \ldots v_n$ *with respect to the point p* and it is denoted by $Sign_p\, s$.

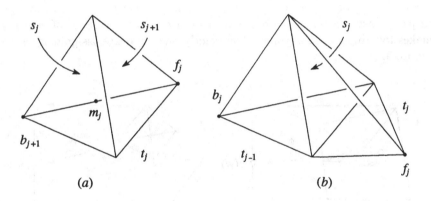

$$(a) \qquad\qquad\qquad (b)$$

Figure II.2. $Sign_{f_j} s_j \neq Sign_{b_{j+1}} s_{j+1}$ and $Sign_{b_j} s_j \neq Sign_{f_j} s_j$

Theorem 1.4 *When the vertices of each n-simplex of a cylindrical simplical linkage in $\phi^{-1}(w)$ are given the ordering of their image vertices $u_0, \ldots, u_n$ in the n-simplex w, the first and last ordered n-simplexes s_1 and s_r in the boundary subcomplex L have opposite sign with respect to their opposing vertices.*

Proof: The claim $Sign_{f_1} s_1 \neq Sign_{b_r} s_r$ follows from an odd number of applications of the following two facts: First, any two consecutive n-simplexes s_j and s_{j+1} of the linkage have opposite signs, $Sign_{f_j} s_j \neq Sign_{b_{j+1}} s_{j+1}$, with respect to the vertices f_j and b_{j+1} opposite them in the $(n+1)$-simplex t_j. Indeed, because the midpoint $m_j = \frac{1}{2}(f_j + b_{j+1})$ is on the same side of the hyperspace through s_j as f_j and also on the same side of the hyperspace through

s_{j+1} as b_{j+1}, then $Sign_{f_j} s_j = Sign_{m_j} s_j$ and $Sign_{b_{j+1}} s_{j+1} = Sign_{m_j} s_{j+1}$ (see Figure 2a). But $Sign_{m_j} s_j \neq Sign_{m_j} s_{j+1}$, by inspection of the matrices $(\dots |f_j| \dots |m_j|)$ and $(\dots |b_{j+1}| \dots |m_j|)$. Second, any intermediate n-simplex s_j $(1 < j < r)$ has opposite signs $Sign_{b_j} s_j \neq Sign_{f_j} s_j$ with respect to the vertex b_j behind it and the vertex f_j in front of it, as these vertices are on opposite sides of the hyperspace through s_j (see Figure 2b). □

We now analyze maps of an $n + 1$-dimensional object onto an $(n - 1)$-dimensional object. The generic case is the projection of $\mathbb{R}^{n+1} = \mathbb{R}^{n-1} \times \mathbb{R}^2$ onto $\mathbb{R}^{n-1}$. The pre-image in $\mathbb{R}^{n+1}$ of the origin $O \in B^{n-1}$ is the surface $\{O\} \times \mathbb{R}^2$. We use simplicial curtains to express the analogous situation that holds for any simplicial map.

Simplicial Curtains   Let (K, L) be any triangulation of any manifold pair $(N^{n+1}, \partial N^{n+1})$. If a simplicial map $\phi : K \to M$ sends an open $(n+1)$-simplex $t = v_0 \dots v_{n+1}$ of K onto an open $(n-1)$-simplex $w = u_0 \dots u_{n-1}$ of M, the surjective assignment of the $n + 2$ vertices of t to the n vertices of w either makes two duplicate assignments or exactly one triplicate assignment, as in Figure 3.

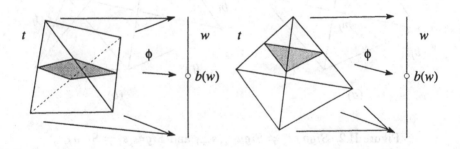

Figure II.3. Simplicial curtain core

So ϕ maps onto w either three or four n-faces of t, as well as the three or four $(n - 1)$-faces in which those n-faces meet. Thus the pre-image under ϕ of the barycenter $b(w)$ contains the convex hull of the barycenters of the $(n-1)$-faces of t mapped onto w, namely, a triangle or quadrilateral whose edges lie in the n-faces of t mapped onto w. Since each n-simplex of the manifold interior $K - L$ is a face of exactly two $(n + 1)$-simplexes of the triangulation K, it follows that each component of the pre-image $\Sigma_w = \phi^{-1}(b(w))$ is an oriented 2-dimensional manifold whose boundary lies in the boundary L of K. We call the pre-image $\Lambda_w = \phi^{-1}(w)$ of the open $(n - 1)$-simplex w of M a *simplicial curtain* in K; it serves as a product neighborhood of the 2-manifold Σ_w in K.

1.2   Combinatorial Maps

The CW complexes of J. H. C. Whitehead [Wh49$_1$] afford a convenience of economy over simplicial complexes, but their cellular (i.e., skeletal preserving) mappings are too loosely structured for our purposes. For convenience, we consider a restricted class of complexes and cellular maps that we call combinatorial.

Combinatorial Maps and Complexes   Let $(D^n, \dot{D}^n)$ denote any CW pair for the ball-sphere pair (B^n, S^{n-1}) with the single interior n-cell $\overset{\circ}{B}{}^n = B^n - \dot{B}^n$; it is unique in dimensions $n = 0, 1$ (for $n = 0$, read $(B^0, S^{-1}) = (\{1\}, \varnothing) = (B^0, \dot{B}^0)$). Let $(D^n \times D^1, \dot{D}^n \times D^1 \cup D^n \times \dot{D}^1)$ be the product complex on

$$(B^{n+1}, S^n) \equiv (B^n \times B^1, \dot{B}^n \times B^1 \cup B^n \times \dot{B}^1)$$

with the single interior $(n+1)$-cell $\overset{\circ}{B}{}^n \times \overset{\circ}{B}{}^1$. A *product $(n+1)$-cell c^{n+1}* in a CW complex L is a cell having a cellular characteristic map $\psi : D^n \times D^1 \to L$ that extends characteristic maps $\phi_j : D^n \times \{j\} \to L$ for (not necessarily distinct) n-cells c_j^n $(j = \pm 1)$ of L. A cellular map $f : L \to K$ *collapses* a product $(n+1)$-cell (c^{n+1}, ψ) if

$$f\,\psi = f\,\phi_j \text{ projection} : D^n \times D^1 \to D^n \times \{j\} \to L \to K (j = \pm 1).$$

A cellular map $f : L \to K$ is called *combinatorial* if each open n-cell $(n \geq 0)$ of L is either carried homeomorphically onto an open n-cell of K or is a product n-cell $(n \geq 1)$ that is collapsed by the map f. In particular, an open 1-cell of L is carried by a combinatorial map either homeomorphically onto an open 1-cell of K or is collapsed to a 0-cell of K.

A CW complex K is called *combinatorial* if each n-cell c^n $(n \geq 0)$ has a combinatorial characteristic map $\phi : (D^n, S^{n-1}) \to (K^n, K^{n-1})$, i.e., the attaching map $\dot{\phi} : S^{n-1} \to K^{n-1}$ is a combinatorial map for some combinatorial complex on S^{n-1}. Notice that according to this inductive definition of combinatorial complexes, every CW complex of dim ≤ 1 is combinatorial and that a 2-dimensional complex is combinatorial provided that each 2-cell attaching map $\dot{\phi} : S^1 \to K^1$ sends each 1-cell of some complex on S^1 either homeomorphically onto an open 1-cell of K or collapses it to a 0-cell of K. These 2-complexes were first considered by Reidemeister [Re32]; see Chapter I, §1.4.

Combinatorial Approximations   The standard cellular approximation theorem [Wh49$_1$, item (L)] states that any map between CW complexes is homotopic to a cellular one, i.e., one that respects the skeletons. The following *combinatorial approximation theorems* (1.5) and (1.6) present refinements for maps of balls and spheres of dimension 1 and 2.

Theorem 1.5 *Any loop $S^1 \to M^1$ in a 1-complex M^1 is homotopic to a combinatorial map $g : L \to M^1$ of some complex L on S^1.*

Proof: Subdivision of the 1-cells into thirds converts the 1-complex M^1 into a simplicial complex N. The combinatorial quotient map $q : N \to M^1$ that collapses the first and last thirds of the 1-cells in M^1, while expanding the middle thirds onto the original 1-cells in M^1, is a deformation of the identity map 1_{K^1}. Then the composition $q\phi : L \to N \to M^1$, for any simplicial approximation $\phi : L \to N$ of the original loop $S^1 \to M^1$, gives a combinatorial map $g : L \to M^1$ homotopic to the original loop. $\qquad\square$

Theorem 1.6 *Skeletal pairs (K^2, K^1) and (M^2, M^1) having the same 1-skeleton $K^1 = M^1$ and 2-cells $\{c_\alpha^2\}$ that are attached via homotopic maps $\{\dot{\varphi}_\alpha \simeq \dot{\lambda}_\alpha : S^1 \to K^1 = M^1\}$ are homotopy equivalent.*

Proof: The technique of proof of Lemma 1 in Chapter I, §2, suffices. Consider a family $H = \{H_\alpha : S_\alpha^1 \times I \to K^1\}$ of homotopies $\{H_\alpha : \dot{\varphi}_\alpha \simeq \dot{\lambda}_\alpha\}$. The adjunction space $W_H = K^1 \cup_H \{B_\alpha^2 \times I\}$ obtained by attaching the solid cylinders $\{B_\alpha^2 \times I\}$ to K^1 via the homotopies in H contains the subspaces

$$K^1 \cup_H \{(B_\alpha^2 \times \{0\}) \cup (S_\alpha^1 \times I)\} \text{ and } K^1 \cup_H \{(S_\alpha^1 \times I) \cup (B_\alpha^2 \times \{1\})\}$$

as strong deformation retracts. Since the latter are homeomorphic to $K = K^1 \cup_{\{\dot{\varphi}_\alpha\}} \{B_\alpha^2\}$ and $M = M^1 \cup_{\{\dot{\lambda}_\alpha\}} \{B_\alpha^2\}$, respectively, we have homotopy equivalences $K \simeq W_H \simeq M$ *rel* K^1. $\qquad\square$

For each $k \geq 1$, let $(P_k, \dot{P}_k)$ be the regular k-sided polygonal complex on (B^2, S^1), with k corner 0-cells, k edge 1-cells, and a single interior 2-cell. By Theorems 1.5 and 1.6, any 2-complex is homotopy equivalent to one K, each of whose 2-cells c^2 has a combinatorial characteristic map $\psi : (P_k, \dot{P}_k) \to (K^2, K^1)$, for some $k \geq 1$. So each 2-cell c^2 has the structure of an open polygon $\overset{\circ}{P}_k = P_k - \dot{P}_k$, each of whose edges is either collapsed to a 0-cell of K or is identified homeomorphically with a 1-cell of K. In other words, K is a *combinatorial 2-complex.*

The following two results show that any map $B^2 \to K$ into a combinatorial 2-complex K is homotopic to a combinatorial map.

Lemma 1.7 *Let $\{P_{k_i} : 1 \leq i \leq m\}$ be any family of disjoint closed polygons in the interior of the unit disc B^2. Any map $F : D \to K^1$ of the punctured disc $D = B^2 - \cup_i \overset{\circ}{P}_{k_i}$ that is combinatorial on the boundary ∂D is homotopic relative ∂D to a combinatorial map $D \to K^1$.*

Proof: The combinatorial map $F : \partial D \to K^1$ is a simplicial map of the simplicial subdivisions $(\partial D)^*$ and $(K^1)^*$ that divide 1-cells into triples of 1-simplexes. Then Theorem 1.3 provides a homotopy relative ∂D of F to a simplicial map $G : D \to K^1$ of some triangulation of D extending $(\partial D)^*$. Each component of the pre-image $G^{-1}(s^1)$ of the middle third simplex s^1 in a 1-cell $c^1 \subseteq K^1$ is either an annular simplicial linkage in $D - \partial D$ or one ending on the middle third simplexes in 1-cells of ∂D, as in Figure 4a.

The collapse of the outer 1-simplexes in the 1-cells in ∂D and K^1, coupled with an expansion of each middle third 1-simplex onto the full closed 1-cell, give deformations of D and K^1. They convert $G : D \to K^1$ into a map for which the pre-image $G^{-1}(c^1)$ of each 1-cell c^1 of K^1 is a *highway system* Λ_{c^1} in this sense: Each component of Λ_{c^1} is either an embedded rectangle $\overset{\circ}{B}{}^1 \times B^1$ whose ends $\overset{\circ}{B}{}^1 \times \{\pm 1\}$ are 1-cells in ∂D or is an embedded annulus $\overset{\circ}{B}{}^1 \times S^1$, in either case, on whose oriented cross-sections $B^1 \times \{t\}$ the null-homotopy G acts like the characteristic map ϕ_{c^1} of the 1-cell c^1.

There is a deformation of D that widens the open highway systems and strong deformation retracts their complement $G^{-1}(K^0)$ onto a graph. This gives a map $D \to K^1$ that is homotopic relative ∂D to the original map and that is combinatorial on a complex C (Figure 4b) formed by the original complex on the ∂D, the graph, and some product cells subdividing the highways. $\square$

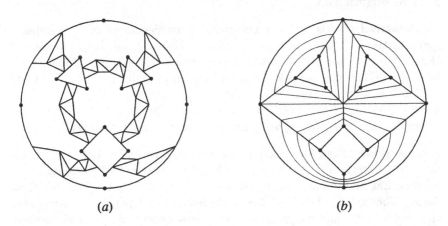

(a) (b)

Figure II.4. Simplicial linkages and combinatorial complex C

Theorem 1.8 *Any map $F : (B^2, S^1) \to (K^2, K^1)$ into a combinatorial complex whose restriction $F|_{S^1}$ is combinatorial on some complex L on S^1 is homotopic relative S^1 to a combinatorial map $(C, L) \to (K^2, K^1)$ of some complex pair (C, L) on (B^2, S^1).*

Proof: The simplicial subdivision K^{1*} of the 1-skeleton K^1 (dividing 1-cells into triples of 1-simplexes) extends to a simplicial subdivision K^* of K, with the center of each 2-cell c^2 of K as the barycenter of some central 2-simplex $t^2 \subseteq c^2$ in K^*. Since the restriction $F|_{S^1}$ is a combinatorial map $L \to K$, it defines a simplicial map $L^* \to K^*$. Then Theorem 1.3 provides a homotopy relative S^1 of F to a map $G : B^2 \to K$ that is simplicial on some triangulation of B^2 extending L^*.

The pre-image $G^{-1}(t^2)$ of a central simplex $t^2 \subset c^2$ is the finite union of 2-simplexes s_j^2 of M that G carries homeomorphically onto t^2. The combinatorial characteristic map $\phi_{c^2} : (P_k, \dot{P}_k) \to (K^2, K^1)$ of an k-sided 2-cell c^2 of K provides a small k-gon $\phi(\epsilon P_k)$ centered in c^2 and contained within the central 2-simplex t^2. There is a deformation of K relative K^1 that radially expands the central k-gon $\phi(\epsilon P_k)$ in each 2-cell c^2 onto the full closed 2-cell, pushing complementary points out to the 1-skeleton K^1.

This deformation of K relative K^1 deforms G relative S^1 into a map $H : B^2 \to K$ for which the pre-image $H^{-1}(c^2)$ of an open k-sided 2-cell c^2 of K is the finite union of small open k-gons in $B^2 - S^1$ on whose closure H acts like the characteristic map $\phi_{c^2} : (P_k, \dot{P}_k) \to (K^2, K^1)$. Then Lemma 1.7 applies to the restriction of H to the punctured disc $H^{-1}(K^1) = B^2 - \cup_{c^2} G^{-1}(c^2)$. This provides a combinatorial map $(B^2, S^1) \to (K^2, K^1)$ whose restriction to S^1 is the original map.   □

Combinatorial Models   For notational convenience, we often collapse a spanning tree T in the 1-skeleton K^1 in a connected 2-complex K to convert (K^2, K^1) into a homotopy equivalent pair with a single 0-cell. This uses the facts that T is a contractible space and the subcomplex inclusion $(T, T) \subseteq (K^2, K^1)$ is a cofibration. When the tree collapse is applied prior to the polygonalization of the 2-cells, the structure of the resulting combinatorial 2-complex can be expressed using group presentation terminology.

A *group presentation* $\mathcal{P} = \langle \mathbf{x} \mid \mathbf{r} \rangle$ consists of a set $X = \{x\}$ of elements called *generators*, together with an indexed set $\mathbf{r} = \{r\}$ of elements called *relators* that are (not necessarily reduced) words in the semigroup $W(\mathbf{x})$ on the alphabet $\mathbf{x} \cup \mathbf{x}^{-1}$. Let $F(\mathbf{x})$ denote the free group $W(\mathbf{x})/\sim$ of words in the alphabet $\mathbf{x} \cup \mathbf{x}^{-1}$ under equivalence relation $\sim$ generated by the elementary relations $xx^{-1} \sim \emptyset \sim x^{-1}x$. Let $N(\mathbf{r})$ denote the smallest normal subgroup of $F(\mathbf{x})$ containing the relation words $r \in \mathbf{r}$. The quotient group $\Pi = F(\mathbf{x})/N(\mathbf{r})$ is called the group presented by $\mathcal{P} = \langle \mathbf{x} \mid \mathbf{r} \rangle$

Let $\vee_x S_x^1$ be the sum of copies of the minimal 1-sphere complex $S_x^1 = c^0 \cup c_x^1$ indexed by the generators $x \in \mathbf{x}$. Each relation word $r = (x_1^{\epsilon_1}, \dots, x_{k(r)}^{\epsilon_{k(r)}})$ of length $k(r)$ spells out a combinatorial loop $\dot{\phi}_r : \dot{P}_{k(r)} \to \vee_x S_x^1$ that is used to attach a 2-cell c_r^2 to the sum $\vee_x S_x^1$. The resulting oriented 2-complex

$$K = c^0 \cup c_x^1 \cup c_r^2 \ (x \in \mathbf{x}, \ r \in \mathbf{r})$$

is called the *model* or *standard complex* $K_{\mathcal{P}}$ of the group presentation $\mathcal{P} = \langle \mathbf{x} \mid \mathbf{r} \rangle$. Each 2-cell c_r^2 ($r \in \mathbf{r}$) acquires a combinatorial characteristic map $\phi_r : (P_{k(r)}, \dot{P}_{k(r)}) \to (K^2, K^1)$, where $k(r)$ is the length of the relator $r \in \mathbf{r}$.

We summarize the preceding discussion with the following theorem, which relates to Chapter I, (15).

Theorem 1.9 *The skeleton pair of a connected 2-complex K is homotopy equivalent to that of the model $K_{\mathcal{P}}$ of a group presentation $\mathcal{P} = \langle \mathbf{x} \mid \mathbf{r} \rangle$.*

Preferred Homotopies: Framed Links and Curtains   The simplicial linkage and curtain techniques apply to refine a homotopy of combinatorial maps of two-dimensional combinatorial complexes. The following statements are offered without details of proof to illustrate the possibilities.

For an inessential combinatorial map $F : S^2 \to K$, there exists a null-homotopy $N : B^3 \to K$ for which the closure of the pre-image $N^{-1}(c^2)$ of each 2-cell c^2 of K is a *framed link* Λ_{c^2} in this sense: each component of Λ_{c^2} is either an embedded polygonal torus $P_n \times S^1$ or a polygonal cylinder $P_n \times B^1$ (embedded except that its polygonal ends are attached to a pair of 2-cells $\{d_+^2, d_-^2\}$ in S^2 by their characteristic maps ψ_+, ψ_-, which according to Theorem 1.4 are oppositely oriented), in either case, on whose oriented polygonal cross-sections $P_n \times \{z\}$ the null-homotopy N acts like the characteristic map $\theta_{c^2} = F\psi_\pm : P_n \to K$ of the 2-cell c^2 of K. The union $\Lambda(N) = \cup \Lambda_{c^2}$ is called the *framed link* for the null-homotopy N.

A null-homotopy N exists for which, in addition, the closure of the pre-image $N^{-1}(c^1)$ of each open 1-cell c^1 of K is a *curtain* Γ_{c^1} in this sense: Each component of Γ_{c^1} is a thickened surface $B^1 \times \Sigma$ (Σ a orientable surface). Each boundary annulus in $B^1 \times \partial\Sigma$ is combinatorially attached along a highway ω_{c^1} of product 2-cells on the boundary of $S^2 \cup N^{-1}(K^2 - K^1)$, on whose oriented cross-sections $B^1 \times \{z\}$ the null-homotopy N acts like the characteristic map $\phi_{c^1} : B^1 \to K$ for the 1-cell c^1. The union $\Gamma(N) = \cup \Gamma_{c^1}$ is called the *curtain system* for the null-homotopy N. The components of the framed link $\Lambda(N)$ and the curtain system $\Gamma(N)$ are spatially separated in B^3, except where attached. The remainder of B^3 belongs to the pre-image $N^{-1}(K^0)$. Such a homotopy N of combinatorial maps is entirely determined by its link and curtain system $\{\Lambda(N), \Gamma(N)\}$.

2   Homotopy Groups for 2-Complexes

In this section, we develop Whitehead's identification of the homotopy theory of 2-dimensional CW complexes with the purely algebraic theory of free crossed modules. In §2.1, we begin with an examination of the long exact homotopy sequence for the skeletal pair (K^2, K^1) of a connected 2-complex K. The algebraic properties of this sequence are then codified in the definition of crossed module. In §2.2, we establish the two-dimensional case of Whitehead's Theorem characterizing maps of CW complexes that are homotopy equivalences. Finally, we deduce a portion of Whitehead's identification of 2-dimensional homotopy theory with the algebraic theory of free crossed modules whose groups of operators are free groups.

2.1   Fundamental sequence for a 2-complex K

We call a space equipped with a basepoint a *based space*, and we call a basepoint preserving map between based spaces a *based map*. We use $1 = <1, 0, \ldots, 0>$ as the basepoint for the $(n+1)$-ball B^{n+1} and n-sphere S^n.

For a based pair (Y, B), we view the homotopy groups $\pi_{n+1}(Y, B)$, $\pi_n(B)$, and $\pi_n(Y)$ in dimensions $n \geq 1$ as sets of homotopy classes of based maps $(B^{n+1}, S^n) \to (Y, B)$, $S^n \to B$, and $S^n \to Y$, respectively. We assume that the reader is familiar with the definition of their group operations and the fact that they are linked by a long exact homotopy sequence:

$$\cdots \xrightarrow{j\#} \pi_{n+1}(Y, B) \xrightarrow{\partial} \pi_n(B) \xrightarrow{i\#} \pi_n(Y) \xrightarrow{j\#} \pi_n(Y, B) \xrightarrow{\partial} \cdots \xrightarrow{i\#} \pi_0(Y).$$

Homotopy Action   Any based map $F : (B^{n+1}, S^n) \to (Y, B)$ can be deformed into another based map $G : (B^{n+1}, S^n) \to (Y, B)$ by *dragging* the image of the basepoint backwards along any loop $\alpha : S^1 \to B$ at the basepoint. More precisely, given F and α, there exists a homotopy $H : G \simeq F$ from G to F mapping $\{1\} \times I$ via $\alpha \circ exp : I \to S^1 \to B$. Such an H is called an α-*homotopy* and is denoted by $H : G \simeq_\alpha F$. For a construction, let

$$R : (B^{n+1}, S^n) \times I \to ((B^{n+1}, S^n) \times \{1\}) \cup (\{1\} \times I)$$

be any retraction and let $< F, \alpha > : ((B^{n+1}, S^n) \times \{1\}) \cup (\{1\} \times I) \to (Y, B)$ be defined by F on $(B^{n+1}, S^n) \times \{1\}$ and by the loop α on $\{1\} \times I$. Then $H = < F, \alpha > \circ R : (B^{n+1}, S^n) \times I \to (Y, B)$ is an α-homotopy. It is an exercise to show that, if $G \simeq_\alpha F$, the based homotopy class $[G]$ depends upon just the based homotopy class $[F]$ and the path-homotopy class $[\alpha] \in \pi_1(B)$. The homotopy class $[G]$ is called the *action* of $[\alpha]$ on $[F]$ and is denoted by $[G] = [\alpha] \cdot [F] \in \pi_{n+1}(Y, B)$.

For each $n \geq 1$, the homotopy action defines a group homomorphism

$$h : \pi_1(B) \to \mathrm{Aut}(\pi_{n+1}(Y, B)), h_{[\alpha]}([F]) = [\alpha] \cdot [F],$$

making the fundamental group $\pi_1(B)$ a group of operators on the group $\pi_{n+1}(Y, B)$. For $n > 1$, the action $h : \pi_1(B) \to \mathrm{Aut}(\pi_{n+1}(Y, B))$ extends linearly to make the abelian group $\pi_{n+1}(Y, B)$ a left module over the integral group ring $\mathbb{Z}(\pi_1(B))$. Similarly, there is an action

$$h : \pi_1(Y) \to \mathrm{Aut}(\pi_n(Y)), h_{[\alpha]}([F]) = [\alpha] \cdot [F],$$

making the fundamental group $\pi_1(Y)$ a group of operators on the group $\pi_n(Y)$, for each $n \geq 1$. For $n = 1$, this action h is conjugation; for $n > 1$, this action h makes the abelian group $\pi_n(Y)$ a left module over $\mathbb{Z}(\pi_1(Y))$.

Here are two important properties of the action that we leave as exercises:

Exercise 1 The action of $\pi_1(B)$ on $\pi_2(Y, B))$ gives $(\partial[G]) \cdot [F] = [G] [F] [G]^{-1}$ for $[F], [G] \in \pi_2(Y, B)$. (Hint: To visualize an α-homotopy $GFG^{-1} \simeq_\alpha F$ where $\alpha = \partial G : S^1 \to B$, let two small oppositely oriented discs in B^2 that are mapped by G under GFG^{-1} coalesce across a path joining the basepoint $1 \in S^1$ to a disc in B^2 that is mapped by F.)

Exercise 2 $\partial : \pi_2(Y, B) \to \pi_1(B)$ respects the actions of $\pi_1(B)$ on both groups in that $\partial([\alpha] \cdot [F]) = [\alpha][\partial F][\alpha]^{-1}$ for $[F] \in \pi_2(Y, B)$ and $[\alpha] \in \pi_1(B)$. (Hint: The boundary of an α-homotopy $G \simeq_\alpha F$ can be sliced open to form a path-homotopy $\partial G \simeq \alpha \cdot \partial F \cdot \alpha^{-1}$.)

Fundamental sequence for 2-complex   We now specialize to the case of a connected 2-complex $K = K^2$. Because the higher homotopy groups $\pi_n(K^1)$ ($n \geq 2$) of the 1-skeleton K^1 are trivial, the long exact homotopy sequence for the skeleton pair (K^2, K^1) breaks down into shorter exact sequences

$$0 \to \pi_n(K^2) \xrightarrow{j\#} \pi_n(K^2, K^1) \to 0 \quad (n \geq 3)$$

and

$$\Pi(K) : 0 \to \pi_2(K^2) \xrightarrow{j\#} \pi_2(K^2, K^1) \xrightarrow{\partial} \pi_1(K^1) \xrightarrow{i_*} \pi_1(K^2) \to 0$$

The four term exact sequence $\Pi(K)$, called the *fundamental sequence* of K, has a surprising rich structure. In §2.2, we shall establish J. H. C. Whitehead's observation [Wh49$_2$, Theorem 6] that the fundamental sequence captures the entire homotopy type of the 2-complex K^2 and the higher homotopy modules of K^2 are superfluous.

Our first task is to interpret the groups and homomorphisms in the fundamental sequence. For convenience, we work, as we may by Theorem 1.9, with the model $K = K_\mathcal{P}$ of a group presentation $\mathcal{P} = \langle \mathbf{x} \mid \mathbf{r} \rangle$. The model K has one-skeleton $K^1 = \vee_x S^1_x$. For each $x \in \mathbf{x}$, the inclusion $i_x : S^1_x \to K^1$ of one of the summands represents a fundamental group element $[i_x] \in \pi_1(K^1)$. As in Section 1, $F(\mathbf{x})$ denotes the free group of words in the alphabet $\mathbf{x} \cup \mathbf{x}^{-1}$.

Lemma 2.1 *For the model $K = K_\mathcal{P}$ of a group presentation $\mathcal{P} = \langle \mathbf{x} \mid \mathbf{r} \rangle$, there is a group isomorphism $\xi : F(\mathbf{x}) \to \pi_1(K^1)$ defined uniquely by the correspondence of each generator $x \in \mathbf{x}$ with the homotopy class $[i_x] \in \pi_1(K^1)$ of the inclusion $i_x : S^1_x \to K^1$.*

Proof: By a universal property of the free group $F(\mathbf{x})$, the homomorphism ξ is well-defined; it sends each reduced word $w = x_1^{\epsilon_1} x_2^{\epsilon_2} \ldots x_m^{\epsilon_m} \in F(\mathbf{x})$ to the homotopy class of the product loop that spells w:

$$f_w = i_{x_1}^{\epsilon_1} \cdot i_{x_2}^{\epsilon_2} \cdot \ldots \cdot i_{x_m}^{\epsilon_m} : S^1 \to K^1 = \vee_x S^1_x.$$

By Theorem 1.5, *the homomorphism ξ is surjective:* each loop $f : S^1 \to K^1$ deforms into a combinatorial loop f_w spelling some word $w \in F(\mathbf{x})$.

By Theorem 1.8, *the homomorphism ξ is injective:* If $\xi(w) = [f_w]$ is trivial, then the map $f_w : S^1 \to K^1$ spelling the word $w = x_1^{\epsilon_1} x_2^{\epsilon_2} \ldots x_m^{\epsilon_m}$ admits a null-homotopy $F : B^2 \to K^1$. The map f_w is combinatorial on the boundary $\dot{P}_m$ of the regular m-gon complex P_m on S^1. As in Theorem 1.8, F deforms relative S^1 into a combinatorial map $G : B^2 \to K^1$, as in Figure 5.

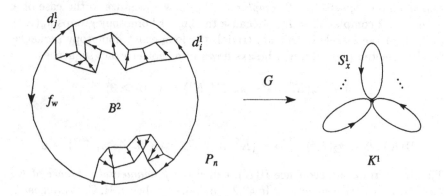

Figure II.5. Free reduction

As there are no 2-cells in K^1, the entire disc B^2 is paved with product 2-cells that are collapsed to 1-cells in K by G. Each non-annular component of the highway system $G^{-1}(c_x^1)$ ends on oppositely oriented open 1-cells d_i^1 and d_j^1 of $\dot{P}_m$ that correspond to entries $x_i^{\epsilon_i}$ and $x_j^{\epsilon_j}$ of the word w with $x_i = x = x_j$. Also $\epsilon_i = -\epsilon_j$, by Theorem 1.5. These components of the highway system are disjoint and provide a complete pairing of all the entries $x_i^{\epsilon_i}$ of the word w. Beginning with an *outermost* component that necessarily ends on adjacent edges of the boundary $\dot{P}_m$, they describe a free reduction of the word $w = x_1^{\epsilon_1} x_2^{\epsilon_2} \dots x_m^{\epsilon_m}$ to 1 in $F(\mathbf{x})$ using only the group axioms and the relations $xx^{-1} = 1 = x^{-1}x$. □

Lemma 2.2 *For the model $K = K_\mathcal{P}$ of a group presentation $\mathcal{P} = \langle \mathbf{x} \mid \mathbf{r} \rangle$, the relative homotopy group $\pi_2(K^2, K^1)$ is generated, up to the homotopy action of $\pi_1(K^1)$, by the homotopy classes $[\phi_r]$ of the characteristic maps $\phi_r : (B^2, S^1) \to (K^2, K^1)$ of the 2-cells $\{c_r^2 : r \in \mathbf{r}\}$.*

Proof: Any based homotopy class $[F] \in \pi_2(K^2, K^1)$ is represented by a combinatorial map $G : (B^2, S^1) \to (K^2, K^1)$, as constructed in Theorem 1.9. The closures of the components of the pre-image $G^{-1}(K^2 - K^1)$ are disjoint polygons $\{P_{k(r_i)} : 1 \le i \le m\}$ in $B^2 - S^1$ on which G acts as signed characteristic maps $\{\phi_{r_i}^{\epsilon_i} : (P_{k(r_i)}, \dot{P}_{k(r_i)}) \to (K^2, K^1)\}$ of certain 2-cells $\{c_{r_i}^2 : 1 \le i \le m\}$ in K. Let the loops $\{\alpha_i : S^1 \to K^1\}$ arise from the restrictions of G to disjoint arcs in $B_2 - \cup P_{k_i}$ joining the base point $1 \in S^1 = \partial B^2$ to the basepoints of these polygons. Since G carries the complement $B^2 - \cup \overset{\circ}{P}_{n_i}$ into the 1-skeleton K^1, the geometry of the configuration of arcs and polygons in B^2 (see Figure 6) shows that $[F] = [G] \in \pi_2(K^2, K^1)$ equals the product $\prod_i [\alpha_i] \cdot [\phi_{r_i}]^{\epsilon_i}$ in counterclockwise order of the arcs. □

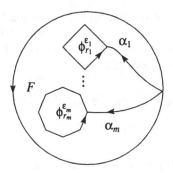

Figure II.6. Polygonal balloons and strings

Theorem 2.3 *The inclusion-induced homomorphism $i_\# : \pi_1(K^1) \to \pi_1(K^2)$ for the model $K = K_\mathcal{P}$ of the group presentation $\mathcal{P} = \langle \mathbf{x} \mid \mathbf{r} \rangle$ is the quotient homomorphism $F(\mathbf{x}) \to F(\mathbf{x})/N(\mathbf{r})$.*

Proof: The homomorphism $i_\#$ is surjective by the cellular approximation theorem, and $\ker i_\# = \operatorname{im} \partial_2$ by exactness of the homotopy sequence for (K^2, K^1). The boundary operator $\partial_2 : \pi_2(K^2, K^1) \to \pi_1(K^1)$ respects the homotopy action by $\pi_1(K^1)$ and therefore it sends the $\pi_1(K^1)$-generators $[\phi_r] \in \pi_2(K^2, K^1)$ to $\pi_1(K^1)$-generators $\partial_2([\phi_r]) = [\dot{\phi}_r]$ for $\operatorname{im} \partial_2 = \ker i_\#$. Since the combinatorial map $\dot{\phi}_r : S^1 \to K^1$ spells $r \in \mathbf{r}$, then $[\dot{\phi}_r] \in \pi_1(K^1)$ and $r \in F(\mathbf{x})$ correspond under the natural isomorphism $\pi_1(K^1) \approx F(\mathbf{x})$ of Lemma 2.1. Therefore, $\ker i_\# = N(\mathbf{r})$. $\qquad\square$

Crossed Modules in Homotopy   Our second task is the description of the rich algebraic structure of the boundary operator $\partial : \pi_2(K^2, K^1) \to \pi_1(K^1)$ in the fundamental sequence for a 2-complex K. This description involves crossed modules and is due to J. H. C. Whitehead [Wh49$_2$], R. Peiffer [Pe49], and K. Reidemeister [Re49]. We express their ideas using an exact sequence associated to the group presentation $\mathcal{P} = \langle \mathbf{x} \mid \mathbf{r} \rangle$ on which K is modeled.

The fundamental group $\pi_1(K^1) = F(\mathbf{x})$ has its homotopy action on the relative homotopy group $\pi_2(K^2, K^1)$ and its conjugation action on itself. As observed earlier, in the discussion of the homotopy action, the boundary operator $\partial : \pi_2(K^2, K^1) \to \pi_1(K^1)$ has these two features:

(1)   First, ∂ is a $\pi_1(K^1)$-homomorphism, that is, $\partial([\alpha] \cdot [F]) = [\alpha]\partial[F][\alpha]^{-1}$ for $[F] \in \pi_2(K^2, K^1)$ and $[\alpha] \in \pi_1(K^1)$.

(2)   Second, the action $(\partial[G]) \cdot [F]$ of $\partial[G] \in \pi_1(K^1)$ on $[F] \in \pi_2(K^2, K^1)$ equals the conjugate $[G][F][G]^{-1}$ of $[F]$ by $[G]$ in $\pi_2(K^2, K^1)$.

Here is the terminology introduced for this situation by J. H. C. Whitehead [Wh49$_2$]. A *G-crossed module* (C, ∂, G) consists of groups C and G, and action of G on the left of C, denoted by $g \cdot c$ for $c \in C$ and $g \in G$, and a homomorphism $\partial : C \to G$ such that

(CM1) $\partial(gc) = g(\partial c)g^{-1}$ for all $g \in G, c \in C$, and

(CM2) $cdc^{-1} = (\partial c) \cdot d$ for all $c, d \in C$.

When the property (CM2) is lacking, (C, ∂, G) is called a *pre-crossed module*. A *morphism* of (pre-) crossed modules $(\eta, \tau) : (C, \partial, G) \to (C', \partial', G')$ consists of group homomorphisms $\eta : C \to C'$ and $\tau : G \to G'$ such that $\tau\partial = \partial'\eta$ and $\eta(gc) = \tau(g)\eta(c)$ for $c \in C$ and $g \in G$. We call η a *τ-homomorphism*. The morphism (η, τ) is called an *isomorphism* if η and τ are group isomorphisms.

Thus, the two features recorded above constitute the $\pi_1(K^1)$-*crossed module structure* of $\partial : \pi_2(K^2, K^1) \to \pi_1(K^1)$. There is the third feature, indicated in Lemma 2.2: $\pi_2(K^2, K^1)$ has a set of $\pi_1(K^1)$-generators in one-to-one correspondence with the indexed set $\mathbf{r}$ of relators of $\mathcal{P} = \langle \mathbf{x} \mid \mathbf{r} \rangle$. All three features of the homotopy crossed module $\partial : \pi_2(K^2, K^1) \to \pi_1(K^1)$ are incorporated in the following algebraic construction associated with $\mathcal{P}$.

(Pre-) Crossed modules from presentations   Let $E(\mathcal{P})$ denote the free group on the set $F(\mathbf{x}) \times \mathbf{r}$ of ordered pairs (w, r) where $w \in F(\mathbf{x})$ and $r \in \mathbf{r}$. Let $F(\mathbf{x})$ act on $E(\mathcal{P})$ by $v \cdot (w, r) = (vw, r)$ for $v, w \in F(\mathbf{x})$ and $r \in \mathbf{r}$. Then $E(\mathcal{P})$ is called the *free operator group* on $\mathbf{r}$ with left operators from $F(\mathbf{x})$. A sequence $\omega = ((w_1, r_1)^{\epsilon_1}, \ldots, (w_m, r_m)^{\epsilon_m})$ in the generators of $E(\mathcal{P})$ and their inverses is called a *word* in $E(\mathcal{P})$ and will be denoted by $\omega = \prod_i (w_i, r_i)^{\epsilon_i}$. Each word ω represents (i.e., freely reduces to) an element $W \in E(\mathcal{P})$.

Let $F(\mathbf{x})$ operates on itself by conjugation and let $\partial : E(\mathcal{P}) \to F(\mathbf{x})$ be the $F(\mathbf{x})$-homomorphism given on the basis elements by $\partial(w, r) = wrw^{-1}$, where $w \in F(\mathbf{x})$ and $r \in \mathbf{r}$. Notice that $\mathrm{im}(\partial) = N(\mathbf{r})$, the normal closure of $\mathbf{r}$ in $F(\mathbf{x})$. The subgroup $I(\mathcal{P}) = \ker(\partial : E(\mathcal{P}) \to F(\mathbf{x}))$ of $E(\mathcal{P})$ is called the *group of identities* for the presentation $\mathcal{P} = \langle \mathbf{x} \mid \mathbf{r} \rangle$.

The operator homomorphism $\partial : E(\mathcal{P}) \to F(\mathbf{x})$ is a pre-crossed module; it duplicates all but the second of the three properties described above for homotopy boundary operator. The action of $\partial(w, r) = wrw^{-1} \in F(\mathbf{x})$ on $(v, s) \in E(\mathcal{P})$ gives $(wrw^{-1}v, s)$. Although $(wrw^{-1}v, s)$ doesn't equal the conjugate $(w, r)(v, s)(w, r)^{-1}$ of (v, s) by (w, r) in $E(\mathcal{P})$, they have the same boundary in $F(\mathbf{x})$. This implies that their difference

$$(w, r)(v, s)(w, r)^{-1}(wrw^{-1}v, s)^{-1}$$

in $E(\mathcal{P})$ belongs to the group of identities $I(\mathcal{P}) = \ker \partial$. These differences measure the failure of (CM 2) for this pre-crossed module.

These elements of $I(\mathcal{P})$ of the form

$$(w, r)(v, s)(w, r)^{-1}(wrw^{-1}v, s)^{-1},$$

where $w, v \in F(\mathbf{x})$ and $r, s \in \mathbf{r}$, are called the *basic Peiffer elements*. More generally, the elements of $E(\mathcal{P})$ of the form $UVU^{-1}(\partial(U) \cdot V)^{-1}$ for $U, V \in E(\mathcal{P})$ are called the *Peiffer elements*. The normal closure $P(\mathcal{P}) \leq I(\mathcal{P})$ in $E(\mathcal{P})$ of the basic Peiffer elements is called the *Peiffer group* for the presentation $\mathcal{P} = \langle \mathbf{x} \mid \mathbf{r} \rangle$. One checks that $P(\mathcal{P})$ is invariant under the action by $F(\mathbf{x})$ on $E(\mathcal{P})$, and that $P(\mathcal{P})$ contains, and is generated by, the Peiffer elements.

Thus, the quotient group $C(\mathcal{P}) = E(\mathcal{P})/P(\mathcal{P})$, provided with the induced action of $F(\mathbf{x})$, and the induced $F(\mathbf{x})$-homomorphism $\partial : C(\mathcal{P}) \to F(\mathbf{x})$, given on the generators by $\partial((w, r)P(\mathcal{P})) = wrw^{-1}$, constitute an $F(\mathbf{x})$-crossed module. For this $F(\mathbf{x})$-crossed module *associated* with the group presentation $\mathcal{P} = \langle \mathbf{x} \mid \mathbf{r} \rangle$, we have $\ker \partial = I(\mathcal{P})/P(\mathcal{P})$ and $\mathrm{im}\, \partial = N(\mathbf{r})$.

Thus, analogous to the fundamental sequence for the 2-complex K is the following *fundamental sequence* for the group presentation $\mathcal{P} = \langle \mathbf{x} \mid \mathbf{r} \rangle$:

$$\Pi(\mathcal{P}) : 0 \to I(\mathcal{P})/P(\mathcal{P}) \to C(\mathcal{P}) \to F(\mathbf{x}) \to F(\mathbf{x})/N(\mathbf{r}) \to 0$$

Representing identities and Peiffer identities   The analysis of the fundamental sequence $\Pi(K)$ of the 2-complex $K = K_{\mathcal{P}}$ is completed below by an identification of the homotopy boundary operator $\partial : \pi_2(K^2, K^1) \to \pi_1(K^1)$ with the crossed module $\partial : C(\mathcal{P}) \to F(\mathbf{x})$. We begin with some preliminary geometric ideas which ultimately identify $\pi_2(K^2, K^1)$ as a quotient of $E(\mathcal{P})$.

We first construct a homomorphism $\eta : E(\mathcal{P}) \to \pi_2(K^2, K^1)$, as follows. For the group presentation $\mathcal{P} = \langle \mathbf{x} \mid \mathbf{r} \rangle$, recall that the action of $v \in F(\mathbf{x})$ on the free group $E(\mathcal{P})$ having basis $\{(w, r) \in F(\mathbf{x}) \times \mathbf{r}\}$ is given by $v(w, r) = (vw, r)$, and that each 2-cell c_r^2 ($r \in \mathbf{r}$) of the model $K = K_{\mathcal{P}}$ has a characteristic map $\phi_r : (B^2, S^1) \to (K^2, K^1)$ whose attaching loop $\partial[\phi_r] = [\dot{\phi}_r]$ represents r under the identification $\tau : F(\mathbf{x}) \equiv \pi_1(K^1)$ of Lemma 2.1. So the correspondence of $(1, r) \in E(\mathcal{P})$ with $[\phi_r] \in \pi_2(K^2, K^1)$ extends uniquely to a homomorphism $\eta : E(\mathcal{P}) \to \pi_2(K^2, K^1)$ that satisfies $\partial\eta = \tau\partial$ and respects the actions of $F(\mathbf{x})$ on $E(\mathcal{P})$ and $\pi_1(K^1)$ on $\pi_2(K^2, K^1)$, under the identification $\tau : F(\mathbf{x}) \equiv \pi_1(K^1)$. In short, η is a τ-homomorphism.

For any word $\omega = \prod_i (w_i, r_i)^{\epsilon_i}$ in $E(\mathcal{P})$ of length m, we construct a *standard representative* $R_\omega : (B^2, S^1) \to (K^2, K^1)$ of $\eta(\omega) \in \pi_2(K^2, K^1)$, as follows. In the 2-ball B^2, we form a descending sequence of disjoint $k(r_i)$-gons $P_{k(r_i)}$ ($1 \leq i \leq m$), centered on the axis $\{0\} \times B^1$ and compatibly oriented with B^2, and a sequence ℓ_i ($1 \leq i \leq m$) of arcs joining the basepoint $1 \in S^1 = \partial B^2$ to those of the polygons $P_{k(r_i)}$ ($1 \leq i \leq m$). The map $R_\omega : (B^2, S^1) \to (K^2, K^1)$ carries the polygons $\{P_{k(r_i)}\}$ via the signed characteristic maps $\{\phi_{r_i}^{\epsilon_i}\}$ and sends their complement into K^1, with the arcs $\{\ell_i\}$ mapped as representative loops $\{\alpha_i\}$ for the elements $\{w_i = [\alpha_i]\}$ in $F(\mathbf{x}) \equiv \pi_1(K^1)$. The map R_ω is captured by the *balloons* and *strings* assignments of Figure 6. Now the *balloons* assignments realize the multiplication in $\pi_2(K^2, K^1)$ and the *strings* assignments realize the action of $\pi_1(K^1)$. So if the word ω freely reduces to $W \in E(\mathcal{P})$, then R_ω represents the homotopy class $\eta(W) \in \pi_2(K^2, K^1)$. In particular, $\partial[R_\omega] = \prod_i w_i r_i^{\epsilon_i} w_i^{-1}$ under the identification $\tau : F(\mathbf{x}) \equiv \pi_1(K^1)$.

By the simplicial linkage techniques of Section 1, any homotopy $H : (B^2, S^1) \times I \to (K^2, K^1)$ of such maps R_ω and $R_{\omega'}$ can be deformed to become one for which the closure of each inverse image $H^{-1}(c_r^2)$ of an attached 2-cell c_r^2 in K is a *framed link* Λ_r in this sense: Each component L of Λ_r is either an embedded polygonal torus $P_{k(r)} \times S^1$ or a polygonal cylinder $P_{k(r)} \times B^1$ whose polygonal ends $P_{k(r)} \times \{\pm 1\}$ lie in $B^2 \times \{0, 1\}$, in either case, on whose oriented polygonal cross-sections $P_{k(r)} \times \{z\}$ the homotopy H acts like the characteristic map ϕ_r for the 2-cell c_r^2. The union $\cup\{\Lambda_r : r \in \mathbf{r}\}$ is called the framed link Λ_H

of the homotopy H. The embedding of the segment $1 \times B^1$ on the cylinder $P_{k(r)} \times B^1$ or the loop $1 \times S^1$ on the torus $P_{k(r)} \times S^1$, gives an index curve λ_L on the cylindrical or toroidal component L that records its twisting.

For inessential maps $(B^2, S^1) \to (K^2, K^1)$, this geometric analysis has an algebraic analogue in a characterization of Peiffer identities due to Reidemeister [Re49] and explicitly stated by Papakyriakopoulos in [Pa63, Theorem 3.1]:

Lemma 2.4 *A word $\omega = \prod_i (w_i, r_i)^{\epsilon_i}$ in $E(\mathcal{P})$ represents a Peiffer identity if and only if ω represents an identity $W \in I(\mathcal{P})$ and there is a pairing (i, j) such that (a) $r_i = r_j$, (b) $\epsilon_i = -\epsilon_j$, and (c) $w_i N(\mathbf{r}) = w_j N(\mathbf{r})$ in $F(\mathbf{x})/N(\mathbf{r})$.*

Proof: The necessity of the pairing condition follows from the definition of the basic Peiffer elements in $P = P(\mathcal{P})$. Now suppose that ω represents an identity $W \in E = E(\mathcal{P})$ and satisfies the pairing condition. Since

$$(wr^\epsilon w^{-1} v, s)^\delta \sim (w, r)^\epsilon (v, s)^\delta (w, r)^{-\epsilon} \bmod P,$$

there is an expansion ω' of the word ω representing an element $W' \sim W \bmod P$ and admitting a pairing (i, j) of its indices such that (a) $r_i' = r_j'$, (b) $\epsilon_i' = -\epsilon_j'$, and (c) $w_i' = w_j'$ in $F(\mathbf{x})$. This element is necessarily an identity $W' \in I = I(\mathcal{P})$ and lies in the commutator subgroup $[E, E]$. It remains to show that

$$I \cap [E, E] \subseteq [I, E] \subseteq P.$$

For the first containment relation, let $\sigma : N(\mathbf{r}) \to E$ denote a splitting homomorphism for $\partial : E \to F(\mathbf{x})$ defined on its free image $N(\mathbf{r}) \trianglelefteq F(\mathbf{x})$. Because $V\sigma(\partial V)^{-1} \in I$ for any $V \in E$ and so

$$UV\sigma(\partial V)^{-1}\sigma(\partial U)^{-1} \sim U\sigma(\partial U)^{-1}V\sigma(\partial V)^{-1} \bmod [I, E],$$

it follows that the assignment $U \to U\sigma(\partial U)^{-1}[I, E]$ defines a homomorphism $\lambda : E \to I/[I, E]$. Because $\lambda[E, E] = 1$, as $[I, I] \le [I, E]$, and because $\lambda(U) = U[I, E]$ when $U \in I$, it follows that $I \cap [E, E] \subseteq [I, E]$.

Finally, $[I, E] \subseteq P$, because

$$UVU^{-1}V^{-1} = UVU^{-1}(\partial(U) \cdot V)^{-1} \in P,$$

when $U \in I$. $\qquad\square$

The geometric analysis of null-homotopic maps and the previous characterization of the Peiffer identities make possible the following lemma.

Lemma 2.5 *The τ-homomorphism $\eta : E(\mathcal{P}) \to \pi_2(K^2, K^1)$ is surjective with $\ker \eta = P(\mathcal{P})$, the group of Peiffer identities.*

Proof: The τ-homomorphism η carries the $F(\mathbf{x})$-generators $(1, r) \in E(\mathcal{P})$ to the $\pi_1(K^1)$-generators $[\phi_r] \in \pi_2(K^2, K^1)$ and so is surjective.

To prove that ker $\eta \subseteq P(\mathcal{P})$, let $W \in$ ker η. Then it is an identity $W \in$ ker $\partial = I(\mathcal{P})$ since $\tau(\partial(W)) = \partial(\eta(W))$ and τ is an isomorphism. For any word ω that represents $W \in$ ker η, the map R_ω represents $\eta(W) = 1$ in $\pi_2(K^2, K^1)$ and so admits a null-homotopy H with framed link Λ_H. Then the ends of a cylindrical component L of Λ_r must be two discs in just the floor $B^2 \times 0$ of $B^2 \times I$, and so they represent two entries $(w_i, r_i)^{\epsilon_i}$ and $(w_j, r_j)^{\epsilon_j}$ of the word ω for which $r_i = r = r_j$ and $\epsilon_i = -\epsilon_j$ by Theorem 1.4. Furthermore, the index curve λ_L on L and the arcs ℓ_i and ℓ_j constitute a null-homotopic loop in $B^2 \times I$. Since H is constant on λ_L and $H|_{B^2 \times \{0\}} = R_\omega$ represents w_i and w_j on ℓ_i and ℓ_j, it follows that $w_i N = w_j N$ in $\pi_1(K^1)$. So the cylindrical components of the framed link Λ_H define a complete pairing of the factors of the word $\omega = \prod_i (w_i, r_i)^{\epsilon_i}$ as in Lemma 2.4. Thus the identity W represented by ω is a Peiffer identity.

To prove that, conversely, $P(\mathcal{P}) \subseteq$ ker η, first observe that η is trivial on the basic Peiffer elements, being a τ-homomorphism to the $\pi_1(K^1)$-crossed module $\partial : \pi_2(K^2, K^1) \to \pi_1(K^1)$. A specific null homotopy H for the map R_ω associated with a basic Peiffer identity

$$\omega = (w, r)(v, s)(w, r)^{-1}(wrw^{-1}v, s)^{-1}$$

is described by a simple unknotted framed link Λ_H (viewed in Figure 7 from the basepoint $1 \in S^1$) in which the cylindrical component ending on (w, r) and $(w, r)^{-1}$ crosses in front of that ending on (v, s) and $(wrw^{-1}v, s)^{-1}$. Since $P(\mathcal{P})$ is the normal closure in $E(\mathcal{P})$ of the basic Peiffer elements and ker η is normal in $E(\mathcal{P})$, the containment relation $P(\mathcal{P}) \subseteq$ ker η follows. $\square$

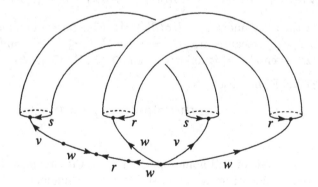

Figure II.7. Framed link for $\omega = (w, r)(v, s)(w, r)^{-1}(wrw^{-1}v, s)^{-1}$

Theorem 2.6 *The $F(\mathbf{x})$-crossed module $\partial : C(\mathcal{P}) \to F(\mathbf{x})$ associated with the presentation $\mathcal{P} = \langle \mathbf{x} \mid \mathbf{r} \rangle$ is isomorphic to the $\pi_1(K^1)$-crossed module $\partial : \pi_2(K^2, K^1) \to \pi_1(K^1)$ for the model $K = K_{\mathcal{P}}$.*

Proof: By Lemma 2.5, the τ-homomorphism $\eta : E(\mathcal{P}) \to \pi_2(K^2, K^1)$ induces a τ-isomorphism $\eta : C(\mathcal{P}) = E(\mathcal{P})/P(\mathcal{P}) \to \pi_2(K^2, K^1)$.   □

Theorems 2.3 and 2.6 identify the fundamental sequences for the model $K = K_{\mathcal{P}}$ of a group presentation $\mathcal{P} = \langle \mathbf{x} \mid \mathbf{r} \rangle$:

$$0 \;\to\; I(\mathcal{P})/P(\mathcal{P}) \;\to\; E(\mathcal{P})/P(\mathcal{P}) \;\to\; F(\mathbf{x}) \;\to\; F(\mathbf{x})/N(\mathbf{r}) \;\to\; 0$$

$$\| \qquad\qquad \eta\| \qquad\qquad \tau\| \qquad\qquad \|$$

$$0 \;\to\; \pi_2(K^2) \;\overset{j_\#}{\to}\; \pi_2(K^2, K^1) \;\overset{\partial}{\to}\; \pi_1(K^1) \;\overset{i_\#}{\to}\; \pi_1(K^2) \;\to\; 0.$$

In particular, this includes the following observation of Reidemeister:

Theorem 2.7 (Reidemeister [Re49]) *The second homotopy group $\pi_2(K^2)$ is isomorphic to the quotient group $I(\mathcal{P})/P(\mathcal{P})$ of identities modulo Peiffer identities.*

There is more to be said about the algebraic structure of the $\pi_1(K^1)$-crossed module $\partial : \pi_2(K^2, K^1) \to \pi_1(K^1)$. The abstract crossed modules that can arise as a homotopy crossed modules are the free crossed modules.

Free Crossed Modules   A G-crossed module $\partial : C \to G$ is called a *free G-crossed module with indexed basis* $\{c_i : i \in \mathcal{I}\} \subseteq C$ if it satisfies this universal property: given a G'-crossed module $\partial' : C' \to G'$, indexed subset $\{c'_i : i \in \mathcal{I}\} \subseteq C'$, and homomorphism $\tau : G \to G'$ such that $\tau(\partial(c_i)) = \partial'(c'_i)$ for each $i \in \mathcal{I}$, then there is a unique homomorphism $\eta : C \to C'$ such that $\eta(c_i) = c'_i$ for each $i \in \mathcal{I}$ and $(\eta, \tau) : (C, \partial, G) \to (C', \partial', G')$ is a G-crossed module homomorphism.

By construction, the $F(\mathbf{x})$-crossed module $\partial : C(\mathcal{P}) \to F(\mathbf{x})$ associated with a group presentation $\mathcal{P} = \langle \mathbf{x} \mid \mathbf{r} \rangle$ is a free $F(\mathbf{x})$-crossed module with indexed basis $\{(1, r) : r \in \mathbf{r}\}$. Hence, the analogous statement holds for the isomorphic $\pi_1(K^1)$-crossed module $\partial : \pi_2(K^2, K^1) \to \pi_1(K^1)$. The converse holds:

Theorem 2.8 *Any free crossed module over a free group has a topological realization as the homotopy crossed module for a 2-complex*

Proof: By the universal property, any free G-crossed module $\partial : C \to G$ where G is a free group $F(\mathbf{x})$ with free basis $\mathbf{x}$ is isomorphic to the $F(\mathbf{x})$-crossed module $\partial : C(\mathcal{P}) \to F(\mathbf{x})$ associated with some group presentation $\mathcal{P} = \langle \mathbf{x} \mid \mathbf{r} \rangle$. Simply take the set $\mathbf{r}$ of relators to be the boundary values in $G = F(\mathbf{x})$ of the indexed basis of C. So Theorem 2.6 says that the given free crossed module over a free group has a topological realization as the homotopy crossed module for the 2-complex $K = K_{\mathcal{P}}$. $\qquad\square$

The techniques employed in Theorem 2.6 may be used to verify this generalization due to J. H. C. Whitehead [Wh49$_2$, Section 16]:

Theorem 2.9 *If X is obtained from A by attaching 2-cells $\{c_i^2\}$ via based maps $\{\phi_i : S^1 \to A\}$, then $\partial : \pi_2(X, A) \to \pi_1(A)$ is a free $\pi_1(A)$-crossed module with indexed basis $\{[\phi_i] \in \pi_2(X, A)\}$.* $\qquad\square$

2.2 $\Pi(K)$ and the homotopy type of a 2-complex K

Whitehead's Theorem for 2-complexes   We first give a direct proof of the 2-dimensional version of Whitehead's result (see the progression, [Wh39, Theorems 15, 17], [Wh48, Theorem 1], and [Wh49$_1$, Theorem 1]) that a map between CW complexes is a homotopy equivalence if and only if induces isomorphism on their homotopy groups.

The next two lemmas concern the models K and L of two presentations $\mathcal{P} = \langle \mathbf{x} \mid \mathbf{r} \rangle$ and $\mathcal{Q} = \langle \mathbf{x} \mid \mathbf{s} \rangle$ that have the same generator set $\mathbf{x}$ and have relator sets $\mathbf{r}$ and $\mathbf{s}$ with the same normal closure $N(\mathbf{r}) = N = N(\mathbf{s})$ in the free group $F = F(\mathbf{x})$. By Lemma 2.3, K and L have identical fundamental group $\Pi = F/N$

Lemma 2.10 *Any map $G : K \to L$ that induces the identity isomorphism on the fundamental group Π is based homotopic to one that is the identity map on the common 1-skeleton $\vee_x S_x^1$ of K and L.*

Proof: If $G_\# : \pi_1(K) \to \pi_1(L)$ is the identity on Π, then $G_\#[i_x] = [G\, i_x] = [j_x]$, for the inclusions $i_x : S_x^1 \subseteq K$ and $j_x : S_x^1 \subseteq L$. So $G|_{\vee_x S_x^1} : \vee_x S_x^1 \to L$ is based homotopic to the inclusion $\vee_x S_x^1 \subseteq L$. As $(B^2 \cup \{0\}) \cup (S^1 \times I)$ is a strong deformation retract of $B^2 \times I$, then $(K \cup \{0\}) \cup (K^1 \times I)$ is a strong deformation retract of $K \times I$. Using this retraction, we may extend to $K \times I$ the homotopy already defined on the 1-skeleton. Thus, G is based homotopic to a map $K \to L$ that is the identity on $\vee_x S_x^1$.

Lemma 2.11 *Any extension* $G : K \to L$ *of the identity map on the common 1-skeleton of* K *and* L *that induces an isomorphism* $G_\# : \pi_2(K) \to \pi_2(L)$ *is a homotopy equivalence.*

Proof: We first show how to construct a right homotopy inverse for G that is also an extension of the identity on the common 1-skeleton. The map $G : (K^2, K^1) \to (L^2, L^1)$ induces a ladder of homomorphisms between the fundamental sequences of K and L:

$$0 \to \pi_2(K^2) \xrightarrow{j_\#} \pi_2(K^2, K^1) \xrightarrow{\partial} \pi_1(K^1) \xrightarrow{i_\#} \pi_1(K^2) \to 0$$

$$\Gamma \downarrow \qquad\qquad \Lambda \downarrow \qquad\qquad \| \qquad\qquad \|$$

$$0 \to \pi_2(L^2) \xrightarrow{j_\#} \pi_2(L^2, L^1) \xrightarrow{\partial} \pi_1(L^1) \xrightarrow{i_\#} \pi_1(L^2) \to 0.$$

Because $G_\# = \Gamma : \pi_2(K^2) \to \pi_2(L^2)$ is an isomorphism, so also is the homomorphism $G_\# = \Lambda : \pi_2(K^2, K^1) \to \pi_2(L^2, L^1)$.

For the characteristic map $\phi_s : (B^2, S^1) \to (L^2, L^1)$ of each 2-cell c_s^2 in L, we select any based map $\psi_s : (B^2, S^1) \to (K^2, K^1)$ representing the pre-image $\Lambda^{-1}([\phi_s]) \in \pi_2(K^2, K^1)$. Since $G_\# = \Lambda$, then $G \psi_s \simeq \phi_s : (B^2, S^1) \to (L^2, L^1)$. Because G is the identity on $K^1 = L^1$ and $(B^2 \times \{0\}) \cup (S^1 \times I) \cong B^2$, then the map ψ_s applied to $B^2 \times \{0\}$ and the restriction of the homotopy $G \psi_s \simeq \phi_s$ to $S^1 \times I$ combine to define a new map $\psi_s : (B^2, S^1) \to (K^2, K^1)$, one that makes $\dot\psi_s = \dot\phi_s : S^1 \to K^1 = L^1$ and makes possible a new homotopy $G \psi_s \simeq \phi_s : (B^2, S^1) \to (L^2, L^1)$ *relative* S^1. Then the extension $J : L \to K$ of the identity on $K^1 = L^1$ given by $J\phi_s = \psi_s$ on each 2-cell c_s^2 in L satisfies $GJ \simeq 1_L : (L^2, L^1) \to (L^2, L^1)$ *relative* L^1. In particular, G has J as a right homotopy inverse.

Since $J_\#([\phi_s]) = [\psi_s] = \Lambda^{-1}[\phi_s]$ for the $\pi_1(L^1)$-generators $[\phi_s] \in \pi_2(L^2, L^1)$ of Lemma 2.2, then $J_\# = \Lambda^{-1} : \pi_2(L^2, L^1) \to \pi_2(K^2, K^1)$, the inverse of the module isomorphism $G_\# = \Lambda$. By the technique of the previous paragraph J has a right homotopy inverse D. Then $D \simeq GJD \simeq G$, so G and J are homotopy inverses. $\square$

In order to deduce Whitehead's theorem on homotopy equivalences, we resort to a technique of expanding presentations of isomorphic groups.

Let $\mathcal{P} = \langle \mathbf{x} \mid \mathbf{r} \rangle$ and $\mathcal{Q} = \langle \mathbf{y} \mid \mathbf{s} \rangle$ be disjoint presentations of isomorphic groups. Let $\Psi : \Omega \leftrightarrow \Xi : \Phi$ be inverse isomorphisms given by the assignments $x \to W_x$ and $y \to V_y$, where W_x and V_y denote words in alphabets $\mathbf{y}$ and $\mathbf{x}$, respectively. There are the *expanded presentations*

$$\mathcal{P}(\Phi) = \langle \mathbf{x}, \mathbf{y} \mid \mathbf{r}, V_y \, y^{-1} (y \in \mathbf{y}) \rangle$$

and

$$Q(\Psi) = \langle y, x \mid s, W_x \ x^{-1}(x \in x)\rangle$$

that have the same generators and whose relators have the same normal closure N in the free group $F = F(x, y)$ because Ψ and Φ are inverse homomorphisms. So $\mathcal{P}(\Phi)$ and $\mathcal{Q}(\Psi)$ present the same group $\Pi = F/N$.

Theorem 2.12 (Whitehead's Theorem [Wh49$_1$]) *A map $F : K \to L$ of connected 2-complexes is a based homotopy equivalence if and only if it induces isomorphisms $F_{\#} : \pi_1(K) \to \pi_1(L)$ and $F_{\#} : \pi_2(K) \to \pi_2(L)$.*

Proof: The direct implication is trivial; it remains to prove the converse. The 2-complexes K and L may be assumed to be the models of arbitrary presentations $\mathcal{P} = \langle x \mid r \rangle$ and $\mathcal{Q} = \langle y \mid s \rangle$ of groups Ω and Ξ. Let $\Psi : \pi_1(K) \leftrightarrow \pi_1(L) : \Phi$ denote the isomorphism $F_{\#}$ and its inverse. The expanded presentations $\mathcal{P}(\Phi)$ and $\mathcal{Q}(\Psi)$ have the same generators and present the same group Π. Their models $K(\Phi)$ and $L(\Psi)$ of $\mathcal{P}(\Phi)$ and $\mathcal{Q}(\Psi)$ have identical 1-skeleton and identical fundamental group Π. The structure of the expanded presentations shows that the inclusions $K \subseteq K(\Phi)$ and $L \subseteq L(\Psi)$ are homotopy equivalences (elementary cellular expansions) that convert the map $F : L \to M$ inducing Ψ on π_1 into a map $K(\Phi) \to L(\Psi)$ inducing the identity automorphism of Π (and an isomorphism on π_2). By Lemmas 1 and 2, $K(\Phi) \to L(\Psi)$ is a homotopy equivalence; hence, so also is F.   □

Maps of fundamental sequences   For a G-crossed module $\partial : C \to G$, let $\pi_1 = \text{coker } \partial$ and $\pi_2 = \text{ker } \partial$. Then $\pi_2 = \text{ker } \partial$ is an abelian group by (CM2) and it is invariant under the action of G by (CM1). Further, this action induces an action of $\pi_1 = \text{coker } \partial$ on the left of $\pi_2 = \text{ker } \partial$, in view of (CM2), making $\pi_2 = \text{ker } \partial$ into a module over $\mathbb{Z}\pi_1$.

Any morphism of crossed modules $(\eta, \tau) : (C, \partial, G) \to (C', \partial', G')$ induces group homomorphisms $\bar\tau : \pi_1 \to \pi_1'$ and $\bar\eta : \pi_2 \to \pi_2'$ by projection and restriction. The latter is an operator homomorphism associated with the former in that $\bar\eta(\bar g \cdot c) = (\bar\tau(\bar g)) \cdot \bar\eta(c)$, for $\bar g \in \pi_1 = \text{coker } \partial$ and $c \in \pi_2 = \text{ker } \partial$. We call (η, τ) an *equivalence* of the crossed modules (C, ∂, G) and (C', ∂', G') when $\bar\eta$ and $\bar\tau$ are group isomorphisms.

Now any cellular map $F : K \to L$ between connected 2-complexes K and L induces ladder of homomorphisms between the exact homotopy sequences of their skeletal pairs. In particular, this yields a morphism between the free crossed modules $\partial : \pi_2(K^2, K^1) \to \pi_1(K^1)$ and $\partial : \pi_2(L^2, L^1) \to \pi_1(L^1)$. And by Whitehead's Theorem 2.12 , the map F is a homotopy equivalence if and only if the induced morphism is an equivalence.

Conversely, a given pair of free crossed modules with free operator groups may be viewed as homotopy crossed modules for a skeletal pairs of 2-complex K and L, by Theorem 2.8. In this case, any morphism

$$0 \to \pi_2(K^2) \xrightarrow{j_\#} \pi_2(K^2, K^1) \xrightarrow{\partial} \pi_1(K^1) \xrightarrow{i_\#} \pi_1(K^2) \to 0$$

$$\downarrow \eta \qquad\qquad \downarrow \tau$$

$$0 \to \pi_2(L^2) \xrightarrow{j_\#} \pi_2(L^2, L^1) \xrightarrow{\partial} \pi_1(L^1) \xrightarrow{i_\#} \pi_1(L^2) \to 0.$$

is realized by a map $(K^2, K^1) \to (L^2, L^1)$ of the skeletal pairs. A mapping between the bouquet of 1-cells in K and L is described by τ; an extension over the attached 2-cells is described by η.

Thus we have established the two dimensional version of Whitehead's observation [Wh49$_2$, Theorems 4 and 6]:

Theorem 2.13 *The homotopy classification of 2-dimensional complexes is identical to the equivalence classification of free crossed modules, whose groups of operators are free groups.*

Whitehead went further and introduced a homotopy relation for maps of crossed modules and showed that "the homotopy theory of 2-dimensional complexes, including the homotopy classification of mappings, is equivalent to the purely algebraic theory of free crossed modules, whose groups of operators are free groups" ([Wh49$_2$, page 468]).

3 Equivariant World for 2-Complexes

The difficult nature of free crossed homotopy modules limits the applicability of the 2-dimensional homotopy classification described in Section 2. The cellular chain complex of the universal coverings of two-dimensional complexes offers an abelianized version of the classification that is much more practical.

3.1 Hurewicz Isomorphism Theorems

We present the first two Hurewicz isomorphism theorems, with very direct proofs that utilize the combinatorial techniques of Sections 2 and 3. The Hurewicz homomorphisms connect the homotopy groups with the more easily computed homology groups. We assume that the reader is familiar with the basic definitions and properties of singular homology theory.

Hurewicz Homomorphisms   By use of the excision and homotopy properties for singular homology theory, we can inductively select generators $s_n \in H_n(S^n)$ and $b_{n+1} \in H_{n+1}(B^{n+1}, S^n)$ of these infinite cyclic singular homology groups for $n \geq 1$ so that $\partial(b_{n+1}) = s_n$, and $q_*(b_n) = s_n$, where $q : B^n \to S^n$ is the quotient map collapsing the boundary sphere S^{n-1}.

The *absolute Hurewicz map* $h_n : \pi_n(K) \to H_n(K)$ $(n \geq 1)$ is defined by $h_n([f]) = f_*(s_n)$ and the *relative Hurewicz map* $h'_{n+1} : \pi_{n+1}(K, L) \to H_{n+1}(K, L)$ is defined by $h'_{n+1}([F]) = F_*(b_{n+1})$.

We leave the following facts as exercises: (1) The Hurewicz maps are homomorphisms. (2) By the care exercised in the inductive choice of the generators s_n and b_{n+1}, the Hurewicz maps form a commutative ladder of homomorphisms between the long exact homotopy and homology sequences for a based pair (K, L). (3) Because the action of $\pi_1(L)$ on $\pi_n(K, L)$ and $\pi_n(L)$ relates based homotopy classes whose representatives are freely homotopic and because freely homotopic maps induce the same homology homomorphisms, then the Hurewicz homomorphisms trivialize the homotopy actions.

Cellular Homology   For convenience, we work with cellular homology hereafter. A brief review of this subject follows. Let K be a CW complex.

The *cellular chain complex* $C(K) = (C_n(K), \partial_n)$ has chain groups $C_n(K) = H_n(K^n, K^{n-1})$ $(n \geq 0)$ and boundary operators $(n \geq 0)$:

$$
\begin{array}{ccc}
C_{n+1}(K) & \xrightarrow{\partial_{n+1}} & C_n(K) \\
\parallel & & \parallel \\
H_{n+1}(K^{n+1}, K^n) & \xrightarrow{\partial_{n+1}} H_n(K^n) \xrightarrow{j_*} & H_n(K^n, K^{n-1}).
\end{array}
$$

By the excision property of singular homology H_*, the characteristic maps $\{\phi_{c^n} : (B^n, S^{n-1}) \to (K^n, K^{n-1})\}$ for the n-cells $\{c^n\}$ of K determine an external direct sum decomposition of nth cellular chain group $C_n(K)$:

$$< \phi_{c^n\,*} >: \oplus_{c^n} H_n(B^n, S^{n-1}) \to H_n(K^n, K^{n-1})$$

There results a free abelian basis $\{\phi_{c^n\,*}(b_n)\}$ for $C_n(K)$ in one-to-one correspondence with the set $\{c^n\}$ of n-cells of K. Each generator $\phi_{c^n\,*}(b_n)$ is usually abbreviated by the corresponding n-cell symbol c^n itself.

In terms of these generators, the boundary operator $\partial_{n+1} : C_{n+1}(K) \to C_n(K)$ $(n \geq 0)$ is given as a linear combination $\partial_{n+1}(c^{n+1}) = \sum_{c^n} [c^{n+1}, c^n] \, c^n$, where the integral coefficient $[c^{n+1}, c^n]$ is called the *incidence number* of the $n+1$-cell c^{n+1} on the n-cell c^n.

One can show that the incidence number $[c^{n+1}, c^n]$ is the degree of the map:

$$ S^n \xrightarrow{\dot{\phi}_{c^{n+1}}} K^n \xrightarrow{q} K^n/(K^n - c^n) \xleftarrow{\phi_{c^n}} B^n/S^{n-1} \equiv S^n . $$

that measures how the attaching map $\dot{\phi}_{c^{n+1}}$ for the $n+1$-cell c^{n+1} wraps around the n-cell c^n.

The chain condition $\partial_n \partial_{n+1} \equiv 0$ $(n \geq 1)$ for the cellular chain complex $C(K) = (C_n(K), \partial_n)$ is best checked using the original definition of the boundary operators, in terms of entries from the long exact singular homology sequences of the pairs (K^n, K^{n-1}) and (K^{n-1}, K^{n-2}). By the chain condition, the group of n-boundaries,

$$B_n(K) = \text{im } (\partial_{n+1} : C_{n+1}(K) \to C_n(K)),$$

is a subgroup of the group of n-cycles,

$$\mathbb{Z}_n(K) = \ker (\partial_n : C_n(K) \to C_{n-1}(K)).$$

The quotient group

$$H_n(C(K)) = \ker \partial_n / \text{im } \partial_{n+1},$$

called the nth *cellular homology group*, consists of cosets $\{z_n\} = z_n + B_n(K)$ of n-cycles $z_n \in \mathbb{Z}_n(K)$. The cellular homology groups $H_n(C(K))$ are isomorphic to the singular homology groups $H_n(K)$; see Massey's text [Ma80, Chapter IV] or Schubert's text [Schu64, Chapter IV].

Theorem 3.1 (Hurewicz [Hu35]) *For any connected 2-complex K, the Hurewicz homomorphism $h_1 : \pi_1(K) \to H_1(K)$ is abelianization.*

Proof: In view of Theorem 1.9, we may consider the model $K = K_{\mathcal{P}}$ of some group presentation $\mathcal{P} = \langle \mathbf{x} \mid \mathbf{r} \rangle$. Since $\pi_1(S^1)$ and $H_1(S^1)$ are infinite cyclic groups generated by $[1_{S^1}]$ and τ_1, respectively, and $h_1[1_{S^1}] = 1_{S^1*}(\tau_1) = \tau_1$, then the Hurewicz homomorphism $h_1 : \pi_1(S^1) \to H_1(S^1)$ is an isomorphism. Since the inclusion maps $i_x : S^1_x \subseteq \vee_x S^1_x$ determine a free product decomposition $\{i_{x\#} : \pi_1(S^1_x) \to \pi_1(\vee_x S^1_x)\}$ by Lemma 2.1 and an internal direct sum decomposition $\{i_{x*} : H_1(S^1_x) \to H_1(\vee_x S^1_x)\}$ by excision, it follows from the naturality of h_1 that it is the abelianization homomorphism when applied to $\vee_x S^1_x$.

For the 2-complex K itself, there is this commutative ladder of Hurewicz homomorphisms:

$$\dots \to \pi_2(K^2) \xrightarrow{j_\#} \pi_2(K^2, K^1) \xrightarrow{\partial} \pi_1(K^1) \xrightarrow{i_\#} \pi_1(K^2) \to 1$$

$$\downarrow h_2 \qquad\qquad \downarrow h_2' \qquad\qquad \downarrow h_1 \qquad\qquad \downarrow h_1$$

$$\dots \to H_2(K^2) \xrightarrow{j_*} H_2(K^2, K^1) \xrightarrow{\partial} H_1(K^1) \xrightarrow{i_*} H_1(K^2) \to 0.$$

Now h_1 for $K^1 = \vee_x S_x^1$ is abelianization and h_2' is surjective as the $\pi_1(K^1)$-generators $[\phi_r]$ of $\pi_2(K^2, K^1)$ in Lemma 2.2 are sent to the free abelian basis members $h_2'([\phi_r]) = \phi_{r*}(b_2) = c_r^2$ of $H_2(K^2, K^1)$. A diagram chase proves that $h_1 : \pi_1(K^2) \to H_1(K^2)$ is surjective and has ker h_1 contained in $[\pi_1, \pi_1]$, the commutator subgroup of $\pi_1 = \pi_1(K^2)$. The reverse containment holds because $H_1(K^2)$ is abelian. So $h_1 : \pi_1(K) \to H_1(K)$ is also abelianization. $\square$

Theorem 3.2 (Hurewicz [Hu35]) *For any simply connected 2-complex K, the Hurewicz homomorphism $h_2' : \pi_2(K^2, K^1) \to H_2(K^2, K^1)$ is abelianization and the Hurewicz homomorphism $h_2 : \pi_2(K) \to H_2(K)$ is an isomorphism.*

Proof: We take the model $K = K_{\mathcal{P}}$ of a group presentation $\mathcal{P} = \langle \mathbf{x} \mid \mathbf{r} \rangle$ of the trivial group. There is this commutative ladder of homomorphisms and short sequences that are exact since K is a simply connected 2-complex.

$$
\begin{array}{ccccccccc}
1 & \to & I(\mathcal{P}) & \to & E(\mathcal{P}) & \xrightarrow{\partial} & F(\mathbf{x}) & \to & 1 \\
& & \downarrow \eta & & \downarrow \eta & & \downarrow \tau & & \\
0 & \to & \pi_2(K^2) & \xrightarrow{j_\#} & \pi_2(K^2, K^1) & \xrightarrow{\partial} & \pi_1(K^1) & \to & 1 \\
& & \downarrow h_2 & & \downarrow h_2' & & \downarrow h_1 & & \\
0 & \to & H_2(K^2) & \xrightarrow{j_*} & H_2(K^2, K^1) & \xrightarrow{\partial} & H_1(K^1) & \to & 0
\end{array}
$$

Injectivity of $h_2 : \pi_2(K^2) \to H_2(K^2)$. Let $[g] \in \pi_2(K^2)$ have $h_2([g]) = 0$. Using the surjectivity of η (Lemma 2.5), we select a word $\omega = \prod_i (w_i, r_i)^{\epsilon_i}$ in $E(\mathcal{P})$ representing an identity $W \in I(\mathcal{P})$ for which

$$\eta(W) = [R_\omega] = \prod_i \tau(w_i) \cdot [\phi_{r_i}]^{\epsilon_i}$$

equals $j_\#([g])$. Because $h_2' \, j_\# = j_* \, h_2$ and because h_2' has values $h_2'([\phi_r]) = \phi_{r*}(b_2) = c_r^2$, we have $0 = h_2'([R_\omega]) = \sum_i \epsilon_i c_{r_i}^2$. Because $\{c_r^2 : r \in \mathbf{r}\}$ is a basis for $H_2(K^2, K^1)$, there is a pairing (i, j) of the indices of the sum $\sum_i \epsilon_i c_{r_i}^2$ such that $c_{r_i}^2 = c_{r_j}^2$ and $\epsilon_i = -\epsilon_j$. Because $w_i N(\mathbf{r}) = w_j N(\mathbf{r})$ in the trivial group $F(\mathbf{x})/N(\mathbf{r})$, this same pairing of the indices of the word $\omega = \prod_i (w_i, r_i)^{\epsilon_i}$ shows that the identity W is a Peiffer identity by Lemma 2.4. So $j_\#([g]) = \eta(W) = 1$ by Lemma 2.5. Because $j_\#$ is injective, we must have $[g] = 0$ in $\pi_2(K^2)$.

The kernel of $h_2' : \pi_2(K^2, K^1) \to H_2(K^2, K^1)$. An analogous argument applies to $[G] \in \pi_2(K^2, K^1)$ having $h_2'([G]) = 0$. A word ω representing $W \in E(\mathcal{P})$ with $[G] = \eta(W)$ admits a pairing of its indices, as above, and so can be expanded, as in the proof of Lemma 2.4, to a word ω' representing $W' \sim W \bmod P(\mathcal{P})$ from $[E(\mathcal{P}), E(\mathcal{P})]$. Thus, $[G] = \eta(W')$ belongs $[\pi_2(K^2, K^1), \pi_2(K^2, K^1)]$. Conversely, the latter commutator subgroup contains $\ker h_2'$ because $H_2(K^2, K^1)$ is abelian.

Surjectivity of $h_2 : \pi_2(K^2) \to H_2(K^2)$. We view any element of

$$H_2(K^2) \equiv \ker \left(\partial : H_2(K^2, K^1) \to H_1(K^1) \right)$$

as the image $h_2'([F]) \in H_2(K^2, K^1)$ of some $[F] \in \pi_2(K^2, K^1)$. Then $\partial([F])$ belongs to $\ker h_1 = [\pi_1(K^1), \pi_1(K^1)]$, as $h_1 \partial = \partial h_2'$. Since $\pi_1(K^1) = \operatorname{im} \partial$, there exists $[G] \in [\pi_2(K^2, K^1), \pi_2(K^2, K^1)]$ with $\partial([G]) = \partial([F])$. Then $[F][G]^{-1}$ has image $h_2'([F][G]^{-1}) = h_2'([F])$ in the abelian group $H_2(K^2, K^1)$ and it also belongs to

$$\pi_2(K^2) \equiv \ker \left(\partial : \pi_2(K^2, K^1) \to \pi_1(K^1) \right).$$

This proves that $h_2 : \pi_2(K^2) \to H_2(K^2)$ is surjective. $\qquad\qquad\square$

3.2 Two Dimensional Equivariant World

We assume that the reader is familiar with the basic theory of covering spaces and the construction of a universal (i.e., simply connected) covering complex $\tilde{K}$ for any connected CW complex K; see, e.g., [Si92, Chapter 15]. We now discuss the cellular chain complex of the universal covering complex $\tilde{K}$ and the equivariant structure induced by the group of covering transformations.

Equivariant cellular chain complex   The covering transformations of the covering projection $p_K : \tilde{K} \to K$ are those homeomorphisms $T : \tilde{K} \to \tilde{K}$ of $\tilde{K}$ for which $p_K \circ T = p_K$. They constitute a group under composition which we denote by $\operatorname{Aut}(p_K)$.

The fundamental group $\pi_1(K)$ and the group $\operatorname{Aut}(p_K)$ of covering transformations are related by the *fundamental isomorphism* $\Delta_{*,\tilde{*}} : \pi_1(K) \to \operatorname{Aut}(p_K)$. It is defined, using 0-cell basepoints $* \in K$ and $\tilde{*} \in p_K^{-1}(*)$ in $\tilde{K}$, by $\Delta([\alpha]) = T \Leftrightarrow T(\tilde{*}) = (\alpha \circ exp)^{\sim}(1)$, where $(\alpha \circ exp)^{\sim} : I \to \tilde{K}$ is the unique path lifting initiating at $\tilde{*}$ of the loop $\alpha \circ exp : I \to S^1 \to K$ at $*$. We use $\Delta_{*,\tilde{*}}$ to identify these isomorphic groups $\pi_1(K)$ and $\operatorname{Aut}(p_K)$ and denote them by Π.

Let $C(\tilde{K})$ be the cellular chain complex $(C_n(\tilde{K}), \partial_n)$ of the universal covering complex $\tilde{K}$. Each covering automorphism $\pi : \tilde{K} \to \tilde{K}$ is a cellular map and so determines a chain map $C(\pi) : C(\tilde{K}) \to C(\tilde{K})$. The action of the automorphism group Π on the cellular chain complex $C(\tilde{K})$ via the induced chain maps $C(\pi) : C(\tilde{K}) \to C(\tilde{K})$, $\pi \in \Pi$, makes $C(\tilde{K})$ a chain complex of (left) modules over the integral group ring $\mathbb{Z}\Pi$.

The cells of $\tilde{K}$ that lie over a given one c in K are permuted by the group $\text{Aut}(p_K)$ of covering transformations $\pi \in \Pi$. So *selected cells $\tilde{c}$ in $\tilde{K}$, one over each cell c in K, provide a preferred $\mathbb{Z}\Pi$-basis for $C(\tilde{K})$ as a chain complex of free $\mathbb{Z}\Pi$-modules.* We call $C(\tilde{K})$ the *equivariant cellular chain complex* of the connected CW complex K.

K. Reidemeister originally utilized such equivariant chain complexes to give a combinatorial classification of lens spaces [Re35] (see also [Re34]).

Equivariant Chain Maps   The same construction for a second connected CW complex L gives a universal covering complex $\tilde{L}$, a covering projection $p_L : \tilde{L} \to L$, fundamental isomorphism $\Delta : \pi_1(L) \equiv \Xi \equiv \text{Aut}(p_L)$, and equivariant cellular chain complex $C(\tilde{L})$ over $\mathbb{Z}\Xi$.

Let $\alpha : \Pi \to \Xi$ be any homomorphism. To be able to employ module-theoretic terminology, we consider this *change of ring procedure*: Each $\mathbb{Z}\Xi$-module M can be viewed as a $\mathbb{Z}\Pi$-module $_\alpha M$ in which the operation of $\pi \in \Pi$ on $m \in \;_\alpha M$ is given by $\pi \cdot m = a(\pi) \cdot m$. Each $\mathbb{Z}\Xi$-module homomorphism $h : M \to N$ is a $\mathbb{Z}\Pi$-module homomorphism $h : \;_\alpha M \to \;_\alpha N$, and the ring homomorphism $\mathbb{Z}\alpha : \mathbb{Z}\Pi \to \mathbb{Z}\Xi, \mathbb{Z}\alpha(\sum_\pi n_\pi \pi) = \sum_\pi n_\pi \alpha(\pi)$, is also a $\mathbb{Z}\Pi$-module homomorphism $\mathbb{Z}\alpha : \mathbb{Z}\Pi \to \;_\alpha \mathbb{Z}\Xi$.

A based map $\tilde{F} : \tilde{K} \to \tilde{L}$ is *α-equivariant* if for all covering transformations $\pi \in \Pi$ and $\alpha(\pi) \in \Xi$, the maps $\tilde{F} \pi : \tilde{K} \to \tilde{K} \to \tilde{L}$ and $\alpha(\pi) \tilde{F} : \tilde{K} \to \tilde{L} \to \tilde{L}$ coincide. Alternately, one says that $\tilde{F} : \tilde{K} \to \;_\alpha \tilde{L}$ is *equivariant*.

The uniqueness property for liftings yields: *Any based map $F : K \to L$ with $F_\# = \alpha : \pi_1(K) \to \pi_1(L)$ has an α-equivariant lifting $\tilde{F} : \tilde{K} \to \tilde{L}$. Conversely, any α-equivariant based map $\tilde{F} : \tilde{K} \to \tilde{L}$ covers a based map $F : K \to L$ with $F_\# = \alpha : \pi_1(K) \to \pi_1(L)$.*

When the equivariant cellular chain complex $C(\tilde{L}) = (C_n(\tilde{L}), \partial_n(\tilde{L}))$ over $\mathbb{Z}\Xi$ is viewed as a chain complex $_\alpha C(\tilde{L}) = (_\alpha C_n(\tilde{L}), \partial_n(\tilde{L}))$ over $\mathbb{Z}\Pi$, we have: *The lifting $\tilde{F} : \tilde{K} \to \tilde{L}$ of a cellular based map $F : K \to L$ with $F_\# = \alpha : \pi_1(K) \to \pi_1(L)$ induces a chain map $C(\tilde{F}) : C(\tilde{K}) \to_\alpha C(\tilde{L})$ of $\mathbb{Z}\Pi$-module homomorphisms, called the equivariant chain map induced by F.*

An arbitrary chain map $C(\tilde{K}) \to \;_\alpha C(\tilde{L})$ is called *0-admissible* if it carries the $\mathbb{Z}$-basis elements of $C_0(\tilde{K})$ to the $\mathbb{Z}$-basis elements of $C_0(\tilde{L})$, preserving basepoints, equivalently, it is induced in dimension 0 by a based equivariant map $\tilde{K}^0 \to_\alpha \tilde{L}^0$.

The advantage the universal covering complex $\tilde{K}$ has over the complex K is that homotopical problems for K transform into homological problems for $\tilde{K}$. In particular, the equivariant cellular chain complex $C(\tilde{K})$ of a 2-complex K is rich enough to contain the homotopy module $\pi_2(K)$:

Lemma 3.3 *For any connected 2-complex K, there is a $\mathbb{Z}\Pi$-module isomorphism $\pi_2(K) \to H_2(\tilde{K})$ relating the homotopy action of $\Pi \equiv \pi_1(K)$ on $\pi_2(K)$ and covering automorphism action of $\Pi \equiv \mathrm{Aut}(p_K)$ on $H_2(\tilde{K})$.*

Proof: Because $\tilde{K}$ is simply connected, the homotopy action of $\pi_1(K)$ on the homotopy group $\pi_2(\tilde{K})$ is trivial. It follows that $\pi_2(\tilde{K})$ is identical to the free homotopy set $[S^2, \tilde{K}]_{free}$; so $\pi_2(\tilde{K})$ acquires an action of the automorphism group $\Pi \equiv \mathrm{Aut}(p_K)$ by composition. Any two based maps $G, F : S^2 \to K$ at $*$ lift to based maps $\tilde{G}, \tilde{F} : S^2 \to \tilde{K}$ at $\tilde{*}$. Any presumed α-homotopy from G to F for a loop $\alpha : S^1 \to K$ at $*$ lifts to a free homotopy from $\tilde{G}$ to some lifting of F. By the uniqueness of liftings, this lifting of F is not $\tilde{F}$, but rather its translation $T \circ \tilde{F} : S^2 \to \tilde{K} \to \tilde{K}$, where $\Delta([\alpha]) = T$ under the fundamental isomorphism $\Delta_{*,\tilde{*}} : \pi_1(K) \to \mathrm{Aut}(p_K)$. By definition of the homotopy action in $\pi_2(K)$, we have $[G] = [\alpha] \cdot [F]$; and by way of the lifting of the α-homotopy, we have $[\tilde{G}]_{free} = [T \circ \tilde{F}]_{free}$. This shows that $p_{K\#} : [S^2, \tilde{K}]_{free} \equiv \pi_2(\tilde{K}) \to \pi_2(K)$ is a $\mathbb{Z}\Pi$-module isomorphism from the covering automorphism action to the homotopy action, assuming the identification $\Delta : \pi_1(K) \equiv \Pi \equiv \mathrm{Aut}(p_K)$ by the fundamental isomorphism.

The homology group $H_2(\tilde{K})$ admits an action by the automorphism group $\Pi = \mathrm{Aut}(p_K)$ via the induced homology homomorphisms $\pi_* : H_2(\tilde{K}) \to H_2(\tilde{K})$ ($\pi \in \Pi$). The Hurewicz isomorphism $h_2 : \pi_2(\tilde{K}) \equiv [S^2, \tilde{K}]_{free} \to H_2(\tilde{K})$ is a $\mathbb{Z}\Pi$-module isomorphism, because both groups have covering automorphism actions and the Hurewicz homomorphism is natural. $\square$

Theorem 3.4 *If $F : K \to L$ is a based cellular map of connected 2-complexes with $F_\# = \alpha$ on π_1, then the homomorphism $\tilde{F}_* : H_2(\tilde{K}) \to H_2(\tilde{L})$ induced by the equivariant chain map $C(\tilde{F}) : C(\tilde{K}) \to {}_\alpha C(\tilde{L})$ is identical to the homotopy module homomorphism $F_\# : \pi_2(K) \to {}_\alpha \pi_2(L)$.*

Proof: By the lifting property $p_L \tilde{F} = F p_K$ and the naturality of the Hurewicz homomorphisms, there is this commutative diagram:

$$\pi_2(K) \xleftarrow{\ p_{K\#}\ } \pi_2(\tilde{K}) \xrightarrow{\ h_2\ } H_2(\tilde{K}) = \ker \partial_2(\tilde{K}) \ \leq\ C_2(\tilde{K})$$

$$\downarrow F_\# \qquad\qquad \downarrow \tilde{F}_\# \qquad\qquad \downarrow \tilde{F}_* \qquad\qquad\qquad \downarrow C_2(\tilde{F})$$

$$\pi_2(L) \xleftarrow{\ p_{L\#}\ } \pi_2(\tilde{L}) \xrightarrow{\ h_2\ } H_2(\tilde{L}) = \ker \partial_2(\tilde{L}) \ \leq\ C_2(\tilde{L})$$

involving the identifications of Lemma 3.3. $\square$

Modification Given a based cellular map $F : K \to L$ of 2-complexes with $F_\# = \alpha$ on π_1 and a $\mathbb{Z}\Pi$-module homomorphism $\gamma : C_2(\tilde{K}) \to {}_\alpha\pi_2(L)$ (called a $C_2(\tilde{K})$-*cochain with coefficients in* ${}_\alpha\pi_2(L)$), we define a new map $F^\gamma : K \to L$, called the *spherical modification* of F by γ, as follows. F^γ acts just like F, except that it preliminarily pinches a circle in each 2-cell c^2 of K to create an additional 2-sphere S^2 based at the boundary of c^2, which it maps via a suitably based representative $S^2 \to \tilde{L} \to L$ of $\gamma(\tilde{c}^2) \in {}_\alpha\pi_2(L) \equiv [S^2, \tilde{L}]_{free}$, where $\tilde{c}^2$ is a preferred 2-cell above c^2. This is a map inducing α on π_1 whose based homotopy class is determined by the based homotopy class of F and the cochain γ. By this construction and the definitions of the equivariant chain maps induced by the cellular based maps $F^\gamma, F : K \to L$, we have the following observation.

Lemma 3.5 *The equivariant chain maps* $C(\tilde{F}^\gamma), C(\tilde{F}) : C(\tilde{K}) \to {}_\alpha C(\tilde{L})$ *induced by* $F^\gamma, F : K \to L$ *coincide through dimension 1 and their difference in dimension 2 is the cochain* $\gamma : C_2(\tilde{K}) \to {}_\alpha\pi_2(L) \equiv \ker \partial_2(\tilde{L})$. □

Theorem 3.6 *For connected 2-complexes* K *and* L, *any 0-admissible chain map* $v : C(\tilde{K}) \to {}_\alpha C(\tilde{L})$ *is realized by a based map* $G : K \to L$ *inducing* $G_\# = \alpha$ *on* π_1.

Proof: Using the 0-admissibility and the simple connectedness, we can construct an α-equivariant map $\tilde{F} : \tilde{K} \to \tilde{L}$ inducing v in dimension 0 and 1. Then the difference $\gamma = v - C_2(\tilde{F}) : C_2(\tilde{K}) \to {}_\alpha C_2(\tilde{L})$ takes values in $\ker \partial_2(\tilde{L}) = H_2(\tilde{L}) \equiv \pi_2(L)$. By Lemma 3.5, the modification $G = F^\gamma : K \to L$ realizes the given chain map v. □

For any connected CW complex K, the product $K \times I$ has the same fundamental group as K and admits $\tilde{K} \times I$ as a universal covering space. So the equivariant vocabulary for based maps $K \to L$ applies in a straightforward way to based homotopies $H : K \times I \to L$. The uniqueness property for liftings yields: *Any based homotopy* $H : K \times I \to L$ *between based maps* $F, G : K \to L$ *with* $F_\# = \alpha = G_\# : \pi_1(K) \to \pi_1(L)$ *has an* α-*equivariant lifting to a based homotopy* $\tilde{H} : \tilde{K} \times I \to \tilde{L}$ *between* α-*equivariant based liftings* $\tilde{F}, \tilde{G} : \tilde{K} \to \tilde{L}$. *Conversely, any* α-*equivariant based homotopy* $\tilde{H} : \tilde{F} \simeq \tilde{G}$ *covers a based homotopy* $H : F \simeq G$ *where* $F_\# = \alpha = G_\# : \pi_1(K) \to \pi_1(L)$.

Theorem 3.7 *For connected 2-complexes* K *and* L, *two based maps* $F, G : K \to L$ *with* $F_\# = \alpha = G_\# : \pi_1(K) \to \pi_1(L)$ *are based homotopic if and only if their induced equivariant chain maps* $C(\tilde{F}), C(\tilde{G}) : C(\tilde{K}) \to {}_\alpha C(\tilde{L})$ *are chain homotopic over* $\mathbb{Z}\Pi$.

Proof: For any based homotopy $H : F \simeq G$, the equivariant lifting $\tilde{H} :$ $\tilde{K} \times I \to \tilde{L}$ induces a chain map $C(\tilde{H}) : C(\tilde{K} \times I) \to {}_\alpha C(\tilde{L})$. Then the $\mathbb{Z}\Pi$-module homomorphisms $s_n : C_n(\tilde{K}) \to {}_\alpha C_{n+1}(\tilde{L})$ $(n \geq 0)$, given by

$$s_n(\tilde{c}^n) = C_{n+1}(\tilde{H})((\tilde{c}^n \times (0,1))$$

define the desired chain homotopy $s : C(\tilde{F}) \simeq C(\tilde{G}) : C(\tilde{K}) \to {}_\alpha C(\tilde{L})$.

Conversely, given a chain homotopy s, there is a 0-admissible equivariant chain map $v : C(\tilde{K} \times I) \to {}_\alpha C(\tilde{L})$ defined by $v_n(\tilde{c}^{n-1} \times (0,1)) = s_n(\tilde{c}^{n-1})$, $v_n(\tilde{c}^n \times \{0\}) = C_n(\tilde{F})(\tilde{c}^n)$, and $v_n(\tilde{c}^n \times \{1\}) = C_n(\tilde{G})(\tilde{c}^n)$, for all basic cells of the product complex $\tilde{K} \times I$. By the proof of Theorem 3.6, this chain map over $\mathbb{Z}\Pi$ has a partial realization through dimension 2 by a partial based homotopy $H : (K \times I)^2 \to L$ between $F, G : K \to L$.

Since L has no 3-cells, $C_3(\tilde{L}) = 0$. So the chain map condition implies that $C_2(\tilde{H}) : C_2(\tilde{K} \times I) \to {}_\alpha C_2(\tilde{L})$ is trivial on the boundary of the 3-cells $\tilde{c}^2 \times (0,1)$. And Theorem 3.4 identifies the induced homomorphism $\tilde{H}_* : H_2((\tilde{K} \times I)^2) \to H_2(\tilde{L})$ is with the homotopy module homomorphism $H_\# : \pi_2((K \times I)^2) \to {}_\alpha \pi_2(L)$. Because the identification $H_2((\tilde{K} \times I)^2) \equiv \pi_2((\tilde{K} \times I)^2)$ corresponds the boundary of the 3-cells $\tilde{c}^2 \times (0,1)$ with the homotopy class of its attaching map, it follows that $H : (K \times I)^2 \to L$ is homotopically trivial on the boundary of each 3-cell $c^2 \times (0,1)$. So it extends to the desired homotopy $H : K \times I \to L$. $\square$

Theorems 3.6 and 3.7 show that the homotopy theory of based maps of 2-complexes translates faithfully in the chain homotopy theory of 0-admissible equivariant chain maps. In Section 4, we present Mac Lane and Whitehead's codification of these results [MaWh50] in terms of k-invariants and the cohomology of groups.

Equivariant world for a presentation model   Let $\mathcal{P} = \langle \mathbf{x} \mid \mathbf{r} \rangle$ be a group presentation. The quotient homomorphism

$$\| - \| : F(\mathbf{x}) \to F(\mathbf{x})/N(\mathbf{r}) = \Pi$$

gives the group presented by $\mathcal{P}$. As in §1, the model $K = K_\mathcal{P}$ is a 2-complex $K = c^0 \cup c_x^1 \cup c_r^2$ whose cells are oriented by combinatorial characteristic maps $\psi_x : B^1 \to K$ and $\phi_r : (B^2, S^1) \to (K^2, K^1)$, where $\dot{\phi}_r : S^1 \to K^1 \equiv \vee S_x^1$ spells the relator word $r \in W(\mathbf{x})$. For each group element $\pi \in \Pi$, generator x, and relator $r = \prod_i x_i^{\epsilon_i}$, the covering complex $\tilde{K}$ has a 0-cell $\pi \, \tilde{c}^0$, an oriented 1-cell $(\pi \, \tilde{c}_x^1, \pi \, \tilde{\psi}_x)$ that joins $\pi \, \tilde{c}^0$ to $\pi x \, \tilde{c}^0$, and an oriented 2-cell $(\pi \, \tilde{c}_r^2, \pi \, \tilde{\phi}_r)$ that is attached by the product combinatorial loop

$$(\| x_1^{\delta_1} \| \, \tilde{\psi}_{x_1})^{\epsilon_1} \cdot (\| x_1^{\epsilon_1} x_2^{\delta_2} \| \, \tilde{\psi}_{x_2})^{\epsilon_2} \cdot \ldots \cdot (\| x_1^{\epsilon_1} \cdots x_{m-1}^{\epsilon_{m-1}} x_m^{\delta_m} \| \, \tilde{\psi}_{x_m})^{\epsilon_m},$$

where $\delta_i = 0$ if $\epsilon_i = 1$ and $\delta_i = -1$ if $\epsilon_i = -1$. There is the fundamental isomorphism $\Delta_{c^0, \tilde{c}^0} : \pi_1(K) \equiv \ \equiv \text{Aut}(p_K)$ corresponding to the basepoints c^0 and $\tilde{c}^0 \in p_K^{-1}(c^0)$, and the action of $\Pi = \text{Aut}(p_K)$ on the cells of $\tilde{K}$ is incorporated into their notation: $\pi' \in \Pi$ carries $\pi \ \tilde{c}^0$, $\pi \ \tilde{c}_x^1$, and $\pi \ \tilde{c}_r^2$ into $\pi'\pi \ \tilde{c}^0$, $\pi'\pi \ \tilde{c}_x^1$, and $\pi'\pi \ \tilde{c}_r^2$, respectively. Therefore, the equivariant cellular chain complex $C(\tilde{K})$ has chain modules $C_0(\tilde{K})$, $C_1(\tilde{K})$, and $C_2(\tilde{K})$ with preferred free $\mathbb{Z}\Pi$-module bases $\{\tilde{c}^0\}$, $\{\tilde{c}_x^1\}$, and $\{\tilde{c}_r^2\}$, and with boundary operators defined by

$$\partial_1(c_x^1) = \| \ x \ \| \ \tilde{c}^0 - \tilde{c}^0 = (\| \ x \ \| - 1)\tilde{c}^0,$$

and

$$\partial_2(\tilde{c}_r^2) = \epsilon_1 \ \| \ x_1^{\delta_1} \ \| \ \tilde{c}_1^1 + \epsilon_2 \ \| \ x_1^{\epsilon_1} x_2^{\delta_2} \ \| \ \tilde{c}_2^1 + \ldots + \epsilon_m \ \| \ x_1^{\epsilon_1} \ldots x_{m-1}^{\epsilon_{m-1}} x_n^{\delta_m} \ \| \ \tilde{c}_m^1,$$

where $r = \prod_i x_i^{\epsilon_i}$, and $\delta_i = 0$ if $\epsilon_i = 1$ and $\delta_i = -1$ if $\epsilon_i = -1$.

The equivariant cellular chain complex $C(\tilde{K})$ of the model $K = K(\mathcal{P})$ can be described succinctly in terms of the presentation $\mathcal{P} = \langle \mathbf{x} \mid \mathbf{r} \rangle$ of Π, as follows. We define the *chain complex* $C(\mathcal{P})$ of the presentation $\mathcal{P}$ to be the $\mathbb{Z}\Pi$-complex of free $\mathbb{Z}\Pi$-modules

$$C(\mathcal{P}) : C_2(\mathcal{P}) \xrightarrow{\partial_2(\mathcal{P})} C_1(\mathcal{P}) \xrightarrow{\partial_1(\mathcal{P})} C_0(\mathcal{P}) \ = \ \mathbb{Z}\Pi$$

with preferred bases $\{b_r^2\}$, $\{b_x^1\}$, and $\{b^0\}$ in correspondence with the relators and generators of $\mathcal{P}$. The boundary operators are defined by

$$\partial_2(\mathcal{P})(b_x^1) = (\| \ x \ \| - 1)b^0$$

and

$$\partial_2(\mathcal{P})(b_r^2) = h(r),$$

where $h : F(\mathbf{x}) \to C_1(\mathcal{P})$ is the *crossed homomorphism* uniquely defined by the three properties:

(a) $h(x) = b_x^1$,

(b) $h(x^{-1}) = - \| \ x \ \|^{-1} \ h(x) = - \| \ x \ \|^{-1} \ b_x^1$, and

(c) $h(W_1 \ W_2) = h(W_1) + \| \ W_1 \ \| \ h(W_2)$, for $W_1, W_2 \in F(\mathbf{x})$.

The crossed homomorphism definition is rigged to ensure that *the chain complex $C(\mathcal{P})$ of the presentation $\mathcal{P}$ is identical to the equivariant chain complex $C(\tilde{K})$ of the model $K = K_\mathcal{P}$ of $\mathcal{P}$.* The point is that everything matches when the $\mathbb{Z}\Pi$-bases $\{\tilde{c}^0\}$, $\{\tilde{c}_x^1\}$, and $\{\tilde{c}_r^2\}$ of $C(\tilde{K})$ are identified with the $\mathbb{Z}\Pi$-bases $\{b^0\}$, $\{b_x^1\}$, and $\{b_r^2\}$ of $C(\mathcal{P})$.

The second boundary operator $\partial_2(\mathcal{P})$ can also be expressed using the *Reidemeister-Fox derivative* $\frac{\partial}{\partial x} : F(\mathbf{x}) \to \mathbb{Z}F(\mathbf{x})$ associated with each generator $x \in X$ and the natural extension $\| - \| : \mathbb{Z}F(\mathbf{x}) \to \mathbb{Z}\Pi$ of the quotient homomorphism $\| - \| : F(\mathbf{x}) \to F(\mathbf{x})/N(\mathbf{r}) = \Pi$. The derivative $\frac{\partial}{\partial x}$ is the unique *derivation*, i.e., function satisfying

$$\frac{\partial\, W_1 W_2}{\partial\, x} = \frac{\partial\, W_1}{\partial\, x} + W_1 \frac{\partial\, W_2}{\partial\, x}, \text{ for all } W_1, W_2 \in F(\mathbf{x}),$$

whose value on $x' \in X$ is $\delta_{x,x'}$ (the Kronecker delta). These derivatives are developed in Fox's *free differential calculus* ([Fo53]). Using them, we can express $\partial_2(\mathcal{P})$ this way:

$$\partial_2(\mathcal{P})(b_r^2) = \sum_{x \in X} \left\| \frac{\partial\, r}{\partial\, x} \right\| b_x^1.$$

By Lemma 3.3, we have this description of the second homotopy module in terms of Reidemeister chains:

Theorem 3.8 *The homotopy module $\pi_2(K)$ of the model $K = K_\mathcal{P}$ is isomorphic to the $\mathbb{Z}\Pi$-submodule $\ker(\partial_2(\mathcal{P}) : C_2(\mathcal{P}) \to C_1(\mathcal{P}))$ of the chain complex $C(\mathcal{P})$ of the presentation $\mathcal{P}$.* $\qquad\square$

Example 1 Consider the presentation $\mathcal{P}_m = \langle x \mid x^m \rangle$ of the finite cyclic group $\mathbb{Z}_m$. The Reidemeister-Fox derivative of the single relator $r = x^m$ with respect to the single generator x is

$$\frac{\partial\, x^m}{\partial\, x} = \frac{\partial\, x}{\partial\, x} + x \frac{\partial\, x^{m-1}}{\partial\, x} = \ldots = 1 + x + \ldots + x^{m-1} \in \mathbb{Z}(F(x)).$$

Let $< x, m > = 1 + x + \ldots + x^{m-1} \in \mathbb{Z}(\mathbb{Z}_m)$. Then $\partial_2(\mathcal{P}_m)(b_r^2) = < x, m > b_x^1$ and the chain complex of the presentation $\mathcal{P}_m$ is

$$C(\mathcal{P}_m) : \mathbb{Z}(\mathbb{Z}_m) \xrightarrow{\ < x,\, m >\ } \mathbb{Z}(\mathbb{Z}_m) \xrightarrow{\ x - 1\ } \mathbb{Z}(\mathbb{Z}_m).$$

The universal covering $\tilde{P}_m$ of the model $P_m = c^0 \cup c_x^1 \cup c_r^2$ (called a *pseudo-projective plane*) is a stack of m discs $x^k\, \tilde{c}^2$ ($x^k \in \mathbb{Z}_m$) that share the boundary circle

$$\tilde{c}^0 \cup \tilde{c}_x^1 \cup x\, \tilde{c}^0 \cup\ x\, \tilde{c}_x^1 \cup x^2\, \tilde{c}^0 \cup \ldots \cup x^{m-1}\, \tilde{c}^0 \cup x^{m-1}\, \tilde{c}_x^1 \cup \tilde{c}^0,$$

with the disc $x^k\, \tilde{c}_r^2$ attached beginning at the 0-cell $x^k\, \tilde{c}^0$. This shows that the equivariant cellular chain complex $C(\tilde{P}_m)$, with preferred $\mathbb{Z}(\mathbb{Z}_m)$-module bases $\{\tilde{c}^0\}$, $\{\tilde{c}^1_x\}$, and $\{\tilde{c}^2_r\}$, really does coincide with the chain complex $C(\mathcal{P}_m)$. By Theorem 3.8, the homotopy module $\pi_2(P_m)$ is isomorphic to the $\mathbb{Z}(\mathbb{Z}_m)$-submodule ker $(< x, \ m >: \mathbb{Z}(\mathbb{Z}_m) \to \mathbb{Z}(\mathbb{Z}_m))$. This is precisely the ideal $\mathbb{Z}(\mathbb{Z}_m)(x-1) \leq \mathbb{Z}(\mathbb{Z}_m)$, a free abelian group of rank $m-1$.

Example 2 Consider the presentation $\mathcal{P}_{m,n} = \langle x, y \mid x^m, y^n, xyx^{-1}y^{-1} \rangle$ of the finite abelian group $\Pi = \mathbb{Z}_m \oplus \mathbb{Z}_n$. The chain complex $C(\mathcal{P}_{m,n})$ is

$$
\mathbb{Z}\Pi^3 \xrightarrow{\begin{pmatrix} < x, \ m > & 0 \\ 0 & < y, \ n > \\ 1-y & x-1 \end{pmatrix}} \mathbb{Z}\Pi^2 \xrightarrow{\begin{pmatrix} x-1 \\ y-1 \end{pmatrix}} \mathbb{Z}\Pi
$$

where $< x, \ m > = 1 + x + \ldots + x^{m-1}$ and $< y, \ n > = 1 + y + \ldots + y^{n-1}$ in $\mathbb{Z}\Pi$. The homotopy module $\pi_2(P_{m,n})$ of the model $P_{m,n}$ is isomorphic to the $\mathbb{Z}(\mathbb{Z}_m \oplus \mathbb{Z}_n)$-submodule ker $\partial_2(\mathcal{P}_{m,n})$. By an examination of the 2-dimensional *cellular cycles* of the covering complex $\tilde{P}_{m,n}$, this can be seen to be the submodule of $\mathbb{Z}\Pi^3$ generated by the four triples:

$$(0, x-1, 0), \quad (< x, \ m >, y-1, 0), \quad (0, 0, y-1), \quad \text{and} \quad (< y, \ n >, 0, 1-x).$$

Exercise 3 Consider any four positive integers m, n, p and q, where $(p, \ m) = 1 = (q, \ n)$, and the presentation $\mathcal{P}_{m,n;p,q} = \langle x, y \mid x^m, y^n, x^p y^q x^{-p} y^{-q} \rangle$ of the finite abelian group $\Pi = \mathbb{Z}_m \oplus \mathbb{Z}_n$. Determine the chain complex $C(\mathcal{P}_{m,n;p,q})$ and the homotopy module $\pi_2(P_{m,n;p,q})$ of the model $P_{m,n;p,q}$.

See Chapter V for further techniques in determining generators of second homotopy modules.

The chain complexes $C(\mathcal{P})$ and $C(\mathcal{Q})$ of two group presentations $\mathcal{P} = \langle \mathbf{x} \mid \mathbf{r} \rangle$ and $\mathcal{Q} = \langle \mathbf{x} \mid \mathbf{s} \rangle$ that have the same generators and present the same group Π may be identified through dimensions 0 and 1:

$$
\begin{array}{ccccc}
C_2(\mathcal{P}) & \xrightarrow{\partial_2(\mathcal{P})} & C_1(\mathcal{P}) & \xrightarrow{\partial_1(\mathcal{P})} & C_0(\mathcal{P}) \\
 & & \parallel & & \parallel \\
C_2(\mathcal{Q}) & \xrightarrow{\partial_2(\mathcal{Q})} & C_1(\mathcal{Q}) & \xrightarrow{\partial_1(\mathcal{Q})} & C_0(\mathcal{Q}).
\end{array}
$$

Let K and L denote the models $K_{\mathcal{P}}$ and $K(\mathcal{Q})$ of these presentations.

Theorem 3.9 *There exists a homotopy equivalence $F : K \to L$ inducing the identity isomorphism on the fundamental group Π if and only if there is a $\mathbb{Z}\Pi$-module isomorphism $\Phi : C_2(\mathcal{P}) \to C_2(\mathcal{Q})$ such that $\partial_2(\mathcal{P}) = \partial_2(\mathcal{Q}) \ \Phi$.*

Proof: If $F : K \to L$ is a homotopy equivalence inducing the identity isomorphism on π_1, then the equivariant chain map $C(\tilde{F}) : C(\tilde{K}) \to C(\tilde{L})$ restricts to give the homotopy module isomorphism $F_\#$ on ker ∂_2:

$$\pi_2(K) \equiv H_2(\tilde{K}) = \ker \partial_2(\mathcal{P}) \leq C_2(\mathcal{P}) \xrightarrow{\partial_2(\mathcal{P})} C_1(\tilde{K})$$

$$\downarrow F_\# \qquad\qquad \downarrow \tilde{F}_* \qquad\qquad \downarrow C_2(\tilde{F}) \qquad\qquad \downarrow C_1(\tilde{F})$$

$$\pi_2(L) \equiv H_2(\tilde{L}) = \ker \partial_2(\mathcal{Q}) \leq C_2(\mathcal{Q}) \xrightarrow{\partial_2(\mathcal{Q})} C_1(\tilde{L}).$$

Since $C_1(\tilde{F}) : C_1(\tilde{K}) \to C_1(\tilde{L})$ is the identity, it follows from the 5-lemma that $\Phi = C_2(\tilde{F}) : C_2(\tilde{K}) \to C_2(\tilde{L})$ is a $\mathbb{Z}\Pi$-module isomorphism satisfying $\partial_2(\mathcal{P}) = \partial_2(\mathcal{Q}) \, \Phi$.

Conversely, a $\mathbb{Z}\Pi$-module isomorphism $\Phi : C_2(\mathcal{P}) \to C_2(\mathcal{Q})$ satisfying the relation $\partial_2(\mathcal{P}) = \partial_2(\mathcal{Q}) \, \Phi$ extends by the identity in lower dimensions to a chain map $v : C(\tilde{K}) \to {}_\alpha C(\tilde{L})$. By Theorem 3.6, v can be realized by a map $F : K \to L$ that is the identity on the common 1-skeleton. By Theorem 3.4, the homotopy module homomorphism $F_\# : \pi_2(K) \to \pi_2(L)$ induced by F is the restriction isomorphism $\Phi : \ker \partial_2(\mathcal{P}) \to \ker \partial_2(\mathcal{Q})$. It follow from Whitehead's Theorem 2.12 that $F : K \to L$ is a homotopy equivalence. $\square$

As in the proof of Theorem 2.12, it is possible to expand disjoint presentations of isomorphic groups into presentations that have the same generators and present the same group. But using Theorem 3.9 and that expansion technique to classify 2-complexes with a prescribed fundamental group encounters this critical difficulty: the existence of a homotopy equivalence can depend upon the induced automorphism of the fundamental group.

Exercise 4 Prove that every automorphism of the fundamental group $\mathbb{Z}_m$ of the pseudo-projective plane P_m is induced by a homotopy equivalence $P_m \to P_m$ (Olum [Ol65]). Equivalently, show that if $ss' \equiv 1(\mathrm{mod}\ m)$, there is a homotopy equivalence of the models of the *expanded presentations*

$$\langle x, y \mid x^m, y^s x^{-1} \rangle \quad \text{and} \quad \langle y, x \mid y^m, x^{s'} y^{-1} \rangle$$

inducing the identity automorphism ($x \mapsto x$ and $y \mapsto y$) on $\pi_1 = \mathbb{Z}_m$.

Exercise 5 Consider the models $P_{m,n}$ and $P_{m,n;p,q}$ of the presentations

$$\mathcal{P}_{m,n} = \langle x, y \mid x^m, y^n, [x, y] \rangle \quad \text{and} \quad \mathcal{P}_{m,n;p,q} = \langle x, y \mid x^m, y^n, [x^p, y^q] \rangle.$$

(a) Prove there is a (simple-) homotopy equivalence $P_{m,n} \to P_{m,n;p,q}$ inducing the identity automorphism ($x \mapsto x$ and $y \mapsto y$) on $\pi_1 = \mathbb{Z}_m \oplus \mathbb{Z}_n$ if and only if $pq \equiv \pm 1 \mod (m, n)$ (Schellenberg [Sche73], see [Si76] and Chapter III, §1.3) .

(b) Prove there is always a simple-homotopy equivalence $P_{m,n} \to P_{m,n,p,q}$ inducing the diagonal automorphism ($x \mapsto x^p$ and $y \mapsto y^q$).

4   Mac Lane-Whitehead Algebraic Types

This section contains the Mac Lane-Whitehead theory [MaWh50] of algebraic types of 2-complexes.

4.1   Homology and Cohomology of Groups

H. Hopf initiated the homology theory of groups with his paper ([Ho41]). He observed that, for complexes K with $\pi_1(K) = \Pi$, the quotient group $H_2(K)/\Sigma_2(K)$ is an invariant, where $\Sigma_2(K)$ is the group of spherical 2-cycles in the cell complex K, i.e., the image of the second Hurewicz homomorphism. For 2-complexes, it is very easy to check the invariance of $H_2(K^2)/\Sigma_2(K^2)$ using Tietze operations; see Chapter I, footnote 26. This invariant quotient is called the *second homology group* $H_2(\Pi)$ of the group Π; it can be viewed as a *lower bound* for the second homology group of complexes K with $\pi_1(K) = \Pi$, since it is a quotient of each one.

History has shown that Hopf's discovery is just the two-dimensional portion of the homology theory of groups. We now begin with some homological algebra to provide a framework for the initial concepts in the homology and cohomology theories of a group.

Resolutions   Let R be any ring and let M be any (left) R-module. A *resolution* of M is an exact sequence of R-modules and homomorphisms

$$ \ldots \to C_n \to C_{n-1} \to \ldots \to C_1 \to C_0 \xrightarrow{\epsilon} M \to 0. $$

Alternately, it is a chain map $\epsilon : C \to M$ from a chain complex $C = (C_n, \partial_n)$ to M (viewed as a chain complex with $M_0 = M$ and $\partial = 0$) such that $\epsilon_* : H_0(C) \approx M$. Every module M has a resolution by projective (even, free) R-modules, called a *projective* (or, *free*) resolution over R. Exactness and the projectivity yield this basic *comparison theorem* ([M68]):

Theorem 4.1 *If $\epsilon : C \to M$ is a projective resolution and $\epsilon' : C' \to M'$ is any resolution, then any module homomorphism $h : M \to M'$ extends to a chain map $u : C \to C'$ and any two extensions are chain homotopic.* $\square$

It follows that there is just one chain homotopy type of a projective resolutions $\epsilon : C \to M$ of M over R. Hence, for each *coefficient module* N, these projective resolutions give just one chain homotopy type of associated chain complexes

$$C \otimes_R N : \ldots \to C_n \otimes_R N \to C_{n-1} \otimes_R N \to \ldots \to C_0 \otimes_R N$$

and just one cochain homotopy type of associated cochain complex

$$\operatorname{Hom}_R(C, N) : \operatorname{Hom}_R(C_0, N) \to \ldots \to \operatorname{Hom}_R(C_n, N) \to \ldots .$$

So the derived homology groups $H_n(C \otimes_R N)$ and the derived cohomology $H^n(\operatorname{Hom}_R(C, N))$ depend upon just the ring R, the resolved module M, and the coefficient module N. While the resolution $\epsilon : C \to M$ is exact, so that the chain complex C is acyclic, the complexes $\operatorname{Hom}_R(C, N)$ and $C \otimes_R N$ are usually not acyclic and their derived groups record algebraic features of the resolved module M, the coefficient module N, and the ring R.

When the ring R is an integral group ring $\mathbb{Z}\Pi$ and the module M is the additive group of integers $\mathbb{Z}$, with the trivial action, these homology and cohomology groups are denoted by $H_n(\Pi, N)$ and $H^n(\Pi, N)$. They are called the *homology* and *cohomology groups* of the group Π with coefficients N.

For any $\mathbb{Z}\Pi$-module N, the invariant submodule $\{n \in N : \pi n = n, \forall \pi \in \Pi\}$ is denoted by N^Π. Here are two fundamental lemmas; the first is an exercise and the second is a project.

Lemma 4.2 *For any $\mathbb{Z}\Pi$-module N, $H^0(\Pi, N) \approx N^\Pi$ and the cohomology homomorphism $h^* : H^0(\Pi, K) \to H^0(\Pi, N)$ induced by a module homomorphism $h : K \to N$ is its restriction $h : K^\Pi \to N^\Pi$.* $\quad\square$

Lemma 4.3 [CaEi56, Chapter XII, Proposition 2.2]) *If Π is a finite group, then $H^n(\Pi, P) = 0$ for all $n \geq 1$ and all $\mathbb{Z}\Pi$-projective modules P.* $\quad\square$

Theorem 4.4 *Let Π be a finite group. Consider any projective resolution*

$$\ldots \to C_n \to C_{n-1} \to \ldots \to C_1 \to \mathbb{Z}\Pi \xrightarrow{\epsilon} \mathbb{Z} \to 0$$

of $\mathbb{Z}$ over $\mathbb{Z}\Pi$, with augmentation homomorphism $\epsilon(\sum_\pi n_\pi \pi) = \sum_\pi n_\pi$.

(a) *There are connecting isomorphisms*

$$\mathbb{Z}_{|\Pi|} \approx H^1(\Pi, \ker \epsilon) \approx H^2(\Pi, \ker \partial_1) \approx H^3(\Pi, \ker \partial_2) \approx \ldots$$

under which $1 \in \mathbb{Z}_{|\Pi|}$ corresponds to the cohomology classes of the cochains $\partial_1 : C_1 \to \ker \epsilon, \partial_2 : C_2 \to \ker \partial_1, \partial_3 : C_3 \to \ker \partial_2, \ldots$.

(b) *There are connecting isomorphisms*

$$0 = H^0(\Pi, \ker \epsilon) \approx H^1(\Pi, \ker \partial_1) \approx H^2(\Pi, \ker \partial_2) \approx \ldots$$

Proof: There are the invariant submodules $\mathbb{Z}^\Pi = \mathbb{Z}$, $(\ker \epsilon)^\Pi = 0$ and $(\mathbb{Z}\Pi)^\Pi = N\mathbb{Z}\Pi$, the ideal generated by the norm element $N = \sum_{\pi \in \Pi} \pi$ of $\mathbb{Z}\Pi$. By Lemma 4.3, $H^1(\Pi, \mathbb{Z}\Pi) = 0$. So by Lemma 4.2, the cohomology sequence induced by the short exact coefficient sequence $0 \to \ker \epsilon \to \mathbb{Z}\Pi \to \mathbb{Z} \to 0$ begins with

$$H^0(\Pi, \ker \epsilon) \xrightarrow{\epsilon_*} H^0(\Pi, \mathbb{Z}\Pi) \xrightarrow{\partial_0} H^0(\Pi, \mathbb{Z}) \to H^1(\Pi, \ker \epsilon) \to 0$$

$$\| \qquad\qquad\qquad \| \qquad\qquad\qquad \|$$

$$0 \qquad\qquad\qquad N\mathbb{Z}\Pi \qquad\qquad\qquad \mathbb{Z}$$

Since $\epsilon(N) = |\Pi|$ (the order of the finite group Π), there results the isomorphism $\mathbb{Z}_{|\Pi|} \approx H^1(\Pi, \ker \epsilon)$. By Lemma 4.3, the connecting homomorphisms in the exact cohomology sequences induced by the coefficient sequences

$$0 \to \ker \partial_n \to C_n \to \mathrm{im} \ \partial_n = \ker \partial_{n-1} \to 0$$

yield the other isomorphisms $H^n(\Pi, \ker \partial_{n-1}) \approx H^{n+1}(\Pi, \ker \partial_n)$, $n \geq 1$. The cohomology class of the cochain $\epsilon : C_0 = \mathbb{Z}\Pi \to \mathbb{Z}$, which generates $H^0(\Pi, \mathbb{Z}) = \mathbb{Z}^\Pi = \mathbb{Z}$, corresponds to the cochains in the statement of the theorem under the connecting homomorphisms. The proof of (a) is complete and (b) is established similarly.   □

Aspherical Complexes   The homology and cohomology of groups is intimately linked to homotopy theory via aspherical complexes. An *aspherical complex* M is a connected CW complex whose higher homotopy groups $\pi_n(M)$, $n \geq 2$, are trivial. Here is the connection.

Theorem 4.5 *The following are equivalent for a CW complex M:*

(a) *M is aspherical;*

(b) *$\tilde{M}$ is contractible; and*

(c) *the equivariant cellular chain complex $C(\tilde{M})$ provides a free resolution $\epsilon : C(\tilde{M}) \to \mathbb{Z}$ over $\mathbb{Z}\Pi$, where $\epsilon(\tilde{c}^0) = 1$ for all 0-cells $\tilde{c}^0$ of $\tilde{M}$.*

Proof: By covering space theory, M is aspherical if and only if the universal covering complex $\tilde{M}$ has trivial homotopy groups, or equivalently (by Hurewicz isomorphism theorems), trivial homology groups for $n > 0$. This homological condition is equivalent to the exactness of $\epsilon : C(\tilde{M}) \to \mathbb{Z}$. □

It follows that the chain complex $C(\tilde{M}) \otimes_{\mathbb{Z}\Pi} \mathbb{Z} \approx C(M)$ and cochain complex $\text{Hom}_{\mathbb{Z}\Pi}(C(\tilde{M}), N)$ for an aspherical CW complex with fundamental group Π may be used to calculate the homology and cohomology groups of Π.

Corollary 4.6 *If M is an aspherical complex with fundamental group Π, there are isomorphisms $H_n(M) \approx H_n(\Pi, \mathbb{Z})$ and $H^n(\tilde{M}, N) \approx H^n(\Pi, N)$.*

Since the 3-skeleton of an aspherical complex M with fundmental group Π can be constructed from any complex K with $\pi_1(K) = \Pi$ by attaching 3-cells via representatives of generators of its second homotopy group, one has $H_2(M) \approx H_2(K)/\Sigma_2(K)$, where $\Sigma_2(K)$ is the group of spherical 2-cycles in the cell complex K. So Corollary 4.6 contains Hopf's original definition of $H_2(\Pi)$ ([Ho41]). On the more algebraic side, if one extends a chain complex $C(\mathcal{P})$ of a presentation $\mathcal{P} = \langle \mathbf{x} \mid \mathbf{r} \rangle$ of Π into a free resolution, it is possible to obtain *Hopf's formula*

$$H_2(\Pi) = (N(\mathbf{r}) \cap [F(\mathbf{x}), F(\mathbf{x})])/[N(\mathbf{r}), F(\mathbf{x})].$$

For a proof, see K. Brown's book ([Br82, pp 42-43]).

Algebraic 2-type The Mac Lane-Whitehead theory [MaWh50] of algebraic type of 2-complexes is based upon this philosophy: *an unfamiliar object can be classified by transforming it into a unique object, while recording the steps of the transformation.*

Any 2-complex K is the 2-skeleton of some aspherical complex M with fundamental group Π; one takes $M^2 = K$ and inductively constructs M^{n+1} by attaching $(n + 1)$-cells to M^n via maps representing generators of $\pi_n(M^n)$. While the aspherical complex M is uniquely determined up to homotopy type by its fundamental group, its not clear how to implement the philosophy just advocated. How much of the transformation process need be recorded?

In an equivalent algebraic construction, the augmented equivariant chain complex $\epsilon : C(\tilde{K}) \to \mathbb{Z}$ extends to a free resolution

$$C(\Pi) : \ldots \to C_3(\Pi) \to C_2(\tilde{K}) \to C_1(\tilde{K}) \to C_0(\tilde{K}) \xrightarrow{\epsilon} \mathbb{Z} \to 0$$

of $\mathbb{Z}$ over $\mathbb{Z}\Pi$. The chain homotopy type of $C(\Pi)$ is unique for the fundamental group Π by the comparison theorem (Theorem 4.1). We shall show that the philosophy to classify K is implemented by recording just the first new boundary operator $\partial_3 : C_3(\tilde{K}) \to C_2(\tilde{K})$ in the extension of the augmented chain complex $C(\tilde{K}) \to \mathbb{Z}$ into the free resolution $C(\Pi)$.

By Lemma 3.3 and the exactness of $C(\Pi)$, ∂_3 takes values in ker $\partial_2(\tilde{K}) = H_2(\tilde{K}) \equiv \pi_2(K)$ and so determines a cochain $\kappa_K : C_3(\Pi) \to \pi_2(K)$. The chain property $\partial_3\,\partial_4 = 0$ implies that the cochain κ_K is a $C(\Pi)$-cocycle and therefore it defines a cohomology class $\{\kappa_K\} \in H^3(\Pi, \pi_2(K))$. This cohomology class is called the *k-invariant* of the 2-complex K; according to the comparison theorem, it depends only on the augmented chain complex $C(\tilde{K}) \to \mathbb{Z}$ and not extension $C(\Pi)$ chosen.

Theorem 4.4(1) yields this basic k-invariant calculation:

Theorem 4.7 *Let K be a connected 2-complex with finite fundamental group Π. Then the cohomology group $H^3(\Pi, \pi_2(K))$ is cyclic of order $|\,\Pi\,|$, and the k-invariant $\{\kappa_K\}$ is a generator.*

We shall show that a 2-complex K is determined up to homotopy type by its fundamental group Π, its second homotopy module $\pi_2(K)$, and its k-invariant $\{\kappa_K\} \in H^3(\Pi, \pi_2(K))$. So the k-invariant may be interpreted as a record of how the fundamental group and the second homotopy module are assembled to form the homotopy type of the 2-complex.

For convenience in the proof, we define an abstract *algebraic 2-type* to be a triple $(\Pi, \pi_2, \{\kappa\})$ consisting of a group Π, a $\mathbb{Z}\Pi$-module π_2 and a cohomology class $\{\kappa\} \in H^3(\Pi, \pi_2)$. The triple $T(K) = (\pi_1(K), \pi_2(K), \{\kappa_K\})$ is the algebraic 2-type of the 2-complex K. A *homomorphism* (α, β) of algebraic 2-types $(\Pi, \pi_2, \{\kappa\})$ and $(\Xi, \xi_2, \{\tau\})$ consists of a group homomorphism $\alpha : \Pi \to \Xi$ and a $\mathbb{Z}\Pi$-module homomorphism $\beta : \pi_2 \to {}_\alpha\xi_2$ such that the cohomology classes $\{\kappa\}$ and $\{\xi\}$ correspond under the homomorphisms

$$H^3(\Xi, \xi_2) \xrightarrow{\alpha^*} H^3(\Pi, {}_\alpha\xi_2) \xleftarrow{\beta_*} H^3(\Pi, \pi_2)$$

defined as follows: Let $\epsilon : C(\Pi) \to \mathbb{Z}$ and $\rho : C(\Xi) \to \mathbb{Z}$ be projective resolutions of $\mathbb{Z}$ over $\mathbb{Z}\Pi$ and $\mathbb{Z}\Xi$, respectively. Via the change of module structure induced by the group homomorphism $\alpha : \Pi \to \Xi$, the second resolution becomes a (not necessarily projective) resolution $\rho : {}_\alpha C(\Xi) \to \mathbb{Z}$ of $\mathbb{Z}$ over $\mathbb{Z}\Pi$. By the comparison theorem (Theorem 4.1), the identity $1 : \mathbb{Z} \to \mathbb{Z}$ extends to a chain map $C(\Pi) \to {}_\alpha C(\Xi)$ that is unique up to chain homotopy. So its induced cohomology homomorphisms are determined by α and may be denoted by $\alpha^* : H^*(\Xi, \xi_2) \to H^*(\Pi, {}_\alpha\xi_2)$. Finally, the $\mathbb{Z}\Pi$-module homomorphism $\beta : \pi_2 \to {}_\alpha\xi_2$ induces the *change of coefficient* homomorphisms $\beta_* : H^*(\Pi, \pi_2) \to H^*(\Pi, {}_\alpha\xi_2)$.

Lemma 4.8 *Let K and L be connected 2-complexes with fundamental groups $\Pi = \pi_1(K)$ and $\Xi = \pi_1(L)$. A group homomorphism $\alpha : \pi_1(K) \to \pi_1(L)$ and $\mathbb{Z}\Pi$-module homomorphism $\beta : \pi_2(K) \to {}_\alpha\pi_2(K)$ are induced by a map $F : K \to L$ if and only if they constitute a homomorphism*

$$(\alpha, \beta) : (\Pi, \pi_2(K), \{\kappa_K\}) \to (\Xi, \pi_2(L), \{\kappa_L\})$$

of the algebraic 2-types of the 2-complexes K and L

Proof: The comparison theorem (Theorem 4.1) implies that the equivariant chain map $C(\tilde{F}) : C(\tilde{K}) \to {}_\alpha C(\tilde{L})$ for a map $F : K \to L$ extends over free resolutions $\epsilon : C(\Pi) \to \mathbb{Z}$ and $\rho : C(\Xi) \to \mathbb{Z}$ to a chain map $v : C(\Pi) \to {}_\alpha C(\Xi)$ by which we may calculate $\alpha^* : H^3(\Xi, \xi_2) \to H^3(\Pi, {}_\alpha\xi_2)$. By Theorem 3.4 and the definition of the k-invariants, the chain map v gives the commutative diagram:

$$
\begin{array}{ccccc}
C_3(\Pi) & \xrightarrow{\kappa_K} & \pi_2(K) & \leq & C_2(\tilde{K}) \\
\beta = F_\# \downarrow & & \downarrow & & \downarrow C_2(\tilde{F}) \\
{}_\alpha C_3(\Xi) & \xrightarrow{\kappa_L} & {}_\alpha\pi_2(L) & \leq & {}_\alpha C_2(\tilde{L})
\end{array}
$$

This yields the desired equality $\beta_*(\{\kappa_K\}) = \alpha^*(\{\kappa_L\})$ in $H^3(\Pi, {}_\alpha\xi_2)$.

Conversely, the equality $\beta_*(\{\kappa_K\}) = \alpha^*(\{\kappa_L\})$ in $H^3(\Pi, {}_\alpha\xi_2)$ implies that for any chain map $v : C(\Pi) \to {}_\alpha C(\Xi)$ extending $1 : \mathbb{Z} \to \mathbb{Z}$, the cochains

$$\kappa_L \, v_3 : C_3(\Pi) \to {}_\alpha C_3(\Xi) \to {}_\alpha\pi_2(L)$$

and

$$\beta \, \kappa_K : C_3(\Pi) \to \pi_2(K) \to {}_\alpha\pi_2(L)$$

are cohomologous. This means that there is a module homomorphism

$$\gamma : C_2(\tilde{K}) \to {}_\alpha\pi_2(L)$$

such that

$$\beta \, \kappa_K - \kappa_L \, v_3 = \gamma \, \partial_3(\Pi) : C_3(\Pi) \to C_2(\tilde{K}) \to {}_\alpha\pi_2(L).$$

Then the three homomorphisms v_0, v_1, and $v_2 + \gamma$ constitute a chain map $C(\tilde{K}) \to {}_\alpha C(\tilde{L})$ inducing $\beta : \pi_2(K) \to {}_\alpha\pi_2(L)$ as in Theorem 3.4. By Theorem 3.6, this chain map (when 0-admissible) is realized by a based map $F : K \to L$ with $F_\# = \alpha$ on π_1 and $F_\# = \beta$ on π_2. $\quad\square$

Theorem 4.9 (Mac Lane-Whitehead [MaWh50]) *Connected 2-complexes are based homotopy equivalent if and only if their algebraic 2-types are isomorphic.*

Proof: This follows from Lemma 4.8 and Whitehead's Theorem 2.12.   □

To derive from Theorem 4.9 a homotopy classification of connected 2-complexes with a prescribed fundamental group Π, it is necessarily to determine exactly which abstract algebraic 2-types $(\Pi, \pi_2, \{\kappa\})$ are realized by 2-complexes and how many isomorphism classes are represented by them. An initial step in this determination is provided by statement (40) of Chapter I: *any two finite, connected, CW-complexes K and L with isomorphic fundamental groups become (simple-) homotopy equivalent when summed with suitably many copies of the 2-sphere S^2.* This homotopy version of Tietze's Theorem for finite group presentations turns the homotopy classification into a problem of cancellation of 2-sphere summands. In terms of Theorem 4.9 and algebraic 2-types $(\Pi, \pi_2, \{\kappa\})$, the homotopy classification problem is one of cancellation of free $\mathbb{Z}\Pi$-summands from the second homotopy module π_2, in a manner that preserves k-invariants.

Cockcroft and Swan [CoSw61] employ algebraic 2-types and a classification of $\mathbb{Z}\mathbb{Z}_p$-modules to handle any prime order, cyclic, fundamental group $\pi_1 = \mathbb{Z}_p$. For the general finite cyclic case $\pi_1 = \mathbb{Z}_n$, Dyer and Sieradski [DySi73] invoke the more basic Theorem 3.9 and the Jacobinski cancellation theorem to cancel free $\mathbb{Z}\mathbb{Z}_n$-summands from π_2. These works show that 2-complexes with $\pi_1 = \mathbb{Z}_n$ are classified by their Euler characteristic: *two connected 2-complexes with isomorphic finite cyclic fundamental groups are homotopy or simple-homotopy equivalent if and only if they have the same Euler characteristic.*

The homotopy and simply-homotopy classifications have been completed for finite abelian fundamental groups (see Chapter III); in general, the Euler characteristic alone does not classify these 2-complexes, but their homotopy and simple-homotopy classifications do coincide. More recently, it has been shown independently by Metzler [Me90] and Lustig [Lu91$_1$] that for general fundamental groups, the homotopy classification and the simple homotopy classification disagree; see Chapter VII.

4.2   Maps between 2-complexes

Action of the Cohomology Groups   Let K and L be 2-complexes with fundamental groups $\pi_1(K) \equiv \Pi$ and $\pi_1(L) \equiv \Xi$. For a group homomorphism $\alpha : \Pi \to \Xi$ and a $\mathbb{Z}\Pi$-module homomorphism $\beta : \pi_2(K) \to {}_\alpha\pi_2(L)$, let $[K, L]_\alpha$

and $[K,L]_{\alpha,\beta}$ consist of the based homotopy classes of based maps $F : K \to L$ with $F_\# = \alpha$ on π_1 and $F_\# = \beta$ on π_2.

The modification $F^\gamma : K \to L$ of a based map $F : K \to L$ by a cochain $\gamma : C_2(\Pi) = C_2(\tilde{K}) \to {}_\alpha\pi_2(L)$, as in Lemma 3.5, defines an action of the cochain group

$$C_2(\Pi, {}_\alpha\pi_2(L)) = \mathrm{Hom}_{\mathbb{Z}\Pi}(C_2(\Pi), {}_\alpha\pi_2(L))$$

on the based homotopy set $[K,L]_\alpha$.

Basic homotopy classification results   The following are left as exercises in application of the equivariant techniques of Section 3:

Lemma 4.10 *For* $[F] \in [K,L]_\alpha$ *and* $\gamma \in C^2(\Pi, {}_\alpha\pi_2(L))$, *we have*

(a) $[F]$ *and* $[F^\gamma]$ *induce the same module homomorphism on* π_2 *if and only if* γ *is a* $\mathbb{Z}\Pi$-*cocycle.*

(b) $[F] = [F^\gamma]$ *if and only if* γ *is a* $\mathbb{Z}\Pi$-*coboundary.*

Theorem 4.11 *There exist actions of the cohomology group* $H^2(\tilde{K}, {}_\alpha\pi_2(L))$ *on* $[K,L]_\alpha$ *and the cohomology group* $H^2(\Pi, {}_\alpha\pi_2(L))$ *on* $[K,L]_{\alpha,\beta}$, *both defined by* $[F]^{\{\gamma\}} = [F^\gamma]$, *satisfying the following properties:*

(a) $[F]^{\{\gamma\}} = [F]$ *if and only if* $\{\gamma\} = 0$,

(b) $([F]^{\{\gamma\}})^{\{\lambda\}} = [F]^{\{\gamma\}+\{\lambda\}}$, *and*

(c) $[F]^{\{\gamma\}} = [F]^{\{\lambda\}}$ *if and only if* $\{\gamma\} = \{\lambda\}$.

Theorem 4.12 *The action of any class* $[F] \in [K,L]_{\alpha,\beta}$ *determines bijections*

$$[F]^{(-)} : H^2(\tilde{K}, {}_\alpha\pi_2(L)) \to [K,L]_\alpha$$

and

$$[F]^{(-)} : H^2(\Pi, {}_\alpha\pi_2(L)) \to [K,L]_{\alpha,\beta}.$$

Corollary 4.13 *Homotopy classes* $[F] \in [K,L]_\alpha$ *are uniquely determined by their induced module homomorphisms* $F_\# : \pi_2(K) \to {}_\alpha\pi_2(L)$ *if and only if the cohomology module* $H^2(\Pi, {}_\alpha\pi_2(L))$ *is trivial.*

Proof: This is the second bijection of Theorem 4.12. □

Corollary 4.14 *Based maps $K \to L$ inducing the same isomorphism α of finite fundamental groups are homotopic if and only if they induce the same module homomorphism β on π_2.*

Proof: When α is an isomorphism, we have $H^2(\Pi, {}_\alpha\pi_2(L)) \approx H^2(\Pi, \pi_2(L))$. Because Π is finite, $H^2(\Pi, \pi_2(L))$ is trivial by Theorem 4.4(2). So Corollary 4.13 implies Corollary 4.14.   □

Chapter III

Homotopy and Homology Classification of 2-Complexes

M. Paul Latiolais

When one wishes to classify 2-complexes, the first classification scheme one must consider is homotopy type. In Section 1, we will define an obstruction to homotopy equivalence, called bias, but show that it is actually an obstruction to homology equivalence. In Section 2, we will refine the bias to the Browning obstruction and use it to get a complete classification of 2-complexes with finite abelian fundamental groups. In Section 3, we will give the currently known facts on the homotopy classification for other fundamental groups. Subsections 3.2 and 3.3, as well as the list of problems are co-authored with Cynthia Hog-Angeloni.

1 Bias Invariant & Homology Classification

The concept of bias is based on Whitehead's Theorem (Chapter II, Theorem 2.12). We will define bias using equivariant chain maps on the equivariant chain complexes of the universal covers of 2-complexes, as discussed in Chapter II, §3.2. Theorem 3.6 of Chapter II identifies maps of 2-complexes with chain maps for the universal coverings.

The bias-invariant was first used in [Me76] and then in [Dy76], [Dy79], [Si77], [SiDy79] to show that certain 2-complexes were not homotopy equivalent. Dyer showed in [Dy86] that the bias invariant, in fact, detected integral homology equivalence. In §1.1, we will motivate and define bias as an obstruction to homotopy equivalence. In §1.2, we will show that bias is the complete obstruction to homology equivalence (with induced isomorphism on fundamental groups). In §1.3, we will make specific calculations of the bias to distinguish the homotopy types of important examples.

1.1   Bias as a homotopy obstruction

Recall that a homotopy equivalence of CW-complexes is simply a map that induces an isomorphism on all homotopy groups. For the 2-dimensional version of Whitehead's theorem, see Chapter II, Theorem 2.12,

In order to see what is involved in constructing the bias, let us consider two 2-complexes, K and L, and a map $f : K \to L$. If f is a homotopy equivalence, it induces an isomorphism on all homotopy groups. In particular, $f_* : \pi_1(K) \to \pi_1(L)$ must be an isomorphism. Therefore, in order to start the classification of 2-complexes up to homotopy type, we must first assume that the fundamental groups of K and L are isomorphic.

Our second step is to check the second homotopy groups of two 2-complexes with the same fundamental group. This would seem much harder to accomplish. However, Theorem 3.4 of Chapter II says that the second homotopy groups are the second homology groups of the universal covers. Consequently, we will look at the cellular chain complexes of the universal covers $\tilde{K}$ and $\tilde{L}$ of K and L respectively and compare them. We will also consider the homomorphism $\tilde{f}_2^{\#} : H_2(\tilde{K}) \to H_2(\tilde{L})$ induced by f. This is, in fact, easier than it sounds.

Since $\tilde{K}$ and $\tilde{L}$ are simply connected, the first homology groups vanish. We have exact sequences:

$$0 \to H_2(\tilde{K}) \to C_2(\tilde{K}) \to C_1(\tilde{K}) \to C_0(\tilde{K}) \to \mathbb{Z} \to 0$$

and

$$0 \to H_2(\tilde{L}) \to C_2(\tilde{L}) \to C_1(\tilde{L}) \to C_0(\tilde{L}) \to \mathbb{Z} \to 0.$$

Let us assume for the moment that f is the identity on the 1-skeletons of K and L. After lifting the cell structures of K and L to $\tilde{K}$ and $\tilde{L}$, the map will still be the identity on 1-skeletons. That gives us the following diagram:

$$
\begin{array}{ccccccccccc}
0 & \to & H_2(\tilde{K}) & \to & C_2(\tilde{K}) & \to & C_1(\tilde{K}) & \to & C_0(\tilde{K}) & \to & \mathbb{Z} & \to & 0 \\
 & & \tilde{f}_2^{\#}\downarrow & & \tilde{f}_2\downarrow & & \tilde{f}_1\| & & \tilde{f}_0\| & & \| \\
0 & \to & H_2(\tilde{L}) & \to & C_2(\tilde{L}) & \to & C_1(\tilde{L}) & \to & C_0(\tilde{L}) & \to & \mathbb{Z} & \to & 0.
\end{array}
$$

The Five Lemma implies that $\tilde{f}_2^{\#}$ is an isomorphism if and only if $\tilde{f}_2$ is an isomorphism. Notice also that the homomorphism $\tilde{f}_2$ can be represented as a matrix over $\mathbb{Z}\pi_1(L)$ (which we are identifying with $\mathbb{Z}\pi_1(K)$ via $f_* : \pi_1(K) \cong \pi_1(L)$), since $C_2(\tilde{K})$ and $C_2(\tilde{L})$ are both free modules over the fundamental groups of K and L. Now these module structures over the fundamental groups are somewhat complicated, so let's forget about them, mathematically. That is, in the matrix representing $\tilde{f}_2$, maps each group element to 1. If $\tilde{f}_2$ is an isomorphism, the resulting matrix will be invertible over $\mathbb{Z}$. Consequently, if the matrix over $\mathbb{Z}$ is not invertible, then our original map f is not a homotopy equivalence.

The above idea is the basis of the definition of the bias invariant. To define bias, we will not assume that f is the identity on the 1-skeleton. However, when one actually computes bias, it is most convenient to assume that f is the identity on the 1-skeleton, as we will see in §2.2.

The construction of the bias invariant allows for generalization to higher dimensions. Though this generalization will not be used in this book, it is useful to know that construction work for the more general (G, d)-complexes.

Definition 1.1 *Let G be group and $d \geq 2$ an integer. A (G, d)-complex K is a finite connected CW-complex of dimension d with $\pi_1(K) = G$ and whose universal cover is $(d-1)$-connected.*

In the case of dimension $d = 2$, the condition on the universal cover in the above definition is automatically satisfied. Given two (G, d)-complexes K and L, we would like to know in general whether they are homotopy equivalent.

We first assume that K and L have the same Euler characteristic. Given any homomorphism $\alpha : \pi_1(K) \to \pi_1(L)$, we can construct a map $f : K \to L$ with $\alpha = f_* : \pi_1(K) \to \pi_1(L)$, the induced homomorphism from f. If α is an isomorphism, the map bias, $b(\alpha)$, which will be defined below, will tell us whether there exists a homology equivalence f with $f_* = \alpha$.

So let $f : K \to L$ be a cellular map with $f_* = \alpha : \pi_1(K) \to \pi_1(L)$ an isomorphism. As we learned in Chapter II, § 3, the CW structure of any complex K may be "lifted" to a CW structure on its universal cover $\tilde{K}$. Also, given a particular choice of preferred base point lifting the 0-cell, each k-cell in K together with each group element of $\pi_1(K) = G$ determine a unique k-cell in $\tilde{K}$. This G-structure on $\tilde{K}$ is consistent with the boundary of each cell, so carries over to the algebraic structure on the chain complex of the universal cover $\tilde{K}$.

Consider the following diagram involving the chain map induced by the lifting $\tilde{f} : \tilde{K} \to \tilde{L}$ of the cellular map $f : K \to L$.

$$H_d(\tilde{K}) \overset{\iota}{\hookrightarrow} C_d(\tilde{K})$$

$$\left\downarrow \tilde{f}_d^{\sharp} \qquad\qquad \right\downarrow \tilde{f}_d \qquad\qquad (1)$$

$$_\alpha H_d(\tilde{L}) \overset{\iota}{\hookrightarrow} {}_\alpha C_d(\tilde{L}),$$

where the α subscript refers to the change of ring procedure associated to α, as in Chapter II, § 3.2. We will modify $\tilde{f}_d$ using π_2 elements (Chapter II, Section 3) without altering $\tilde{f}_i, i < d$, to attempt to make $\tilde{f}_d^{\sharp}$ an isomorphism.

Lemma 1.2 *Let K and L be (G,d)-complexes as above, and let $f : K \to L$ be a cellular map which induces an isomorphism α on the fundamental groups. Let $\gamma : C_d(\tilde{K}) \to {}_\alpha H_d(\tilde{L})$ be a module homomorphism. Then there exists a map $g : K \to L$ such that*

(a) $\tilde{g}_i = \tilde{f}_i : C_i(\tilde{K}) \to {}_\alpha C_i(\tilde{L})$, *for* $i < d$

(b) $\tilde{g}_d = \tilde{f}_d + \iota \circ \gamma : C_d(\tilde{K}) \to {}_\alpha C_d(\tilde{L})$, *and*

(c) $\tilde{g}_d^{\sharp} = (\tilde{f}_d^{\sharp} + \gamma \circ \iota) : H_d(\tilde{K}) \to H_d(\tilde{L})$, *where ι is the appropriate inclusion homomorphism.*

This lemma is a generalization of Theorem 3.6 of Chapter II. The technique of the proof is patterned after [DySi73], p. 41. For an older reference, see [Pu58].

Proof: Note that the d-cells of K correspond to the generating set of $C_d(\tilde{K})$ as a free $\mathbb{Z}G$-module. If $D^{(d)}$ is any d-cell of K, the corresponding element of the generating set of $C_d(\tilde{K})$ is denoted by $\tilde{D}^{(d)}$. The image of $\tilde{D}^{(d)}$ under the homomorphism γ is an element of $H_d(\tilde{L}) \cong \pi_d(L)$. Let $\gamma(\tilde{D}^{(d)}) = \Delta$. As an element of $\pi_d(L), \Delta$ is represented by some map $p : S^d \to L$. Note that we are equating $H_d(\tilde{L})$ with $\pi_d(L)$ to avoid too much notation (by the use of a Hurewicz isomorphism and covering projection isomorphism).

Define the map $h : K \to K \vee S^d$ to be the identity on $K \setminus D^{(d)}$ and to map $D^{(d)}$ onto $D^{(d)} \vee S^d$ by collapsing some $(S^{d-1}, e) \subset (D^{(d)}, e_o)$ to e_0 (where e_0 is the wedge point) and the ball bounded by S^{d-1} onto S^d. The rest of $D^{(d)}$ gets stretched to cover $D^{(d)}$.

Now compose

$$K \overset{h}{\longrightarrow} K \vee S^d \overset{f \cup p}{\longrightarrow} L,$$

where $f \cup p$ is the obvious adjunction of f and p on $D^{(d)} \vee S^d$. Call the above composition map g. Notice that $(\tilde{g})_i$ is the same as $\tilde{f}_i$, for $i < d$, and that

$$(\tilde{g})_d(\tilde{D}^{(d)}) = \tilde{f}_d(\tilde{D}^{(d)}) + \iota \circ \gamma(\tilde{D}^{(d)}).$$

Now continue the process by changing g as above with the next d-cell. Since the complex is finite, the process is finite. $\square$

Let $\Sigma_d(K) \subset H_d(K)$ be the spherical elements of $H_d(K)$. That is, $\Sigma_d(K)$ is the image of the Hurewicz homomorphism $\pi_d(K) \to H_d(K)$, or alternately, the image of the projection $p_* : H_d(\tilde{K}) \to H_d(K)$. So diagram (1) projects to

$$
\begin{array}{ccc}
\Sigma_d(K) & \hookrightarrow & C_d(K) \\
\Big\downarrow{\scriptstyle f_d^{\sharp}} & & \Big\downarrow{\scriptstyle f_d} \\
\Sigma_d(L) & \hookrightarrow & C_d(L).
\end{array}
$$

Any homomorphism $\delta : C_d(K) \to \Sigma_d(L)$ may be lifted to some homomorphism $\tilde{\delta} : C_d(\tilde{K}) \to {}_\alpha H_d(\tilde{L})$ as follows: First, compose δ with the projection $p_* : C(\tilde{K}) \to C(K)$. Since $C(\tilde{K})$ is a free module and $q_* : H_d(\tilde{L}) \to \Sigma(L)$ is onto, it follows that $\delta \circ p_* : C(\tilde{K}) \to \Sigma_d(L)$ can be lifted to $\tilde{\delta} : C_d(\tilde{K}) \to H_d(\tilde{L})$, so that $\delta \circ p_* = q_* \tilde{\delta}$. Therefore, by Lemma 1.2, given any homomorphism $\delta : C_d(K) \to \Sigma_d(L)$, there exists a map $g : K \to L$ inducing the isomorphism α, such that $g_d = f_d + \iota \circ \delta$ and $g_d^{\sharp} = f_d^{\sharp} + \delta \circ \iota$ (ι being the appropriate inclusion homomorphism from $H_d \to C_d$).

Exercise 1.3 *Prove that $H_d(K)$ is a direct summand of $C_d(K)$.*

Now let $\beta : H_d(K) \to \Sigma_d(L)$ be some other homomorphism. Since $H_d(K)$ is a direct summand of $C_d(K)$, we can extend β to $C_d(K)$ by letting it be zero on the other factor. We have proved the forward direction of:

Lemma 1.4 *Let K and L be (G, d)-complexes and let $\delta : H_d(K) \to \Sigma_d(L)$ be any homomorphism and $f : K \to L$ any map. Then there exists a map $g : K \to L$ such that*

$$f_* = g_* : \pi_1(K) \to \pi_1(L)$$

and the induced homomorphism $g_d : H_d(K) \to H_d(L)$ is the homomorphism

$$f_d + \iota \circ \delta : H_d(K) \to H_d(L).$$

Conversely, given two maps $f : K \to L$ *and* $g : K \to L$ *with* $f_* = g_* :$ $\pi_1(K) \to \pi_1(L)$, *then* $g_d = f_d + \iota \circ \delta$, *for some* $\delta : H_d(K) \to \Sigma_d(L)$, *as below:*

$$
\begin{array}{ccc}
\Sigma_d(K) & \overset{\iota}{\hookrightarrow} & H_d(K) \\[2mm]
f_d^\sharp \downarrow & \delta \swarrow & \downarrow f_d \\[2mm]
\Sigma_d(L) & \overset{\iota}{\hookrightarrow} & H_d(L)
\end{array}
$$

Proof: The first part follows from the previous lemma. For the converse, consider the commutative diagrams (with horizontal exact sequences):

$$
\begin{array}{ccccccccc}
0 & \to & \Sigma_d(K) & \to & H_d(K) & \to & H_d(\pi_1(K)) & \to & 0 \\[2mm]
 & & \downarrow f_d^\sharp & & \downarrow f_d & & \downarrow H_d(f_*) & & \\[2mm]
0 & \to & \Sigma_d(L) & \to & H_d(L) & \to & H_d(\pi_1(L)) & \to & 0
\end{array}
$$

and

$$
\begin{array}{ccccccccc}
0 & \to & \Sigma_d(K) & \to & H_d(K) & \to & H_d(\pi_1(K)) & \to & 0 \\[2mm]
 & & \downarrow g_d^\sharp & & \downarrow g_d & & \downarrow H_d(g_*) & & \\[2mm]
0 & \to & \Sigma_d(L) & \to & H_d(L) & \to & H_d(\pi_1(L)) & \to & 0
\end{array}
$$

The homomorphism $H_d(f_*) : H_d(\pi_1(K)) \to H_d(\pi_1(L))$ is an isomorphism, whenever f_* is an isomorphism, by a theorem of H. Hopf [Br82]. Note that this isomorphism only depends on the induced isomomorphism f_* on fundamental groups. Consequently $H_d(f_*) = H_d(g_*)$.

Since the two diagrams agree on the right, then the images of f_d and g_d agree up to an element of $H_d(L)$. That is precisely what we are trying to prove. □

Exercise 1.5 *Prove that for any choice of bases, the matrices representing* $f^\sharp$ *and* f_d *have determinants that differ at most by sign.*

We are now ready to construct a map bias, $b(\alpha)$, where $\alpha \in Iso(\pi_1(K), \pi_1(L))$. Let $f : K \to L$ be a map with $f_* = \alpha : \pi_1(K) \to \pi_1(L)$ an isomorphism. From our earlier discussion, we know that f is a homotopy equivalence if and only if $\tilde{f}_d^\sharp : H_d(\tilde{K}) \to H_d(\tilde{L})$ is an isomorphism. If f is a homotopy equivalence, then $f_d^\sharp : \Sigma_d(K) \to \Sigma_d(L)$ will also be an isomorphism. If $\chi(K) = \chi(L)$, then $\mathrm{rank}_{\mathbf{Z}}\Sigma_d K = \mathrm{rank}_{\mathbf{Z}}\Sigma_d L$.

Choose a set of free generators for $\Sigma_d(K)$ and for $\Sigma_d(L)$. Given these bases, the homomorphism $f_d^\sharp : \Sigma_d(K) \to \Sigma_d(L)$ can be expressed as a matrix, $M_f^\sharp$, with entries in $\mathbb{Z}$. So $f_d^\sharp$ is an isomorphism if and only if the determinant of $M_f^\sharp$ is ± 1. In particular, if the determinant is not ± 1, then f cannot be a homotopy equivalence. This $det[M_f^\sharp]$ is independent of bases modulo sign.

Now the obvious question is, "What will be the effect of the modification by π_2 elements on the determinant of the matrix $M_f^\sharp$, where $M_f^\sharp$ represents the induced homomorphism $\Sigma_d(K) \to \Sigma_d(L)$?" Consider, again the diagram:

$$
\begin{array}{ccc}
\Sigma_d(K) & \overset{\iota}{\hookrightarrow} & H_d(K) \\[2mm]
f_d^\sharp \downarrow & \delta \swarrow & \downarrow f_d \\[2mm]
\Sigma_d(L) & \overset{\iota}{\hookrightarrow} & H_d(L)
\end{array}
$$

From the above exercise, $det[M_f^\sharp] = \pm 1$ if and only if the determinant of the matrix M_f representing f_d is also ± 1. Henceforth, we will focus on M_f rather than $M_f^\sharp$. Assume for the moment that the respective bases elements make the inclusion homomorphisms diagonal. That is, if $\sigma_1, \ldots, \sigma_t$ are the generators of $H_d(K)$, then the generators of $\Sigma_d(K)$ can be expressed as $s_1, \ldots, s_r$, for $r \leq t$, and the inclusion homomorphism can be expressed as $s_i \mapsto n_i \cdot \sigma_i$, where $n_i \in \mathbb{Z}$. Similarly, let the generators of $H_d(L)$ be $\gamma_1, \ldots, \gamma_t$, then under our assumption the generators of $\Sigma_d(L)$ will be $g_1, \ldots, g_r$, with the inclusion homomorphism expressed as $g_i \mapsto m_i \cdot \gamma_i$, for $m_i \in \mathbb{Z}$. In this set up, it is easy to see that the modification by π_2 elements will effect the determinant of the matrix representing f_d by multiples of the m_i's. Define $m = gcd\{m_i\}$, then it is clear that if $det[M_f] \neq \pm 1 \bmod m$, then there is no homotopy equivalence $g : K \to L$ with $g_* = \alpha$. In fact, as we will see later, $det[M_f] \neq \pm 1 \bmod m$, if and only if there is no homology equivalence g with $g_* = \alpha$.

Proposition 1.6 *Let $\mathbb{Z}^r \overset{\iota}{\hookrightarrow} \mathbb{Z}^t$ be an injection with matrix (m_{ij}) with respect to some bases of $\mathbb{Z}^r$ and $\mathbb{Z}^t$. If $1 \neq m = gcd\{m_{ij}\}$, then m is also the $gcd\{$torsion coefficients of $\mathbb{Z}^t/\iota(\mathbb{Z}^r)\}$. If $1 = gcd\{m_{ij}\}$ then $\mathbb{Z}^t/\iota(\mathbb{Z}^r)$ can be generated by fewer than t generators.(Compare with p.2 of [Dy86].)*

Proof: Diagonalizing the matrix (m_{ij}) will not change the $gcd\{m_{ij}\}$. $\square$

Definition 1.7 (Map Bias) *Let K and L be two (G, d)-complexes and let $f : K \to L$ be any map with $f_* = \alpha : \pi_1(K) \to \pi_1(L)$ an isomorphism. Let the set $\{m_{ij}\}$ be the integer coefficients of the generators of $\Sigma_d(L)$ expressed* *as elements of $C_d(L)$ with its standard set of generators (from the d-cells of L). Let $m = gcd\{m_{ij}\}$. Let M_f be the matrix representation of the induced homomorphism $f_d : H_d(K) \rightarrow H_d(L)$ with respect to any bases of $H_d(K)$ and $H_d(L)$. Define the map bias $b(\alpha) = \overline{det[M_f]} \in \mathbb{Z}_m / \pm 1$.*

Note that, by Proposition 1.6, $m = gcd\{$torsion coefficients of $C_d(L)/\Sigma_d(L)\}$ $= gcd\{$torsion coefficients $H_d(L)/\Sigma_d(L)\}$, and that $H_d(L)/\Sigma_d(L)$ is the homology invariant $H_d(G)$ of the group G, by [Br82, Theorem 5.2].

Lemma 1.8 *The map bias, $b(\alpha) \in \mathbb{Z}_m / \pm 1$, depends only on K, L and α. That is, $b(\alpha)$ is independent of the choice functions $f : K \rightarrow L$, with $f_* = \alpha$.*

Proof: Given two maps f and g which induce α on π_1, then the induced homomorphisms on Σ_d will differ only by homomorphisms of the form $\iota \circ \delta$, where $\delta : H_d(K) \rightarrow \Sigma_d(L)$ is some homomorphism and $\iota : \Sigma_d(L) \rightarrow H_d(L)$ is the obvious inclusion. But adding in such homomorphisms will only effect the determinant by multiples of the coefficients $\{m_{ij}\}$ of the generators of $\Sigma_d(L)$. This has no effect modulo m and sign, since $m = gcd\{m_{ij}\}$. $\square$

Theorem 1.9 *Given a pair of isomorphisms $\alpha : \pi_1(K) \rightarrow \pi_1(L)$ and $\chi : \pi_1(L) \rightarrow \pi_1(L)$, then $b(\chi \circ \alpha) = b(\chi)b(\alpha)$. (Compare with [Dy86], Lemma 1.)*

Exercise 1.10 *Prove theorem 1.9, and that the bias is a unit in $\mathbb{Z}_m / \pm 1$. Consequently, bias can be measured in the group of units $\mathbb{Z}_m^* / \pm 1$.* $\square$

Definition 1.11 (Bias) *Let $\beta : Aut(\pi_1(L)) \rightarrow \mathbb{Z}_m^* / \pm 1$ be the homomorphism that sends $\gamma \mapsto b(\gamma) \in \mathbb{Z}_m^* / \pm 1$, and let $D \subset \mathbb{Z}_m^* / \pm 1$ be the image of β. Define the pair bias $b(K, L) \equiv [b(\alpha)] \in \mathbb{Z}_m^* / \pm D$, for any isomorphism $\alpha : \pi_1(K) \rightarrow \pi_1(L)$.*

Theorem 1.12 *If $b(K, L) \not\equiv 1 \in \mathbb{Z}_m^* / \pm D$, then K and L are not homotopy equivalent.*

Proof: We prove the contrapositive. Suppose K and L are homotopy equivalent. Let $h : K \rightarrow L$ be our homotopy equivalence, with $h_* = \beta : \pi_1(K) \rightarrow \pi_1(L)$. Let $\alpha : \pi_1(K) \rightarrow \pi(L)$ be another isomorphism. We must show that $[b(\alpha)] = 1$ in $\mathbb{Z}_m^* / \pm D$. Since h is a homotopy equivalence, it will induce an isomorphism on the homology groups. By Theorem 1.9 above,

$$b(\beta)b(\beta^{-1} \circ \alpha) = b(\beta \circ \beta^{-1} \circ \alpha) = b(\alpha).$$

But, $\beta^{-1} \circ \alpha \in Aut(\pi_1(L))$. Consequently, $1 = [b(\beta)] = [b(\alpha)] \in \mathbb{Z}_m^* / \pm D$. $\square$

Two interesting questions immediately arise for a given G and d:

- Is the bias surjective onto $\mathbb{Z}_m^* / \pm D$?

- What does D look like?

Note that Metzler used the idea of bias in [Me76], to give the first known examples of homotopy inequivalent 2-complexes with the same finite fundamental group (namely, $(\mathbb{Z}_5)^3$) and the same Euler characteristic.

1.2 Bias as the complete homology obstruction

A map $f : K \to L$ of (G, d)-complexes is a *homology equivalence*, if f induces isomorphisms $\pi_1(K) \to \pi_1(L)$ and $H_k(K) \to H_k(L)$ for all $k \leq d$. Before we use the bias to distinguish homotopy types of $(G, 2)$-complexes, let us point out that the bias is really detecting homology equivalence:

Theorem 1.13 (Dyer [Dy86], Theorem 2) *Let* $\alpha : \pi_1(K) \to \pi_1(L)$ *be an isomorphism. Then there exists a homology equivalence* $f : K \to L$, *with* $f_* = \alpha$, *if and only if* $b(\alpha) \equiv \pm 1 \mod m$.

Proof: Let $f : K \to L$ be a map such that $f_* = \alpha$ and $det[M_f] \equiv b(\alpha) \not\equiv \pm 1$ mod m. By Lemma 1.4, each other map $K \to L$ inducing α changes f_d only by adding a homomorphism $\delta : H_d(K) \to \Sigma_d(L)$. Such a change only alters the determinant of M_f by multiples of m. So no map $g : K \to L$ with $g_* = \alpha$ has M_g an isomorphism as $det[M_g] \equiv det[M_f] \not\equiv \pm 1$ mod m. So there is no homology equivalence with the given identification of fundamental groups.

Now suppose $b(\alpha) \equiv \pm 1$ mod m, then there exists a map $f : K \to L$ with $f_* = \alpha$ and $det[M_f] \equiv \pm 1$ mod m. Consider the diagram:

$$
\begin{array}{ccccccccc}
0 & \to & \Sigma_d(K) & \to & H_d(K) & \to & H_d(G) & \to & 0 \\
  &   & \downarrow f_d^\sharp & & \downarrow f_d & & \downarrow H_d(\alpha) & & \\
0 & \to & \Sigma_d(L) & \to & H_d(L) & \to & H_d(G) & \to & 0
\end{array}
$$

Remember that $H_d(\alpha)$ is an isomorphism. The above diagram may be expressed as:

$$
\begin{array}{ccccccccc}
0 & \to & \mathbb{Z}^n & \to & \mathbb{Z}^{n+k} & \to & \mathbb{Z}_{m_1} \times \ldots \times \mathbb{Z}_{m_n} \times \mathbb{Z}^k & \to & 0 \\
  &   & \downarrow M_f^\sharp & & \downarrow M_f & & \downarrow H_d(f_*) & & \\
0 & \to & \mathbb{Z}^n & \to & \mathbb{Z}^{n+k} & \to & \mathbb{Z}_{m_1} \times \ldots \times \mathbb{Z}_{m_n} \times \mathbb{Z}^k & \to & 0
\end{array}
$$

By a change of basis in the groups $\mathbb{Z}^n$ and $\mathbb{Z}^{n+k}$, we may assume that the above inclusions are diagonal homomorphisms, and that each basis element $\theta_i \in \mathbb{Z}^n$ is sent to $m_i \cdot \tau_i \in \mathbb{Z}^{n+k}$, where $\{\tau_i\}$ is the corresponding basis of $\mathbb{Z}^{n+k}$ and the $m_i \neq 1$ are the above torsion coefficients of $H_d(G)$. The basis elements of the above groups may also be chosen so that m_1 divides m_2, m_2 divides m_3, etc. Note that $H_d(f_*)$ projected to the free summand $\mathbb{Z}^k$ will be an isomorphism (as will $H_d(f_*)$ restricted to the torsion subgroup $\mathbb{Z}_{m_1} \times \ldots \times \mathbb{Z}_{m_n}$). Consequently, M_f will be invertible if and only if the restricted homomorphism from $\mathbb{Z}^n$ to $\mathbb{Z}^n$ is invertible. We may, therefore, restrict our attention to

$$
\begin{array}{ccccccccc}
0 & \to & \mathbb{Z}^n & \to & \mathbb{Z}^n & \to & \mathbb{Z}_{m_1} \times \ldots \times \mathbb{Z}_{m_n} & \to & 0 \\
& & \downarrow M_f^{\sharp} & & \downarrow M_f & & \downarrow H_d(\alpha) & & \\
0 & \to & \mathbb{Z}^n & \to & \mathbb{Z}^n & \to & \mathbb{Z}_{m_1} \times \ldots \times \mathbb{Z}_{m_n} & \to & 0
\end{array}
$$

Let us assume that $m_1 \neq 1$. We will leave the case where $m_1 = 1$ as an exercise below. The proof will continue by induction on n, the rank of $\Sigma_d = \mathbb{Z}^n$ as an abelian group. Suppose $n = 1$, i.e. $\Sigma_d = \mathbb{Z}$, then $det[M_f] \equiv \pm 1 \bmod m$, where m is the single torsion coefficient of $H_d(G)$. This implies that $M_f = (\delta)$, where $\delta = \pm 1 + rm$, for some $r \in \mathbb{Z}$. We want to try to modify f via a modification by π_2 elements (Lemma 1.4), so that the resulting function will be an isomorphism on H_d (or equivalently, an isomorphism on Σ_d). Consider the diagram:

$$
\begin{array}{ccccc}
\mathbb{Z} & \xrightarrow{\times m} & \mathbb{Z} & \xrightarrow{q} & \mathbb{Z}_m \\
\times\delta \downarrow & \gamma \swarrow & \downarrow \times\delta & & \downarrow \times\delta \\
\mathbb{Z} & \xrightarrow{\times m} & \mathbb{Z} & \xrightarrow{q} & \mathbb{Z}_m
\end{array}
$$

Notice that $\gamma \circ (\times m)$ is multiplication by m followed by multiplication by some chosen $s \in \mathbb{Z}$. Consequently, $\gamma \circ (\times m)$ is multiplication by sm, with s chosen arbitrarily. So $M_f + \gamma \circ (\times m) = (\delta + sm) = (\pm 1 + rm + sm)$. If we choose $s = -r$, we are done.

Now suppose $\Sigma_d \equiv \mathbb{Z}^n$, for some n. Consider, again, the diagram:

$$
\begin{array}{ccccccccc}
0 & \to & \mathbb{Z}^n & \to & \mathbb{Z}^n & \to & \mathbb{Z}_{m_1} \times \ldots \times \mathbb{Z}_{m_n} & \to & 0 \\
& & \downarrow M_f^{\sharp} & & \downarrow M_f & & \downarrow H_d(\alpha) & & \\
0 & \to & \mathbb{Z}^n & \to & \mathbb{Z}^n & \to & \mathbb{Z}_{m_1} \times \ldots \times \mathbb{Z}_{m_n} & \to & 0
\end{array}
$$

This time we will use γ to modify M_f, to get M_f to be invertible. By a lemma of A. Sieradski (See [Dy85], Appendix B), we may assume that $H_d(\alpha)$ is a diagonal homomorphism, and since $H_d(\alpha)$ is an isomorphism, the matrix of $H_d(\alpha)$ is a diagonal matrix of the form:

$$\begin{pmatrix} d_1, & \cdots, & 0 \\ \vdots & \ddots & \vdots \\ 0, & \cdots, & d_n \end{pmatrix}$$

where each d_i is relatively prime to each m_i. (Note: the following argument is copied from the proof of Theorem 6.9, p.145 of [Dy86]). The above matrix may be expressed as:

$$\begin{pmatrix} d_1, & 0, & \cdots, & 0 \\ \vdots & \ddots & & \vdots \\ 0, & \cdots, & d_{n-1}, & 0 \\ 0, & \cdots, & 0, & d_n \end{pmatrix} = \begin{pmatrix} d_1, & 0, & \cdots, & 0 \\ \vdots & \ddots & & \vdots \\ 0, & \cdots, & d_{n-1} \cdot d_n, & 0 \\ 0, & \cdots, & 0, & 1 \end{pmatrix} \cdot \begin{pmatrix} 1, & 0, & \cdots, & 0 \\ \vdots & \ddots & & \vdots \\ & & 1 & \\ 0, & \cdots, & 0 & d'_n, & 0 \\ 0, & \cdots, & 0, & 0 & d_n \end{pmatrix}$$

where $d'_n \cdot d_n \equiv 1 \bmod m_n$ (and hence mod m_{n-1}). Both matrices on the right have determinant $\pm 1 \bmod m$. Consequently, we have reduced the problem to the 2×2 case. Since we are assuming that M_f has determinant ± 1, we may now assume that M_f is of the form

$$\begin{pmatrix} d_1 + m_1 \cdot a & m_1 \cdot b \\ m_2 \cdot c & d_2 + m_2 \cdot d \end{pmatrix}$$

Consequently, $H_d(\alpha)$ will have determinant $d_1 d_2 \equiv \pm 1 \bmod m_1$. Note that a suitable homomorphism γ in the modification by π_2 elements will change the matrix

$$\begin{pmatrix} d_1 + m_1 \cdot a & m_1 \cdot b \\ m_2 \cdot c & d_2 + m_2 \cdot d \end{pmatrix}$$

into the form

$$\begin{pmatrix} d_1 & 0 \\ 0 & d_2 \end{pmatrix}$$

So we may assume that f_d is of the form

$$\begin{pmatrix} d_1 & 0 \\ 0 & d_2 \end{pmatrix}$$

Now we want to find a γ so that the determinant of f_d is ± 1. So we want now to find new $a, b, c,$ and d, so that

$$\begin{pmatrix} d_1 + m_1 \cdot a & m_1 \cdot b \\ m_2 \cdot c & d_2 + m_2 \cdot d \end{pmatrix}$$

will have determinant ± 1. Our problem is now merely combinatorial. We need to find an a, b, c, and d, so that $d_1 d_2 + m_1 m_2 (ad-bc) + m_1 a d_2 + m_2 d_1 = \pm 1$. Let $d_1 d_2 = \pm 1 + k m_1$, and choose a and c so that $-k = cm_2 + ad_2$ (possible since $(m_2, d_2) = 1$). Choose $d = 0$ and $b = -1$, then

$$det \begin{pmatrix} d_1 + m_1 a & m_1 \\ m_1 c & d_2 \end{pmatrix} =$$

$d_1 d_2 + m_1 a d_2 + m_1 m_2 c = \pm 1 + k m_1 + m_1 a d_1 + m_1 m_2 (-d_2 a - k)/m_2 = \pm 1. \square$

Exercise 1.14 *Prove Theorem 1.13 in the case where* $gcd\{m_i\} = 1$.

Corollary 1.15 *Two (G, d)-complexes K, and L are homology equivalent if and only if* $b(K, L) = 1$ *in* $\mathbb{Z}_m^* / \pm D$.

1.3   Homotopy distinction of twisted presentations

In the homotopy classification of $(G, 2)$-complexes with finite abelian G, we must first deal with recognizable presentations of such groups. We can express G as $\mathbb{Z}_{m_1} \times \ldots \times \mathbb{Z}_{m_n}$ with $m_i | m_{i+1}$ and $m_1 > 1$. Its standard presentation

$$\mathcal{P} = \langle\, a_1, \ldots, a_n \mid a_1^{m_1}, a_2^{m_2}, \ldots, a_n^{m_n}, [a_i, a_j], \ i < j \,\rangle,$$

has commutator relators $[a_i, a_j] = a_i a_j a_i^{-1} a_j^{-1}$. Note that if r is a number relatively prime to m_1, then $\{a_1^r, a_2, \ldots, a_n\}$ also generates G. Consequently, we can define a *twisted presentation* of G as

$$\mathcal{P}_r = \langle\, a_1, \ldots, a_n \mid a_1^{m_1}, a_2^{m_2}, \ldots, a_n^{m_n}, [a_1^r, a_2], [a_i, a_j], \ i < j, \ (i, j) \neq (1, 2) \,\rangle.$$

We could *twist* all the generators independently, but we will discover later that this gains no new homotopy types.

Remark: Let K_1 be the model of the standard presentation $\mathcal{P}$ and K_r the model of the *twisted presentation* $\mathcal{P}_r$ above. In Definition 1.11, the modulus for the bias $b(K_s, K_r)$ for a pair of twisted complexes is gcd{torsion coefficients of $H_2(G)$}, which equals $m_1 = gcd\{m_1, \ldots, m_n\}$ in this setting.

Theorem 1.16 *For K_r and K_s, the bias* $b(K_s, K_r) \equiv [s]^{-1}[r] \in \mathbb{Z}_{m_1}^* / \pm D$.

Proof: Let s' be an integer such that $s \cdot s' \equiv 1 \ mod \ m_1$. In particular, let $s \cdot s' = 1 + k \cdot m_1$. We can assume that k is positive. Check that

$$\left(\sum_{i=0}^{s-1} a_1^i\right) \cdot \left(\sum_{j=0}^{s'-1} a_1^{sj}\right) = \left(\sum_{i=0}^{1+k \cdot m_1} a_1^i\right) = k \cdot \left(\sum_{i=0}^{m_1-1} a_1^i\right) + 1.$$

Using Fox's free differential calculus (Chapter II, Section 3.2), the respective homomorphisms $C_2(\tilde{K}_s)\xrightarrow{\partial_2}C_1(\tilde{K}_s)$ and $C_2(\tilde{K}_r)\xrightarrow{\partial_2}C_1(\tilde{K}_r)$ are represented by the matrices

$$
\begin{array}{c|ccccc}
 & \tilde{a}_1 & \tilde{a}_2 & \cdots & \tilde{a}_{n-1} & \tilde{a}_n \\
\tilde{R}(s)_1 & \sum\limits_{i=0}^{m_1-1} a_1^i & 0 & \cdots & 0 & 0 \\
\tilde{R}(s)_2 & 0 & \sum\limits_{i=0}^{m_2-1} a_2^i & \cdots & 0 & 0 \\
\vdots & \vdots & \vdots & \ddots & \vdots & \vdots \\
\tilde{R}(s)_{n-1} & 0 & 0 & \cdots & \sum\limits_{i=0}^{m_{n-1}-1} a_{n-1}^i & 0 \\
\tilde{R}(s)_n & 0 & 0 & \cdots & 0 & \sum\limits_{i=0}^{m_n-1} a_n^i \\
\tilde{R}(s)_{12} & (\sum\limits_{i=0}^{s-1} a_1^i)(1-a_2) & (\sum\limits_{i=0}^{s-1} a_1^i)(a_1-1) & \cdots & 0 & 0 \\
\vdots & \vdots & \vdots & \ddots & \vdots & \vdots \\
\tilde{R}(s)_{(n-1)n} & 0 & 0 & \cdots & (1-a_n) & (a_{n-1}-1)
\end{array}
$$

and

$$
\begin{array}{c|ccccc}
 & \tilde{a}_1 & \tilde{a}_2 & \cdots & \tilde{a}_{n-1} & \tilde{a}_n \\
\tilde{R}(r)_1 & \sum\limits_{i=0}^{m_1-1} a_1^i & 0 & \cdots & 0 & 0 \\
\tilde{R}(r)_2 & 0 & \sum\limits_{i=0}^{m_2-1} a_2^i & \cdots & 0 & 0 \\
\vdots & \vdots & \vdots & \ddots & \vdots & \vdots \\
\tilde{R}(r)_{n-1} & 0 & 0 & \cdots & \sum\limits_{i=0}^{m_{n-1}-1} a_{n-1}^i & 0 \\
\tilde{R}(r)_n & 0 & 0 & \cdots & 0 & \sum\limits_{i=0}^{m_n-1} a_n^i \\
\tilde{R}(r)_{12} & (\sum\limits_{i=0}^{r-1} a_1^i)(1-a_2) & (\sum\limits_{i=0}^{r-1} a_1^i)(a_1-1) & \cdots & 0 & 0 \\
\vdots & \vdots & \vdots & \ddots & \vdots & \vdots \\
\tilde{R}(r)_{(n-1)n} & 0 & 0 & \cdots & (1-a_n) & (a_{n-1}-1)
\end{array}
$$

Consider the homomorphism $\mu : C_2(\tilde{K}_s) \to C_2(\ddot{K}_r)$ with matrix:

$$
\begin{array}{c}
\\
R(r)_1 \\
\vdots \\
R(r)_n \\
R(r)_{12} \\
\\
\vdots \\
R(r)_{(n-1)n}
\end{array}
\begin{array}{cccccc}
R(s)_1 & \cdots & R(s)_n & R(s)_{12} & \cdots & R(s)_{(n-1)n} \\
\left(\begin{array}{c} 1 \end{array}\right. & \cdots & 0 & 0 & \cdots & 0 \\
\vdots & \ddots & \vdots & \vdots & & \vdots \\
0 & \cdots & 1 & 0 & \cdots & 0 \\
k \cdot (\sum_{i=0}^{r-1} a_1^i) \cdot (a_2 - 1) & \cdots & 0 & (\sum_{i=0}^{r-1} a^i)(\sum_{i=0}^{s'-1} a_1^{s \cdot i}) & \cdots & 0 \\
\vdots & & \vdots & \vdots & \ddots & \vdots \\
\left. 0 \vphantom{\sum} \right. & \cdots & 0 & 0 & \cdots & 1
\end{array}\right)
$$

Note that the determinant of the augmented matrix is $s' \cdot r$. We leave it to the reader to check that the diagram

$$C_2(\tilde{K}_s) \xrightarrow{\partial_2} C_1(\tilde{K}_s)$$

$$\mu \downarrow \qquad\qquad \|$$

$$C_2(\tilde{K}_r) \xrightarrow{\partial_2} C_1(\tilde{K}_r)$$

commutes. Now use Lemma 1.2 with f any map which is the identity on the 1-skeleton and $\gamma = \tilde{f}_2 - \mu$. $\qquad\qquad\qquad\qquad\qquad\qquad\qquad\square$

For the group $G = \mathbb{Z}_{m_1} \times \ldots \times \mathbb{Z}_{m_n}$ with torsion coefficients $m_1 | m_2 \ldots | m_n$, Sieradski [Si77] showed that $D = \pm(\mathbb{Z}_{m_1}^*)^{n-1}$. So the bias can be explicitly computed for twisted presentations. So the number of distinct homotopy types of twisted models is at least the order of $\mathbb{Z}_{m_1}^* / \pm (\mathbb{Z}_{m_1}^*)^{n-1}$. In §2, we will see that this is the number of homotopy types of $(G, 2)$-complexes of minimal Euler characteristic. Simple number theoretic arguments allow us to compute this order directly, in terms of the Euler function ϕ ([Si77], p. 137).

For two twisted presentations where the bias does vanish, Sieradski was able to construct 3-deformations between them. Thus, we have this theorem:

Theorem 1.17 (($\Rightarrow$) Metzler [Me76]; ($\Leftarrow$) Sieradski [Si77])   *The standard two complexes K_r and K_s of two twisted presentations of a finite abelian group are homotopy equivalent, simple-homotopy equivalent and 3-deformation equivalent if and only if the bias of the pair vanishes:*

$$b(K_r, K_s) = 1 \in \mathbb{Z}_{m_1}^* / \pm (\mathbb{Z}_{m_1}^*)^{n-1}$$

The above classification will be useful to us in Section 2, when we classify all $(G, 2)$-complexes, where G is a finite abelian group.

2   Classifications for Finite Abelian π_1

The next step in the homotopy classification of $(G, 2)$-complexes for G a finite abelian group is to construct the Browning obstruction theory, due to Wesley Browning ([Br78],[Br79$_1$], [Br79$_2$], [Br79$_3$]). In some sense, this obstruction theory is merely an extension of the idea of bias, but the algebra used is much more sophisticated. Consequently, we will merely define the Browning obstruction group and the obstruction elements. We will not include the lengthy and complicated proofs.

In §2.1, we will define the Browning obstruction elements and the Browning obstruction group. In §2.2, we will prove that, for finite abelian fundamental groups, the Browning obstruction is equivalent to the bias obstruction.

2.1   The Browning obstruction group

Wesley Browning [Br78] defined an obstruction to two finite (G, d)-complexes with the same Euler characteristic being homotopy equivalent, provided the fundamental group satisfied Eichler's condition. For alternate treatments, see [GuLa91] and [Gr91]. We will not define Eichler's condition here. Suffice it to say that groups that do not have binary polyhedral quotients satisfy Eichler's condition. Consequently, finite abelian groups satisfy Eichler's condition, and Browning's theorem holds for such groups.

Let G be a finite group satisfying Eichler's condition. Let K and L be (G, d)-complexes with $\chi(K) = \chi(L)$. We will assume the Euler characteristic is minimal with respect to G, since otherwise the Browning obstruction group vanishes [Br78]. Browning actually proved that, above the minimum Euler characteristic, all (G, d)-complexes with the same Euler characteristic are homotopy equivalent, for any finite group G.

If we assume $f : K \to L$ is the identity on the $(d-1)$-skeleton, then f is a homotopy equivalence if and only if the homomorphism $\tilde{f}_d : C_d(\tilde{K}) \to C_d(\tilde{L})$ is an isomorphism, as we have said before.

Definition 2.1 *Let u be the set of primes that divide the order of the finite group G. Let $\mathbb{Z}_u = \{ \frac{a}{b} \mid a, b \in \mathbb{Z}$ and $(b, p) = 1$, for every $p \in u \}$. The localization at u of a module or a homomorphism between modules results by tensoring with $\mathbb{Z}_u$. (E.g., the localization M_u of a module M means $M \otimes \mathbb{Z}_u$).*

Lemma 2.2 *Given a map $f : K \to L$ which is the identity on the $(d-1)$-skeleton, then there exists a map g such that $g^{(i)} = f^{(i)}$, for $i < d$ and the matrix representing $\tilde{g}_d : C_d(\tilde{K}) \to C_d(\tilde{L})$ localizes to an invertible matrix.*

As the proof makes free use of homological algebra and the theory of pointed modules, we refer the reader to Browning's thesis [Br78, (2.10.3)], [GuLa91, Lemma 3.11], or [Gr91].

Let R be any ring with unit. Let $Gl_n(R)$ be the group of invertible $n \times n$-matrices with entries in the ring R. We can consider the group $GL_n(R)$ as a subgroup of $GL_{n+1}(R)$ by adding a just 1 as a new diagonal entry:

$$A \mapsto \begin{pmatrix} A & 0 \\ 0 & 1 \end{pmatrix}.$$

Let $GL_\infty(R)$ represent the direct limit $\varprojlim_{n \to \infty} GL_n(R)$. That is, $GL_\infty(R)$ is the group of infinite invertible matrices which are the identity matrix accept in a finite number of entries. Define

$$K_1(R) \equiv GL_\infty(R)/[GL_\infty(R), GL_\infty(R)],$$

the abelian quotient of $GL_\infty(R)$. Whitehead showed that the commutator subgroup $[GL_\infty(R), GL_\infty(R)]$ is actually the subgroup generated by the elementary matrices (See [Co73]). So, in fact, $K_1(R)$ is the group of matrices modulo row and column operations.

Given a matrix $M \in GL_n(R)$, let $\tau(M)$ represent the class of M in $K_1(R)$. Given a chain equivalence between two free chain complexes over a ring R, the torsion of the chain equivalence can be computed as the equivalence class in $K_1(R)$ of the invertible matrix representing the chain map (see [Co73]). If the chain equivalence is the identity below the top level, d, then its torsion will be plus or minus the torsion of the matrix on the d-level. By adjusting the lower dimensional skeleta, Lemma 2.2 gives for any finite group G:

Lemma 2.3 *Given two (G,d)-complexes K and L with $\chi(K) = \chi(L)$ and a group isomorphism $\alpha : \pi_1(K) \to \pi_1(L)$, then there exists a map $f : K \to L$, with $f_* = \alpha$, such that the chain map $\tilde{f}_* : C_*(\tilde{K}) \to C_*(\tilde{L})$ localizes to a chain equivalence: $\tilde{f}_* \otimes \mathbb{Z}_u : C_*(\tilde{K}) \otimes \mathbb{Z}_u \to C_*(\tilde{L}) \otimes \mathbb{Z}_u$.*

The local torsion of f, $\tau(f)$ will be defined using the torsion of the matrix representing the chain equivalence $(f_*)_u$, with respect to the standard bases, as in [Co73, (14.2)]. If F is the identity on the $(d-1)$-skeleton, this matrix is equivalent up to sign to the torsion of the matrix for the homomorphism $C_d(\tilde{K})_u \to C_d(\tilde{L})_u$. The following lemma helps us compute this torsion.

Lemma 2.4 *For a finite group G and a finite set u of primes dividing the order of G, the homomorphism from the units of $\mathbb{Z}_u G$ to $K_1(\mathbb{Z}_u G)$, $\iota : (\mathbb{Z}_u G)^* \to K_1(\mathbb{Z}_u G)$, is onto. If G is commutative and M is in $GL_n(\mathbb{Z}_u G)$ then $\tau(M) = \iota(Det[M])$.*

Proof: By [Sw70], Lemma 9.2, $\mathbb{Z}_u G$ is a semi-local ring. By [Sil81], Chapter 6, K_1 of a semi-local ring is represented by units. The second part of this lemma follows from direct calculation.   □

Note that if f_d were an isomorphism, then the induced matrix M_f would be invertible over $\mathbb{Z}G$, i.e., $\tilde{M}_f$ would represent an element of $K_1(\mathbb{Z}G)$. Consequently, such invertible matrices should represent zero in any obstruction group. So we will quotient out by the image of $K_1(\mathbb{Z}G)$ in $K_1(\mathbb{Z}_u G)$. It should be clear that if $\tilde{M}_f$ is not zero in $K_1(\mathbb{Z}_u G)/\iota_*(K_1(\mathbb{Z}G))$, where i_* is induce by inclusion, then f cannot be a homotopy equivalence. Define $K_1(\mathbb{Z}G, u) \equiv K_1(\mathbb{Z}_u G)/\iota_*(K_1(\mathbb{Z}G))$.

In determining whether there exists a homotopy equivalence inducing α, we must also consider the effect of modifying f by composing it with a map $h : L \to L$ which is the identity on the fundamental group of L. Even though a given map may not be a homotopy equivalence, some other map, inducing the same isomomorphism on fundamental groups may be a homotopy equivalence. The homotopy inverse composed with the original map would be a map which induces the identity on the fundamental group. We must therefore quotient out by the effects on the obstruction via maps on L that induce the identity on the fundamental group. This will be dealt with in the following definition:

Definition 2.5 *Let $Aut_u(L)$ refer to the maps $h : L \to L$ which induces the identity on the $\pi_1(L)$ and induces a chain equivalence on the chain complex $C_*(L)_u$. Define the* Browning Obstruction Group *to be*

$$Br_L(G) \equiv K_1(\mathbb{Z}G, u)\big/\tau(Aut_u(L))$$

where $\tau(Aut_u(L))$ represents the subgroup of all elements of the form $\tau(h)$ in $K_1(\mathbb{Z}G, u)$, $h \in Aut_u(L)$.

Define the Browning obstruction element *of K and L with respect to α as*

$$B_\alpha(K, L) \equiv \tau(\tilde{f}_*) \in Br_L G$$

where $f : K \to L$ is a map which induces a given isomorphism α on the fundamental groups and localizes to a chain equivalence $\tilde{f}_* : C_*(\tilde{K})_u \to C_*(\tilde{L})_u$.

We note that $B_\alpha(K, L)$ is independent of the choice of f. Also, though the obstruction element depends on α, the group it lies in will not depend on α (See [GuLa91]). The following theorems are stated without proof.

Theorem 2.6 (Browning's Theorem [Br79₃], [GuLa91]) *Let G be a finite group satisfying Eichler's condition. Given two (G, d)-complexes K and L with $\chi(K) = \chi(L)$, there exists a homotopy equivalence $g : K \to L$ which induces $\alpha : \pi(K) \cong \pi_1(L)$ if and only if $B_\alpha(K, L) = 0$.*

The transitivity of the Browning obstruction follows from the definition:

Theorem 2.7 (Transitivity) *For (G,d)-complexes K, L and T and isomorphisms $\alpha : \pi_1(K) \to \pi_1(L)$ and $\beta : \pi_1(L) \to \pi_1(T)$, then*

$$B_{\beta \circ \alpha}(K,T) = \beta_*(B_\alpha(K,L)) \cdot B_\beta(L,T),$$

where $\beta_* : Br_L(G) \to Br_T(G)$ is the obvious homomorphism induced by β.

2.2   Homotopy classification for finite abelian π_1

Let G be a finite abelian group $\mathbb{Z}_{m_1} \times \ldots \times \mathbb{Z}_{m_n}$, with torsion coefficients $m_1|m_2|\ldots|m_n$. Let $N_1, N_2, \ldots, N_n$ be the respective sums $N_i = \sum_{j=1}^{m_i - 1} a_i^j$, where a_i generates $\mathbb{Z}_{m_i}$. Let $N = \sum_{g \in G} g$, the *norm* element of $\mathbb{Z}G$. The following exercise may be easily established by induction:

Exercise 2.8 *Let u be the set of prime divisors of G. Prove that any element $\mu \in \mathbb{Z}_u G$ with augmentation $\epsilon(\mu) = 1$ may be written as a product of elements $\mu = \mu_1 \cdot \mu_2 \cdot \ldots \cdot \mu_n$, with $\mu_i \cdot N_i = N_i$.* $\square$

Proposition 2.9 (Browning [Br79_3])   *Let K be the standard complex of the presentation*

$$\mathcal{P} = \langle\, a_1, a_2, \ldots, a_n \mid a_1^{m_1}, a_2^{m_2}, \ldots, a_n^{m_n},\ [a_i, a_j]\ (i < j) \,\rangle.$$

Then the sets $\{1 + rN \mid r \in \mathbb{Z}\}$ and $\{\mu \in (\mathbb{Z}_u G)^* \mid \epsilon(\mu) = 1\}$ are contained in $\tau(Aut_u(K)) \subset K_1(\mathbb{Z}G, u)$.

Proof: To realize the elements that augment to 1, we merely note that Exercise 2.8 tells us that the matrix:

$$\begin{pmatrix}
\mu_1 & 0 & \cdots & 0 & 0 & \cdots & 0 \\
0 & \mu_2 & \cdots & 0 & 0 & \cdots & 0 \\
\vdots & \ddots & \ddots & \vdots & \vdots & \ddots & \vdots \\
0 & 0 & \cdots & \mu_n & 0 & \cdots & 0 \\
0 & 0 & \cdots & 0 & 1 & \cdots & 0 \\
\vdots & \vdots & \ddots & \vdots & \vdots & \ddots & \vdots \\
0 & 0 & \cdots & 0 & 0 & \cdots & 1
\end{pmatrix}$$

will commute with the boundary homomorphism $C_2(\tilde{K}) \xrightarrow{\partial_2} C_1(\tilde{K})$,

$$
\begin{array}{c}
 \\
\tilde{R}_1 \\
\tilde{R}_2 \\
\vdots \\
\tilde{R}_{n-1} \\
\tilde{R}_n \\
\tilde{R}_{12} \\
\vdots \\
\tilde{R}_{(n-1)n}
\end{array}
\begin{array}{ccccc}
\tilde{a}_1 & \tilde{a}_2 & \cdots & \tilde{a}_{n-1} & \tilde{a}_n \\
\left(\sum\limits_{i=0}^{m_1-1} a_1^i \right. & 0 & \cdots & 0 & 0 \\
0 & \sum\limits_{i=0}^{m_2-1} a_2^i & \cdots & 0 & 0 \\
\vdots & \vdots & \ddots & \vdots & \vdots \\
0 & 0 & \cdots & \sum\limits_{i=0}^{m_{n-1}-1} a_{n-1}^i & 0 \\
0 & 0 & \cdots & 0 & \sum\limits_{i=0}^{m_n-1} a_n^i \\
(1-a_2) & (a_1-1) & \cdots & 0 & 0 \\
\vdots & \vdots & \ddots & \vdots & \vdots \\
0 & 0 & \cdots & (1-a_n) & \left. (a_{n-1}-1) \right)
\end{array}
$$

Consequently, by modifications of the identity map $id : K \to K$ by π_2 elements, we can construct a geometric map, $f : K \to K$, which induces the identity on the 1-skeleton and whose torsion is the torsion of the above matrix, which is μ.

Similarly, to realize $1 + rN$, we merely need to point out that $N \cdot \partial(\tilde{S}_{i,j}) = 0$ where $S_{i,j}$ is the algebraic boundary in $C_1(\tilde{K})$ of the 2-cell corresponding to the commutator relator $[a_i, a_j]$. Therefore, if we start with the identity matrix on $C_2(\tilde{K})$, and replace the 1 corresponding to the $\tilde{S}_{i,j}$ entry with $1 + kN$, we see that the new matrix again commutes with the boundary homomorphism.$\Box$

Corollary 2.10 *The elements of* $K_1(\mathbb{Z}G, u)/\tau(Aut_u(K))$ *are representable by integers in* $(\mathbb{Z}/|G|\mathbb{Z})^*/\pm 1$, *where* $|G|$ *is the order of the group.*

Proof: Since $\mathbb{Z}_u G$ is semi-local, $K_1(\mathbb{Z}_u G)$ is representable by units, [Sil81], Section 6.5. So let $[\kappa] \in K_1(\mathbb{Z}G, u)/\tau(Aut_u(K))$, where $\kappa \in (\mathbb{Z}_u G)^*$. Let $k = \epsilon(\kappa) \in \mathbb{Z}_u$. Since $\kappa \cdot \frac{1}{k}$ augments to 1, it is in $\tau(Aut_u(K))$. Consequently, κ is equivalent to $k \in (\mathbb{Z}_u)^*$.

By this argument, $1 + rN$ is equivalent to $1 + r|G| \in K_1(\mathbb{Z}G, u)/\tau(Aut_u(K))$. So any integer congruent to 1 mod $|G|$, is equivalent to 1. In particular, if $\bar{b} \in \mathbb{Z}$ is the multiplicative inverse of b in $\mathbb{Z}/|G|\mathbb{Z}$, then $\bar{b} \equiv \frac{1}{b} \in K_1(\mathbb{Z}G, u)/\tau(Aut_u(K))$. Therefore, if $r = \frac{a}{b} \in \mathbb{Z}_u$, then r is equivalent in $K_1(\mathbb{Z}G, u)/\tau(Aut_u(K))$ to $a\bar{b}$. $\Box$

Theorem 2.11 *Let K be a finite 2-complex with minimal Euler characteristic and with $\pi_1(K) = G$, where G is a finite abelian group with torsion coefficients $m_1|m_2|\ldots|m_n$. Then K is homotopy equivalent to the standard complex K_r the model of some twisted presentation $\mathcal{P}_r$ (see, §1.3) of G.*

Proof: First, let K_1 is the standard 2 complex of the usual presentation of G and K_r is the standard complex of the above presentation. We will construct a map $f : K_1 \to K_r$ which is the identity on the 1-skeleton and whose torsion is $[\sum_{i=0}^{r-1} a_1^i] \in K_1(\mathbb{Z}_u G)$. Note that $\sum_{i=0}^{r-1} a_1^i \cdot \partial \tilde{S}_{1,2} = \partial \tilde{R}_{1,2}$, where $\tilde{R}_{1,2}$ is the 2-cell in $\tilde{K}_r$ corresponding to the relator $a_1^r a_2 a_1^{-r} a_2^{-1}$, and $\tilde{S}_{1,2}$ is the 2-cell in $\tilde{K}_1$ corresponding to the relator $[a_1, a_2]$. Note also that $\sum_{i=0}^{r-1} a_1^i$ is a unit in $\mathbb{Z}_u G$. For, if $rs = 1 + qm_1$ for some s and q, then

$$\sum_{i=0}^{r-1} a_1^i \Big(\sum_{i=0}^{s-1} (a_1^r)^i - \frac{q}{r} \sum_{i=0}^{m_1-1} a_1^i \Big) = 1.$$

Also notice that $\sum_{i=0}^{r-1} a_1^i$ augments to $r \in \mathbb{Z}$. $B_\alpha(K_1, K_r) = [r]$, where $\alpha : \pi_1(K_1) \to \pi_1(K_r)$ is the homomorphism induced from the identification of the generators. Consequently all of the elements of $K_1(\mathbb{Z}G, u)/\tau(Aut_u(K))$ are realizable by obstructions to complex K being homotopy equivalent to a standard complex of a twisted presentation.

Now, suppose K is some $(G, 2)$-complex with minimal Euler characteristic. Let $\alpha : \pi_1(K_1) \to \pi_1(K)$ be an isomorphism. By Corollary 2.10, $B_\alpha(K_1, K) = [s]$ for some integer s relatively prime to $|G|$. Now, let r be an integer which is the multiplicative inverse of s mod m_1. Notice $B_\beta(K_r, K_1) = [s]^{-1}$, where β is induced by the identity on the generators of G. Using the transitivity of the Browning invariant, we see that $B_{\alpha \circ \beta}(K_r, K) = \beta_*(B_\beta(K_r, K_1)) \cdot B_\alpha(K_1, K)$. Since β is the identity on generators, it acts as the identity on the obstruction group. Thus $B_{\alpha \circ \beta}(K_r, K) = [r] \cdot [s] = 1$. So K is homotopy equivalent to K_r, by Browning's Theorem.

Therefore, each 2-complex with minimal Euler characteristic and fundamental group G is homotopy equivalent to one of the standard 2-complex of a twisted presentation. Furthermore, the number of their homotopy types is that of the twisted models, the order of the quotient group $\mathbb{Z}_{m_1}^* / \pm (\mathbb{Z}_{m_1}^*)^{n-1}$.   □

The results of the above theorem can be extended to a simple-homotopy classification. Using the results of [La86], all minimal $(G, 2)$-complexes (with G finite, abelian) are simple-homotopy equivalent to a standard complex of a twisted presentation. Of course we know, from Sieradski (see Theorem

1.17), that if the Browning obstruction vanishes for standard complexes of twisted presentations, then they 3-deform to one another. We still don't know whether two arbitrary $(G,2)$-complexes that are homotopy equivalent (hence simple-homotopy equivalent) must 3-deform to one another. This is a case of the Generalized Andrews-Curtis Conjecture discussed in Chapter I.

The techniques used in this section can also be used on non-abelian groups that are semi-direct products of cyclic groups (see [La91], and [Gu-La93]).

3 Classifications for Non-Finite π_1

This section gives techniques and facts currently known for 2-complexes with non-finite fundamental groups. In §3.1, we generalize the Browning obstruction elements and give examples. In §3.2 and §3.3, we give techniques and examples for free products of groups.

3.1 Infinite groups; generalized Browning invariant

We generalize the results from homotopy equivalence to partial homotopy equivalence, i.e. equivariant homology equivalence of finite covers.

In this section, we will work with finite quotients Q of our fundamental group G and the localizations at the set u of primes dividing the order of Q. If L is a 2-complex with fundamental group G and $\theta : G \to Q$ is a surjection onto a finite group, we construct a group $Br_L(Q)$. Given a 2-complex K and an identification

$$\begin{array}{ccc} \pi_1(K) & \stackrel{\alpha}{\cong} & \pi_1(L) \\ \downarrow \theta & & \downarrow \theta' \\ Q & = & Q. \end{array}$$

We define a total obstruction to homology equivalence of finite covers in $Br_L(Q)$ which is defined for all 2-complexes L with fundamental group G and Euler characteristic $\chi(L)$. We are able to locate the obstruction as a torsion element in a quotient of $K_1(\mathbb{Z}Q, u)$, which is defined to be the quotient of $K_1(\mathbb{Z}_u Q)$ by $\iota_*(K_1(\mathbb{Z}Q))$.

There are two technicalities. The first condition is that $H_2(\ker \theta)$ must be a module which, as a group, is free abelian. This does not appear to be a serious setback. For example, if G is a finite free product of finite abelian groups, its commutator subgroup has free abelian second homology. The second condition is that Q satisfy Eichler's condition. Again, this is not a serious constraint, since "most" finite groups satisfy Eichler's condition.

Let G be a fixed group. As before, we consider 2-dimensional CW-complexes K with only one vertex $K^{(0)}$ (the zero skeleton). In general, we look at pairs (K, ϕ), where ϕ is an isomorphism, $\phi : \pi_1(K, K^0) \to G$. Given any two such complexes K, L, there is a map $f : K \to L$ such that f_* is an isomorphism of fundamental groups making the diagram

$$\pi_1(K, K^0) \xrightarrow{f_*} \pi_1(L, L^0)$$

$$\phi_K \searrow \qquad \swarrow \phi_L$$

$$G$$

commute. Thus, we can identify the fundamental groups of K and L via f_*.

Let $\theta : G \to Q$ be a surjection of groups with kernel N. Let $\bar{K}$ be the covering of K associated to N (strictly speaking to $\phi^{-1}(N)$). Given K and L and f as above, there is a unique lift $\bar{f} : \bar{K} \to \bar{L}$ of f sending the preferred base point of $\bar{K}$ to that of $\bar{L}$.

Definition 3.1 *The map* $f : K \to L$ *above is said to be a* partial homotopy equivalence *with respect to* Q *if* $\bar{f}$ *induces an isomorphism of integral homology.*

Remarks: (1) Among the pairs (K, ϕ) as above, this is an equivalence relation: If L is partially homotopy equivalent (with respect to Q) to M via a map $g : L \to M$, then the lift of $g \circ f$ is $\bar{g} \circ \bar{f}$, and it clearly induces an isomorphism $H_*(\bar{K}) \to H_*(\bar{M})$.

(2) Clearly, the isomorphism of homology in the definition is Q-equivariant.

(3) A partial homotopy equivalence with respect to G is just a homotopy equivalence, since in that case $\bar{K}$ is the universal cover $\tilde{K}$. At the other extreme, a partial homotopy equivalence with respect to the trivial group is a homology equivalence (with an isomorphism of π_1). Thus partial homotopy equivalence is necessary for homotopy equivalence and sufficient for homology equivalence.

Consider $u, \bar{K}$ and Q as usual with $\theta : \pi_1 K \to Q$ onto, Q finite of order n, u the set of primes of n, $H_2(\ker(\theta))$ free abelian, and $\bar{K}$ the cover of K associated to $\ker(\theta)$.

Definition 3.2 *Let* $Aut_u(\bar{L})$ *be the set of maps* $g : L \to L$ *that induce the identity on* $\pi_1(L)$ *and a localized chain equivalence* $(\bar{g}_*)_u : C_*(\bar{L})_u \cong C_*(\bar{L})_u$, *where* $\bar{L}$ *is the cover of* L *associated to* Q.

Definition 3.3 *Let* $\theta : \pi_1(L) \to Q$ *be a homomorphism onto a finite group. let* $\alpha : \pi_1(K) \to \pi_1(L)$ *be an isomorphism and let* $f : K \to L$ *be a map with* $f_* = \alpha$. *Let* $\bar{f} : \bar{K} \to \bar{L}$, *the lift of* f *corresponding to* $\ker(\theta)$, *induce a local chain equivalence* $(\bar{f}_*)_u : C_*(\bar{K}) \to C_*(\bar{L})$, *then define*

$$B_{\alpha,Q}(K,L) \equiv \tau(\bar{f}) \bmod Aut_u(\bar{L}) \in K_1(\mathbb{Z}Q, u)/Aut_u(\bar{L}).$$

Theorem 3.4 (Browning's Theorem for partial homotopy [GuLa91]) *Given two finite 2-complexes* K *and* L *with* $\alpha : \pi_1(K) \cong \pi_1(L)$. *Let* $\theta : \pi_1(K) \to Q$ *a homomorphism onto a finite group with* $H_2(\ker\theta)$ *free abelian. There exists a partial homotopy equivalence* $f : K \to L$ *with respect to* Q *with* $f_* = \alpha$ *if and only if* $B_{\alpha,Q}(K,L) = 0$ *in* $K_1(\mathbb{Z}Q, u)/Aut_u(\bar{L})$.

We now consider the case when $G = H * J$, where H and J are finite groups.

Definition 3.5 *For any finitely presentable group* G, *let* $\chi_{min}(G)$ *be the minimum Euler characteristic for a 2-complex with fundamental group* G.

Proposition 3.6 *If* $\theta : H * J \to Q$ *is a surjective homomorphism onto a finite group with the restrictions to* H *and* J *injective, then* $\ker \theta$ *is a free group. In particular,* $H_2(\ker(\theta))$ *is trivial, hence, free.*

Proof: Let $N = \ker \theta$. Let K and L be finite 2-complexes with $\pi_1(K) = H, \pi_1(L) = J, \chi(K) = \chi_{min}(H)$ and $\chi(L) = \chi_{min}(J)$. $K \vee L$ will have fundamental group $H * J$. Let $\overline{K \vee L}$ be the finite lift of $K \vee L$ whose fundamental group is N, with covering map $p : \overline{K \vee L} \to K \vee L$. Since H and J inject into Q, then the covers of K and L in $\overline{K \vee L}$ will be unions of universal covers, $p^{-1}(K) = \sqcup_i \tilde{K}$, $p^{-1}(L) = \sqcup_j \tilde{L}$, where $\sqcup$ is disjoint union and $\tilde{K}$ and $\tilde{L}$ are universal covers. Now replace the 'wedge', $\vee$, in $K \vee L$ by an arc connecting a point of K with a point of L, then $\overline{K \vee L} = p^{-1}(K) \cup p^{-1}(arc) \cup p^{-1}(L)$. So off of the covers of K and L, $\overline{K \vee L}$ is 1-dimensional. Since the components of $p^{-1}(K)$ and $p^{-1}(L)$ are simply-connected, $\overline{K \vee L}$ is the fundamental group of a graph. Therefore $N = \pi_1(\overline{K \vee L})$ is free. $\qquad\square$

Corollary 3.7 *If* $H = \mathbb{Z}_{m_1} \times \mathbb{Z}_{n_1}$, $J = \mathbb{Z}_{m_2} \times \mathbb{Z}_{n_2}$, *and* $Q = H \times J$ *satisfy the conditions of* (3.6), *then there is a unique partial homotopy type for* $(H * J, 2)$- *complexes* K *having a specific Euler characteristic* $\chi(K) > \chi_{min}(H * J)$.

Proof: Let L be the 2-complex with minimal Euler characteristic and with $\pi_1(L) = H * J$. Since $\chi(L) < \chi(K)$, then $\chi(K) = \chi(L \vee nS^2)$ for $n > 0$. We will compute the Browning obstruction to partial equivalence using $L \vee nS^2$. We may always construct a map $h : L \vee nS^2 \to L \vee nS^2$ which is the identity on the 1-skeleton and multiplication by μ on one of the S^2's for any $\mu \in \mathbb{Z}G$. In particular if $\mu = r$ is any integer relatively prime to the order of Q, $[r] \in Aut_u(\overline{L \vee nS^2})$. Given any unit in $\mathbb{Z}_uQ$, multiply it by an r to get it into $\mathbb{Z}Q$, then lift it to $\mathbb{Z}G$. It will then be realizable. Therefore, all torsion elements of $K_1(\mathbb{Z}Q, u)$ are actually in $\tau(Aut_u(\overline{L \vee nS^2}))$.   □

The above corollary is of particular interest, since there are examples where $\chi_{min}(H * J) < \chi_{min}(H) + \chi_{min}(J)$, [HoLuMe85]. This is the case if we let m_1, m_2, r_1, r_2, n_1, and n_2 be integers with $r_i > 1$, $r_i^{m_i} - 1 = n_i \cdot q_i$, $(q_1, q_2) = 1$, $r_i \equiv 1 \bmod n_i$, and $(m_i, n_i) \neq 1$. For example, let $m_1 = n_1 = 2$, $m_2 = n_2 = 3$, $r = 9$. The examples of [HoLuMe85] have been shown to have the same homology type (Metzler, private communication). Partial homotopy type may be able to distinguish homotopy types.

Remark 3.8 *Let G be any finite group satisfying Eichler's condition. Let $\rho : G \to Gl(n, \mathbb{Z})$ be a representation. The representations are in one-to-one correspondence with the semi-direct products $G \ltimes \mathbb{Z}^n$. If $\theta : G \ltimes \mathbb{Z}^n \to G$, then $ker(\theta)$ will have free second homology. Therefore, we may use the theory of partial homotopy equivalences to try to distinguish the homotopy types of 2-complexes with fundamental group $G \ltimes \mathbb{Z}^n$.*

3.2   Results when π_1 is a free product of cyclic groups

Among infinite groups, homotopy classification has, as far as we know, only been achieved for free groups:

Theorem 3.9 *Every compact, connected 2-complex with free fundamental group is homotopy equivalent to a finite bouquet of 1- and 2-dimensional spheres* (see [Wa65], Proposition 3.3).

It is not within the bounds of possibility of this section to give all the arguments for Theorem 3.9, but we are going to give a guide to the framework for a proof.

It clearly suffices to establish the assertion for polyhedra which arise by varying given 2-complexes within their 3-deformation types. Hence by Chapter I, without loss of generality, we may deal with standard complexes K of finite presentations of a free group.

There is a bouquet L of 1-spheres and 2-spheres having the Euler characteristic of K and a map $f : K \to L$ inducing an isomorphism of fundamental groups. Moreover, suitable 3-deformations augment L by a sum of 2-discs and modify K to yield (Exercise):

- The 1-skeleta of K and L coincide.

- There are maps $f : K \to L$ and $g : L \to K$ whose restriction to the 1-skeleta is the identity map.

That gives us the commutative diagram for the chain complexes of the universal coverings:

$$
\begin{array}{ccccccccccc}
0 & \to & H_2(\tilde{K}) & \to & C_2(\tilde{K}) & \to & C_1(\tilde{K}) & \to & C_0(\tilde{K}) & \to & \mathbb{Z} & \to & 0 \\
(1) & \| & & \downarrow\uparrow & & \tilde{f}_2 \downarrow\uparrow \tilde{g}_2 & & \| & & \| & & \| & & \| \\
0 & \to & H_2(\tilde{L}) & \to & C_2(\tilde{L}) & \to & C_1(\tilde{L}) & \to & C_0(\tilde{L}) & \to & \mathbb{Z} & \to & 0
\end{array}
$$

Now L^2 is the standard complex of the presentation

$$
\mathcal{P} = \langle a_1, \ldots, a_n \mid a_1, \ldots, a_k, 1, \ldots, 1 \rangle,
$$

with m relations. Its second boundary operator is just the projection onto $(\mathbb{Z}G)^k$, the direct summand of $C_1(\tilde{L})$ generated by those 1-chains that are associated to the trivial generators of $\mathcal{P}$. The second chain group splits into the sum $C_2(\tilde{L}) = (\mathbb{Z}G)^k \oplus (\mathbb{Z}G)^{m-k}$ where the second summand equals $H_2(\tilde{L})$.

By commutativity of diagram (1), the second boundary operator in the upper row is also the projection onto $(\mathbb{Z}G)^k$, and the sequence splits. Hence

(2) $H_2(\tilde{K})$ is a finitely generated stably free module.

Were $H_2(\tilde{K})$ actually free, it would have rank $m - k$. Then the sum of the isomorphism of the $(\mathbb{Z}G)^k$ summands of the second chain groups with any isomorphism $H_2(\tilde{K}) \cong (\mathbb{Z}G)^{m-k}$ would define a chain map $C_2(\tilde{K}) \to C_2(\tilde{L})$, yielding the situation of diagram (1) with $\tilde{f}_2$ an isomorphism. This would suffice to show that K and L are homotopy equivalent.

So far, we have reduced the proof of Theorem 3.9 to the question whether

(3) stably free modules over the group ring $\mathbb{Z}G$ of the free group are free.

Although in Section 2, *localization* of the group ring was the right tool, in proving (3) *tensoring with field coefficients* turns out to be helpful: The analogous statement to (3) for the group ring QG of the free group G with coefficients in a field Q can be established (see, [Ho-An90$_1$] for a topological proof and the bibliography there), and then lifted to $\mathbb{Z}$-coefficients [Ba64]. So the sketch of the proof of Theorem 3.9 is complete.

For general free products G of cyclic groups, there are the standard presentations $\mathcal{P} = \langle a_1, \ldots, a_n \mid a_1^{m_1}, \ldots, a_k^{m_k}, 1, \ldots, 1 \rangle$ with associated standard complex L^2. Let K^2 be any compact connected 2 - complex with $\pi_1(K) = G$. Based on work of G. Bergman [Be74] about free products of rings, Hog-Angeloni [Ho-An90$_2$] proved that tensoring with field coefficients still yields the situation of diagram (1) with $\tilde{f}_2$ an isomorphism. Furthermore, $\tilde{f}_2$ can be chosen to have trivial Whitehead torsion. Whether the result can be lifted, as in the case of a free group, to $\mathbb{Z}$-coefficients or whether there is an integer obstruction, remains a topic for further research. For some partial results, see [Lat86].

3.3   Trees of homotopy types, simple-homotopy types, and 3-deformation types

The concluding subsection of this chapter will illuminate what has been built up further on the chain of implications (41) of Chapter I: For finite, connected, CW-complexes K^2 and L^2,

$$K^2 \overset{3}{\wedge} L^2 \underset{?}{\overset{?}{\rightleftarrows}} K^2 \wedge L^2 \rightleftarrows K^2 \simeq L^2 \rightrightarrows$$

$$\pi_1(K^2) \approx \pi_1(L^2) \text{ and } \chi(K^2) = \chi(L^2).$$

So far we have dealt only with a systematic study of the obstructions to reversing the last implication.

For a fixed group G, consider the (directed) tree of homotopy types (see [DySi73]) whose vertices consist of homotopy types of 2-complexes where the type of a 2-complex K is joined by an edge to the type of its sum (one-point union) $K \vee S^2$, with the 2-sphere S^2. The trees of simple-homotopy and 3-deformation types are defined analogously. Of special interest in each of these trees are the roots and the junctions. The roots are the types that do not admit a factorization involving an S^2 summand. They generate the rest of the types above them in the tree by the operation of forming sums with S^2. The junctions are the types that admit two or more inequivalent factorizations involving an S^2 summand. Each junction is a 2-dimensional instance of non-cancellation of the 2-sphere S^2 with respect to the sum operation.

For example, by Theorem 3.9, the tree of homotopy types for a free group is a bamboo stalk with no junctions and with a single root determined by the type of the sum $\vee S^1$ of copies of the 1-sphere S^1, see Fig 1a. Moreover, since the Whitehead group of a free group is trivial, the trees of homotopy type and simple-homotopy type coincide.

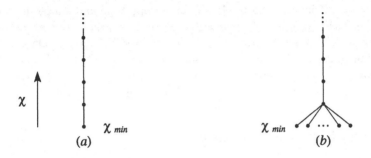

Figure III.1. Homotopy trees

By Section 2, the tree for the finite abelian group $G \approx \mathbb{Z}_{m_1} \times \ldots \times \mathbb{Z}_{m_n}$ has the form of Figure 1 with $|\mathbb{Z}_{m_1}^* / \pm (\mathbb{Z}_{m_1}^*)^{n-1}|$ roots represented by the twisted presentations. Again, the tree of homotopy types coincides with the tree of simple-homotopy types, see [La86].

Actually, Figure 1b shows the shape of the tree of homotopy types for all finite groups: From the Jordan-Zassenhaus-theorem [Sw70], it follows that there are only finitely many types at each level, and Browning's Theorem [Br79$_3$] proves that there is a unique homotopy type at each level above χ_{min}. As for the tree of simple-homotopy types, it is an open problem whether the number of types at the minimum Euler characteristic is finite, but Kreck-Hambleton [HaKr92$_1$] show that there is only one simple-homotopy type above χ_{min}; see Chapter IX, §1.

For infinite groups, rather different phenomena arise:

- Dunwoody [Du76] exhibited an example of a root above χ_{min} - level.

- Lustig [Lu93] (also compare [LuPr92]) discovered infinitely many distinct homotopy types at the same level of χ.

- Metzler [Me90] and Lustig [Lu91] give examples of homotopy equivalent, but simple-homotopy inequivalent 2-complexes; compare Chapter VII, §5.6, and Chapter XII, §3.2.

For a study, to which extent the tree of (3-deformation, simple- homotopy, respectively) homotopy types of the free product is determined by the corresponding trees of the factors, see Section 3 of Chapter XII. For a survey on trees of homotopy types of general (π, m) - complexes see [Dy79].

What about the trees of 3-deformation types? The converse of the first implication in (41) of Chapter I behaves differently from the others. While for $\pi_1 = \{1\}$, it is not hard to classify homotopy type and simple-homotopy type, there are examples of contractible 2-complexes that are conjectured to represent exotic 3-deformation types (see Chapter XII, §1.1). On the other hand, the generalized Andrews-Curtis Conjecture (see Chapter I, (30)) states that simple-homotopy type and 3-deformation type coincide. Chapter XII is dedicated to a discussion of the status of research on this problem.

3.4   Problems for Chapter III

1. Complete the (simple-) homotopy classification for $\pi_1 =$ free product of cyclic groups.

2. (Problem a) of [Ho-AnLaMe90]) For other classes of finite groups (see above for the abelian class), give a complete set of representatives for the vertices at the minimum Euler characterisitic of the tree of (simple-) homotopy types

3. (Problem b) of [Ho-AnLaMe90]) The examples of [Me90] and [Lu91] had infinite fundamental group. Does there exist a pair of homotopy equivalent 2-complexes with finite fundamental group that are not simple-homotopy equivalent?

4. (Problem c) of [Ho-AnLaMe90]) Is it possible to get *infinitely* many compact connected 2-complexes which are pairwise simple-homotopy distinct, but which all have the same homotopy type?

5. (compare Problem b) of [Dy79]) Add to the list of groups with a single minimal root in the (3-deformation-, simple)-homotopy tree.

6. (compare Problem c) of [Dy79]) Do there exist 2-complexes K^2, L^2 such that $K \vee 2S^2 \simeq L \vee 2S^2$ but $K \vee S^2 \not\simeq L \vee S^2$?

Chapter IV

Crossed Modules and Π_2 Homotopy Modules

Micheal N. Dyer

1 Introduction

This chapter is partly an introduction to crossed modules, with emphasis on the role that they play in the study of 2-complexes, and an introduction to various identity properties.

Recall from Chapter II that, if $X \subseteq Y$ are topological spaces, then the boundary map $\partial : \pi_2(Y, X) \to \pi_1(X)$ is well known to be a $\pi_1(X)$-crossed module. We will call this the *crossed module associated with the pair* (X, Y).

In Section 2, we will study projective and free crossed modules, in particular J. Ratcliffe's characterization of these modules. We also show that, if (Y, X) is a pair of 2-complexes, then the crossed module associated to it is projective. This is definitely not true if the pair is not 2-dimensional. Further, we characterize when the kernel of a projective crossed module is trivial in homological terms.

In Section 3, we study the coproduct of crossed modules. The purpose here is to demonstrate how the second homotopy module of a 2-complex can be built up from subcomplexes.

In Chapter II, it was shown that a 2-complex is aspherical iff it satisfies the identity property. Let X be a 2-complex with fundamental group G. If N is a subgroup of G, let X_N denote the covering of X corresponding to N. In

3.3, we study a generalized N-identity property and show how it is equivalent to the vanishing of the Hurewicz map $h : \pi_2(X) \to H_2(X_N)$, the so-called N-Cockcroft property.

2   Crossed and Precrossed Modules

A G-crossed module consists of a triple (C, ∂, G), where C and G are groups with G acting on C (on the left, with action denoted by $g \cdot c$ for $g \in G$ and $c \in C$) and $\partial : C \to G$ a homomorphism of groups satisfying

(CM1) $\partial(gc) = g(\partial c)g^{-1}$ for all $g \in G, c \in C$, and

(CM2) $cdc^{-1} = (\partial c) \cdot d$ for all $c, d \in C$.

Thus, if G acts on itself by conjugation action, then (CM1) says that ∂ is a G-homomorphism.

It follows easily from the definition that $K = \ker \partial$ is contained in the center of G and that $N = \mathrm{im}\, \partial$ is a normal subgroup of G. Let $Q = G/N$. Then, because N acts on C via conjugation action by a pre-image under ∂ (by (CM2)) and because the conjugation action induces the trivial homomorphism on the homology groups $H_i C$, we see that these homology groups are $\mathbb{Z}Q$-modules. Thus, C_{ab}, the abelianization of C, is a $\mathbb{Z}Q$-module . So we have the central extension

$$0 \to K \to C \xrightarrow{\partial} N \to 1.$$

Obvious and important special cases are: (1) the case where C is a $\mathbb{Z}G$-module (so $\partial = 0$) and (2) the case where C is a normal subgroup of G (so ∂ is the inclusion).

As in Example 1 in §3.2, Chapter II, let m be a positive integer, $\mathcal{P}$ denote the presentation $\langle x \mid x^m \rangle$ of the cyclic group C_m, and X be its corresponding 2-complex. Let S denote the element of $\mathbb{Z}C_m$ given by the sum $S = 1 + x + \cdots + x^{m-1}$ and $IC_m = \mathbb{Z}Q(x-1)$ be the so-called augmentation ideal in $\mathbb{Z}Q$, which in this case is generated by the single element $(x-1)$. Then the cellular chain complex of the universal cover $\tilde{X}$ looks like $C_i \tilde{X} = \mathbb{Z}C_m$ for $i = 0, 1, 2$ with $\partial_2 : C_2 \tilde{X} \to C_1 \tilde{X}$ given by multiplication by S and ∂_1 multiplication by $(x - 1)$. In this case $\pi_2(X) \approx IC_m$. Furthermore, if $X^{(1)}$ denotes the one-skeleton of X, then $\pi_2(X, X^{(1)}) \approx \mathbb{Z}C_m$. This is a rare occurrence of an abelian relative homotopy group $\pi_2(Y, Z)$.

A *morphism* (α, β) from the crossed module (C, ∂, G) to (C', ∂', G') is a pair of group homomorphisms $\alpha : C \to C'$ and $\beta : G \to G'$ so that the following diagram commutes :

$$
\begin{array}{ccc}
C & \xrightarrow{\alpha} & C' \\
\partial \downarrow & & \downarrow \partial' \\
G & \xrightarrow{\beta} & G'
\end{array}
$$

for which $\alpha(g \cdot c) = \beta(g) \cdot \alpha(c)$ $(g \in G, c \in C)$. Let $\mathcal{CM}$ denote the category of crossed modules. .

If, in the above diagram, β is the identity on $G = G'$, we say that α is a *G-crossed module homomorphism* or *G-morphism* and denote this category by $\mathcal{CM}_G$.

A *pre-crossed module* is a triple (C, ∂, G) as above satisfying just (CM1). There are similar categories $\mathcal{PCM}$ (respectively, $\mathcal{PCM}_G$) of pre-crossed modules (respectively, pre-crossed modules over G).

Let (E, ∂, G) be a pre-crossed module. Then we can define the *identity* subgroup of E to be $I = \ker \partial$. Furthermore, we let P (the so-called *subgroup of Peiffer identities*) be the normal closure (in E) of the set

$$
W = \{ xyx^{-1}\partial(x) \cdot y \,|\, x, y \in E \}.
$$

The elements of W are called Peiffer elements . Of course, Peiffer elements are identities and the triple $(E/P, \partial, G)$ is a crossed module.

2.1 Free crossed modules

A *G*-crossed module $\partial : C \to G$ is called a *free* crossed module with indexed basis $\{c_\alpha : \alpha \in \mathcal{A}\} \subseteq C$ if it satisfies the following universal property: given a G'-crossed module $\partial' : C' \to G'$, an indexed subset $\{c_\alpha' : \alpha \in \mathcal{A}\} \subseteq C'$ and homomorphism $\tau : G \to G'$ such that $\tau(\partial(c_\alpha)) = \partial'(c_\alpha')$ for each $\alpha \in \mathcal{A}$, then there is a unique homomorphism $\eta : C \to C'$ such that $\eta(c_\alpha) = c_\alpha'$ for each $\alpha \in \mathcal{A}$ and the pair (η, τ) is a morphism of crossed modules.

Examples of free crossed modules are ordinary free modules over a group, free groups, knot groups, and higher dimensional link groups (all with the obvious bases). Indeed, because a knot group G has an element $c \in G$ such that the normal closure $\ll c \gg_G$ of c in G is all of G , the group G is a free G-crossed module with basis $\{c\}$.

J. H. C. Whitehead proved in [Wh49$_2$] (see Chapter II, Theorem 2.9) that if X is obtained from A by attaching only 2-cells, then the crossed module associated with the pair (X, A) is free with basis corresponding to the homotopy classes of the attached 2-cells. He also showed that any free crossed module can be constructed this way.

For the reader's convenience, we repeat the standard construction of the (unique, up to isomorphism) free crossed module associated to the data: let G be a group and $\{g_\alpha\}_{\alpha \in A}$ be an indexed set of elements in G . Let E be the free group on $G \times A$. The group G acts on E via $g \cdot (x, \alpha) = (gx, \alpha)$. We let the map $\hat{\partial} : E \to N = \ll\{g_\alpha\}\gg_G$ be defined by $\hat{\partial}(x, \alpha) = xg_\alpha x^{-1}$. Notice that the triple $(E, \hat{\partial}, G)$ is a pre-crossed module, called the *free pre-crossed module determined by* G *and the set* $\{g_\alpha\}$. Let P be the subgroup of Peiffer identities in E; that is to say, P is the normal closure in E of the set W of *basic* Peiffer elements:

$$W = \{(x, \alpha)(y, \beta)(x, \alpha)^{-1}(xg_\alpha x^{-1}y, \beta)^{-1} | x, y \in G \text{ and } \alpha, \beta \in A\}.$$

Let the group $C = E/P$ and the homomorphism $\partial : C \to G$ be defined by the map $\hat{\partial}$. Let $p : E \to C$ be the natural projection and, for each $\alpha \in A$, let $c_\alpha = p(1, \alpha)$. Then (C, ∂, G) is a free crossed module with basis $\{c_\alpha\}$.

The subgroup $I = \ker \hat{\partial}$ are identities among the boundaries $\{g_\alpha\}$. A typical element

$$\prod_{i=1}^{n}(x_i, \alpha_i)^{\epsilon_i}$$

of E is an identity if and only if

$$\prod_{i=1}^{n} x_i g_{\alpha_i}^{\epsilon_i} x_i^{-1} = 1$$

in G. Recall that $P \subseteq I$. A slight generalization of Theorem 2.6 in Chapter II shows that $K = \ker \partial = I/P$.

The following lemma is the key to studying free crossed modules.

Lemma 2.1 (Ratcliffe [Ra80]) $P \cap [E, E] = [E, I]$.

Proof: : Suppose that $e \in E$ and $i \in I$. Then $p(i) \in \ker \partial$ which is in the center of C. Hence, $p[e, i] = [p(e), p(i)] = 0$, so $[e, i] \in \ker p$. This shows that $[E, I] \subset P \cap [E, E]$.

To see the reverse inclusion, suppose that $q \in P \cap [E, E]$. We write

$$q = \prod_{i=1}^{n} t_i w_i^{\epsilon_i} t_i^{-1},$$

where $w_i \in W, t_i \in E, \epsilon_i = \pm 1$.

Write
$$w_i = (x_i, \alpha_i)(y_i, \beta_i)(x_i, \alpha_i)^{-1}(x_i g_i x_i^{-1} y_i, \beta_i)^{-1}, \quad g_i = g_{\alpha_i}.$$

Modulo $[E, P] \subset [E, I], q \equiv q'$ where $q' = \prod_{i=1}^n w_i^{\epsilon_i}$. Since $q \in [E, E]$, the totality of the factors $(y_i, \beta_i)^{\epsilon_i}$ and $(x_i g_i x_i^{-1} y_i, \beta_i)^{-\epsilon_i}$ must fall into inverse pairs. First, suppose that for no proper subset of $\{w_1, \ldots, w_n\}$ do the $(y_i, \beta_i)^{\epsilon_i}$ and $(x_i g_i x_i^{-1} y_i, \beta_i)^{-\epsilon_i}$ fall into inverse pairs.

Given $u = (x, \alpha)^\epsilon$ with $\epsilon = \pm 1$ and $v = (y, \beta)$, let $v^* = (x g_\alpha^\epsilon x^{-1} y, \beta)$ and $[u; v] = uvu^{-1}(v^*)^{-1}$. The element $[u; v]$ is called a *crossed commutator*. Observe that $[u; v]^{-1} = v^* u v^{-1} u^{-1}$, so

$$u^{-1}[u; v]^{-1} u = u^{-1} v^* u v^{-1} = [u^{-1}; v^*];$$

hence, $[u; v]^{-1} \equiv [u^{-1}; v^*]$ modulo $[E, P]$. Therefore, $q' \equiv q''$ mod $[E, P]$, where $q'' = \prod_{i=1}^n [u_i; v_i]$ with

$$u_i = (x_i, \beta_i)^{\epsilon_i}, \quad v_i = (z_i y_i, \beta_i), \quad v_i^* = (x_i g_i^{\epsilon_i} x_i^{-1} z_i y_i, \beta_i),$$

and

$$z_i = \begin{cases} 1 & \text{if } \epsilon_i = 1 \\ x_i g_i x_i^{-1} & \text{if } \epsilon_i = -1 \end{cases}$$

Modulo $[E, E]$, $q'' \equiv \prod_{i=1}^n v_i \cdot \prod_{j=1}^n (v_j^*)^{-1}$. But $q'' \in [E, E]$ implies that for each i, $v_i = v_j^*$ for some j. Modulo $[E, P]$, we may rearrange the terms of q''. This allows us to reindex so that $v_i = v_{i+1}^*$ (i modulo n). Therefore, all the β_i are the same, say $\beta_i = \beta$, and $z_i y_i = x_{i+1} g_{i+1}^{\epsilon_{i+1}} x_{i+1}^{-1} z_{i+1} y_{i+1}$ (i modulo n). Hence,

$$y_n = z_n^{-1} \left(\prod_{i=1}^n x_i g_i^{\epsilon_i} x_i^{-1} \right) z_n y_n$$

and

$$\prod_{i=1}^n x_i g_i^{\epsilon_i} x_i^{-1} = 1.$$

Thus, $\imath = \prod_{i=1}^n (x_i, \alpha_i)^{\epsilon_i}$ is an identity.

To say again what we have done, we have

$$q'' = \prod_{i=1}^n [u_i; v_i]$$

and

$$[u_i; v_i] = (x_i, \alpha_i)^{\epsilon_i} (z_i y_i, \beta)(x_i, \alpha_i)^{-\epsilon_i} (x_i g_i^{\epsilon_i} x_i^{-1} z_i y_i, \beta)^{-1}.$$

Let
$$r_i = (x_1, \alpha_1)^{\epsilon_1} \cdots (x_{i-1}, \alpha_{i-1})^{\epsilon_{i-1}}$$
and
$$q_i = r_i[u_i; v_i]r_i^{-1};$$
then $q'' \equiv q_n \ldots q_1 \bmod [E, P]$. Observe that

$$q_i = r_{i+1}(z_i y_i, \beta) r_{i+1}^{-1} r_i (x_i g_i^{\epsilon_i} x_i^{-1} z_i y_i, \beta)^{-1} r_i^{-1}.$$

The product $q_n \cdots q_1$ telescopes, since $x_i g_i^{\epsilon_i} x_i^{-1} z_i y_i = z_{i-1} y_{i-1}$, giving

$$q_n \cdots q_1 = \imath(z_n y_n, \beta) \imath^{-1} (x_1 g_1^{\epsilon_1} x_1^{-1} z_1 y_1, \beta)^{-1} = [\imath, (z_n y_n, \beta)].$$

Therefore, $q_n \cdots q_1 \in [E, I]$. Hence, $q \equiv q_n \cdots q_1 \bmod [E, P]$ implies that $q \in [E, I]$.

In the general case we may partition $\{w_1, \cdots, w_n\}$ into subsets $\{S_j\}_{j=1}^k$ with the property that the totality of the factors $(y_i, \beta_i)^{\epsilon_i}$ and $(x_i g_i x_i^{-1} y_i, \beta_i)^{-\epsilon_i}$ fall into inverse pairs within some S_j, and which are minimal with respect this property. Then apply the above argument to show that $q \equiv \prod_{j=1}^k [\imath_j, t_j] \bmod [E, P]$ with $\imath_j \in I$ and $t_j \in E$. This proves that $[E, I] \supset P \cap [E, E]$.   □

We note that, in the special case that there is a splitting of the homomorphism $\partial : C \to N$, Ratcliffe's lemma follows from an argument given in [Re49]. See the proof of Lemma 2.4 in Chapter II. Another proof of Lemma 2.1 is due to G. Ellis and T. Porter [ElPo86].

2.2   A characterization of free crossed modules

The following theorem of John Ratcliffe [Ra80] gives a useful homological characterization of free crossed modules.

Theorem 2.2 *If (C, ∂, G) is a crossed module, $N = \operatorname{Im} \partial, Q = G/N$, and $\{c_\alpha\}$ is an indexed subset of C, then C is a free crossed module with basis $\{c_\alpha\}$ if and only if*

1. *C_{ab} is a free $\mathbb{Z}Q$-module with basis $\{\bar{c}_\alpha\}$,*

2. *N is the normal closure of $\{\partial c_\alpha\}$, and*

3. *$\partial_* : H_2(C) \to H_2(N)$ is trivial.*

Proof: Suppose that C is a free crossed module with basis $\{c_\alpha\}$. Properties (1) and (2) were observed by Whitehead [Wh49$_2$] and follow easily from the construction of the free crossed module on the set $\{\partial c_\alpha\}$. This is because, if C is free, then $C \approx E/P$, so clearly (2) is true. The exact sequence of groups $1 \to P \to E \to C \to 1$ gives rise to the Stallings-Stammbach 5-term exact sequence [HiSt71] in homology:

$$H_2(E) \to H_2(C) \to \mathbb{Z} \otimes_{\mathbb{Z}C} H_1(P) \to H_1(E) \to H_1(C) \to 0.$$

Note that $H_1(E) \approx \mathbb{Z}G^{\mathcal{A}}$, the direct sum of $|\mathcal{A}|$-copies of the group ring, and that we may, from the definition of P, identify the image of $\mathbb{Z} \otimes_{\mathbb{Z}C} H_1(P)$ in $H_1(C)$ as isomorphic to $IN \cdot \mathbb{Z}G^{\mathcal{A}}$. Hence,

$$H_1(C) \approx (\mathbb{Z}G^{\mathcal{A}})/(IN \cdot \mathbb{Z}G^{\mathcal{A}}) \approx \mathbb{Z}Q^{\mathcal{A}}.$$

This shows (1) and (2).

To see that (3) holds, we use the 5-term exact sequence

$$H_2(C) \to H_2(N) \to K \to H_1(C) \to H_1(N) \to 0.$$

coming from the central extension $0 \to K \to C \to N \to 1$. Noting that the exact sequence above is a presentation for C and that the extension

$$1 \to I \to E \to N \to 1$$

is a presentation for N, we may identify, by the formula of Hopf, $H_2(C) \approx P \cap [E,E]/[E,P]$ and $H_2(N) \approx I \cap [E,E]/[E,I]$. Furthermore, $K \approx I/P$. Then, by Lemma 2.1, the homomorphism $\partial_* : H_2(C) \to H_2(N)$ is trivial.

Conversely, suppose that C and $\{c_\alpha\}$ satisfy (1) - (3). Let C' be the free crossed module with basis $\{c'_\alpha\}$ so that $\partial c'_\alpha = \partial c_\alpha$ for all α. By (2), Im $\partial' = N$. Because C' is free, there is a G-morphism $\phi : C' \to C$ such that $\phi(c'_\alpha) = c_\alpha$ for each α. The equality $\partial' = \partial \circ \phi$ shows that ϕ induces a homomorphism $\phi_0 : K' \to K$.

The 5-term homology sequence is natural, so the following diagram commutes

$$
\begin{array}{ccccccccc}
0 & \to & H_2(N) & \to & K' & \to & H_1(C') & \to & H_1(N) & \to & 0 \\
& & \| & & \downarrow \phi_0 & & \downarrow \bar{\phi} & & \| & & \\
0 & \to & H_2(N) & \to & K & \to & H_1(C) & \to & H_1(N) & \to & 0.
\end{array}
$$

The first zeros in the above diagram come from (3).

By (1), $\bar{\phi}$ is an isomorphism; hence, by the 5-lemma ϕ_0 is an isomorphism. Observe that the following diagram commutes

$$
\begin{array}{ccccccccc}
0 & \to & K' & \to & C' & \to & N & \to & 1 \\
& & \downarrow \phi_0 & & \downarrow \phi & & \| & & \\
0 & \to & K & \to & C & \to & N & \to & 1.
\end{array}
$$

Hence, ϕ is an isomorphism by the short 5-lemma. Therefore, C is free with basis $\{c_\alpha\}$.   □

2.3   Projective crossed modules

Fix a group G. A G-crossed module C is said to be *projective* if it is projective in the category $\mathcal{C}\mathcal{M}_G$, that is to say, for any surjective morphism of G-crossed modules $f : A \to B$ and any G-morphism $\mathcal{C}\mathcal{M}_G$ $g : C \to B$, there exists a morphism $h : C \to A$ such that $f \circ h = g$.

Suppose that C is a G-crossed module and that A is a $\mathbb{Z}Q$-module, where $Q = G/N$, with N equal to the image of the boundary of C. Regard A as a G-module via the canonical projection $G \to Q$. Then a *G-crossed extension* of A by C is an extension

$$E : 0 \to A \overset{\imath}{\to} B \overset{\eta}{\to} C \to 1$$

where B is a G-crossed module, $\imath$ is a G-homomorphism, and η is a G-morphism. Clearly E is a central extension.

Note that $A \times C$ is a G-crossed module with G acting diagonally and with boundary $\hat{\partial} : A \times C \to G$ given by $\hat{\partial}(a,c) = \partial(c)$. We call

$$E_0 : 0 \to A \to A \times C \to C \to 1$$

the trivial G-crossed extension of A by C.

Theorem 2.3 (Ratcliffe)   *If C is a G-crossed module, then the following are equivalent:*

1. *C is projective;*

2. *every G-crossed extension $0 \to A \to B \to C \to 1$ operator splits;*

3. *there is a projective $\mathbb{Z}Q$-module P and a free G-crossed module B such that $P \times C$ is G-isomorphic to B; and*

4. *$H_1(C)$ is a projective $\mathbb{Z}Q$-module and $\partial_* : H_2(C) \to H_2(N)$ is trivial.*

Proof: The equivalence of (1) and (2) is proved in the standard way.

To see that (2) implies (3), let $\{c_\alpha\}$ be a set of operator generators of C, and let B be a free crossed module with basis $\{b_\alpha\}$ such that $\partial b_\alpha = \partial c_\alpha$ for all α. Because B is free, there is a G-morphism $\eta : B \to C$ such that $\eta(c_\alpha) = \eta(b_\alpha)$

for each α. It is clear that η is an epimorphism, because $\{c_\alpha\}$ generates C under the action of G.

Let $P = \ker \eta$. By (2), the G-crossed extension $0 \to P \to B \to C \to 1$ splits with a G-morphism. Therefore, B is G-isomorphic to $P \times C$. Observe that $P \oplus H_1(C) \approx H_1(B)$ as $\mathbb{Z}Q$-modules, so P is a projective $\mathbb{Z}Q$-module.

To see that (3) implies (4), observe that $P \oplus H_1(C) \approx H_1(B)$ implies that $H_1(C)$ is a projective $\mathbb{Z}Q$-module. Let $i : C \to P \times C$ be the natural inclusion. Then $\hat{\partial} \circ i = \partial$ yields the following commutative diagram:

$$
\begin{array}{ccc}
H_2(C) & & \partial_* \\
\downarrow i_* & \searrow & \\
H_2(P \times C) & \xrightarrow{\hat{\partial}_*} & H_2(N).
\end{array}
$$

By Theorem 2.2 (3), $\hat{\partial}_*$ is trivial; hence, ∂_* is trivial. To see that (4) implies (2), suppose that $\phi : C' \to C$ is a G-epimorphism. We claim that $\phi_* : H_2(C') \to H_2(C)$ is also an epimorphism (in fact all we will use is that $H_2(C) \to H_2(N)$ is trivial). Let $K = \ker \partial$ and $K' = \ker \partial'$. Because $\partial \circ \phi = \partial'$, ϕ induces a homomorphism $\phi_0 : K' \to K$. According to T. Ganea in [Ga68], there is a 6-term exact homology sequence (extending the usual 5-term sequence)

$$ K \otimes H_1(C) \to H_2(C) \xrightarrow{\partial_*} H_2(N) \to K \to H_1(C) \cdots . $$

According to U. Stammbach in [St73], the sequence is natural, so the following diagram is commutative :

$$
\begin{array}{ccccccccccc}
K' \otimes H_1(C') & \to & H_2(C') & \to & H_2(N) & \to & K' & \to & H_1(C') & \to & H_1(N) \\
\downarrow \phi_0 \otimes \bar{\phi} & & \downarrow \phi_* & & \| & & \downarrow \phi_0 & & \downarrow \bar{\phi} & & \| \\
K \otimes H_1(C) & \to & H_2(C) & \to & H_2(N) & \to & K & \to & H_1(C) & \to & H_1(N).
\end{array}
$$

Because ϕ is an epimorphism, $\bar{\phi}$ is an epimorphism. The 5-lemma implies that ϕ_0 is a epimorphism. Hence, $\phi_0 \otimes \bar{\phi}$ is an epimorphism. By (4) ∂_* is trivial. Hence the Ganea homomorphism $K \otimes H_1(C) \to H_2(C)$ is an epimorphism. It follows that $\phi_* : H_2(C') \to H_2(C)$ is an epimorphism.

By considering the 5-term exact sequence associated with the (central) extension $0 \to \ker \phi \to C' \xrightarrow{\phi} C \to 1$, we see that $\ker \phi$ injects into $H_1(C')$; therefore, the sequence $0 \to \ker \phi \to H_1(C') \xrightarrow{\bar{\phi}} H_1(C) \to 0$ is short exact. The sequence splits, because $H_1(C)$ is projective. Therefore, the sequence $0 \to \ker \phi \to C' \xrightarrow{\phi} C \to 1$ operator splits. $\qquad\square$

Projective modules over $\mathbb{Z}G$ form examples of projective G-crossed modules. Also, if $N \triangleleft G$ is a normal subgroup of G, then N is a projective (conjugation) crossed module if and only if $H_2(N) = 0$ and $H_1(N)$ is a projective $\mathbb{Z}Q$-module, where $Q = G/N$. A group N is said to be *superperfect* if both $H_1(N)$ and $H_2(N)$ are trivial. Thus, any superperfect normal subgroup N of a group G is a projective crossed module, which is not free as a crossed module. In the next section, we will see that projective crossed modules abound.

2.4   Two-complexes and projective crossed modules

The purpose of this section is to demonstrate that projective crossed modules occur in a very natural setting. Namely, if X is a connected subcomplex of a connected 2-dimensional CW-complex Y, then the boundary map $\partial :$ $\pi_2(Y,X) \to \pi_1(X)$ has the structure of a *projective* $\pi_1(X)$-crossed module. This theorem generalizes a special case of the theorem of J. H. C. Whitehead (Theorem 2.9 in Chapter II), which states that if Y is obtained from X by adding only 2-cells, then the crossed module (C, ∂, G) associated with the pair (Y,X) is a free crossed module. If X is not 2-dimensional and Y is obtained from X by adding 1-cells *and* 2-cells, then it is *not* necessarily true that the crossed module (C, ∂, G) is projective. An example of this is obtained by letting $X = \mathbb{R}P^3$ be the real projective 3-space and Y be obtained from $X \vee S^1$ by attaching a 2-cell so that the 2-skeleton of Y is the realization of the presentation $\langle x, y \mid x^2, [x, y] \rangle$ ([Dy87$_1$]).

Theorem 2.4 (Dy87$_1$) *Let X be a connected subcomplex of a connected 2-complex Y. Then the crossed module associated with the pair (Y, X) is projective in the category of G-crossed modules.*

Proof: Let $\pi_1(X) = G, \pi_1(Y) = H$, and let $i : X \to Y$ be the inclusion map. Let $C = \pi_2(Y, X), Q = \mathrm{im}\ \{\pi_1(i) : G \to H\}$, and $N = \ker\ \{\pi_1(i) : G \to H\}$. We will show that (1) $\partial_* : H_2(C) \to H_2(N)$ is trivial and that (2) $H_1(C)$ is a projective $\mathbb{Z}Q$-module.

(1) Let $Y^{(1)}$ be the 1-skeleton of Y and let

$$\partial' : C' = \pi_2(Y, X \cup Y^{(1)}) \to \pi_1(X \cup Y^{(1)}) = G * F$$

be the associated crossed module, where F is a free group. Let $N' = \mathrm{im}\ \partial'$ and $\alpha : C \to C'$, $\beta : G \to G * F$, and $\delta : N \hookrightarrow N'$ be the maps induced by the inclusion $X \hookrightarrow X \cup Y^{(1)}$. Note that β is the inclusion $G \hookrightarrow G * F$ and that δ is the restriction of β to N. Then we see that $\partial' \circ \alpha = \beta \circ \partial$.

We claim that $N = N' \cap G$, if we identify G with βG. Clearly, $N \subset N' \cap G$. To see the reverse inclusion, observe that N' is the normal closure of words

$[r_\alpha] = \partial e_\alpha$ corresponding to the boundaries ∂e_α of the 2-cells in Y outside those of X. Thus an element $[\omega] \in N' \cap G$ can be written as

$$[\omega] = \prod_{j=1}^{k} a_j r_j^{\epsilon_j} a_j^{-1} \ (a_j \in G * F, \ \epsilon_j = \pm 1)$$

where the loop ω is homotopic, relative to the base point, to a loop ω' in X', the homotopy taking place inside $X \cup Y^{(1)}$.

Then we may build a map $\eta : (B^2, S^1) \to (Y, X)$ as follows: picture the disk B^2 with k pairwise disjoint disks $D_1, \ldots, D_k$ inside, each disk disjoint from the boundary and each disk joined to the basepoint $*$ by a straight line. The picture is that of k balloons (or lollypops) inside B^2 on strings (or sticks) being held by one hand at the basepoint on the boundary of B^2. We build the map η by putting ω' clockwise around the boundary of B^2, and running, from left to right, the loop representing a_j along each of the strings, and mapping each disk D_j to the corresponding 2-cell e_j, with the orientation being determined by ϵ_j. See, for example, Figure 1 of Chapter 6 (where $R \leftrightarrow r$ and $u \leftrightarrow a$). The map on $B^2 - \{lollypops\}$ is that of the homotopy between ω and ω' in $X \cup Y^{(1)}$. Thus, $[\eta]$ is a member of $\pi_2(Y, X)$ whose boundary $\partial[\eta] = [\omega'] \in N$. Thus the claim is proved.

By the theorem of Kurosch [Se80], N is then a free summand of N' and $\delta : N \to N'$ is the inclusion onto that free summand. Applying the functor $H_2(-)$ to the commutative diagram represented by $\partial' \circ \alpha = \beta \circ \partial$ we have the commutative square:

$$\begin{array}{ccc} H_2(C) & \overset{\partial_*}{\to} & H_2(N) \\ \alpha_* \downarrow & & \downarrow \delta_* \\ H_2(C') & \overset{\partial_*'}{\to} & H_2(N'). \end{array}$$

By Theorem 14.2 of [HiSt71] δ_* is a monomorphism. Because $(C', \delta', G * F)$ is a free crossed module, we have $\partial_*' = 0$. Hence, $\delta_* \partial_* = 0$; so $\partial_* = 0$.

(2) Let X_N be the covering of X corresponding to the subgroup N and $\tilde{Y}$ be the universal covering of Y. Consider the pair $(\tilde{Y}, X_N)$, where X_N may be identified with a connected subcomplex of $\tilde{Y}$. Note that both $\tilde{Y}$ and X_N are stable under the action of $Q \leq H = \pi_1(Y)$.

We may identify $C = \pi_2(Y, X)$ with $\pi_2(\tilde{Y}, p^{-1}X)$ using the covering map $p : \tilde{Y} \to Y$. This in turn may be identified with $\pi_2(\tilde{Y}, X_N)$, where X_N is the component of $p^{-1}X$ containing the base point. We will show that $H_2(\tilde{Y}, X_N)$ is the abelianization of $\pi_2(\tilde{Y}, X_N)$ (under the Hurewicz map) by the following

Lemma 2.5 *Let (W, A) be a topological pair with W simply connected and A path connected. Then the Hurewicz map $h : \pi_2(W, A) \to H_2(W, A)$ is*

surjective with the kernel of h equal to the commutator subgroup π_2' of $\pi_2 = \pi_2(W, A)$.

Proof: The map h can be shown to be surjective by a simple diagram chase on the Hurewicz ladder between the exact sequences (in homotopy and homology) for the pair (W, A), using the fact that W is 1-connected. In order to see that $\ker h = \pi_2'$, we use the relative Hurewicz Theorem, which says that $\ker h$ is the normal closure in π_2 of $\{(g \cdot c)c^{-1} | c \in \pi_2, g \in \pi_1(A)\}$. Consider the surjection $\partial_2 : \pi_2(W, A) \to \pi_1(A)$. For any $g \in \pi_1(A)$ there is a $d \in \pi_2$ with $\partial_2 d = g$. Thus $(g \cdot c)c^{-1} = ((\partial_2 d) \cdot c)c^{-1} = dcd^{-1}c^{-1} \in \pi_2'$. Thus, $\ker h \subseteq \pi_2'$; the reverse inclusion is obvious because $H_2(W, A)$ is abelian. $\square$

Clearly, $H_2(\tilde{Y}, X_N)$ has a $\mathbb{Z}Q$-module structure. We will show that this module is in fact a projective $\mathbb{Z}Q$-module. Consider the cellular chain complex $C_*(\tilde{Y}, X_N)$ of free $\mathbb{Z}Q$-modules given by

$$C_2(\tilde{Y})/C_2(X_N) \xrightarrow{\bar{\partial}_2} C_1(\tilde{Y})/C_1(X_N) \xrightarrow{\bar{\partial}_1} C_0(\tilde{Y})/C_0(X_N).$$

Now, X_N is *connected* implies that $H_0(\tilde{Y}, X_N) = 0$, hence $\bar{\partial}_1$ is surjective. Also, $H_1(\tilde{Y}) = 0$, so im $\bar{\partial}_2 = \ker \bar{\partial}_1$. In addition, the $\mathbb{Z}Q$-module $C_0(\tilde{Y})/C_0(X_N)$ is a free module, so the above exact sequence is split over $\mathbb{Z}Q$. Thus, $H_2(\tilde{Y}, X_N) = \ker \bar{\partial}_2$ is a projective $\mathbb{Z}Q$-module, being a direct summand of the free $\mathbb{Z}Q$-module $C_2(\tilde{Y})/C_2(X_N)$. This completes the proof that $\partial : C \to G$ is a projective $\mathbb{Z}Q$-module. $\square$

2.5   The kernel of a projective crossed module

Let the group homomorphism $\partial : C \to G$ denote a G-crossed module (C, ∂, G). Let $K = \ker \partial$. The main result of this section characterizes the triviality of K in terms of the first and second homology of $N = \operatorname{im} \partial$ and the second homology of C. This result is then used to prove the following: Let Y be a 2-complex that does not have the homotopy type of S^2, $Y^{(1)}$ be the 1-skeleton of Y, and let $C = \pi_2(Y, Y^{(1)})$. Then $\pi_2(Y) = 0$ if and only if $H_2(C) = 0$.

A group G is *perfect* if the abelianization $H_1(G) = 0$; G is *superperfect* if both $H_1(G)$ and $H_2(G)$ are trivial.

Theorem 2.6 ([Dy91₂])   *Let (C, ∂, G) be a projective G-crossed module, with $K = \ker \partial, N = \operatorname{im} \partial$, and $Q = G/N$. Let the conditions (i) $H_2(C) = 0$, (ii) $H_2(N) = 0$, and (iii) $H_1(N)$ is a free abelian group, be denoted as hypothesis (**H**). Then*

1. *The kernel $K = 0$ implies hypothesis* (**H**).

2. *If hypothesis* (**H**) *holds, then either the $\mathbb{Z}Q$-module K is trivial or the underlying abelian group K^0 of K is isomorphic to the integers $\mathbb{Z}$.*

3. *If hypothesis* (**H**) *holds, then $K^0 \approx \mathbb{Z}$ if and only if the following are true:*

    (a) *the quotient group $Q = 1$,*

    (b) *the group G is superperfect, and*

    (c) *the abelianization $H_1(C) \approx \mathbb{Z}$.*

Proof: To see (1) recall that the following sequence is exact and central:

$$0 \to K \to C \to N \to 1$$

If $K = 0$, then $C = N$. By Ratcliffe's characterization (2.3) of projective crossed modules, we see that $H_2(C) = H_2(N) = 0$. Similarly, $H_1(C)$ is a projective $\mathbb{Z}Q$-module implies that it is free abelian. Thus, hypothesis (**H**) holds.

To prove (2) and (3), we need three lemmas.

Lemma 2.7 *If $H_2(N) = 0$, then K is a free abelian group.*

Proof: Consider the Stallings-Stammbach sequence associated with the above central extension:

$$H_2(N) \to K \to H_1(C) \to H_1(N) \to 0.$$

Then $H_2(N) = 0$ implies that K is a subgroup of $H_1(C)$, which is a free abelian group.   □

Lemma 2.8 *Let A be any free abelian group. Then $H_2(A) = 0$ if and only if $A = 0$ or $\mathbb{Z}$.*

Proof: Let $\{e_\alpha | \alpha \in I\}$ be an ordered basis for A. The second homology of A is isomorphic to the symmetric product $A \wedge A$ with basis $\{e_\alpha \wedge e_\alpha | \alpha < \beta\}$ (see [Br82]). Thus, $H_2(A) = 0$ iff $A = 0$ or $\mathbb{Z}$. A more elementary proof would be to notice that if A is not 0 or $\mathbb{Z}$ then A contains $\mathbb{Z} \oplus \mathbb{Z}$ as a direct summand. Hence, $H_2(A)$ contains $\mathbb{Z} = H_2(\mathbb{Z} \oplus \mathbb{Z})$ as a direct summand.   □

Lemma 2.9 *Let M be a $\mathbb{Z}Q$-module whose underlying abelian group $M^0 = \mathbb{Z}$. Then M is a projective $\mathbb{Z}Q$-module if and only if $Q = \{1\}$.*

Proof: If Q is infinite, then no module M with $M^0 = \mathbb{Z}$ can (even) be a submodule of a free $\mathbb{Z}Q$-module. For let $m \in M \subseteq F$, where F is a free $\mathbb{Z}Q$-module with basis B, $m = \sum_{i=1}^{n}\{m_i e_i | m_i \in \mathbb{Z}Q, e_i \in B\} \neq 0$. Let the carrier of $n \in \mathbb{Z}Q$ be $\mid n \mid$, which is the finite set of all $q \in Q$ with non-zero coefficients in n. We may then choose $x \in Q$ so that $\mid m_i \mid \neq \mid xm_i \mid$ for some $i \in \{1, \ldots, n\}$. This shows that M cannot be a submodule of a free module, provided Q is infinite.

If Q is finite and M is a projective $\mathbb{Z}Q$-module with $M^0 = \mathbb{Z}$, then a theorem of R. Swan ([Br82], page 239) says that the order of Q must divide the $\mathbb{Z}$-rank of M. Thus $\mid Q \mid = 1$. □

In order to prove (2), we assume that hypothesis **(H)** is true. Thus, we have that $H_1(N)$ is free abelian and $H_2(N) = 0$. Notice that the action of N on K is trivial. This implies that the second cohomology group

$$H^2(N, K) \approx Hom(H_2(N), K) \oplus Ext(H_1(N), K) = 0.$$

(See [HiSt71], page 222, for the universal coefficient theorem.) So the central extension $0 \to K \to C \to N \to 1$ splits, as groups only, yielding $C \approx N \times K$. Then $H_2(C) = 0 = (H_1(N) \otimes K) \oplus H_2(K)$ (See [HiSt71], page 222, for the Künneth theorem). If $H_1(N) \neq 0$, then $H_1(N)$ is a free abelian group implies that $H_1(N) \otimes K$ is isomorphic to a direct sum of copies of K, which implies that $K = 0$. If $H_1(N) = 0$, then $H_2(K) = 0$ together with the second lemma imply that $K = 0$ or $\mathbb{Z}$. This proves (2).

To see (3), assume hypothesis **(H)** together with (3a, b, c). We see that $G = N$ so $H_1(N) = 0$. Then, using the Stallings-Stammbach sequence above, $H_1(N) = 0 = H_2(N)$ shows that $K \approx H_1(C) \approx \mathbb{Z}$.

For the converse, assume hypothesis **(H)** together with $K = \mathbb{Z}$. First note that this yields $H_1(N) = 0$ (for otherwise $K = 0$). It follows that $H_1(C) \approx \mathbb{Z}$. By theorem 2.4, $H_1(C)$ is a projective $\mathbb{Z}Q$-module. Lemma 2.9 then shows that $Q = 1$. This completes the proof of (3) in 2.6. □

Now for the application to aspherical 2-complexes. Recall that a connected 2-complex X is aspherical iff $\pi_2(X) = 0$.

Corollary 2.10 *Let Y be a connected 2-dimensional CW-complex with 1-skeleton $Y^{(1)}$. Let $C = \pi_2(Y, Y^{(1)})$, and assume that Y does not have the homotopy type of the 2-sphere S^2. Then the following are equivalent:*

1. Y is aspherical;

2. $H_2(C) = 0$; and

3. C is a free group.

Proof: We let (C, ∂, G) denote the free crossed module

$$\partial_2 : \pi_2(Y, Y^{(1)}) \to \pi_1(Y^{(1)}).$$

Because $N = \text{im } \partial$ is a subgroup of the free group G, then $H_1(N)$ is free abelian, and $H_2(N) = 0$ are automatic. Because $Y^{(1)}$ is a 1-complex, we have $K = \ker \partial = \pi_2(Y)$. Then, part (1) of the above theorem shows that $K = 0$ implies that $H_2(C) = 0$. If $H_2(C) = 0$ and $\pi_2(Y) = 0$, then, by part (3) of the above theorem, we have G is free and superperfect, so $G = \{1\}$. Hence, $Q = \pi_1(Y)$ is trivial. But then Y is a simply connected 2-complex with $\pi_2(Y) = \mathbb{Z}$, hence $Y \simeq S^2$, which was forbidden. Clearly (1) implies (3), because $C = N$. If C is free and Y is not aspherical, then an argument similar to the above shows that $Y \simeq S^2$. $\square$

We contrast this with the **identity property** encountered in Lemma 2.4 of Chapter II (see also Section 4 of this chapter). Assume that Y is the realization $K(\mathcal{P})$ of the presentation $\mathcal{P} = \langle \mathbf{x} \mid \mathbf{r} \rangle$ of the group Q. Here the group $G = F(\mathbf{x}) = \pi_1(Y^{(1)})$, where $F(\mathbf{x})$ is the free group on the set $\mathbf{x}$, and $Q = \pi_1(Y) = G/N$, where N is the normal closure of the set $\mathbf{r}$. It is a simple fact (see Chapter II) that any connected 2-complex has the homotopy type of a 2-complex which is the realization of a presentation. Consider the free crossed module (C, ∂, G) associated to the pair $(Y, Y^{(1)})$. Let $K = \pi_2(Y) = \ker \partial$. It follows from the Stallings-Stammbach sequence associated to $0 \to K \to C \to N \to 1$ that Y is aspherical iff $H_1(C) = H_1(N)$ is isomorphic to the free $\mathbb{Z}\pi_1(Y)$-module with basis in one-to-one correspondence to the 2-cells of Y. This is the traditional way of detecting asphericity, known as the *identity property*: For each product of conjugates of elements of $\mathbf{r} \cup \mathbf{r}^{-1}$,

$$\prod_{i=1}^{n} \{g_i(R_i^{\epsilon_i})g_i^{-1}|g_i \in G, R_i \in \mathbf{r}, \epsilon_i = \pm 1\},$$

that freely reduces to $1 \in G$, there is a pairing $i \leftrightarrow j$ of the indices such that $R_i = R_j$, $\epsilon_i = -\epsilon_j$, and $g_i N = g_j N$. This is succinctly encapsulated by the statement that $H_1(N)$ is a free $\mathbb{Z}\pi_1(Y)$-module on the basis $\{R[N, N] | R \in \mathbf{r}\}$.

Thus, the hypothesis **(H)** in 2.6 (with $H_1(N)$ a free abelian group) is an apparent weakening of the identity property, or perhaps, better said, trades one hard problem for another. It is not expected that showing that $H_2(C) = 0$ will be any easier in general, but it might allow one to bring the extensive machinery surrounding the homology of groups to this problem.

3   On the Second Homotopy Module of a 2-Complex

Let $M \to G$ and $N \to G$ be G-crossed modules. In this section we will study the coproduct $M \bowtie N \to G$ of crossed modules. We will prove the theorem of M. Gutierrez and J. Ratcliffe [GuRa81] showing how the second homotopy group of a 2-complex is determined by its subcomplexes. We will also discuss a 2-dimensional version of the Brown-Higgins theorem ([BrHi78], [Br84]) which shows that in many cases the following is true: If K is a connected 2-complex, K_1 and K_2 are connected subcomplexes such that $K = K_1 \cup K_2$ and $K_1 \cap K_2$ is a connected subcomplex K_0 of K, then the relative group $\pi_2(K, K_0)$ is a coproduct $\pi_2(K_1, K_0) \bowtie \pi_2(K_2, K_0)$. The coproduct of two crossed modules was first defined by J. H. C. Whitehead in [Wh41$_1$]. We follow the development of W. Bogley and M. Gutierrez in [BoGu92].

3.1   Coproducts of crossed modules

First, some notation. If M is a G-group, we write $[m, g] = m(g \cdot m^{-1})$ and $[g, m] = (g \cdot m)m^{-1}$. For $R \subset G$, $[M, R]$ is the smallest G-stable subgroup containing the $[m, r]$, and $M_R = M/[M, R]$. Similarly, if K is a $\mathbb{Z}G$-module and $R \subset G$, let $[R, K]$ be the submodule generated by the elements $(r - 1)k$, where $r \in R$ and $k \in K$, and let $K_R = K/[R, K]$.

We will define the coproduct in the category $\mathcal{CM}_G$, due to J. H. C. Whitehead. Let $d : M \to G$ and $d' : N \to G$ be G-crossed modules. To establish notation, we deal with elements $m, m_k, \mu, \ldots$, and $n, n_k, \nu, \ldots$, lying in M and N, respectively. If no confusion arises, we will write dn for $d'n$.

We define the *coproduct* $M \bowtie N$ of two G-crossed modules M and N to be the quotient of the free product $M * N$ by the smallest G-invariant normal subgroup containing the Peiffer relations

$$r(m, n) = mnm^{-1}(dm \cdot n^{-1}) = [m, n][n, dm]$$

and

$$s(m, n) = nmn^{-1}(dn \cdot m^{-1}) = [n, m][m, dn].$$

In other words, in $M \bowtie N$, $[m, n] = [dm, n] = [dn, m]$ for all $m \in M$ and $n \in N$. The map $dd' : M \bowtie N \to G$, given by $dd'(m * n) = dmd'n$, where $m * n \in M * N$, defines a G-crossed module structure on $M \bowtie N$. There are inclusion induced G-maps $i : M \to M \bowtie N$ and $i' : N \to M \bowtie N$.

Lemma 3.1 *The sequence* $1 \to [iM, i'N] \to M \bowtie N \to M_{dN} \times N_{dM} \to 1$ *is exact in the category of groups.*

In particular $M \bowtie N$ abelianizes to

$$H_1(M)_{dN} \oplus H_1(N)_{dM} \approx \mathbb{Z} \otimes_{\mathbb{Z}dN} H_1(M) \oplus \mathbb{Z} \otimes_{\mathbb{Z}dM} H_1(N).$$

We note that $M \bowtie N$ is the push-out in the category $\mathcal{CM}_G$.

The relations $r(m, n)$ and $s(m, n)$ suggest a kind of double semi-direct product (first noted in [Br84]). Consider the semi-direct product

$$M \triangleleft N = M * N / \ll s(m, n), \forall m, n \gg .$$

There is the defining split exact sequence $1 \to M \to M \triangleleft N \to N \to 1$ with the action of N on M given by the identity $s(m, n)$. One then argues that $M \bowtie N$ is the quotient of $M \triangleleft N$ by the normal closure of the identities $r(m, n)$. Note that both M and N embed in $M \triangleleft N$ (so we may refer to m and n as elements of $M \triangleleft N$ without ambiguity) and that dd' induces a G-map $dd' : M \triangleleft N \to G$.

The first four parts of the following lemma shows that all the elements of $M \triangleleft N$ can be written as $\mu\nu$. In particular, in $M \triangleleft N$, $r(m, n) = [m, dn][n, dm]$. We denote the latter by $\{m, n\}$, and the smallest G-normal subgroup containing the $\{m, n\}$ by $\{M, N\}$.

Lemma 3.2 *If* $m, m_k, \mu \in M$, $n, n_k, \nu \in N$, $g \in G$ *and* $\epsilon = \pm 1$ *then the following equalities hold, the first one in N and the remaining in $M \triangleleft N$.*

1. *If* $x \in G$ *we have* $xdn \cdot n^\epsilon = x \cdot n^\epsilon$,

2. $[dm, dn] \cdot \mu = [m, dn](\mu)[dn, m]$,

3. $[dn, dm] \cdot \mu = [dn, m](\mu)[m, dn]$,

4. $nm = (dn \cdot m)n$,

5. $g \cdot \{m, n\} = \{g \cdot m, g \cdot n\}$,

6. $\mu\{m, n\}\mu^{-1} = \{m, n\}$,

7. $\nu\{m, n\}\nu^{-1} = \{d\nu \cdot m, d\nu \cdot n\}$,

8. $\{m, n\}^{-1} = \{dn \cdot m, n^{-1}\}$, *and*

9. $\prod_{k=1}^p \{m_k, n_k\} = (\prod_{k=1}^p [dn_k, m_k])^{-1}(\prod_{k=1}^p [n_k, dm_k])$.

Proof: The first four follow easily from the definitions. Equations (5)-(7) are left to the reader, they are proved in [Br84]. They show that $\{M, N\}$ is the subgroup generated by the $\{m, n\}$. To see (9), it is enough to show the case $p = 2$. By (3) and (4) $\{m, n\}\{\mu, \nu\}$

$$= [m, dn]([dn, dm] \cdot [\mu, d\nu])[n, dm][\nu, d\mu]$$

$$= [m, dn][dn, m][\mu, d\nu][m, dn][n, dm][\nu, d\mu]$$

$$= [\mu, d\nu][m, dn][n, dm][\nu, d\mu]$$

$$= ([dn, m][d\nu, \mu])^{-1}[n, dm][\nu, d\mu].$$

Formula (8) is trickier: $\{dn \cdot m, n^{-1}\}\{m, n\}$

$$= [dn, m]n^{-1}((dndmdn^{-1}) \cdot n)[m, dn][n, dm]$$

by definition. By (1) and (3), $(dndmdn^{-1}) \cdot n = n(dm \cdot n)n^{-1}$ so the expression reduces to $[dn, m][dm, n][m, dn][n, dm]$

$$= [dn, m]([dm, dn] \cdot [m, dn])$$

$$= [dn, m][m, dn][m, dn][dn, m] = 1$$

by (2). Formulas (8) and (9) were first proved in [BoGu92].   □

The importance of (8) is that it shows that all elements of $\{M, N\}$ are products of the $\{m, n\}$, which products are described by (9).

We note that
$$[\ker d, dN] \subset \ker i \subset \ker d \cap [M, dN].$$

These inclusions appear in [Gi92] as well as [BoGu92].

3.2   A special case

In general it is hard to determine $\ker i$ or even whether $[\ker d, dN] \subset \ker i$ is zero. An interesting special case is the following: let M and N be sub-G-crossed modules of a G-crossed module X. Then $M \bowtie N \to X$ has image MN and i is monic. We may thus omit i without ambiguity. The following result is contained in [Br84], (2.7) and (2.11), and [BoGu92],(top of page 47).

Theorem 3.3 *If* M *and* N *are* G-subgroups *of a* G-group X, *then* $\ker (M \bowtie N \to MN) = M \cap N/[M, N].$

Note that $M \cap N$ means intersection in X; in $M \bowtie N, iM \cap i'N = [iM, i'N]$.

Proof: As G/dN-groups, and thus as G-groups, $M \bowtie N/N \approx M/[M,N]$. The following commutative diagram shows that α is an isomorphism.

$$
\begin{array}{ccccccccc}
 & & & & 1 & & 1 & & \\
 & & & & \downarrow & & \downarrow & & \\
 & & & & N & = & N & & \\
 & & & & \downarrow & & \downarrow & & \\
0 & \to & K & \to & M \bowtie N & \to & MN & \to & 1 \\
 & & \downarrow \alpha & & \downarrow & & \downarrow & & \\
0 & \to & M \cap N/[M,N] & \to & M/[M,N] & \to & M/M \cap N & \to & 1 \\
 & & \downarrow & & \downarrow & & & & \\
 & & 1 & & 1 & & & &
\end{array}
$$

$\square$

As pointed out in [Br84], an explicit formula for α_{-1}, deduced from the snake lemma, is $\alpha^{-1}(\mu[M,N]) = i(\mu^{-1})i'(\mu)$.

3.3   On the kernel of a coproduct of crossed modules

If $f : M \to P$ and $g : N \to Q$ are G-morphisms, we have G-morphisms $f_0 : M \to P \to P \bowtie Q$ and $g_0 : N \to Q \to P \bowtie Q$ and so the map $f \bowtie g : M \bowtie N \to P \bowtie Q$ is called the *direct sum map* . Recall that ker $(f \bowtie g)$ must be a $\mathbb{Z}G$-module.

Theorem 3.4 ([BoGu92], Lemma 1.4)   *If f and g are onto, then*

$$\text{ker } (f \bowtie g) = i(\text{ker } f) + i'(\text{ker } g).$$

Proof: If $f(m)g(n) = 1$ in $P \bowtie Q$, then the multi-valued Lemma 3.3 and the hypothesis imply that we can find m_k, n_k so that $f(m) = (\prod[dg(n_k), f(m_k)])^{-1}$ and $g(n) = \prod[g(n_k), df(m_k)]$ because $dg(n) \cdot f(m) = dn \cdot f(m) = f(dn \cdot m)$ and similarly $df(m) \cdot g(n) = g(dm \cdot n)$. There are elements $\sigma \in \text{ker } f$ and $\tau \in \text{ker } g$ so that $m = (\prod[dn_k, m_k])^{-1}\sigma$ and $n = \tau \prod[n_k, dm_k]$ and $m \bowtie n = i(\sigma) + i'(\tau)$ in $M \bowtie N$. $\square$

Note that normal subgroups R and S of G can be considered as G-crossed modules with boundary the inclusion map and G action by conjugation.

We define $I = i(\text{ker } d) \cap i'(\text{ker } d')$.

Theorem 3.5 ([BoGu92], Proposition 1.5)

1. $I = ker(dd') \cap [iM, i'N]$.

2. *If* $I = 0$, *then* ker $i =$ ker $d \cap [M, dN]$ *and* ker $i' =$ ker $d' \cap [N, dM]$ *so* $i(\text{ker } d) + i'(\text{ker } d')$ *is the direct sum*

$$\text{ker } d/(\text{ker } d \cap [M, dN]) \oplus \text{ker } d'/(\text{ker } d' \cap [N, dM]).$$

3. *If* $dd' : M \bowtie N \to G$ *splits as a group homomorphism or if* $H_2(M \bowtie N) \to H_2(\text{im } dd')$ *is onto, then* $I = 0$.

Proof: To prove (1), we observe that $i(\text{ker } d) = $ ker $dd' \cap iM$ and that $i'(\text{ker } d') = $ ker $dd' \cap i'N$. This together with the fact that $iM \cap i'N = [iM, i'N]$ completes the proof.

For (2), we see that if $I = 0$, then ker dd' embeds in $M \bowtie N/[iM, i'N] = M_{dN} \times N_{dM}$ by Lemma 3.1 so we have a short exact sequence

$$0 \to \text{ker } i \to \text{ker } d \to i(\text{ker } d) \subset M_{dN},$$

which shows that ker $i = $ ker (ker $d \to M_{dN}) = $ ker $d \cap [M, dN]$.

To see (3), if $dd' : M \bowtie N \to $ im dd' splits, then it is easy to see that ker $dd' \cap [M \bowtie N, M \bowtie N] = 0$. More generally, if $H_2(M \bowtie N) \to H_2(\text{im } dd')$ is onto, the central extension

$$0 \to \text{ker } dd' \to M \bowtie N \to \text{im } dd' \to 1$$

is a commutator extension and the kernel of dd' is algebraically disjoint from $[M \bowtie N, M \bowtie N]$. $\square$

The following theorem is due to M. Gutierrez (unpublished). An argument of this sort was used in [BoGu92], Theorem 2.3. An example where the sum is not direct appears in [BoGu92], Example 1.6.

Theorem 3.6 *Let* $d : M \to G$ *and* $d' : N \to G$ *be two* G-*crossed modules. Write* $R = dM$, $S = d'N$, *and* $I = i(\text{ker } d) \cap i'(\text{ker } d')$. *We have a sequence*

$$0 \to i(\text{ker } d) + i'(\text{ker } d') \to \text{ker } dd' \to R \cap S/[R, S] \to 0$$

of $\mathbb{Z}G/RS$-*modules which is exact and natural. In general the sum in the left hand term is not direct. If* $H_2(M \bowtie N) \to H_2(\text{im } dd')$ *is onto, then* $I = 0$ *and the sum is direct. In that case,* ker $i = $ ker $d \cap [M, dN]$.

Proof: Consider the commutative diagram of G-crossed modules:

$$
\begin{array}{ccccccccc}
 & & & & & & 0 & & \\
 & & & & & & \downarrow & & \\
 & & 0 & & & & R \cap S/[R,S] & & \\
 & & \downarrow & & & & \downarrow & & \\
0 & \to & i(\ker d) + i'(\ker d') & \to & M \bowtie N & \overset{d \bowtie d'}{\to} & R \bowtie S & \to & 1 \\
 & & \downarrow & & \| & & \downarrow & & \\
0 & \to & \ker dd' & \to & M \bowtie N & \overset{dd'}{\to} & RS & \to & 1 \\
 & & & & & & \downarrow & & \\
 & & & & & & 1 & &
\end{array}
$$

obtained from Lemmas 3.5 and 3.6. The snake lemma then gives the sequence.

$\square$

Finally, we note that if $d : M \to G$ and $d' : N \to G$ are two projective G-crossed modules, then $dd' : M \bowtie N \to G$ is also a projective crossed module. This can be proved using using Ratcliffe's characterization of projective crossed modules (Section 2).

3.4 Computing π_2 from subcomplexes

The following theorem shows how the second homotopy group of a 2-complex is determined by those of its subcomplexes. This was first proved by M. Gutierrez and J. Ratcliffe in [GuRa81], in the case that K_0 is the 1-skeleton of K. The present version contains elements of results from [GuRa81], [Br84], and [BoGu92, Theorem 2.3].

Theorem 3.7 *Let K be a connected 2-complex, K_1 and K_2 be connected subcomplexes such that $K = K_1 \cup K_2$ and $K_1 \cap K_2$ is a connected subcomplex K_0 of K, and let $i_j : K_j \to K$ $(j = 0, 1, 2)$ be the inclusion maps. Assume further that $\pi_2(K, K_0) \approx \pi_2(K_1, K_0) \bowtie \pi_2(K_2, K_0)$ as a $\pi_1(K_0)$-crossed module. Then there is an exact sequence of $\pi_1(K)$-modules*

$$0 \to i_{1*}\pi_2(K_1) + i_{2*}\pi_2(K_2) \overset{\zeta}{\to} \pi_2(K) \overset{\eta}{\to} N_1 \cap N_2/[N_1, N_2] \to 0$$

where ζ is induced by inclusion, N_i is the kernel of $\pi_1(K_0) \to \pi_1(K_i)$, and the action of $\pi_1(K)$ on $N_1 \cap N_2/[N_1, N_2]$ is induced by conjugation in $\pi_1(K_0)$.

Proof: This is a corollary to the above Theorem 3.8. $\square$

The key question is then: Is $\pi_2(K, K_0) \approx \pi_2(K_1, K_0) \bowtie \pi_2(K_2, K_0)$ as a $\pi_1(K_0)$-crossed module?

If K is obtained from K_0 by adding only cells in dimension 2, then the answer is clearly "yes", because everything in sight is free ([BrHi78], [Br84], [GuRa81], [BoGu92]). It is shown in [BrHi78] and [Br84] that the answer is "yes" if the pairs (K_1, K_0) and (K_2, K_0) are 1-connected, giving the so-called Brown-Higgins theorem. In [BoGu92], if (K, K_1) is 1-connected, necessary and sufficient conditions are given for a positive answer.

We content ourselves with the following simple characterization (see [BoGu92], in the argument of Theorem 4.5).

Let $M = \pi_2(K_1, K_0)$, $N = \pi_2(K_2, K_0)$, $C = \pi_2(K, K_0)$, and $G = \pi_1(K_0)$. The inclusion induced maps $M \to C$ and $N \to C$ induce a G-morphism $\alpha : M \bowtie N \to C$, which has kernel P, which is a $\mathbb{Z}G$-module. Then α is an isomorphism if and only if

1. α is surjective, and

2. α induces an isomorphism $\alpha_* : H_1(M \bowtie N) \approx H_1(C)$.

To see this, observe that the conditions are clearly necessary. Assuming (1) and (2), we see that, because P is central in $M \bowtie N$, the projectivity of C and the surjectivity of α imply that $M \bowtie N \approx C \times P$ as G-groups. Thus, $H_1(C) \oplus P \approx H_1(M \bowtie N)$, so (2) implies that $P = 0$.   □

Open problem: In the light of the previous section and the above theorem, it is clearly important to compute $H_2(M \bowtie N)$ in terms of $H_2(M)$ and $H_2(N)$.

Because of the projectivity of C, the surjectivity of α implies that of $H_2(M \bowtie N) \to H_2(C)$. Hence, if the former is trivial then the latter is as well, and so the triviality of the ker $\{C \to G\}$ can be detected by using Theorem 2.6.

As an indication of the power of (non-2-dimensional) version of Theorem 3.9, we state a lovely theorem of R. Brown [Br84]. This has been generalized (in a slightly different context) to H_3 in [BoGu92] and to H_4 in [DuElGi92]. A direct proof is given in [BoGu92].

Theorem 3.8 *Let M and N be normal subgroups of a group and let $L = M \cap N$. Then there is the exact sequence*

$$H_2(MN) \to H_2(M/L) \oplus H_2(N/L) \to L/[M, N]$$

$$\to H_1(MN) \to H_1(M/L) \oplus H_1(N/L) \to 0.$$

As an example of Theorem 3.9, (taken from [GuRa81]) consider the 2-complex K modeled on the standard presentation $\{x, y\, ; \, x^2, [x, y]\}$ for the group $\mathbb{Z}_2 \times \mathbb{Z}$.

Let K_1 be the real projective plane wedged with a copy of S^1 modeled on $\{x, y \,; x^2\}$ and let K_2 be the torus, modeled on $\{x, y \,; [x, y]\}$. Because $\pi(K_1)$ is generated by $(x - 1)$ in $\mathbb{Z}(\mathbb{Z}_2 * \mathbb{Z})$, we have $i_{1*}\pi_2(K_1) = \mathbb{Z}(\mathbb{Z}_2 \times \mathbb{Z})(x - 1)$, which is isomorphic to $\mathbb{Z}(\mathbb{Z}) = \mathbb{Z}[t, t^{-1}]$ with $(x - 1)$ corresponding to 1, and the action of $\mathbb{Z}_2 \times \mathbb{Z}$ on $\mathbb{Z}(\mathbb{Z})$ given by $x \cdot u = -u$ and $y \cdot u = tu$. As K_2 is the torus, we have $\pi_2(K_2) = 0$.

If f is in the free group $F(x, y)$, let $\ll f \gg$ denote the normal closure of f in $F(x, y)$. An element of $\ll x^2 \gg$ is of the form $\prod_{i=1}^n f_i x^{2\epsilon_i} f_i^{-1}$ with $\epsilon_i = \pm 1$. Such an element is in $\ll [x, y] \gg$, the commutator subgroup of $F(x, y)$, iff $\sum_{i=1}^n \epsilon_i = 0$. It is now clear that $\ll x^2 \gg \cap \ll [x, y] \gg = \ll [y, x^2] \gg$. One sees that x acts trivially on $\ll [y, x^2] \gg / [\ll x^2 \gg, \ll [x, y] \gg]$. Therefore, $[\ll x \gg, \ll [y, x^2] \gg] \subset [\ll x^2 \gg, \ll [x, y] \gg]$. A calculation shows that an element of the form $[fx^2f^{-1}, [x, y]]$ is in $[\ll x \gg, \ll [y, x^2] \gg]$. Therefore,

$$N_1 \cap N_2/[N_1, N_2] = \ll [y, x^2] \gg / [\ll x \gg, \ll [y, x^2] \gg],$$

which is the relation module of the group $\langle x, y \mid [y, x^2] \rangle$ modulo the action of x. Hence, $N_1 \cap N_2/[N_1, N_2] \approx \mathbb{Z}(\mathbb{Z}) \approx \mathbb{Z}[t, t^{-1}]$ with $[y, x^2]$ corresponding to 1, and the action of $\mathbb{Z}_2 \times \mathbb{Z}$ on $\mathbb{Z}(\mathbb{Z})$ given by $x \cdot u = u$ and $y \cdot u = tu$.

Thus the exact sequence of the theorem becomes

$$0 \to \mathbb{Z}(\mathbb{Z}) \to \pi_2(K) \to \mathbb{Z}(\mathbb{Z}) \to 0.$$

Although the sequence splits as a sequence of $\mathbb{Z}(\mathbb{Z})$-modules, we claim that it does not split as $\mathbb{Z}(\mathbb{Z}_2 \times \mathbb{Z})$-modules. Observe that the element $[y, x^2]$ yields the identity

$$(yx^2y^{-1})x^{-2}x[x, y]x^{-1}[x, y] = 1.$$

Therefore, $\pi_2(K)$ is determined by $(x - 1, 0)$ and $(y - 1, x + 1)$ in $\mathbb{Z}(\mathbb{Z}_2 \times \mathbb{Z}) \oplus \mathbb{Z}(\mathbb{Z}_2 \times \mathbb{Z})$. Note that

$$x(y - 1, x + 1) = (y - 1)(x - 1, 0) + (y - 1, x + 1).$$

Hence the action of x on $\pi_2(K) \approx \mathbb{Z}(\mathbb{Z}) \oplus \mathbb{Z}(\mathbb{Z})$ is given by $x(1, 0) = (-1, 0)$ and $x(0, 1) = (t - 1, 1)$.

Suppose that the sequence $0 \to \mathbb{Z}(\mathbb{Z}) \to \mathbb{Z}(\mathbb{Z}) \oplus \mathbb{Z}(\mathbb{Z}) \to \mathbb{Z}(\mathbb{Z}) \to 0$ splits via $\sigma : \mathbb{Z}(\mathbb{Z}) \to \mathbb{Z}(\mathbb{Z}) \oplus \mathbb{Z}(\mathbb{Z})$. The $\sigma(1) = (u, 1)$ for some $u \in \mathbb{Z}(\mathbb{Z})$. Because x acts trivially on the second copy of $\mathbb{Z}(\mathbb{Z})$, the element $(u, 1)$ must be invariant under the action of x in order that σ respect the action of x ; but $x \cdot (u, 1) = (-u + t - 1, 1)$ and $-u + t - 1 \neq u$, because $(t - 1)$ is not divisible by 2 in $\mathbb{Z}(\mathbb{Z})$. Therefore, the sequence does not split as a sequence of $\mathbb{Z}(\mathbb{Z}_2 \times \mathbb{Z})$-modules.

This example is especially nice because one can easily visualize the universal cover of K and verify the calculations. We leave this as an exercise for the reader.

4   Identity Properties

Let $\mathcal{P} = \langle \mathbf{x} \mid \mathbf{r} \rangle$ be a presentation of the group G. As usual, we let $K = K_{\mathcal{P}}$ be the 2-complex canonically associated with $\mathcal{P}$ and $\tilde{K}$ be the universal cover of K. The second homotopy group of K can be thought of as a submodule of the free left $\mathbb{Z}G$-module $C_2(\tilde{K}) \approx \oplus_{R \in r} \mathbb{Z}Ge_R$ by identifying $\pi_2(K)$ as $H_2(\tilde{K}) = \ker (C_2(\tilde{K}) \to C_1(\tilde{K}))$.

If L is a subgroup of G, let IL be the augmentation ideal in $\mathbb{Z}L$. Let $1 \to N \to F(\mathbf{x}) \to G \to 1$ be the presentation of G, where $F(\mathbf{x})$ is the free group on the set $\mathbf{x}$ and $N = \ll \mathbf{r} \gg$ is the normal closure of the set of relators $\mathbf{r}$ inside $F(X)$. Let $\phi : F(X) \to G$ be the natural quotient, L be any subgroup of G, and $K = \phi^{-1}L$. We say that the presentation $\mathcal{P}$ has the *right* (resp. *left*) *L-identity property* if, for each product of conjugates of elements of $\mathbf{r} \cup \mathbf{r}^{-1}$, $\prod_{i=1}^{n} f_i R_i^{\epsilon_i} f_i^{-1}$, where $f_i \in F(\mathbf{x}), R_i \in \mathbf{r}$, and $\epsilon_i = \pm 1$, that reduces to the identity element $1 \in F(\mathbf{x})$, there is a pairing $i \leftrightarrow j$ of the indices such that $R_i = R_j$, $\epsilon_i = -\epsilon_j$, and $Kf_i = Kf_j$ (resp. $f_i K = f_j K$).

The terminology "right identity property" comes from the fact that $Kf_i = Kf_j$ as right cosets of K.

The identity property has been studied in various guises in the these places: [BaHoPr92], [BoPr92], [BrHu82], [ChCoHu81], [Co54], [Dy91$_2$], [GiHo92], [Gi93], [Ha91], [Pa63], [Pr92$_2$], [Pr91].

We show in the first subsection that the presentation $\mathcal{P}$ has the right (resp. left) L-identity property if and only if, for $K = K_{\mathcal{P}}$, each element $\xi \in \pi_2(K) \subset C_2(\tilde{K})$ has all its coordinates lying in $IL \cdot \mathbb{Z}G$ (resp. $\mathbb{Z}G \cdot IL$).

When L is normal in G, then the left and right identity properties coincide, and we can refer simply to the L-*identity property*.

A presentation that has the $\{1\}$-identity property is *aspherical*; i. e., $\pi_2(K) = 0$, (see Chapter II and Section 2 of this chapter).

For any subgroup L of G, let K_L denote the covering of K corresponding to the subgroup L. A 2-complex K is said to be L-*Cockcroft* if the Hurewicz map $\pi_2(K) \to H_2(K_L)$ is trivial. It is said to be *Cockcroft* if it is G-Cockcroft. A presentation is said to be L-Cockcroft if the corresponding 2-complex is also. We will show in the next section that the presentation $\mathcal{P}$ is L-Cockcroft if and only if $\mathcal{P}$ satisfies the right L-identity property. This property was first noticed in [Co54] in connection with the question of J. H. C. Whitehead. In particular, if X is a connected 2-complex that is a subcomplex of an aspherical 2-complex Y and $L = \ker \{\pi_1(X) \to \pi_1(Y)\}$, then X is L-Cockcroft. This is easily seen to be true by looking at X_L as a subcomplex of the contractible space $\tilde{Y}$.

On the other hand, the left identity property arises quite naturally if one thinks about formulating a generalization of the 1-identity property by using pairings of terms of identity sequences (or, equivalently, pairings of disks of spherical pictures).

Just as an example of the utility of the Cockcroft property in group theory, we state the following theorem from [Dy87$_1$]. If L is a normal subgroup of the group G, the *weight of L in G* (denoted by $wt_G L$) is the minimal number of elements whose normal closure in G is L. A group Q is said to be a *Rosset group* if Q contains a non-trivial normal abelian torsion-free subgroup.

Theorem 4.1 *Suppose* $1 \to L \to G \to Q \to 1$ *is an exact sequence of groups with Q a Rosset group, G finitely presented, $H_1(L)$ finitely generated as an abelian group, and $wt_G L < \infty$. Let X be any finite 2-complex with fundamental group isomorphic to G. Then the Euler characteristic $\chi(X) \geq 0$, with $\chi(X) = 0$ iff X is L-Cockcroft and $H_2(L) = 0$.*

4.1 H-Cockcroft complexes

Let X be a 2-complex with fundamental group $\pi_1(X) = G$; such a complex is called a $(G, 2) - complex$. For any subgroup $L \leq G$ we say that X is L-*Cockcroft* if the Hurewicz map $\pi_2(X) \to H_2(X_L)$ is trivial.

There are many ways to characterize this property. Let us describe several. For any group G the augmentation ideal in $\mathbb{Z}G$ is denoted by IG. For any $(G, 2)$-complex X, we let $C_*(\tilde{X}) \to \mathbb{Z}$ denote the augmented cellular chain complex of the universal cover $\tilde{X}$ of X (considered as free left $\mathbb{Z}G$-modules with a preferred basis determined by the cells of X). From the definition it follows that if we choose *any* basis whatever for the free $\mathbb{Z}G$-module $C_2(\tilde{X})$, then X is L-Cockcroft iff the *coordinates* of each element of $\pi_2(X) \subseteq C_2(\tilde{X})$ are contained in the *right* ideal $IL \cdot \mathbb{Z}G$ of $\mathbb{Z}G$. This follows because we may identify $C_2(X_L)$ as $\mathbb{Z} \otimes_{\mathbb{Z}L} C_2(\tilde{X}) = C_2(\tilde{X})/(IL \cdot C_2(\tilde{X}))$ as an abelian group. If L is a normal subgroup of G, then, of course, $IL \cdot \mathbb{Z}G = \mathbb{Z}G \cdot IL$ and $C_2(X_L)$ is a free $\mathbb{Z}G/L$-module.

For any (left) $\mathbb{Z}G$-module M that is a submodule of a free $\mathbb{Z}G$-module F, define the *Fox ideal* of the inclusion $M \subseteq F$, $\mathcal{F}(M \subseteq F) = \mathcal{F}M$, to be the two sided ideal generated by all the coordinates of elements of M (for some choice of a basis for F). The ideal $\mathcal{F}M$ is independent of the choice of the basis for F because it is 2-sided. Most of the following theorem was proved in [BrDy81] and [Dy91$_1$].

Theorem 4.2 *The following are equivalent for the $(G,2)$-complex X:*

1. *The complex X is L-Cockcroft.*

2. *The coordinates of each element of $\pi_2(X)$ are in the ideal $IL \cdot \mathbb{Z}G$.*

3. *The Fox ideal $\mathcal{F}(\pi_2(X))$ is contained in the ideal $IL \cdot \mathbb{Z}G$.*

4. *If $i : X \to Y$ is the inclusion of X into the $[H,2]$-complex Y so that $L \subseteq \ker \{\pi_1(X) \to \pi_1(Y)\}$, then $\pi_2(i)$ is trivial.*

5. *If X is the realization of the presentation $cal P = \langle \mathbf{x} \mid \mathbf{r} \rangle$ of G and N is the normal closure of the set of relators $\mathbf{r}$, then*

$$\mathbb{Z} \otimes_{\mathbb{Z}L} H_1(N) \approx \bigoplus_{R \in \mathbf{r}} (\mathbb{Z}(L \setminus G)_R$$

where $L \setminus G$ is the set of right cosets of L in G, and the former group is generated by the set $\bar{\mathbf{r}} = \{R[K,N] \mid R \in \mathbf{r}\}$.

6. *If X is the realization of the presentation $\mathcal{P} = \langle \mathbf{x} \mid \mathbf{r} \rangle$ of G, then $\mathcal{P}$ has the right L-identity property.*

Proof: The equivalence of (1) and (2) is given in the paragraph preceding the statement of the theorem. The equivalence (2) $\Leftrightarrow$ (3) follows because $\pi_2(X)$ is a left module and the coordinates are contained in the right ideal $IL \cdot \mathbb{Z}G$.

The equivalence (1) $\Leftrightarrow$ (4) is most useful. It shows that (for normal subgroups L) that the L-Cockcroft property is the answer to the question: when can you add 1-cells and 2-cells to X to "kill" the map on π_2? Let $K = \ker \{\pi_1(X) \to \pi_1(Y)\}$. This follows because the map $\pi_2(i)$ factors into the maps $\beta \circ h$, where h is the Hurewicz map $\pi_2(X) \to H_2(X_K)$ and $\beta : H_2(X_K) \hookrightarrow \pi_2(Y)$ is the map induced by the inclusion $X_K \to \tilde{Y}$.

In order to prove (1) $\Leftrightarrow$ (5), recall that the presentation $\mathcal{P} = \{\mathbf{x}; \mathbf{r}\}$ gives rise to an exact sequence $1 \to N \to F(\mathbf{x}) \to G \to 1$, where $N = \ll \mathbf{r} \gg$ is the normal closure of the set of relators $\mathbf{r}$ inside $F(X)$. Furthermore, recall that any $(G,2)$-complex has the homotopy type of one which is the realization of a presentation.

Now $\pi_2(X)$ is defined by the sequence

$$0 \to \pi_2(X) \to C_2(\tilde{X}) = \bigoplus_{R \in \mathbf{r}} \mathbb{Z}G_R \to H_1(N) \to 0.$$

By tensoring this sequence with $\mathbb{Z} \otimes_{\mathbb{Z}L} -$ we see that

$$\mathbb{Z} \otimes_{\mathbb{Z}L} \pi_2(X) \to (\bigoplus_{R \in \mathbf{r}} \mathbb{Z}G_R)/(IL \cdot (\bigoplus_{R \in \mathbf{r}} \mathbb{Z}G_R)) \to \mathbb{Z} \otimes_{\mathbb{Z}L} H_1(N) \to 0$$

is exact. Note that $\mathbb{Z} \otimes_{\mathbb{Z}L} H_1(N) \approx N/[K, N]$. Thus, X is L-Cockcroft iff

$$\mathbb{Z} \otimes_{\mathbb{Z}L} H_1(N) \approx \bigoplus_{R \in \mathbf{r}} (\mathbb{Z}G/IL \cdot \mathbb{Z}G)_R$$

which in turn is isomorphic to

$$\bigoplus_{R \in \mathbf{r}} \mathbb{Z}(L \setminus G)_R \approx \bigoplus_{R \in \mathbf{r}} (\mathbb{Z}(K \setminus F))_R,$$

where $L \setminus G$ denotes the set of right cosets of L.

Now for the equivalence of (5) and (6). So assume that

$$\bigoplus_{R \in \mathbf{r}} \mathbb{Z}(K \setminus F) e_R = \mathbb{Z}(K \setminus F)^{\mathbf{r}} \approx \mathbb{Z} \otimes_{\mathbb{Z}L} H_1(N)$$

via the basis assignment $e_R \mapsto R[K, N]$ for each $R \in \mathbf{r}$. We will verify the right L-identity property. If $\prod_{i=1}^{n} f_i R_i^{\epsilon_i} f_i^{-1} = 1$ ($f_i \in F(\mathbf{x}), R_i \in \mathbf{r}$, and $\epsilon_i = \pm 1$), then $\sum_{i=1}^{n} \epsilon_i \bar{f}_i \bar{R}_i \bar{f}_i^{-1} = 0$ in $H_1(N)$, which in turn implies that $\sum_{i=1}^{n} \epsilon_i (1 \otimes \bar{f}_i \bar{R}_i \bar{f}_i^{-1}) = 0$ in $\mathbb{Z} \otimes_{\mathbb{Z}L} H_1(N)$. Thus the hypothesis implies that $\sum_{i=1}^{n} \epsilon_i (Kg_i) e_{R_i} = 0$ in $\sum_{R \in \mathbf{r}} (\mathbb{Z}(K \setminus F) e_R$. Thus for each R the partial sum $\sum_{R_i = R} \epsilon_i (Kg_i) e_{R_i}$ is a member of $IK \cdot \mathbb{Z}F$. This implies the existence of the pairing with the necessary properties.

To see the converse, we build the group E, the free group on the set $F \times \mathbf{r}$, and the exact sequence $1 \to I \to E \xrightarrow{\bar{\partial}} N \to 1$, where the map $\bar{\partial}$ sends $(f, R) \mapsto fRf^{-1}$ and I is the group of identities. Recall that the normal subgroup P of I is the subgroup of Peiffer identities, and that the relative homotopy group $\pi_2(X, X^{(1)}) = E/P$, while $\pi_2(X) = I/P$. Thus we have the following commutative diagram with exact horizontal rows:

$$
\begin{array}{ccccccccc}
1 & \to & I & \to & E & \xrightarrow{\bar{\partial}} & N & \to & 1 \\
 & & \alpha \downarrow & & \downarrow & & \| & & \\
0 & \to & I/P & \to & \pi_2(X, X^{(1)}) & \to & N & \to & 1 \\
 & & \eta \approx \downarrow & & \downarrow & & \downarrow & & \\
0 & \to & \pi_2(X) & \to & \mathbb{Z}(G)^{\mathbf{r}} & \xrightarrow{\rho} & H_1(N) & \to & 0 \\
 & & s \downarrow & & f \downarrow & & \downarrow & & \\
0 & \to & W & \to & \mathbb{Z} \otimes_{\mathbb{Z}L} \mathbb{Z}(G)^{\mathbf{r}} & \xrightarrow{1 \otimes \rho} & \mathbb{Z} \otimes_{\mathbb{Z}L} H_1(N) & \to & 0.
\end{array}
$$

All the vertical arrows except s are clearly surjective. We claim that s is surjective as well. Let $\ell \in \ker 1 \otimes \rho = W$. Then there is an $x \in \mathbb{Z}(G)^{\mathbf{r}}$ such that $f(x) = \ell$. Because $\mathbb{Z} \otimes_{\mathbb{Z}L} H_1(N) = H_1(N)/IL \cdot H_1(N)$ we must have $\rho(x) \in IL \cdot H_1(N)$; hence, $\exists y \in IL \cdot \mathbb{Z}(G)^{\mathbf{r}}$ so that $\rho(x) = \rho(y)$. Thus, $\rho(x - y) = 0$ and $s(x - y) = f(x - y) = f(x) - f(y) = \ell - 0$. Therefore, s is surjective.

Thus, to each $\ell \in W$, there corresponds an identity $\imath = \prod(f_i, R_i)^{\epsilon_i} \in I$; that is, $\bar{\rho}(\imath) = \prod(f_i R_i^{\epsilon_i} f_i^{-1}) = 1 \in F$ with $\alpha \eta s(\imath) = \ell$. Now we can easily see that $\mathcal{P}$ has the right L-identity property implies that $W = 0$. For let $\ell \in W$ and let $\imath = \prod(f_i, R_i)^{\epsilon_i} \in I$ with $\alpha \eta s(\imath) = \ell$. Hence, the existence of a pairing shows that $\ell = \sum \epsilon_i (K f_i) e(R_i) = 0$. Hence, $W = 0$.   □

For example, let Y be the realization of a one-relator presentation $\langle \mathbf{x} \mid R \rangle$ of a group G. We write the relator $R = S^q$ in the free group $F(\mathbf{x})$, with S not a proper power, and $q \geq 1$. Let H be the normal subgroup generated by the image $\bar{S}$ of S in G. Then, Y is L-Cockcroft if and only if L contains H. This follows because $\pi_2(Y) \approx \mathbb{Z}G(\bar{S} - 1)$, as a submodule of $\mathbb{Z}G$.

4.2   Characterizing Cockcroft complexes

In this section, we characterize when a finite presentation $\mathcal{P} = \langle \mathbf{x} \mid \mathbf{r} \rangle$ of the group G is Cockcroft; that is to say, when the realization K of $\mathcal{P}$ has trivial Hurewicz map $\pi_2(K) \to H_2(K)$. We have seen from the above theorem that K is Cockcroft iff K has the G-identity property. To restate this in detail: for each product $\prod_{i=1}^{n} f_i R_i^{\epsilon_i} f_i^{-1}$, where $f_i \in F(\mathbf{x}), R_i \in \mathbf{r}$, and $\epsilon_i = \pm 1$, which reduces to the identity element $1 \in F(\mathbf{x})$, there is a pairing $i \leftrightarrow j$ of the indices such that $R_i = R_j$ and $\epsilon_i = -\epsilon_j$. This is sometimes called the *weak identity property* [Dy91$_1$].

The *directed deficiency* of the finite presentation $\mathcal{P}$ is *dir def* $\mathcal{P} = |\mathbf{r}| - |\mathbf{x}|$. The directed deficiency is useful because it "goes in the same direction" as the Euler characteristic of the realization of $\mathcal{P}$; i. e., *dir def* $\mathcal{P} + 1 = \chi(|\mathcal{P}|)$. The *directed deficiency* of a finite presentable group G, denoted *dir def* G, is the minimum of the directed deficiencies *dir def* $\mathcal{P}$ for all finite presentations $\mathcal{P}$ of G.

The finite presentation $\mathcal{P}$ of the group G is said to be *efficient* if *dir def* $G = sH_2(G) - rank\ H_1(G)$, where sA denotes the minimum number of generators of the finitely generated abelian group A. A group G is said to be *efficient* if it admits an efficient presentation.

Theorem 4.3 *Let* $\mathcal{P} = \langle \mathbf{x} \mid \mathbf{r} \rangle$ *be a finite presentation of a group* G *and* K *be the realization of* $\mathcal{P}$. *Let* N *be the normal closure of the set of relators* $\mathbf{r}$ *in the free group* $F(\mathbf{x})$. *Then the following are equivalent.*

  1. K *is Cockcroft.*

  2. $\mathcal{P}$ *has the* G-*identity property.*

  3. $\mathbb{Z} \otimes_{\mathbb{Z}G} H_1(N) \approx \mathbb{Z}^{\mathbf{r}}$.

4. $\mathcal{P}$ is an efficient presentation and $H_2(G)$ is free abelian.

Proof: We have already shown the equivalence of (1), (2), and (3). We show now the equivalence of (3) and (4). Assume that (4) is true and consider the following ladder of exact sequences:

$$
\begin{array}{ccccccccc}
0 & \to & H_2(K) & \to & \mathbb{Z}^r & \to & \mathbb{Z}^x & \to & H_1(K) & \to & 0 \\
  &   & \alpha\downarrow & & \beta\downarrow & & \| & & \| \\
0 & \to & H_2(G) & \to & \mathbb{Z}\otimes_{\mathbb{Z}G} H_1(N) & \overset{\rho}{\to} & H_1(F) & \to & H_1(G) & \to & 0.
\end{array}
$$

where the upper sequence is derived from the cellular chain complex $C_*(K)$ of K and the lower sequence from the 5-term exact sequence of Stallings-Stammbach [HiSt71]. Notice that the maps α and β are surjective. Now $\mathcal{P}$ is efficient and $H_2(G)$ is free abelian implies that *dir def* $\mathcal{P} = |\mathbf{r}| - |\mathbf{x}| = rank\, H_2(G) - rank\, H_1(G)$. By rank arguments, we have $|\mathbf{r}| - |\mathbf{x}| = rank\, H_2(G) - rank\, H_1(K)$. Thus, $rank\, H_2(G) = rank\, H_2(K)$. But α is surjective implies that α is an isomorphism. The five lemma then implies that β is an isomorphism, so that (3) is true.

Assuming (3) is true, we see that β is an isomorphism; hence, so is α. So $H_2(G)$ is free abelian and $|\mathbf{r}| - |\mathbf{x}| = sH_2(G) - rank\, H_1(G)$. Hence, $\mathcal{P}$ is efficient. $\qquad\square$

As an example, we consider the group G of [BrGe84] having presentation

$$\mathcal{P} = \langle a, b \mid r, s \rangle$$

where $r = [ab^{-1}, a^{-1}ba]$, and $s = [ab^{-1}, b^{-1}a^{-1}bab]$. It is shown in [BrGe84], Theorem 7.1, that $H_i(G) \approx \mathbb{Z} \oplus \mathbb{Z}$ for all $i \geq 1$. Thus, $K = |\mathcal{P}|$ is Cockcroft.

The following two theorems are due to W. Bogley in [Bo92]. The first is a way of building a Cockcroft complex from two Cockcroft subcomplexes.

Theorem 4.4 *Let X be the realization of the presentation $\mathcal{P} = \langle \mathbf{x} \mid \mathbf{r}, \mathbf{s} \rangle$, $\rho = \ll \mathbf{r} \gg_F$, and $\sigma = \ll \mathbf{s} \gg_F$, where F is the free group on the set $\mathbf{x}$. Let M be any subgroup of F containing $\sigma\rho$, with images $N = M\sigma\rho$ in $G/\sigma\rho$, $N_\rho = M\rho$ in G/ρ, and $N_\sigma = M\sigma$ in G/σ. Let X_r be the realization of $\langle \mathbf{x} \mid \mathbf{r} \rangle$, X_s of $\langle \mathbf{x} \mid \mathbf{s} \rangle$. Then, X is Cockcroft if and only if X_r is N_ρ-Cockcroft, X_s is N_σ-Cockcroft, and $\sigma \cap \rho \subseteq [\rho, M] \cap [\sigma, M]$.*

Theorem 4.5 *Let X be the realization of the presentation $\mathcal{P} = \langle \mathbf{x} \mid \mathbf{r} \rangle$ and let F be the free group on the set $\mathbf{x}$. Let F_n denote the lower central series of F, defined by $F_1 = F$ and $F_{i+1} = [F, F_i]$ for all $i \geq 1$. If, for some $n \geq 1$, the image of the relator set $\mathbf{r}$ in the free abelian group F_n/F_{n+1} is linearly independent, then X is Cockcroft.*

For example, if $\mathcal{P} = \langle a, b \mid r, s \rangle$, where $r = aba^{-1}b^2$ and $s = bub^{-1}u^3$, then the image of the set $\{r, s\}$ is linearly independent in $H_1(F)$; hence, $|\mathcal{P}|$ is Cockcroft. This also follows because clearly, $H_2(|\mathcal{P}|) = 0$.

As another example, we consider a group G having a presentation $\mathcal{P} = \langle \mathbf{x} \mid \mathbf{r} \rangle$ for which the image of the set $\mathbf{r}$ in $H_1(F)$ is linearly independent. Assume further that $H_1(G)$ is free abelian. Then, G is an E-group [St74] and $X = |\mathcal{P}|$ is not only Cockcroft, it is *very* Cockcroft. Let $G(\alpha)$ denote the transfinite derived series of G, where α is any ordinal; this is defined as $G(1) = G$, $G(\alpha + 1) = [G(\alpha), G(\alpha)]$, and $G(\beta) = \cap\{G(\alpha) \mid \alpha < \beta\}$, for any limit ordinal β. If we let $P(G)$ denote the intersection of the transfinite derived series, then X is $P(G)$-Cockcroft.

4.3   Minimal subgroups

Let X be a $(G, 2)$-complex. It is immediate that, if X is L-Cockcroft and if $L \leq H$, then X is H-Cockcroft. Further, if X is L-Cockcroft and $g \in G$, then X is gLg^{-1}-Cockcroft, because the coverings X_L and $X_{gLg^{-1}}$ are equivalent, or algebraically because $I(gLg^{-1}) = g(IL)g^{-1}$ and so

$$I(gLg^{-1}) \cdot \mathbb{Z}G = g(IL)g^{-1} \cdot \mathbb{Z}G = gIL \cdot \mathbb{Z}G.$$

Thus $\pi_2(X) \subseteq IL \cdot C_2(\tilde{X})$ implies that $\pi_2(X) = g\pi_2(X) \subseteq gIL \cdot C_2(\tilde{X}) = I(gLg^{-1}) \cdot C_2(\tilde{X})$. Given a Cockcroft $(G, 2)$-complex X it is therefore of interest to determine minimal subgroups L of G such that X is L-Cockcroft. The *existence* of such minimal subgroups has been proved independently by J. Harlander [Ha91] and N. Gilbert and J. Howie [GiHo92].

Theorem 4.6 *Let X be a Cockcroft $(G, 2)$-complex. Then there exists a subgroup L of G which is minimal with respect to the property that X is L-Cockcroft.*

We will give both proofs, for reasons that will soon become clear.

Proof: ([GiHo92]) Let $L_0 \geq L_1 \geq \cdots$ be a descending chain of subgroups of G with intersection L_∞, and suppose that for all $i \geq 0$, the 2-complex X is L_i-Cockcroft. We have the Fox ideal $\mathcal{F}(\pi_2(X)) = \mathcal{F}(X) \subseteq IL_i \cdot \mathbb{Z}G$ for all $i \geq 0$, and so $\mathcal{F}(X) \subseteq \cap_{i \geq 0} IL_i \cdot \mathbb{Z}G$. We claim that $\cap_{i \geq 0} IL_i \cdot \mathbb{Z}G = IL_\infty \cdot \mathbb{Z}G$, so that X is L_∞-Cockcroft. An application of Zorn's lemma completes the proof.

Consider the descending chain $IL_0 \cdot \mathbb{Z}G \geq IL_1 \cdot \mathbb{Z}G \geq \cdots$ of right ideals of $\mathbb{Z}G$. It is clear that $\cap_{i \geq 0} IL_i \cdot \mathbb{Z}G \supseteq IL_\infty \cdot \mathbb{Z}G$. Now suppose that $\alpha = \sum_{j=1}^{m} \lambda_j g_j \in$

$\bigcap_{i \geq 0} IL_i \cdot \mathbb{Z}G$. Considering the image of α in $\mathbb{Z}(L_i \setminus G)$ we see that for all $i \geq 0$, and all $g \in G$,

$$\sum_{g_i \in L_i g} \lambda_j = 0.$$

We obtain a partition of the indexing set $\{1, \ldots, m\}$ such that the sum of those λ_j indexed by each part is zero, and if $i \leq k$ then L_k produces a finer partition than L_i. Let $J_1 \cup \cdots \cup J_p$ be the finest partition of $\{1, \ldots, m\}$ so produced. Renumber if necessary so that $J_1 = \{1, \ldots, q\}$. Then for all $i \geq 0$, $\{g_1, \ldots, g_q\} \subseteq L_i h_i$ for some $h_i \in G$, and therefore $g_1^{-1} g_j \in L_i$ for $j = 1, \ldots, q$. It follows immediately that $g_1^{-1} g_j \in L_\infty$ for $j = 1, \ldots, q$. A similar argument applies to each of the other parts of the partition. We conclude that α has trivial image in $\mathbb{Z}(L_\infty \setminus G)$, so that $\alpha \in IL_\infty \cdot \mathbb{Z}G$, as required. $\qquad\square$

Proof: ([Ha91]) (Sketch) Again, Let $L_0 \geq L_1 \geq \cdots$ be a descending chain of subgroups of G with intersection L_∞, and suppose that for all $i \geq 0$, the 2-complex X is L_i-Cockcroft. This chain of subgroups induces a sequence of coverings of X:

$$\cdots \to X_{L_i} \to X_{L_{i-1}} \to \cdots \to X_{L_0},$$

which in turn induces sequences on the cellular chain complexes and on the corresponding homologies. This gives rise to the following commutative square:

$$
\begin{array}{ccc}
\pi_2(X) & \to & H_2(\lim_{\leftarrow i} C_2(X_{L_i})) \\
\downarrow & & \downarrow \\
H_2(X_{L_\infty}) & \to & \lim_{\leftarrow i} H_2(X_{L_i}).
\end{array}
$$

The top horizontal and the left vertical maps are induced from the corresponding Hurewicz maps, while the right vertical map is defined because each cycle in the inverse limit of the chain groups is a sequence of cycles in the respective chain groups. The bottom map is induced by the covering maps. It is an easy calculation that the right vertical map is an isomorphism, because everything in sight is 2-dimensional. Finally, it can be shown directly that the lower horizontal map is injective. This finishes the second proof. $\qquad\square$

A subgroup L of G that is minimal such that X is L-Cockcroft is called a *Cockcroft threshold* of X. (The terminology was suggested by W. Bogley.) The above theorem says nothing of uniqueness; if the Cockcroft threshold is unique, then it must be normal. As noted in the introduction, S. Pride [Pr92$_2$] has constructed examples for which the Cockcroft thresholds are not unique. If the Cockcroft threshold of X is $\pi_1(X)$ itself, we say that X is *absolutely Cockcroft*. Examples of absolutely Cockcroft 2-complexes are spines of aspherical, closed, orientable 3-manifolds [GiHo92].

Both proofs show that if we restrict ourselves to *normal* subgroups, then there exists minimal normal subgroups that are L-Cockcroft. As was observed in the introduction, the first proof may be modified to show that the left L-identity property (as well as the right L-identity property) admits threshold subgroups. The second proof shows that if we restrict ourselves to the (non-empty) family $\{L\}$ of (resp. normal) subgroups of G for which $H_2(X_L) = 0$, then this family also admits (resp. normal) minimal subgroups L having the property that $H_2(X_L) = 0$. We call such a subgroup of $\pi_1(X)$ a $null -$ H_2 threshold for X. This yields the following corollary, which generalizes a theorem of R. Strebel [St74] on E-groups. Note that this result is related to Proposition 2.4 of [Ge83].

Corollary 4.7 *Let X be a 2-complex. If L is a null $- H_2$ threshold for X, then $H_1(L)$ is a torsion abelian group, $H_2(L) = 0$, and, if L is a normal subgroup, then the cohomological dimension $cd_{\mathbf{Q}}G/L$ of G/L over $\mathbf{Q}$, the rationals, is ≤ 2.*

Proof: Because $H_2(X_L) = 0$, it follows that $H_2(L) = 0$. We factor $H_1(L)$ as follows:

$$0 \to T(H_1(L)) \to H_1(L) \to H_1(L)/T(H_1(L)) \to 0,$$

where $T(H_1(L))$ is the torsion subgroup of $H_1(L)$ and $H_1(L)/T(H_1(L))$ is torsion free. Because $H_1(L)/T(H_1(L))$ is a torsion free abelian group, it is a D-group, in the sense of Strebel [St74], and, if non-trivial, we would have (by the properties of D-groups) that $H_2(X_W) = 0$ for some subgroup W strictly contained in L. This is not possible by the minimality of L.

Finally, one sees from the above data that the chain complex $C_*(\tilde{X})$ tensored with $\mathbf{Q}$

$$0 \to \mathbf{Q} \otimes C_2(\tilde{X}) \to \mathbf{Q} \otimes C_1(\tilde{X}) \to \mathbf{Q} \otimes C_0(\tilde{X}) \to \mathbf{Q} \to 0$$

is exact.   □

Chapter V

Calculating Generators of Π_2

W. A. Bogley and S. J. Pride

This article discusses combinatorial geometric techniques that determine explicit generators for the second homotopy module of a 2-complex in terms of its cell structure. Applications of the techniques are also presented. The discussion focuses on the theory of *pictures*. Pictures have been used for many purposes; we shall be concerned only with their application to π_2 calculations.

There are three sections, each of which is divided into several subsections. The first section provides an overview of the theory of pictures from a homotopy-theoretic perspective. The second section deals with generalities related to the generation of π_2. Some proofs are included in these first two sections, and there are several exercises. The third section is devoted to a summary description without proofs of various calculations and applications that have been obtained in the study of π_2.

This paper may be taken as a companion and sequel to [Pr91], where pictures are treated within the purely combinatorial context of identity sequences.

1 The Theory of Pictures

As is customary, 2-complexes will be specified by means of group presentations. Let $\mathcal{P} = \langle \mathbf{x} \mid \mathbf{r} \rangle$ be a presentation for a group G. Thus, $G = F/N$ where F is the free group with basis $\mathbf{x}$ and N is the normal closure in F of the set $\mathbf{r}$ of (not necessarily reduced) words in $\mathbf{x} \cup \mathbf{x}^{-1}$. We allow the occurrence of trivial and repeated relators $r \in \mathbf{r}$, so $\mathbf{r}$ should be treated as an indexed set of words. We will say that two words u and v in $\mathbf{x} \cup \mathbf{x}^{-1}$ are *identically equal* if they represent the same element in the free semigroup on $\mathbf{x} \cup \mathbf{x}^{-1}$. The words u and v are *freely equal* if they represent the same element of F.

The *cellular model* of $\mathcal{P}$ is the 2-complex $K_{\mathcal{P}}$ that has a single 0–cell, one 1–cell for each generator $x \in \mathbf{x}$ and one 2–cell for each relator. An orientation of the cells in the one-skeleton $K_{\mathcal{P}}^{(1)}$ determines an isomorphism $\pi_1 K_{\mathcal{P}}^{(1)} \cong F$. The 2–cell2–cell corresponding to a relator $r \in \mathbf{r}$ is attached along a based loop in the one-skeleton that spells the word r. The inclusion $K_{\mathcal{P}}^{(1)} \subseteq K_{\mathcal{P}}$ induces a surjection $F \to \pi_1 K_{\mathcal{P}}$ with kernel N. In particular, $\pi_1 K_{\mathcal{P}}$ is canonically isomorphic to G, and so $\pi_2 K_{\mathcal{P}}$ is a left $\mathbf{Z}G$-module under the homotopy action of $\pi_1 K_{\mathcal{P}}$. We are interested in determining $\mathbf{Z}G$-module generators of $\pi_2 K_{\mathcal{P}}$ in terms of the presentation $\mathcal{P}$. The 0–cell of $K_{\mathcal{P}}$ will be used as basepoint for all homotopy groups, and we will often write $\pi_2 \mathcal{P}$ for $\pi_2 K_{\mathcal{P}}$.

1.1   Pictures

Pictures were introduced in [Ig79$_1$, Ro79]. There are many references for the basic theory, including [BrHu82, BoPr92, CoHu82, DuHo92$_1$, DuHo92$_2$, Fe83, Ho89, Ho90, Hu81, Ig79$_1$, Ig79$_2$, Pr91, Ro79, Wa80]. In this section, we review without proofs those aspects of the theory that pertain to π_2 calculations.

A *picture* $\mathbf{P}$ is a finite collection of pairwise disjoint closed discs $\Delta_1, \ldots, \Delta_m$ in a closed disc D^2, together with a finite collection of pairwise disjoint compact one-manifolds $\alpha_1, \ldots, \alpha_n$ properly embedded in $D^2 - \bigcup_i \mathrm{int}\Delta_i$ (where "int" denotes interior). By the *discs* of $\mathbf{P}$ we mean the discs $\Delta_1, \ldots, \Delta_m$ and not the ambient disc D^2; $\alpha_1, \ldots, \alpha_n$ are the *arcs* of $\mathbf{P}$. An arc can be either a circle or an interval. The *boundary* ∂D^2 will be denoted $\partial \mathbf{P}$. The *corners* of a disc Δ of $\mathbf{P}$ are the closures of the connected components of $\partial \Delta - \bigcup_j \alpha_j$. The *regions* of $\mathbf{P}$ are the closures of the connected components of $D^2 - (\bigcup_i \Delta_i \cup \bigcup_j \alpha_j)$. The *components* of $\mathbf{P}$ are the connected components of $\bigcup_i \Delta_i \cup \bigcup_j \alpha_j$; $\mathbf{P}$ is *connected* if it has at most one component. The picture $\mathbf{P}$ is *spherical* if it has at least one disc and no arc of $\mathbf{P}$ meets $\partial \mathbf{P}$.

One can think of a spherical picture $\mathbf{P}$ as being supported on the two-sphere simply by identifying $\partial \mathbf{P}$ to a point. When $\mathbf{P}$ is connected, a tessellation of S^2 is then obtained by shrinking each disc of $\mathbf{P}$ to a vertex. The discs (resp. arcs, regions) of $\mathbf{P}$ correspond to the vertices (resp. edges, faces) of the tessellation. The geometry of the sphere therefore restricts the structure of connected spherical pictures as follows. An *angle function* on a picture $\mathbf{P}$ is a real-valued function θ on the set of corners of $\mathbf{P}$. Associated to θ is a *curvature function* γ defined on the discs Δ of $\mathbf{P}$ by

$$\gamma(\Delta) = 2\pi - \sum_{c \subseteq \partial \Delta} \theta(c)$$

and on the regions F of $\mathbf{P}$ by

$$\gamma(F) = 2\pi - \sum_{c \subseteq \partial F} (\pi - \theta(c)).$$

In both of these sums, c denotes a corner. Noting that $\mathbf{P}$ has twice as many corners as arcs, an Euler characteristic count reveals that

$$\sum_{\Delta} \gamma(\Delta) + \sum_{F} \gamma(F) = 2\pi\chi(S^2) = 4\pi,$$

where the sums are taken over all discs Δ and regions F of $\mathbf{P}$. The following result is an immediate consequence.

Lemma 1.1 *For any angle function on any connected spherical picture, some disc or region has positive curvature.*   □

The real number $\theta(c)$ is thought of as the radian angle measure of the corner c. A picture $\mathbf{P}$ is *flat* at a disc Δ if the sum of the angles of all corners of Δ is exactly 2π, which is to say that $\gamma(\Delta) = 0$. A region F containing d corners is *positively curved* (that is, $\gamma(F) > 0$) if the sum of the angle measures of the corners in F is greater than $(d - 2)\pi$.

There are other ways to formulate curvature. One can exchange the roles of the discs and regions in the definition of the curvature function. This simply amounts to a consideration of the dual tessellation of the sphere. One can also replace the constant π by any positive real number; popular choices for this are 1 and 180. None of these variations provides information beyond that which is contained in Lemma 1.1. Curvature considerations appear in [Ed91, Ho89, Ho90, DuHo92$_1$, DuHo92$_2$], among many others. The following exercise is fundamental to small cancellation theory [Ly66, LySc77], and indicates some basic restrictions on the structure of connected spherical pictures.

Exercise 1.2 *Show that if the boundary of every region (resp. disc) of a connected spherical picture contains at least p (resp. q) corners, then $1/p + 1/q > 1/2$. Prove this in two ways, first by assigning angles so that discs are flat, and then again by assigning angles so that regions are flat.*

Assume that a group presentation $\mathcal{P} = \langle \mathbf{x} \mid \mathbf{r} \rangle$ is given. Let $\mathbf{P}$ be a picture, and fix an orientation of the ambient disc, thereby determining a sense of positive rotation (e.g. clockwise). Assume that the discs and arcs of $\mathbf{P}$ are labeled by elements of $\mathcal{P}$ as follows.

- Each arc of **P** is equipped with a normal orientation (indicated by an arrow transverse to the arc), and is labeled by an element of $\mathbf{x} \cup \mathbf{x}^{-1}$.

- Each disc Δ of **P** is equipped with a sign $\epsilon(\Delta) = \pm 1$ and is labeled by a relator $R(\Delta) \in \mathbf{r}$.

For a corner c in a disc Δ of **P**, $W(c)$ denotes the word in $\mathbf{x} \cup \mathbf{x}^{-1}$ obtained by reading in order the (signed) labels on the arcs that are encountered in a walk around $\partial\Delta$ in the positive direction, beginning and ending at an interior point of c. The oriented and labeled picture **P** is a *picture over* $\mathcal{P}$ if:

- For each corner c in each disc Δ of **P**, $W(c)$ is identically equal to a cyclic permutation of $R(\Delta)^{\epsilon(\Delta)}$.

Note that if no relator of $\mathcal{P}$ is a cyclic permutation of its inverse, nor of any other relator or its inverse, then the labelings $R(\Delta)$ and signs $\epsilon(\Delta)$ are actually determined by the labelings and normal orientations on the arcs of **P**. Examples are displayed in Figure 1.

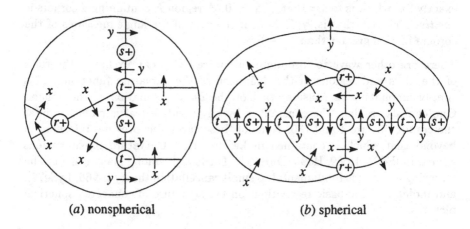

(a) nonspherical   (b) spherical

Figure V.1. $\mathcal{P} = \langle x, y \mid r, s, t \rangle; r = x^4, s = y^2, t = (xy)^2$

A corner c is a *basic corner* of the ambient disc Δ of **P** if $W(c)$ is identically equal to $R(\Delta)^{\epsilon(\Delta)}$. When $R(\Delta)$ is nonempty, $R(\Delta)$ is identically equal to a unique word of the form Q^p where $p \geq 1$ and Q is not a proper power; Q is the *root* of $R(\Delta)$ and p is the *period*. The disc Δ then has exactly p basic corners.

A picture **P** over $\mathcal{P}$ becomes a *based* picture over $\mathcal{P}$ when it is equipped with basepoints as follows.

- Each disc Δ has one basepoint, which is a selected point in the interior of a basic corner of Δ.

- **P** has a global basepoint, which is a selected point in $\partial \mathbf{P}$ that does not lie on any arc of **P**.

Figure 2 shows three based spherical pictures that are all supported by the same spherical picture over $\mathcal{P} = \langle x \mid x^4 \rangle$.

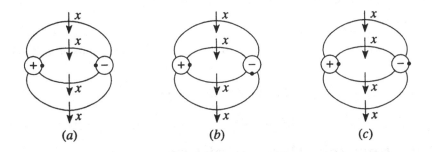

(a)   (b)   (c)

Figure V.2. Based pictures over $\mathcal{P} = \langle x \mid x^4 \rangle$

Let **P** be a based picture over $\mathcal{P}$. The *boundary label* on **P** is the word $W(\mathbf{P})$ obtained by reading in order the (signed) labels on the arcs of **P** that are encountered in a walk around $\partial \mathbf{P}$ in the positive direction, beginning and ending at the global basepoint. In Figure 1(a), if we place the global basepoint in the "southwest" region of the picture, the boundary label is $x^2 y x^{-2} y^{-1} = [x^2, y]$. Alteration of the global basepoint of **P** or of the orientation of the ambient disc changes the boundary label of **P** only up to cyclic permutation and inversion.

A path β in **P** that does not meet the interior of any disc of **P** will be called *transverse* if (i) whenever β meets $\partial \mathbf{P}$ or a disc of **P**, it does so only in its endpoints, (ii) no endpoint of β touches any arc, and (iii) β meets the arcs in just finitely many transverse intersections. A traverse of an oriented transverse path β determines a labeling word $W(\beta)$ in the alphabet $\mathbf{x} \cup \mathbf{x}^{-1}$.

When a transverse path β in **P** is actually an embedded circle, then it encloses a *subpicture* **B** of **P**. This subpicture consists of the discs and (portions of) arcs that are separated from $\partial \mathbf{P}$ by β. When **P** is spherical, the *complement* of **B** in **P** is defined as follows. Delete the interior of **B** to form an oriented annulus. Identification of $\partial \mathbf{P}$ to a point produces an oriented disc that has boundary β, and which supports a new picture over $\mathcal{P}$. The complement of **B** in **P** is obtained from this new picture by a planar reflection. The complement has the same boundary label as **B** and its discs are those of **P − B**, taken with opposite signs. See Figure 3.

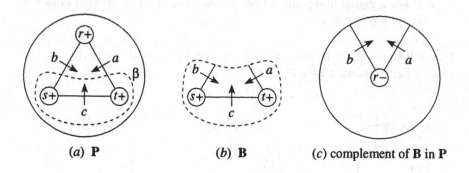

      (a) **P**   (b) **B**   (c) complement of **B** in **P**

Figure V.3. Complement

1.2   Homotopy theory of pictures

Let $K = K_{\mathcal{P}}$ be the cellular model of $\mathcal{P} = \langle \mathbf{x} \mid \mathbf{r} \rangle$. Each based picture **P** over $\mathcal{P}$ determines a map $f_{\mathbf{P}} : (D^2, \partial D^2) \to (K, K^{(1)})$ of based pairs as follows. Thicken each arc α of **P** to a product $\alpha \times D^1$ in **P** so that no thickened arc meets any basepoint of **P** and so that each slice $\alpha \times \{z\}(z \in D^1)$ is properly embedded parallel to α. The orientation and labeling on α show how to collapse $\alpha \times D^1$ onto D^1 by the projection and then map D^1 characteristically onto a 1–cell of K. The basepoint, orientation and labeling on each disc Δ determine a based characteristic mapping of Δ onto a 2–cell of K. The remaining material in D^2 is mapped to the 0–cell of K. The resulting map $f_{\mathbf{P}}$ is continuous since **P** is a picture over $\mathcal{P}$, and is well-defined up to based homotopy of pairs. Further, the homotopy class of $f_{\mathbf{P}}$ is unchanged by isotopic deformations of D^2.

The *homotopy class* of **P** is the element $[f_{\mathbf{P}}] \in \pi_2(K, K^{(1)})$. In case **P** is spherical, the map $f_{\mathbf{P}}$ canonically induces a based spherical map $f_{\mathbf{P}} : S^2 \to K$, and we think of $[f_{\mathbf{P}}]$ as an element of $\pi_2\mathcal{P} = \pi_2 K$. The simplicial techniques in Chapter II, §1, of this volume serve to establish the following result, which can be traced to [Wh41$_1$]. See also [Fe83, Hu81, Si80].

Theorem 1.3 *Every element of $\pi_2(K, K^{(1)})$ has the form $[f_{\mathbf{P}}]$ for some based picture over $\mathcal{P}$. Every element of $\pi_2\mathcal{P}$ has the form $[f_{\mathbf{P}}]$ for some based spherical picture **P** over $\mathcal{P}$.*   □

For a nonspherical picture **P** over $\mathcal{P}$, varying choices of global basepoint can change the homotopy class of **P** only up to the homotopy action by an element of $F \cong \pi_1 K^{(1)}$. In case $\mathcal{P}$ has relators with period greater than one, the map $f_{\mathbf{P}}$ depends intrinsically on the choice of disc basepoints. Consider the three based spherical pictures in Figure 2. Each of these gives rise to a lifted spherical map $\tilde{f}_{\mathbf{P}} : S^2 \to \tilde{K}$ to the universal covering complex of K. In Figure 2(a), the map $\tilde{f}_{\mathbf{P}}$ (and hence $f_{\mathbf{P}}$) "folds" the two discs of **P** together mirrorwise across an equator of S^2, and then maps these identified discs to a single 2–cell of $\tilde{K}$. In particular, one finds that $[f_{\mathbf{P}}] = 0$ in $\pi_2 K \cong H_2 \tilde{K}$. On the other hand, the based pictures in Figures 2(b) and (c) determine homotopically nontrivial spherical maps. For each of these, the lifted map $\tilde{f}_{\mathbf{P}}$ carries the two discs of **P** to distinct 2–cells of $\tilde{K}$.

There are well-known alternative procedures for producing homotopy elements from based pictures. To each based (spherical) picture over $\mathcal{P}$ is associated the Peiffer equivalence class of an (identity) sequence over $\mathcal{P}$. A homotopy class in $\pi_2(K, K^{(1)})$ then arises from Whitehead's description of that group as a *free crossed F-module*. (See Chapters II and IV of this volume.) From a based picture one can read off homological chains in $C_2 \tilde{K}$ and in $C_2 K$, giving "pictorial" views of the Hurewicz homomorphisms $\pi_2(K, K^{(1)}) \to H_2(\tilde{K}, \tilde{K}^{(1)})$ and $\pi_2(K, K^{(1)}) \to H_2(K, K^{(1)})$. For spherical pictures, these associated chains are actually cycles, and determine the isomorphism $\pi_2 K \to H_2 \tilde{K}$ and the homomorphism $\pi_2 K \to H_2 K$. As an example, the homotopy classes arising from the pictures in Figures 2(b) and (c) are easily seen to be distinct upon consideration of the associated cycles in $C_2 \tilde{K} = H_2(\tilde{K}, \tilde{K}^{(1)})$. While these perspectives are very useful, we shall not pursue them here. See [BrHu82, Fe83, Pr91, Ra83] for further details.

The algebraic operations in second homotopy groups are easily visualized in terms of pictures. Let **P** and **Q** be based pictures over $\mathcal{P}$, represented schematically in Figure 4.

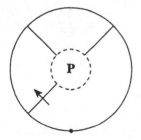 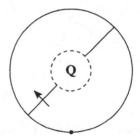

Figure V.4. **P** and **Q**

Two new pictures $\mathbf{P} \cdot \mathbf{Q}$ and $\mathbf{P}^{-1}$ are constructed as in Figure 5.

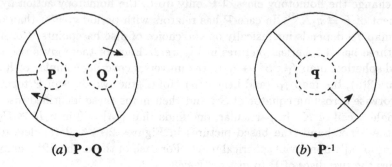

(a) $\mathbf{P} \cdot \mathbf{Q}$ (b) $\mathbf{P}^{-1}$

Figure V.5. Products and inverses

Thus $\mathbf{P} \cdot \mathbf{Q}$ is a certain union of $\mathbf{P}$ and $\mathbf{Q}$ and $\mathbf{P}^{-1}$ is a mirror image of $\mathbf{P}$ obtained by a planar reflection and by changing the signs on all discs of $\mathbf{P}$. For spherical $\mathbf{P}$ and $\mathbf{Q}$, these pictures will be denoted $\mathbf{P}+\mathbf{Q}$ and $-\mathbf{P}$, respectively. The point is that $[f_{\mathbf{P} \cdot \mathbf{Q}}] = [f_{\mathbf{P}}] \cdot [f_{\mathbf{Q}}]$ and $[f_{\mathbf{P}^{-1}}] = [f_{\mathbf{P}}]^{-1}$ in $\pi_2(K, K^{(1)})$, with analogous formulae holding in the (additive abelian) group $\pi_2 K$ for spherical $\mathbf{P}$ and $\mathbf{Q}$. The homotopy boundary of $[f_{\mathbf{P}}]$ under $\pi_2(K, K^{(1)}) \to \pi_1 K^{(1)} \cong F$ is determined by the boundary label $W(\mathbf{P})$. For a word u in $\mathbf{x} \cup \mathbf{x}^{-1}$, the associated element of F acts on elements of $\pi_2(K, K^{(1)})$, while the associated element of $G = F/N$ acts on elements of $\pi_2 K$. Representative pictures for these actions are denoted $u \cdot \mathbf{P}$, and are displayed in Figure 6 (for nonspherical $\mathbf{P}$ and spherical $\mathbf{R}$).

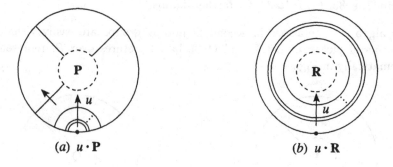

(a) $u \cdot \mathbf{P}$ (b) $u \cdot \mathbf{R}$

Figure V.6. Actions

If a word u in $\mathbf{x} \cup \mathbf{x}^{-1}$ represents an element in the normal closure N of $\mathbf{r}$ in F, then it is easy to construct a based picture over $\mathcal{P} = \langle \mathbf{x} \mid \mathbf{r} \rangle$ with boundary label u. There is the following pictorial version of the "van Kampen lemma" [vK33][BrHu82, p.190][LySc77, V.1.1,V.1.2].

Theorem 1.4 *A word u in $\mathbf{x} \cup \mathbf{x}^{-1}$ represents an element of N if and only if there is a based picture over $\mathcal{P}$ with boundary label u. If β and γ are oriented transverse paths with the same endpoints in a picture $\mathbf{P}$ over $\mathcal{P}$, then $W(\beta)N = W(\gamma)N$.* □

This result can be proved by noting that each loop in $\mathbf{P}$ lifts to a loop in the universal cover $\tilde{K}$, and so determines an element of N. As an example, the picture in Figure 1(a) shows that x^2 determines a central element in the group presented by $\mathcal{P} = \langle x, y \mid x^4, y^2, (xy)^2 \rangle$.

Certain configurations in a based picture contribute nothing to the homotopy class. A *floating arc* in a based picture $\mathbf{P}$ is an arc of $\mathbf{P}$ that separates the ambient disc into two components, one of which contains the global basepoint of $\mathbf{P}$ and all remaining arcs and discs of $\mathbf{P}$. A *folding pair* in $\mathbf{P}$ is a connected spherical subpicture of $\mathbf{P}$ that contains exactly two discs such that (i) the two discs are labeled by the same relator and have opposite signs, (ii) the basepoints of the discs lie in the same region, and (iii) each arc in the subpicture has an endpoint on each disc. The picture in Figure 2(a) is a folding pair, but those in Figures 2(b) and (c) are not.

Let X be a set of based spherical pictures over $\mathcal{P}$. By an *X-picture* we mean either a picture $\mathbf{P}$ from X or its mirror image $-\mathbf{P}$. The $\mathbf{Z}G$-submodule of $\pi_2 \mathcal{P}$ generated by the homotopy classes $[f_{\mathbf{P}}](\mathbf{P} \in X)$ will be denoted $J(X)$. We say that X *generates* $\pi_2 \mathcal{P}$ if $J(X) = \pi_2 \mathcal{P}$.

The following operations can be applied to based pictures over $\mathcal{P}$.

BRIDGE: Bridge move; see Figure 7.

FLOAT: Insert or delete a floating arc.

FOLD: Insert of delete a folding pair.

REPLACE(X): Replace a subpicture of a given picture by the complement of that subpicture in an X-picture.

Figure V.7. Bridge move

The first three of these operations have no effect on homotopy classes of based pictures; suitable homotopies are described in [Fe83, pp. 56-59][Hu81, Proposition 3]. We shall say that based pictures **P** and **Q** are *equivalent* if **P** can be transformed into **Q** (up to planar isotopy) by a finite sequence of operations BRIDGE, FLOAT, FOLD. Based spherical pictures **P** and **Q** are *X-equivalent* if **P** can be transformed into **Q** (up to planar isotopy) by a finite sequence of operations BRIDGE, FLOAT, FOLD, REPLACE(X).

The operation REPLACE has the following effect. Suppose that $\mathbf{P}_1$ and $\mathbf{P}_2$ are based spherical pictures over $\mathcal{P}$, each of which contains as a subpicture an isomorphic copy of a picture **B**. (While **B** itself is a based picture over $\mathcal{P}$, the boundary and selected global basepoint of **B** are not actually part of $\mathbf{P}_1$ or $\mathbf{P}_2$.) For $i = 1, 2$, let β_i be a transverse path in $\mathbf{P}_i$ from the global basepoint of $\mathbf{P}_i$ to that of **B**. Let **R** be the result when the subpicture **B** of $\mathbf{P}_1$ is replaced by the complement of **B** in $\mathbf{P}_2$.

Lemma 1.5 *In* $\pi_2\mathcal{P}$,

$$[f_{\mathbf{R}}] = [f_{\mathbf{P}_1}] - W(\beta_1)W(\beta_2)^{-1}[f_{\mathbf{P}_2}].$$

In particular, if $\mathbf{P}_2$ *is an X-picture, then* $[f_{\mathbf{R}}] - [f_{\mathbf{P}_1}] \in J(X)$.

Proof: A sequence of bridge moves applied to $W(\beta_2)^{-1} \cdot \mathbf{P}_2$ yields a picture $\mathbf{P}_2'$ containing **B**, and where the global basepoint of **B** is "exposed", lying in the boundary region of $\mathbf{P}_2'$. See Figure 8.

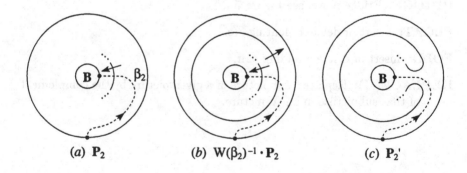

(a) $\mathbf{P}_2$ (b) $W(\beta_2)^{-1} \cdot \mathbf{P}_2$ (c) $\mathbf{P}_2'$

Figure V.8. Exposed basepoint

Let $\mathbf{P}_1'$ be the picture obtained from $\mathbf{P}_1$ by inserting a copy of $-\mathbf{P}_2'$ so that the global basepoint of $-\mathbf{P}_2'$ is adjacent to that of the subpicture **B** of $\mathbf{P}_1$. A sequence of bridge moves "around" the inserted material yields the picture

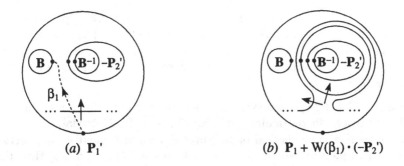

Figure V.9. Insert and bridge

$P_1 + W(\beta_1) \cdot (-P_2')$, showing that $[f_{\mathbf{P}_1'}] = [f_{\mathbf{P}_1}] - W(\beta_1)W(\beta_2)^{-1}[f_{\mathbf{P}_2}]$. See Figure 9.

On the other hand, the oppositely oriented and adjacent copies of $\mathbf{B}$ in $\mathbf{P}_1'$ can be "cancelled" by a sequence of bridge moves and deletions of folding pairs. The resulting picture is exactly $\mathbf{R}$, and so $[f_{\mathbf{P}_1'}] = [f_{\mathbf{R}}]$. $\qquad\square$

Theorem 1.6 *Let* $\mathbf{P}$ *and* $\mathbf{Q}$ *be based pictures over* $\mathcal{P}$.

1. $[f_{\mathbf{P}}] = [f_{\mathbf{Q}}]$ *in* $\pi_2(K(\mathcal{P}), K_{\mathcal{P}}^{(1)})$ *if and only if* $\mathbf{P}$ *and* $\mathbf{Q}$ *are equivalent.*

2. *For spherical* $\mathbf{P}$ *and* $\mathbf{Q}$, $[f_{\mathbf{P}}] - [f_{\mathbf{Q}}] \in J(X)$ *in* $\pi_2\mathcal{P}$ *if and only if* $\mathbf{P}$ *and* $\mathbf{Q}$ *are* X-*equivalent.* $\qquad\square$

A homotopy-theoretic proof of 1 is outlined in [Hu81, Proposition 3]. For a combinatorial proof see [Pr91, Theorem 2.5*]. This result implies 2 in the case where $X = \emptyset$. Note that the operation REPLACE(X) includes the following operation.

INSERT(X): Insert or delete an X-picture.

With 1.5, the result 2 follows from 1 and the proof of [Pr91, Theorem 2.6].

2 Generation of Π_2

2.1 Asphericity

The 2-complex $K_\mathcal{P}$ is aspherical (i.e., has contractible universal covering) if and only if $\pi_2 K_\mathcal{P} = 0$. More generally, $K_\mathcal{P}$ (or rather $\mathcal{P}$) is said to be *combinatorially aspherical* (CA) if $\pi_2 K_\mathcal{P} = \pi_2\mathcal{P}$ is generated by the set of based spherical pictures over $\mathcal{P}$ that contain exactly two discs. (Such presentations are also called *aspherical* in [Hu79, Identity Theorem and Theorem 1].) As a fundamental example, the Simple Identity Theorem of R. C. Lyndon [Ly50] implies that if the relator of a one-relator presentation $\mathcal{P}$ is not freely trivial, then $\mathcal{P}$ is (CA).

Note that if $\mathcal{P}$ is (CA), then every spherical picture over $\mathcal{P}$ contains an even number of discs. In particular, no relator of $\mathcal{P}$ is freely trivial, for such a relator gives rise to a spherical picture over $\mathcal{P}$ with just one disc. In general, if a spherical picture over $\mathcal{P}$ contains just two discs Δ_1 and Δ_2, then the relator label $R(\Delta_1)$ is freely conjugate to $R(\Delta_2)^{\pm 1}$. Consider the following *Relator Hypothesis* for $\mathcal{P}$.

RH: No relator of $\mathcal{P}$ is freely trivial, nor is a conjugate of any other relator or its inverse.

It is clear that if $K_{\mathcal{P}}$ is aspherical, then $\mathcal{P}$ is (CA). For the converse, $K_{\mathcal{P}}$ is aspherical if and only if (i) $\mathcal{P}$ satisfies the Relator Hypothesis, (ii) each relator of $\mathcal{P}$ has period one, and (iii) $\mathcal{P}$ is (CA) (see [Hu81, Proposition 5]).

Discarding freely trivial and repeated relators (up to conjugacy), any presentation $\mathcal{P} = \langle \mathbf{x} \mid \mathbf{r} \rangle$ contains a subpresentation $\mathcal{P}_0 = \langle \mathbf{x} \mid \mathbf{r}_0 \rangle$ where $\mathcal{P}_0$ satisfies the Relator Hypothesis. Further, there is a homotopy equivalence

$$K_{\mathcal{P}} \simeq K(\mathcal{P}_0) \vee \bigvee_{\mathbf{r} - \mathbf{r}_0} S^2.$$

In particular, $\pi_1 K(\mathcal{P}_0) \cong \pi_1 K_{\mathcal{P}} \cong G$ and

$$\pi_2 K_{\mathcal{P}} \cong \pi_2 K(\mathcal{P}_0) \oplus \bigoplus_{\mathbf{r} - \mathbf{r}_0} \mathbf{Z}G.$$

A basis for the free summand $\oplus_{\mathbf{r} - \mathbf{r}_0} \mathbf{Z}G$ is easily given in terms of based spherical pictures over $\mathcal{P}$ having just one or two discs. In considering the structure of π_2, there is therefore no loss of generality if one works only with presentations that satisfy the Relator Hypothesis. In addition, one may also assume that each relator is cyclically reduced, as cyclic reduction of relators of a presentation $\mathcal{P}$ does not affect the homotopy type of $K_{\mathcal{P}}$.

We remark that covering space topology naturally leads to the consideration of 2-complexes having more that just a single 0–cell, and of presentations that contain repeated relators. For example, the canonical cell structure on the universal cover of the model of $\langle x \mid x^4 \rangle$ has four 0–cells, and this covering space is homeomorphic to the model of $\langle x \mid x, \ x, \ x, \ x \rangle$.

A *dipole* in a picture over $\mathcal{P}$ consists of an arc which meets two corners c_1, c_2 in distinct discs such that (i) the two discs are labeled by the same relator and have opposite signs, (ii) c_1 and c_2 lie in the same region of the picture, and (iii) $W(c_1) = W(c_2)^{-1}$. See Figure 10.

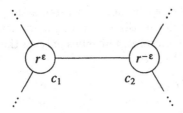

Figure V.10. Dipole

By a *complete dipole* over $\mathcal{P}$, we mean a connected based spherical picture over $\mathcal{P}$ that contains just two discs, and where each arc of the picture constitutes a dipole. Note that a complete dipole is just a folding pair except possibly with a "twist", in that the disc basepoints need not lie in the same region (unless the relator that labels the two discs has period one). All the pictures of Figure 2 are complete dipoles. A complete dipole will be called *primitive* if the relator labeling the discs has root Q and period $p > 1$, and there is a transverse path joining the disc basepoints with label Q^f, where $\gcd(f, p) = 1$. In Figure 2, only (b) is a primitive dipole. The following lemma shows that, modulo primitive dipoles, one need not be concerned with choices of disc basepoints.

Lemma 2.1 *Suppose that X contains a primitive dipole for each proper power relator in $\mathcal{P}$. If a based spherical picture $\mathbf{P}$ over $\mathcal{P}$ is obtained from $\mathbf{Q}$ simply by changing disc basepoints, then $[f_\mathbf{P}] - [f_\mathbf{Q}] \in J(X)$.*

Proof: Using the operation REPLACE(X), primitive dipoles can be used to alter disc basepoints in any prescribed manner. □

Combinatorially aspherical presentations that satisfy the Relator Hypothesis are called *almost aspherical* in [GuRa81], and are said to have the *Identity Property* in [HoSc79]. They are precisely the presentations $\mathcal{P}$ for which $\pi_2\mathcal{P}$ is generated by primitive dipoles. Combinatorial asphericity of a presentation has strong consequences for the group presented. Outstandingly, the maximal finite subgroups are cyclic and are determined (up to conjugacy) by the roots of the relators [Hu79]. In particular, no non-cyclic finite group admits a (CA) presentation. An extensive catalogue of (CA) presentations appears in [Pr91].

We will say that $\mathcal{P}$ is *combinatorially reducible* (CR) if each connected spherical picture over $\mathcal{P}$ contains a dipole. Note that if $\mathcal{P}$ is (CR) then each spherical picture over $\mathcal{P}$ has a dipole, and that $\mathcal{P}$ satisfies the Relator Hypothesis. Further, if a picture **P** contains a dipole, then a series of bridge moves splits off a complete dipole. It follows that (CR) $\Rightarrow$ (CA). An advantage of the (CR) property is that it can often be detected using a simple combinatorial test. The power of this test lies in the fact that its application does not require the consideration of any pictures, but relies only on the structure of a certain graph that is easily read off $\mathcal{P}$.

The *star graph* of $\mathcal{P} = \langle \mathbf{x} \mid \mathbf{r} \rangle$ is the geometric graph $\mathcal{P}^{st}$ with vertex set $\mathbf{x} \cup \mathbf{x}^{-1}$ and with one geometric edge with endpoints x^ϵ and $y^{-\delta}$ for each cyclic syllable $x^\epsilon y^\delta$ of the root of each relator. For example, for $n \geq 1$, $\langle x \mid x^n \rangle^{st}$ has two vertices connected by a single edge. Closely related is the *link graph* Lk $\mathcal{P}$; it has an edge for each cyclic syllable of each relator. In the link graph, proper power relators give rise to multiple parallel edges. The link graph of $\langle x \mid x^n \rangle$ has two vertices joined by n edges. Topologically, the link is the boundary of a basic open neighborhood of the 0–cell in $K_{\mathcal{P}}$. See [Pr88, Pr91, Si81] for examples and alternate descriptions of link and star graphs.

The star graph is related to pictures as follows. Let c be a corner of a disc Δ in a picture **P** over $\mathcal{P}$. Writing $W(c)^{\epsilon(\Delta)} = y^\delta V x^\epsilon$, the corner c determines the cyclic syllable $x^\epsilon y^\delta$ of (the root of) $R(\Delta)$, and hence determines an edge of $\mathcal{P}^{st}$. If F is a region of **P**, then traversing a component of ∂F that does not meet $\partial \mathbf{P}$ determines a sequence of corners, which in turn defines a cycle (i.e., a closed loop) in $\mathcal{P}^{st}$. A spherical picture **P** has no dipoles if and only if all cycles arising this way are cyclically reduced (i.e. contain no backtracking).

A *weight function* on $\mathcal{P}^{st}$ is a real-valued function on the edges of $\mathcal{P}^{st}$. Note that each syllable of each relator acquires a weight; the weight on a path in $\mathcal{P}^{st}$ is the sum of the weights of its constituent edges, counted with multiplicities and without regard to the direction of traversal. The following *weight test* evolved from the *coloring test* of A. Sieradski [Si81] through the work of S. Gersten [Ge87$_1$] and S. Pride [Pr88].

Theorem 2.2 *Suppose that* $\mathcal{P}^{st}$ *admits a weight function such that the following two conditions hold.*

1. *The sum of the weights of the syllables of the root of any relator is not more than* $n - 2/p$, *where* p *is the period of the relator and* n *is the length of the root.*

2. *Each cyclically reduced loop in* $\mathcal{P}^{st}$ *has weight at least two.*

Then $\mathcal{P}$ is (CR).   □

Proof: Let **P** be a connected spherical picture over $\mathcal{P}$. The dual picture **P*** has a single disc in the interior of each region of **P**, and has the same number of arcs as **P**. Each arc of **P*** meets exactly one arc of **P** in exactly one transverse intersection, and **P*** is a connected spherical picture. There is an obvious bijective correspondence between the corners of **P*** and the corners of **P**. To each corner of **P*** there is therefore associated an edge of $\mathcal{P}^{st}$, and so we may define an angle function on **P*** by letting the angle on a corner of **P*** be the product of π times the weight on the corresponding edge of $\mathcal{P}^{st}$. The condition 1 implies that each region of **P*** is non-positively curved. If we suppose that **P** has no dipoles, then the condition 2 implies that each disc of **P*** is non-positively curved, contrary to Lemma 1.1.   □

A refined version of the weight test is given in the *cycle test* of G. Huck and S. Rosebrock [HuRo92]. Applications to decision problems are discussed in Chapter VI of this volume. Huck has recently devised another form of the weight test (unpublished), in which the condition 2 of the weight test is weakened to require only that each simple cycle in $\mathcal{P}^{st}$ has weight at least two. (A cycle in a graph is *simple* if it touches each vertex and edge of the graph at most once.) Presentations satisfying Huck's weight test are (CA), and are in fact *diagrammatically aspherical* (DA) in the sense of [ChCoHu81, CoHu82], but they need not be (CR). For example, the presentations $\langle x \mid xx^{-1}x \rangle$ and $\langle x, y \mid x^2y^{-1}, xy^{-1} \rangle$ both satisfy Huck's weight test, and yet each admits a based spherical picture without dipoles. The (DA) property is characterized in terms of pictures in [CoHu82, Lemma 6] and in [Hu81, Proposition 8]. Diagrammatically aspherical presentations are called "aspherical" in [LySc77, p. 156] and in [Si80], where a homotopy-theoretic interpretation of (DA) is given. For completeness, we record here the fact that (CR) $\Rightarrow$ (DA) $\Rightarrow$ (CA), and that neither of these implications is reversible. See [ChCoHu81, p. 8], [CoHu82], [Hu79, Note added in proof], [Si80], [Ge87$_1$], and Chapter X of this volume for further discussion.

Finally, a (CR) presentation in which each relator has period one is said to be *diagrammatically reducible* (DR). The model of a (DR) presentation is aspherical. The (DR) property was first described in [Si81] by Sieradski. A homotopy-theoretic characterization of (DR) appears in [Ge87$_2$]. Applications of diagrammatic reducibility to the study of equations over groups were discovered independently by Sieradski (unpublished) and Gersten [Ge87$_1$]. At present it is unknown to what extent analogous results on equations over groups can be proved in the more general setting involving (CR). See also [Ge86, Ge87$_2$] in this regard.

Exercise 2.3 *Use the weight test to show that* $\langle x, y, z \mid x^{m_1} y^{n_1} z^{p_1}, x^{m_2} y^{n_2} z^{p_2} \rangle$ *is (DR) if* $m_1 m_2, n_1 n_2$ *and* $p_1 p_2$ *are all negative. What other hypotheses on the exponents will ensure diagrammatic reducibility?*

2.2   A Dehn algorithm for π_2

A procedure for determining generators of $\pi_2 \mathcal{P}$ is the following pictorial analogue of the *Dehn algorithm*. First, look for a set X of "obvious" based connected spherical pictures over $\mathcal{P}$. Next, devise a measure μ of *complexity* for based pictures over $\mathcal{P}$. One is free to choose whatever will work, though the function μ should take values in an ordered monoid that has no strictly decreasing infinite sequence, and should be *additive*, in that whenever **B** is a subpicture with complement **B**' in a based spherical picture **Q**, then $\mu(\mathbf{Q}) = \mu(\mathbf{B}) + \mu(\mathbf{B}')$. Suppose that such a complexity function has been selected.

Theorem 2.4 *Assume that for each connected based spherical picture* **P** *over* $\mathcal{P}$, *there is a based spherical picture* $\mathbf{P}_0$ *over* $\mathcal{P}$ *such that*

1. **P** *and* $\mathbf{P}_0$ *are* X-*equivalent,*

2. $\mu(\mathbf{P}_0) \leq \mu(\mathbf{P})$ *and*

3. $\mathbf{P}_0$ *contains a subpicture* **B** *such that* **B** *is also a subpicture of an* X-*picture* **Q**, *and such that* $\mu(\mathbf{B}) > \mu(\mathbf{B}')$, *where* **B**' *is the complement of* **B** *in* **Q**.

Then $\pi_2 \mathcal{P}$ *is generated by* X *together with the collection of based spherical pictures over* $\mathcal{P}$ *having minimal complexity.*

(Of course, one hopes that the pictures of minimal complexity are trivial, or at least are easily described.)

Proof: By Theorem 1.3, each element of $\pi_2 \mathcal{P}$ is of the form $[f_\mathbf{P}]$ for some (not necessarily connected) based spherical picture **P** over $\mathcal{P}$. By considering a connected component of **P**, there is a picture $\mathbf{P}_0$ satisfying the conditions 1-3 of the theorem. We can replace **B** by **B**' and transform $\mathbf{P}_0$ into a new picture **P**' with smaller complexity. (This is because μ is additive.) By Theorem 1.6.2, $[f_\mathbf{P}] - [f_{\mathbf{P}'}] \in J(X)$. Proceeding inductively, we find a based spherical picture **P**'' where $[f_\mathbf{P}] - [f_{\mathbf{P}''}] \in J(X)$ and **P**'' has minimal complexity. $\square$

As a simple application of this technique, one can easily show that $\pi_2 \mathcal{P}$ is generated by the set of all connected based spherical pictures over $\mathcal{P}$: Take the complexity of a picture to be the number of components. For a general example of an additive complexity function on $\mathcal{P} = \langle \mathbf{x} \mid \mathbf{r} \rangle$, suppose that $f :$ $\mathbf{x} \cup \mathbf{r} \to \mathbf{N}$ is a function with values in the non-negative integers. Each based picture $\mathbf{P}$ over $\mathcal{P}$ then acquires a complexity $\mu(\mathbf{P})$, given as the following sum over the discs and arcs of $\mathbf{P}$. Each disc contributes the f-value of its relator label. Each arc having both endpoints on discs of $\mathbf{P}$ contributes the f-value of its generator label, each arc having one endpoint on a disc of $\mathbf{P}$ contributes half the f-value of its generator label, and each are that does not meet any disc of $\mathbf{P}$ contributes zero. The measure μ is then additive. One is interested in spherical pictures with $\mu = 0$, and in subpictures $\mathbf{B}$ in X-pictures $\mathbf{Q}$ for which $\mu(\mathbf{B}) > \mu(\mathbf{Q})/2$.

To illustrate the method, consider the presentation $\mathcal{P} = \langle x, y \mid r, s, t \rangle$ for the dihedral group of order eight, where $r = x^4, s = y^2$ and $t = (xy)^2$. Let X consist of a primitive dipole for each of r, s, t, together with the picture $\mathbf{Q}$ of Figure 1(b), which can be equipped with basepoints in any allowable fashion. (See Lemma 2.1.) Define $f(x) = f(y) = 0, f(r) = 4, f(s) = f(t) = 1$, and let μ be the corresponding complexity function. Let $\mathbf{P}$ be a spherical picture over $\mathcal{P}$. If $\mathbf{P}$ contains a dipole, then a series of bridge moves splits off a complete dipole, and so as in the proof of Lemma 2.1, $\mathbf{P}$ is X-equivalent to a picture $\mathbf{P}_0$ that contains an entire X-picture, and where $\mu(\mathbf{P}) = \mu(\mathbf{P}_0)$. Suppose then that $\mathbf{P}$ contains no dipole. It is easy to check that $\langle x, y \mid s, t \rangle$ is (CR), and so $\mathbf{P}$ must contain at least one disc Δ labeled by r. Replacing $\mathbf{P}$ by $-\mathbf{P}$ if necessary, we may assume that Δ has sign $+1$. Since $\mathbf{P}$ has no dipoles, there is a subpicture of $\mathbf{P}$ of the form shown in Figure 11.

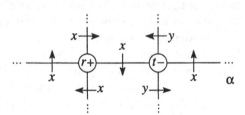

Figure V.11. A subpicture

Note that in $\mathbf{P}$, the arc α does not touch the disc Δ, for otherwise $\mathbf{P}$ would contain a subpicture with boundary label of the form $(yx^k)^{\pm 1}$ $(k = 0$ or $-1)$, in contradiction to the fact that G is not cyclic. (See Theorem 1.4.) It follows that $\mathbf{P}$ has a subpicture $\mathbf{B}$ of the form shown in Figure 12.

Now, $\mu(\mathbf{B}) = 9$ and $\mathbf{B}$ is also a subpicture of $\mathbf{Q}$, while $\mu(\mathbf{Q}) = 16$. Since pictures with zero complexity are homotopically trivial, Theorem 2.4 implies that X generates $\pi_2\mathcal{P}$.

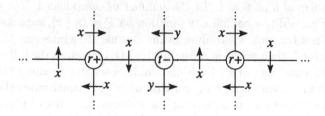

Figure V.12. **B**

Exercise 2.5 *Determine generators of* $\pi_2 P$ *where* $P = \langle x, y \mid x^a, y^b, (xy)^c \rangle$ *presents the* (a, b, c)- *triangle group* $(a, b, c \geq 2)$. (Hint: If $1/a + 1/b + 1/c \leq 1$, then P is (CR); see [BoGu92, Example 5.4] for help in building pictures when $1/a + 1/b + 1/c > 1$.)

2.3 Subpresentations

Suppose that $\{P_i : i \in I\}$ is a collection of subpresentations of P, and let X_i denote the collection of all based spherical pictures over P_i $(i \in I)$. We shall say that a set Y of based spherical pictures over P *generates* $\pi_2 P$ *over* $\{P_i : i \in I\}$ if $Y \cup \bigcup_{i \in I} X_i$ generates $\pi_2 P$. Note that one may equivalently take X_i to be a set of generators for $\pi_2 P_i$ $(i \in I)$.

The value of this concept is that we can often isolate certain subpresentations which we know little about, or which in some way are arbitrary; then, relative to these subpresentations we can often determine a "nice" set of generators of $\pi_2 P$. This point of view is very useful when considering the second homotopy module of "canonical" presentations arising from group constructions. For example, suppose $G = G_1 \times G_2$. Choose presentations $P_i = \langle \mathbf{x}_i \mid \mathbf{r}_i \rangle$ for G_i $(i = 1, 2)$. Then $P = \langle \mathbf{x}_1, \mathbf{x}_2 \mid \mathbf{r}_1, \mathbf{r}_2, x_1 x_2 x_1^{-1} x_2^{-1} (x_i \in \mathbf{x}_i) \rangle$ is a presentation of G. Since G_1 and G_2 are arbitrary, we cannot say anything about generators of $\pi_2 P_i$. However, it is not hard to find generators for $\pi_2 \mathbf{P}$ over $\{P_1, P_2\}$. See Section 10 below.

We will say that P is (CA) *over* $\{P_i : i \in I\}$ if $\pi_2 P$ is generated over $\{P_i : i \in I\}$ by primitive dipoles. Similarly, P is (A) *over* $\{P_i : i \in I\}$ if $\pi_2 P$ is generated by the empty set over $\{P_i : i \in I\}$. Note that P is (A) over $\{P_i : i \in I\}$ if and only if the inclusion-induced map $\oplus_{i \in I} \pi_2 P_i \to \pi_2 P$ is surjective. These formulations of (CA) and (A) both imply that no relator of $P - \bigcup_{i \in I} P_i$ is a consequence of the other relators of P.

Consider a presentation $P = \langle \mathbf{x} \mid \mathbf{r}_1, \mathbf{r}_2, \ldots, \mathbf{r}_n \rangle$. For $i = 1, \ldots, n$, let R_i

be the normal closure of r_i in the free group F with basis $\mathbf{x}$, let $\mathcal{P}_i = \langle \mathbf{x} \mid r_i \rangle$, and let $N_i = \prod_{j \neq i} R_j$. Set $R = \prod_{i=1}^{n} R_i, G = F/R$ and $G_i = F/R_i$. The family $\{R_1, R_2, \ldots, R_n\}$ is said to be *independent* if $R_i \cap N_i = [R_i, N_i]$ for $i = 1, \ldots, n$. This and related notions have been studied in [Bo91, DuElGi92, Gi93, GuRa81, Hu81]. For example [DuElGi92, GuRa81, Hu81], if $\{R_1, R_2, \ldots, R_n\}$ is independent, then $\mathcal{P}$ is (A) over $\{\mathcal{P}_i : i = 1, \ldots, n\}$. The converse holds in case $n = 2$. It is shown in [DuElGi92, Theorem 2.1] that $\{R_1, R_2, \ldots, R_n\}$ is independent if and only if the inclusions $R_i \to R$ induce an isomorphism $\oplus_{i=1}^{n}(\mathbf{Z}G \otimes_{\mathbf{Z}G_i} H_1 R_i) \to H_1 R$ of (induced) relation modules.

Consider a pair of presentations $\mathcal{Q} \subseteq \mathcal{P}$ having the same set of generating symbols, say

$$\mathcal{Q} = \langle \mathbf{x} \mid \mathbf{s} \rangle \subseteq \langle \mathbf{x} \mid \mathbf{s}, \mathbf{r} \rangle = \mathcal{P}.$$

In this case, $K_{\mathcal{P}}$ is obtained from $K(\mathcal{Q})$ by attaching 2–cells. We will say that $\mathcal{P}$ is *reducible over* $\mathcal{Q}$ if for each spherical picture over $\mathcal{P}$ that is not a picture over $\mathcal{Q}$, there is a transverse path γ joining corners c_1, c_2 in distinct discs Δ_1, Δ_2 such that (i) Δ_1 and Δ_2 are labeled by the same relator $r \in \mathbf{r}$ and have opposite signs, (ii) $W(c_1) = W(c_2)^{-1}$, and (iii) $W(\gamma)$ represents the identity in the group presented by $\mathcal{Q}$. See Figure 13. This notion compares closely with [CoHu82, p. 179], [Hu81, Proposition 8] and [Si80, (5)].

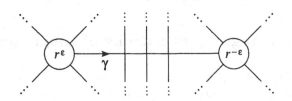

Figure V.13. Reducible

Lemma 2.6 *If* $\mathcal{P}$ *is reducible over* $\mathcal{Q}$, *then* $\mathcal{P}$ *is (CA) over* $\mathcal{Q}$.

Proof: Let X be a complete set of primitive dipoles over $\mathcal{P}$, and let $\mathbf{P}$ be a spherical picture over $\mathcal{P}$. We show by induction on the number of discs in $\mathbf{P}$ which are labeled by elements of $\mathbf{r}$ that $\mathbf{P}$ is X-equivalent to a spherical picture over $\mathcal{Q}$. By hypothesis, $\mathbf{P}$ contains a subpicture of the sort depicted in Figure 13. By (iii), there is a picture $\mathbf{A}$ over $\mathcal{Q}$ with boundary label $W(\gamma)$, and so there is a spherical picture $\mathbf{Q}$ over $\mathcal{P}$ as shown in Figure 14.

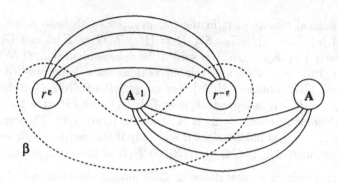

Figure V.14. A spherical picture **Q**

The two discs labeled r in **Q** support a complete dipole in **Q**, while the oppositely oriented copies of **A** can be cancelled as in the proof of 1.5, so that $[f_{\mathbf{Q}}] \in J(X)$. The subpicture **B** of **Q** enclosed by the loop β is also a subpicture of **P**. Replacing the subpicture **B** in **P** by the complement of **B** in **Q**, it follows that **P** is X-equivalent to a picture having two fewer discs labeled by elements of **r**.   □

3   Applications and Results

3.1   Universal group of a family of subgroups

Abels and Holz [AbHo92] have considered the concept of a family of subgroups of a group being "n-generating". In particular, the 3-generating property is intimately connected with the second homotopy module. We give an account of their ideas in terms of pictures.

Let $\mathcal{H} = \{H_i : i \in I\}$ be a family of subgroups of a group G. We can consider the "free product of this family amalgamated along their intersections". This is a group $H = \bigsqcup_{\cap} H_i$ together with a homomorphism $\phi_i : H_i \to H$ for each $i \in I$ such that $\phi_i|_{H_i \cap H_j} = \phi_j|_{H_i \cap H_j}$ for each $i, j \in I$, and where H is universal with this property.

Choose a presentation $\mathcal{P}_i = \langle \mathbf{x}_i \mid \mathbf{s}_i \rangle$ for H_i. We thus have an isomorphism $\alpha_i : H_i \to F_i/N_i$, where F_i is the free group on $\mathbf{x}_i$ and N_i is the normal closure of $\mathbf{s}_i$ in F_i. For any (unordered) pair of elements $i, j \in I$ choose a generating set $\{h_\lambda : \lambda \in \Lambda\{i,j\}\}$ of $H_i \cap H_j$. For each $\lambda \in \Lambda\{i,j\}$, select $a_\lambda^i \in F_i$ and $a_\lambda^j \in F_j$ such that $\alpha_i(h_\lambda) = a_\lambda^i N_i$ and $\alpha_j(h_\lambda) = a_\lambda^j N_j$. Then

$$\mathcal{P} = \langle \mathbf{x}_i \ (i \in I) \mid \mathbf{s}_i \ (i \in I), a_\lambda^i = a_\lambda^j \ (i,j \in I, \lambda \in \Lambda\{i,j\}) \rangle$$

is a presentation of H. We let $\mathbf{r}_{\{i,j\}} = \{a_\lambda^i = a_\lambda^j : i,j \in I, \lambda \in \Lambda\{i,j\}\}$.

We consider certain "obvious" spherical pictures over $\mathcal{P}$. First, for $i \in I$, let X_i denote the set of spherical pictures over $\mathcal{P}_i$, and let $X^1 = \bigcup_{i \in I} X_i$.

Second, for $i,j \in I$, let V be some relation among the generators h_λ ($\lambda \in \Lambda\{i,j\}$) of $H_i \cap H_j$, say

$$V = h_{\lambda_1}^{\epsilon_1} h_{\lambda_2}^{\epsilon_2} \dots h_{\lambda_n}^{\epsilon_n} \quad (\lambda_\xi \in \Lambda\{i,j\}, \epsilon_\xi = \pm 1, \xi = 1, \dots, n).$$

Then,

$$V_i = (a_{\lambda_1}^i)^{\epsilon_1} (a_{\lambda_2}^i)^{\epsilon_2} \dots (a_{\lambda_n}^i)^{\epsilon_n} \in N_i$$

and similarly we have a word V_j in $a_{\lambda_\xi}^j$ representing an element of N_j. By 1.4 there are pictures $\mathbf{D}_V^i, \mathbf{D}_V^j$ over $\mathcal{P}_i, \mathcal{P}_j$ respectively with boundary labels $W(\mathbf{D}_V^i) = V_i^{-1}$ and $W(\mathbf{D}_V^j) = V_j$. We obtain a spherical picture $\mathbf{P}_V$ over $\mathcal{P}$ as shown in Figure 15.

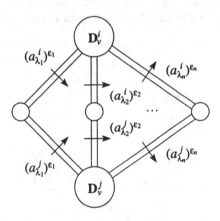

Figure V.15. $\mathbf{P}_v \in X^2$

We let X^2 denote the collection of all such pictures $\mathbf{P}_V$ for all pairs $i,j \in I$.

Third, for each triple $i,j,k \in I$ choose a generating set $\{h_\gamma : \gamma \in \Gamma\{i,j,k\}\}$ of $H_i \cap H_j \cap H_k$. Each h_γ can be written as a word $b_\gamma^{ij}(h_{\lambda(ij)})$ ($\lambda(ij) \in \Lambda\{i,j\}$) in the generators of $H_i \cap H_j$. Similarly there are words $b_\gamma^{jk}(h_{\lambda(jk)}), b_\gamma^{ki}(h_{\lambda(ki)})$ representing h_γ in the generators of $H_j \cap H_k, H_k \cap H_i$, respectively. Then we have, for instance, $\alpha_i(b_\gamma^{ij}(h_{\lambda(ij)})) = \alpha_i(b_\gamma^{ki}(h_{\lambda(ki)}))$, and it follows that

$$b_\gamma^{ij}(a_{\lambda(ij)}^i)N_i = b_\gamma^{ki}(a_{\lambda(ki)}^i)N_i$$
$$b_\gamma^{jk}(a_{\lambda(jk)}^j)N_j = b_\gamma^{ij}(a_{\lambda(ij)}^j)N_j$$
$$b_\gamma^{jk}(a_{\lambda(jk)}^k)N_k = b_\gamma^{ki}(a_{\lambda(ki)}^k)N_k.$$

By 1.4 there are pictures $\mathbf{D}_\gamma^i, \mathbf{D}_\gamma^j, \mathbf{D}_\gamma^k$ over $\mathcal{P}_i, \mathcal{P}_j, \mathcal{P}_k$ with clockwise boundary labels

$$W(\mathbf{D}_\gamma^i) = b_\gamma^{ij}(a_{\lambda(ij)}^i)b_\gamma^{ki}(a_{\lambda(ki)}^i)^{-1}$$

and similarly for $\mathbf{D}_\gamma^j, \mathbf{D}_\gamma^k$. Also, there are pictures $\mathbf{A}_\gamma^{ij}$ (respectively $\mathbf{A}_\gamma^{jk}, \mathbf{A}_\gamma^{ki}$) containing only discs labeled by relators in $\mathbf{r}_{\{i,j\}}$ (respectively $\mathbf{r}_{\{j,k\}}, \mathbf{r}_{\{k,i\}}$), and having clockwise boundary labels

$$W(\mathbf{A}_\gamma^{ij}) = b_\gamma^{ij}(a_{\lambda(ij)}^j)b_\gamma^{ij}(a_{\lambda(ij)}^i)^{-1}$$

and similarly for $\mathbf{A}_\gamma^{jk}, \mathbf{A}_\gamma^{ki}$. We can join these six pictures together to form a spherical picture $\mathbf{P}_\gamma$ as shown schematically in Figure 16.

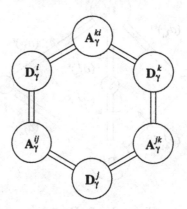

Figure V.16. $\mathbf{P}_v \in X^3$

We let X^3 denote the collection of all such $\mathbf{P}_\gamma$ for all triples $i, j, k \in I$. Finally, L denotes the submodule of $\pi_2 \mathcal{P}$ generated by $X^1 \cup X^2 \cup X^3$.

We now discuss the work of Abels and Holz mentioned earlier. We have the family $\mathcal{H} = \{H_i : i \in I\}$ of subgroups of the group G. Let $\mathcal{N}$ denote the nerve of the cover of G that consists of all the left cosets of the H_i ($i \in I$) in

G. Abels and Holz define the family $\mathcal{H}$ to be *n-generating* if the simplicial complex $\mathcal{N}$ is $(n-1)$-connected, i.e., if the geometric realization of $\mathcal{N}$ has trivial homotopy groups in dimensions less than n. It turns out that

- $\mathcal{H}$ is 1-generating $\Leftrightarrow \bigcup_i H_i$ generates G (folklore), and

- $\mathcal{H}$ is 2-generating $\Leftrightarrow$ the natural map $H = \bigsqcup_\cap H_i \to G$ is an isomorphism [Be75, So73].

Abels and Holz show that if $\mathcal{H}$ is 2-generating, and if we choose a presentation $\mathcal{P}$ for $H = \bigsqcup_\cap H_i$ as above, then

$$\pi_2(\mathcal{N}) \cong \pi_2(\mathcal{P})/L.$$

Thus, $\mathcal{H}$ is 3-generating if and only if L generates $\pi_2\mathcal{P}$.

3.2 Generalized graphs of groups

We consider a much more general concept of graphs of groups than the standard concept [Se80]. Let Γ be a directed graph with vertex set $\mathbf{v}$ and oriented edge set $\mathbf{e}$, and for each $v \in \mathbf{v}, e \in \mathbf{e}$, let there be assigned groups G_v, G_e. Also, for each oriented edge $e \in \mathbf{e}$ with initial (resp. terminal) vertex $i(e)$ (resp. $t(e)$), let there be assigned subgroups $H_e \leq G_{i(e)}, \bar{H}_e \leq G_{t(e)}$, and an isomorphism $\phi_e : H_e \to \bar{H}_e$. Finally, suppose that for each $e \in \mathbf{e}$ there is chosen a fixed element $\xi_e \in G_{i(e)} * G_e * G_{t(e)}$ (or $\xi_e \in G_{i(e)} * G_e$ if $i(e) = t(e)$). We allow ξ_e to be trivial; if ξ_e is not trivial, then we require that the first and last terms of ξ_e (in normal form) belong to G_e.

The *fundamental group* of the above system defined to be the quotient group G of the free product $*_{z \in \Gamma} G_z$ by the elements

$$h\xi_e \phi_e(h)^{-1} \xi_e^{-1} \quad (h \in H_e, e \in \mathbf{e}).$$

We obtain a presentation for G as follows. Choose a presentation $\mathcal{P}_z$ for G_z for each $z \in \Gamma$. For $e \in \mathbf{e}$, let $\{h_{i,e} : i \in I(e)\}$ be a set of generators of H_e, and for $i \in I(e)$ choose words $a_{i,e}, \bar{a}_{i,e}$ in the generators of $\mathcal{P}_{i(e)}, \mathcal{P}_{t(e)}$ which represent $h_{i,e}, \phi_e(h_{i,e})$ respectively; let T_e be a word in the generators of $\mathcal{P}_{i(e)} \cup \mathcal{P}_e \cup \mathcal{P}_{t(e)}$ which represents ξ_e. A presentation $\mathcal{P}$ for G results from combining all the presentations $\mathcal{P}_z (z \in \Gamma)$ and adding the additional relations

$$a_{i,e} T_e \bar{a}_{i,e}^{-1} T_e^{-1} \quad (e \in \mathbf{e}, i \in I(e)).$$

Now suppose that

$$U : h_{i_1,e}^{\epsilon_1} h_{i_2,e}^{\epsilon_2} \ldots h_{i_r,e}^{\epsilon_r} = 1$$

$(i_j \in I(e), \epsilon_j = \pm 1, i = 1, \ldots, r)$ is a relation amongst the generators of H_e. Then we have a picture $\mathbf{D}_U$ over $\mathcal{P}_{i(e)}$ with anti-clockwise boundary label

$$W(\mathbf{D}_U)^{-1} = a_{i_1,e}^{\epsilon_1} a_{i_2,e}^{\epsilon_2} \ldots a_{i_r,e}^{\epsilon_r},$$

and a picture $\bar{\mathbf{D}}_U$ over $\mathcal{P}_{t(e)}$ with (clockwise) boundary label

$$W(\bar{\mathbf{D}}_U) = \bar{a}_{i_1,e}^{\epsilon_1} \bar{a}_{i_2,e}^{\epsilon_2} \ldots \bar{a}_{i_r,e}^{\epsilon_r}.$$

A spherical picture $\mathbf{P}_U$ over $\mathcal{P}$ is then constructed as in Figure 17

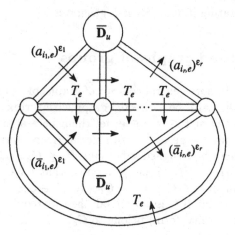

Figure V.17. $\mathbf{P}_u$

Theorem 3.1 ([BaPr])   *Suppose that the following two conditions hold.*

1. *The set of edges e of Γ for which ξ_e is trivial spans a forest in Γ.*

2. *For each $e \in \mathbf{e}$ all the terms of ξ_e (in normal form) have infinite order.*

Then, the pictures $\mathbf{P}_U$ generate $\pi_2 \mathcal{P}$ over $\{\mathcal{P}_z : z \in \Gamma\}$. $\square$

A proof and numerous applications of this result are given in [BaPr]. The proof relies on a new result due to Klyachko [Kl92] about graphs on spheres. We remark that in the setting of Theorem 3.1 one can show that the groups G_z naturally embed into G [Kl92].

3.3 Split extensions

Let K and H be groups and let $\psi : K \to \mathrm{Aut}(H)$ $(k \mapsto \psi_k)$ be a homomorphism. Consider the split extension $G = H \times_\psi K$ consisting of all pairs $hk(h \in H, k \in K)$, where multiplication is defined by $(hk)(h'k') = (h\psi_k(h'))(kk')$. Choose presentations $\mathcal{H} = \langle \mathbf{x} \mid \mathbf{s} \rangle, \mathcal{K} = \langle \mathbf{y} \mid \mathbf{t} \rangle$ for H and K respectively. If F is the free group on $\mathbf{y}$, then there is an epimorphism $F \to K$ $(y \mapsto k_y)$ whose kernel is the normal closure of $\mathbf{t}$ in F. There is the composite homomorphism $\psi^* : F \to \mathrm{Aut}(H)$ and split extension $G^* = H \times_{\psi^*} F$. A presentation for G^* is given by

$$\mathcal{P}^* = \langle \mathbf{x}, \mathbf{y} \mid \mathbf{s}, xy^{-1}a_{xy}^{-1}y(x \in \mathbf{x}, y \in \mathbf{y}) \rangle.$$

Here, for $x \in \mathbf{x}, y \in \mathbf{y}, a_{xy}$ is a selected word on $\mathbf{x}$ representing $\psi_{k_y}(h_x)$, where h_x is the element of H represented by x. A presentation $\mathcal{P}$ for G is obtained by adjoining the relators $\mathbf{t}$ to $\mathcal{P}^*$.

Now G^* is the fundamental group of a (standard) graph of groups, and so generators of $\pi_2(\mathcal{P}^*)$ can be obtained from 3.1. We will describe a set of pictures which generate $\pi_2\mathcal{P}$ over $\mathcal{P}^*$.

Let $\mathbf{t}^\#$ denote the set of all cyclic permutations of elements of $\mathbf{t} \cup \mathbf{t}^{-1}$ that end with an element of $\mathbf{y}$ (rather than with an element of $\mathbf{y}^{-1}$). Let $T \in \mathbf{t}^\#$, say $T = Uy$, and let $x \in \mathbf{x}$. Now $Ua_{xy}U^{-1}x^{-1}$ defines the identity of G^* (since $TxT^{-1}x^{-1}$ does), so there is a picture $\mathbf{B}_{T,x}$ over $\mathcal{P}^*$ with boundary label $Ua_{xy}U^{-1}x^{-1}$. We thus obtain a spherical picture $\mathbf{P}_{T,x}$ over $\mathcal{P}$ as shown in Figure 18.

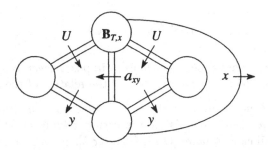

Figure V.18. $\mathbf{P}_{T,x}$

Theorem 3.2 ([BaPr]) *The pictures* $\mathbf{P}_{T,x}$ $(T \in \mathbf{t}^\#, x \in \mathbf{x})$ *generate* $\pi_2\mathcal{P}$ *over* $\mathcal{P}^*$. □

This theorem has recently been used by Lysionok and Pride [LyPr] to describe the structure of $\pi_2 \mathcal{P}$ in terms of modules associated with H and K.

3.4   Iterative constructions

Ol'shanskii [Ol91] has devised a method for constructing groups with strange properties by considering certain types of ascending chains of presentations

$$\mathcal{P}_0 \subseteq \mathcal{P}_1 \subseteq \dots$$

and looking at the group G_∞ defined by $\mathcal{P}_\infty = \bigcup_{i=0}^\infty \mathcal{P}_i$. For the chains considered by Ol'shanskii, it turns out that for each i, $\mathcal{P}_{i+1}$ is reducible over $\mathcal{P}_i$. Thus $\mathcal{P}_\infty$ is (CA) over $\mathcal{P}_0$ (Lemma 3.3).

We describe Ol'shanskii's basic construction; for more details see [Ol91]. Let n be a very large ($n > 10^{10}$) odd integer, and let $\mathcal{P}_0 = (\mathbf{x} : -)$. Suppose that $\mathcal{P}_i = \langle \mathbf{x} \mid \mathbf{r}_i \rangle$ has been constructed, and let G_i be the group defined by $\mathcal{P}_i$. We say that a word on $\mathbf{x}$ is *simple in rank* i if it is not conjugate in G_i to a power of a shorter word, and it is not conjugate in G_i to a power of a root of a relator in $\mathbf{r}_i$. Let $\mathbf{u}_{i+1}$ denote a maximal set of words of length $i + 1$ which are simple in rank i with the property that if $A, B \in \mathbf{u}_{i+1}$ and $A \neq B$, then A is not conjugate in G_i to $B^{\pm 1}$.

Now define a set $\mathbf{s}_{i+1}$ as follows. First, we include in $\mathbf{s}_{i+1}$ relators of the form A^n ($A \in \mathbf{u}_{i+1}$) (relators of the first type). Also for each $A \in \mathbf{u}_{i+1}$ we include in $\mathbf{s}_{i+1}$ a set $\mathbf{s}_A$ of words of the form

$$T_1 A^{n_1} T_2 A^{n_2} \dots T_k A^{n_k}$$

(relators of the second type). Ol'shanskii imposes seven conditions R1-R7 [Ol91, pp. 271-272] concerning the form of the relators of the second type.

When there are no relators of the second type the conditions R1-R7 are vacuous and the theory in this case is easier, relying on the geometry of what Ol'shanskii calls A-*maps* [Ol91, Chapter 5]. In this case G_∞ is the free Burnside group on $\mathbf{x}$ of exponent n [Ol91, Theorem 19.7]. When there are relators of the second type the geometry becomes much more complicated and one has to analyse what Ol'shanskii calls B-*maps* [Ol91, Chapter 7]. The presence of relators of the second type gives considerable flexibility of construction. In particular, one can impose relators of the second type in such a way that G_∞ is a so-called "Tarski monster" [Ol91, Theorem 28.1].

The geometry of A- and B-maps ensures that $\mathcal{P}_{i+1}$ is reducible over $\mathcal{P}_i$ [Ol91, Lemma 18.2, Lemma 25.1], and so $\mathcal{P}_\infty$ is (CA).

The above can be varied by choosing $\mathcal{P}_0$ differently. In particular, one can take $\mathcal{P}_0$ to be a presentation of a free product of groups or of a hyperbolic group [Ol91, Chapter 11].

In connection with free Burnside groups, one should mention the earlier work of Novikov-Adian [Ad75]. Recently, S. Ivanov [Iv92] and I. G. Lysionok [Ly92, Ly93] have investigated free Burnside groups of large even exponent. In particular, Lysionok has determined generators of the second homotopy module of presentations of these groups. The paper of Rips [Ri82] has similarities with the work of Ol'shanskii, though Rips does not consider π_2.

3.5 The Steinberg group and K_3 of a ring

Let A be a ring. The *Steinberg group* $St(A)$ is the group defined by the presentation $\mathcal{St}(A)$ with generators

$$x_{ij}(a) \qquad (i, j \in \mathbf{N}, i \neq j, a \in A)$$

and with relations

$$x_{ij}(a)x_{ij}(b) = x_{ij}(a + b) \qquad \text{(all } i, j, a, b)$$

$$[x_{ij}(a), x_{kl}(b)] = 1 \qquad (i \neq l, \ j \neq k, \text{ all } a, b)$$

$$[x_{ij}(a), x_{jk}(b)] = x_{ik}(ab) \qquad (i, j, k \text{ distinct, all } a, b).$$

There is a homomorphism $St(A) \to GL(A)$ that maps the generator $x_{ij}(a)$ to the elementary matrix with a in the (i, j)-position. The kernel of this map is denoted $K_2(A)$; it measures the 'nonobvious' relations amongst the elementary matrices over A.

Now Igusa [Ig79$_2$] has described certain pictures over the presentation $\mathcal{St}(A)$. Igusa's pictures are valid for each ring A, and so can be regarded as 'obvious'. These pictures are exhibited explicitly in [Wa80]. (Note however that a different convention is used for drawing pictures there.) Igusa has proved that if M is the submodule of $\pi_2\mathcal{St}(A)$ that is generated by these 'obvious' pictures, then $\pi_2\mathcal{St}(A)/M$ is isomorphic to $K_3(A)$. Thus $K_3(A)$ measures the 'nonobvious' pictures over $\mathcal{St}(A)$.

3.6   Presentations defined using simplicial complexes

Let $\mathcal{K} = (\mathbf{v}, \Sigma)$ be an abstract simplicial complex. For each vertex $v \in \mathbf{v}$ let $\mathbf{x}_v$ be a collection of generating symbols. An element of $\mathbf{x}_v$ will be said to be of *type v*. For each simplex $\sigma \in \Sigma$ let

$$\mathbf{x}_\sigma = \bigcup_{v \in \sigma} \mathbf{x}_v$$

and let $\mathbf{r}_\sigma$ be a set of cyclically reduced words on $\mathbf{x}_\sigma$, where each element of $\mathbf{r}_\sigma$ involves at least one generator of every type $v \in \sigma$. We allow $\mathbf{r}_\sigma$ to be empty; σ is *essential* if $\mathbf{r}_\sigma$ is non-empty. We let

$$\mathcal{P}_\sigma = (\mathbf{x}_\sigma : \bigcup_{\tau \leq \sigma} \mathbf{r}_\tau)$$
$$\mathcal{P}_\sigma^\circ = (\mathbf{x}_\sigma : \bigcup_{\tau < \sigma} \mathbf{r}_\tau)$$

and let G_σ, G_σ° be the groups defined by $\mathcal{P}_\sigma, \mathcal{P}_\sigma^\circ$ respectively. Thus G_σ is the "group corresponding to the simplex σ" and G_σ° is the "group corresponding to the proper faces of σ". We have the natural epimorphism $G_\sigma^\circ \to G_\sigma$. We let $\hat{\mathbf{r}}_\sigma$ denote the set of all words on $\mathbf{x}_\sigma$ which represent nontrivial elements in the kernel of this epimorphism.

If W is a word on $\mathbf{x}_\sigma$ then a *k-factorization* of W is an expression of W as a product $W_1 W_2 \dots W_k$ where W_λ $(1 \leq \lambda \leq k)$ does not involve generators of every type $v \in \sigma$ (that is, W_λ is a word on $\mathbf{x}_{\sigma_\lambda}$ for some proper face σ_λ of σ). We will say that $\mathcal{P}_\sigma$ has *property-B_k* if no element of $\hat{\mathbf{r}}_\sigma$ has a k-factorization. Roughly speaking, this gives a measure of the amount of collapse in passing from G_σ° to G_σ—the larger the k, the less the collapse.

Consider the following graph Γ: the vertices are the essential simplices of $\mathcal{K}$; there is an edge joining σ, τ if $\sigma \cap \tau$ is nonempty and is a proper face of both σ and τ. We will say that the *triangle condition* holds if Γ has no triangles.

Theorem 3.3 *Let* $\mathcal{P} = \langle \mathbf{x}_v \ (v \in \mathbf{v}) \mid \mathbf{r}_\sigma \ (\sigma \in \Sigma) \rangle$, *and let* G *be the group defined by* $\mathcal{P}$. *Suppose that one of the following two conditions holds.*

(I) *Each* $\mathcal{P}_\sigma$ *has property* B_5.

(II) *Each* $\mathcal{P}_\sigma$ *has property* B_3 *and the triangle condition holds.*

Then,

1. *the natural mappings* $G_\sigma \to G$ *are injective for each* $\sigma \in \Sigma$, *and*

2. $\mathcal{P}$ *is* (CA) *over* $\{\mathcal{P}_\sigma : \sigma$ *is a maximal simplex in* $\mathcal{K}\}$. □

The proof of 1 in the case when each $\mathbf{x}_v$ is a singleton is due to Edjvet [Ed88, Theorem 1]; the proof can easily be adapted to the more general situation considered here. The proof of 2 is an unpublished result of the second author. (A proof of 2 in the case when $\mathcal{K}$ is one-dimensional can be found in [Pr92$_1$].)

The above theorem had its origins in the papers [Pr87, PrSt89, PrSt90, Pr92$_1$] which considered the case when $\mathcal{K}$ is one-dimensional. In this case there are many nice examples (for example, certain Coxeter groups [PrSt90]) when the conditions (I) and (II) are easily verified. For higher dimensional situations it is difficult to verify the conditions (I) and (II), and it would be of interest to find nice examples.

3.7 One-relator products

Let H_1, H_2 be two nontrivial groups, and let R be a cyclically reduced element of the free product $H_1 * H_2$ of free product length at least two. The quotient G of $H_1 * H_2$ by the normal closure of R is called a *one-relator product* of H_1 and H_2, which we denote by $(H_1, H_2; R)$. If we choose presentations $\mathcal{Q}_1 = \langle \mathbf{x}_1 \mid \mathbf{s}_1 \rangle, \mathcal{Q}_2 = \langle \mathbf{x}_2 \mid \mathbf{s}_2 \rangle$ for H_1, H_2 respectively, then we obtain a presentation

$$\mathcal{P} = \langle \mathbf{x}_1, \mathbf{x}_2 \mid \mathbf{s}_1, \mathbf{s}_2, \tilde{R} \rangle$$

for $G = (H_1, H_2; R)$, where $\tilde{R}$ is a word on $\mathbf{x}_1 \cup \mathbf{x}_2$ representing R.

Two major issues that have been addressed for one-relator products are the following.

1. When are the natural maps $H_1, H_2 \to (H_1, H_2; R)$ injective?

2. Determine a "nice" set of generators of $\pi_2 \mathcal{P}$ over $\{\mathcal{Q}_1, \mathcal{Q}_2\}$.

For the most part it suffices to restrict discussion to the case when one of the factors is infinite cyclic. In fact, in the above setting, let $H = H_1 * H_2$, and let C be an infinite cyclic group generated by t. There is an automorphism ψ of $H * C$ given by

$$h_1 \mapsto h_1 \ (h_1 \in H_1), h_2 \mapsto th_2t^{-1} \ (h_2 \in H_2), t \mapsto t.$$

Let R_0 be the image of R under ψ, and let $G_0 = (H, C \mid R_0)$. Let $\phi : H \to G, \phi_0 : H * C \to G_0$ be the natural surjections. Then there is an isomorphism $\psi_* : G * C \to G_0$ such that $\phi_0 \psi = \psi_*(\phi * id)$. It follows from this that if ϕ_0 is injective on H, then ϕ is injective on each of H_1 and H_2. Moreover, from the presentation $\mathcal{P}$ of G we can obtain a presentation

$$\mathcal{P}_0 = \langle \mathbf{x}_1, \mathbf{x}_2, t \mid \mathbf{s}_1, \mathbf{s}_2, \tilde{R}_0 \rangle$$

of G_0, where $\tilde{R}_0$ is obtained from $\tilde{R}$ by replacing each symbol $x_2^{\pm 1}$ in R_0 by $t x_2^{\pm 1} t^{-1}$ for all $x_2 \in \mathbf{x}_2$, and then cyclically reducing the result. Suppose we have determined a set X_0 of spherical pictures over $\mathcal{P}_0$ that generate $\pi_2\mathcal{P}$ over $\{\mathcal{Q}_1, \mathcal{Q}_2\}$. Let X be the set of pictures over $\mathcal{P}$ obtained from the pictures in X_0 by erasing all arcs labeled $t^{\pm 1}$. Then X generates $\pi_2\mathcal{P}$ over $\{\mathcal{Q}_1, \mathcal{Q}_2\}$. To prove this, one converts a given spherical picture $\mathbf{P}$ over $\mathcal{P}$ to a spherical picture $\mathbf{P}_0$ over $\mathcal{P}_0$ using the process described in [BoPr92, proof of Lemma 3.3]. This process has the property that upon erasing all arcs of $\mathbf{P}_0$ labeled by t, one returns to $\mathbf{P}$. Let Y_i be the set of all spherical pictures over $\mathcal{Q}_i$ ($i = 1, 2$). Since X_0 generates $\pi_2\mathcal{P}_0$ over $\{\mathcal{Q}_1, \mathcal{Q}_2\}$, $\mathbf{P}_0$ is $(X_0 \cup Y_1 \cup Y_2)$-equivalent to the empty picture. Erasing all arcs labeled t in this transformation, one finds an $(X \cup Y_1 \cup Y_2)$-equivalence from $\mathbf{P}$ to the empty picture.

From now on we will consider a one-relator product $G = (H, t : R)$ where H is a nontrivial group and t generates an infinite cyclic group. We choose a presentation $\mathcal{Q} = \langle \mathbf{x} \mid \mathbf{s} \rangle$ for H, and obtain a presentation $\mathcal{P} = \langle \mathbf{x}, t \mid \mathbf{s}, \tilde{R} \rangle$ for G. We will then be concerned with the following two issues.

1. When is the natural map $H \to G$ injective?

2. Find a "nice" set of generators of $\pi_2\mathcal{P}$ over $\mathcal{Q}$.

The main results in this direction have been obtained under hypotheses on either (i) H, or (ii) the "shape" of R. As regards (i), it has been shown that if H is locally indicable, then $H \to G$ is injective [BrS80, Ge83, Ho81$_2$, Kr85, Sh81] and $\mathcal{P}$ is (CA) over $\mathcal{Q}$ [Ho84]. As regards (ii), most work has been done when R is a proper power: $R = S^n$ ($n > 1$). (Ordinary one-relator group theory is easier when the relator is a proper power, so one would expect a similar phenomenon in the relative case considered here.) The root S is said to be *exceptional* if some cyclic permutation of S has the form

$$U h U^{-1} k$$

where h, k are elements of H of finite orders p, q, and

$$1/p + 1/q + 1/n > 1.$$

If S is exceptional, then for each distinct cyclic permutation of S of the above form, there is an "obvious" spherical picture over $\mathcal{P}$ (see [DuElGi92, Ho89, Ho90]) and we can consider the set X of all these spherical pictures, together with a primitive dipole. If S is not exceptional, then X contains only the primitive dipole.

Theorem 3.4 ([Ho89, Ho90])   *If $n \geq 4$, then $H \to G$ is injective and $\pi_2\mathcal{P}$ is generated by X over $\mathcal{Q}$.* □

This theorem is not too difficult to prove when $n \geq 6$ using small cancellation theory (see [CoPe85, GoSh86, Ly66]). When $n = 4, 5$, the proof is much more difficult and is due to Howie [Ho89, Ho90]. Some partial results for $n = 3$ have been obtained by Duncan and Howie [DuHo92$_1$].

The question of what happens when $n = 2$ is far from clear. Juhasz (unpublished) has claimed that if $n = 2$ and H is torsion-free, then $H \to G$ is injective and $\mathcal{P}$ is (CA) over $\mathcal{Q}$. Duncan and Howie [DuHo92$_2$] have pointed out that there are examples with $n = 2$ and S having only one exceptional form (so that X consists of a single picture), but where $\pi_2\mathcal{P}$ is not generated by X over $\mathcal{Q}$. For example, take either (i) $H = (a, b : a^2, b^3)$ and $S = (atbt^{-1})^4(atb^2t^{-1})^2$ or (ii) $H = (a, b : a^3, b^3)$ and $S = atbt^{-1}atb^2t^{-1}$.

For relators that are not proper powers, one can try to get an insight into this case by considering relators of small t-length. The first case of real interest is t-length three, so that up to cyclic permutation, R has the form $th_1th_2t^\epsilon h_3$ where $h_i \in H$ and $\epsilon = \pm 1$. When $\epsilon = 1$ (resp. $\epsilon = -1$), then it follows from a theorem of Levin [Le62] (resp. Howie [Ho83$_1$]) that $H \to G$ is injective. As regards $\pi_2\mathcal{P}$, we have the following results.

Theorem 3.5 [BoPr92])   *Suppose that $R = th_1th_2th_3$ where $h_1, h_2, h_3 \in H$ are not all equal. Then, $\mathcal{P}$ is (A) over $\mathcal{Q}$ if and only if neither of the following two conditions holds.*

1. *For each $i = 1, 2, 3$, $h_i h_{i+1}^{-1}$ has finite order p_i (subscripts mod 3) and $1/p_1 + 1/p_2 + 1/p_3 > 1$, or*

2. *there exist integers $j \in \{1, 2, 3\}$, $p > 2$, and $0 \leq k < p$ such that $\text{sgp}\{h_i h_{i+1}^{-1} : i = 1, 2, 3\}$ is finite cyclic with generator $h_j h_{j+1}^{-1}$ of order p, and $h_{j+1}h_{j+2}^{-1} = (h_j h_{j+1}^{-1})^k$ where either $k = 1$, $p = k + 2$, $p = 2k + 1$, or $p = 6$ and $k = 2$ or 3.* □

The proof of 3.5 uses a relative form of the weight test [BoPr92, Theorem 2.1] to detect the (A) property. In case 1 holds, explicit spherical pictures are constructed and it can be shown that the homotopy classes of these pictures do not lie in the submodule of $\pi_2\mathcal{P}$ that is generated by the image of $\pi_2\mathcal{Q} \to \pi_2\mathcal{P}$. When 2 holds, G is shown to have finite subgroups that are not conjugate to subgroups of H. It follows from a theorem of Serre (see [BoPr92, Theorem 1.4]) that $\mathcal{P}$ is not (A) over $\mathcal{Q}$.

Theorem 3.6 ([BoPr92])   *Suppose that $R = tatbt^{-1}c \, (a, b, c \in H)$.*

1. *If one of b, c has infinite order in H, then $\mathcal{P}$ is (A) over $\mathcal{Q}$.*

2. *If b, c have finite orders p, q, then $\mathcal{P}$ is (A) over $\mathcal{Q}$ except possibly if $1/p + 1/q > 1/2$ or $a^{-1}ba = c^k$ for some k, or $aca^{-1} = b^k$ for some k.*

$\square$

Edjvet [Ed91] has greatly improved the result 3.6. Suppose b, c have finite orders p, q. Assume first that $1/p + 1/q > 1/2$. Edjvet uses the relative form of the weight test and a "curvature adjustment" technique based on 1.1 to prove that $\mathcal{P}$ is (A) over $\mathcal{Q}$ unless a, b, c satisfy any one of nine special families of relations in H. In these nine cases, Edjvet constructs explicit spherical pictures over $\mathcal{P}$ that are not pictures over $\mathcal{Q}$, and which have no dipoles. It is reasonable to expect that the homotopy classes of these pictures do not lie in the submodule of $\pi_2\mathcal{P}$ that is generated by the image of $\pi_2\mathcal{Q} \to \pi_2\mathcal{P}$. The same sort of results are proved in the case when $1/p + 1/q \leq 1/2$, except here there are six special families of relations, and the cases $(p, q) = (8, 4), (9, 3)$ are left unresolved.

The t-length four case becomes even more complicated. In the case when all powers of t are positive, $\pi_2\mathcal{P}$ is studied in [BaBoPr]. (The injectivity of $H \to G$ follows from the result of Levin mentioned above.) When the powers of t are not all positive, the injectivity of $H \to G$ has been discussed in [EdHo91]. Very little has been done concerning $\pi_2\mathcal{P}$.

We remark that as this article was being written, new work of Klyachko [K192] came to our attention. He proves a new result concerning diagrams on spheres and uses this to investigate the injectivity of $H \to G$ in some significant cases. Results about $\pi_2\mathcal{P}$ are not mentioned explicitly, but it is clear that Klyachko's ideas will have considerable future use in computations of π_2.

Chapter VI

Applications of Diagrams to Decision Problems

Günther Huck and Stephan Rosebrock

In this chapter, classical decision problems such as the word and conjugacy problem are introduced and methods are given for solving them in certain cases. All the methods we present involve Van-Kampen diagrams as one of the most powerful tools when dealing with the classical decision problems.

1 Introduction

In 1912, Max Dehn formulated in his article „Über unendliche diskontinuierliche Gruppen" ("On infinite discontinuous groups") three fundamental problems for infinite groups given by finite presentations: the *identity problem*, the *transformation problem*, and the *isomorphism problem*. The following is a translation of Dehn's definition of the first two problems called in modern terms the *word problem* and the *conjugacy problem*:

The identity problem (word problem): Let an arbitrary element of the group be given as a product of the generators. Find a method to decide in a finite number of steps whether or not this element equals the identity element.

The transformation problem (conjugacy problem): Any two elements S and T of the group are given as a product of the generators. Find a method to decide whether or not S and T are conjugate, i.e., whether or not there exists an element U of the group that satisfies the equation $S = UTU^{-1}$.

In this chapter we wish to give an account of some recent developments that use Dehn's original geometric ideas in extending partial solutions of the word and conjugacy problems, in particular some recent generalizations of small-cancellation theory.

2   Decidability and Dehn's Algorithm

Dehn's definitions of the word and conjugacy problems represent quite accurately what we call in modern terms decision problems. In a *decision problem* one considers a question with possible answers 'yes' or 'no' about the elements of a countable set. If there exists an *algorithm*, i.e., a well defined set of directions for a sequence of mathematical steps, which when applied to any element of the set will, after finitely many steps, produce the correct answer, we say the problem is *decidable* or *(recursively) solvable*; equivalently the subset for which the answer to the question is yes (respectively no) is called *recursive*. When Dehn formulated his fundamental problems for infinite groups, "algorithm" was an intuitively well understood notion which had not yet been formalized. The formalizations of the notions algorithm and computability by Turing (1936) [Tur37], Church (1941) [Chu41], and Markov (1954) [Ma54] were proved to be equivalent and led in the 1950s to a generally accepted theory of computability and thereby to a precise definition of "decidable" or "recursive set". One can briefly say that a question is decidable if there exists a Turing machine that halts on all inputs producing the right answer. The precise definition of algorithm made it possible to prove many decision problems to be unsolvable in general. Novikov [No55] and Boone [Bo55] constructed in 1955 the first examples of finitely presented groups with unsolvable word problem, deriving them from Turing machines with unsolvable halting problem. For an account of these results and the definition of Turing machine, see [Ro73, Chapter 12] and [Sti82].

The word problem as stated by Dehn is clearly equivalent to the problem of deciding whether or not two given words represent the same group element; it is also a special case of the conjugacy problem. Although the word problem and conjugacy problem are usually stated for a finite presentation of a group, they apply more generally to *recursive presentations*, i.e., presentations with a finite set of generators and a recursive set of relations. One could easily allow countably many generators in this context; however, the following result does not carry over to the class of countably generated presentations: For finitely

generated recursively presented groups the solvability of the word problem
or the conjugacy problem are algebraic invariants, i.e., independent of the
specific finitely generated presentation (see [Co89] sect. 9.3, and [Mi92]).
For simplicity we will restrict ourselves in the following to finitely presented
groups.

In many cases the word problem or conjugacy problem is solved by a *normal
form*, i.e., a unique representation for each group element as a word in the
generators, which mostly includes an algorithm that transforms a group element represented by some word into its normal form. The simplest examples
are free groups and free abelian groups. The word problem for free abelian
groups is trivial; in the standard presentation of a free group the normal
form for the word problem (resp. the conjugacy problem) is the set of freely
reduced words (resp. cyclically reduced words). Algebraic classes of finitely
presented groups with solvable word problem include linear groups, residually free groups, residually finite groups, and residually nilpotent groups. A
rather comprehensive account of results on decision problems in group theory
can be found in [Mi92].

Definition 2.1 *Given a finite presentation* $\mathcal{P} = <x_1, \ldots, x_n \mid R_1, \ldots, R_m>$
for a group G, we say $\mathcal{P}$ *is a* Dehn presentation *or satisfies Dehn's algorithm
if every non-trivial word w which represents the identity element in G has a
subword v that is equal to a subword of a cyclic conjugate of a defining relator
or its inverse* $R_i^{\pm 1}$ *and has length* $|v| > |R_i|/2$.

If we assume, as is customary in small-cancellation theory, that the set of
defining relations is *symmetrized*, i.e., closed with respect to inversion and
cyclic permutation, the subword v becomes a prefix of a relator R (we will
generally use "relator" or "relation" to mean defining relation). Dehn's algorithm for solving the word problem of a symmetrized Dehn presentation
is now obvious: A non-trivial word represents the identity element in G if
and only if it can be reduced to the trivial word by the following elementary
reductions:

1. free reductions, and

2. if v is a subword of w such that $|v| > |u|$ and $vu = R$ is a relator, then
   replace v by the shorter subword u^{-1}.

It is not difficult to convert this into a well defined algorithm. Dehn's proof
in [De12] that this algorithm applies to the standard presentations of closed
orientable surface groups of genus $n \geq 2$ is a very simple argument, using

the regular tessellation of the (hyperbolic) plane that represents the universal cover of the standard cell decomposition of the surface:

Consider this regular tessellation by $4n$-gons with $4n$ regions coming together at each vertex. We select one region of this tessellation and call it the central region. Dehn divides this tessellation into concentric rings around the central region, where the first ring consists of all regions that touch the boundary of the central region, the second ring consists of the regions that touch the outer boundary of the first ring etc.. One can now easily observe, by looking at a picture of such a tessellation, that, except for the central region, $4n - 2$ or $4n - 3$ successive edges of the boundary of each region lie in the outer boundary of the ring to which the region belongs, two edges are *radial* edges, i.e., connect the inner and outer boundary of the ring, and at most one edge of the region lies on the inner boundary of the ring, we will call the boundaries of the rings "circles". Now let the word $w = 1$ be represented by a closed edge path that starts w.l.o.g. at a vertex of the central region. Let a be a vertex on the largest circle which is reached by the path, more precisely let a be the first vertex along the path that lies on this circle, i.e., the edge before a is a radial edge that crosses a ring (unless the entire path lies on the boundary of the central region, but then the proof is trivial). Then, either the path continues from a on this largest circle for at least $4n - 3$ edges or there is some backtracking. In the first case an elementary reduction of type 2 is possible (homotope the path across a region replacing the $4n - 3$ or $4n - 2$ boundary edges of the region by the opposite 3 or 2 edges); in the second case w allows a free reduction.

Dehn's algorithm for the conjugacy problem for fundamental groups of closed orientable surfaces, which is of similar simplicity (although the proof is not quite as simple), does not carry over to the general case of a Dehn presentation. Nevertheless, by Gromov's result that hyperbolic groups (in the sense of Gromov) are exactly the groups that allow a Dehn presentation (compare Theorem 4.2 in this chapter) and by the fact that hyperbolic groups have solvable conjugacy problem (see [CoDePa90]) we know that groups with a Dehn presentation always have solvable conjugacy problem.

3   Cayley Graph and van Kampen Diagrams

Let G be a finitely generated group and $X = \{x_1, \dots, x_n\}$ a set of generators. Denote by $X^{\pm 1}$ the set of generators and their inverses.

Definition 3.1 *The* Cayley graph $\Gamma(G, X)$ *is a directed labeled graph defined as follows: the set of vertices is the set of group elements G and for every* *pair $(g, x_i) \in G \times X$ there is a directed edge labeled x_i from the vertex g to the vertex gx_i.*

Another customary definition of the Cayley graph uses $G \times X^{\pm 1}$ as edge set. This means that for each edge from g to gx_i labeled x_i there is a parallel but oppositely oriented edge from gx_i to g labeled x_i^{-1}. In our definition we identify these two parallel edges and think of the *inverse* of the edge from g to gx_i as the same edge traversed against its orientation, i.e., from gx_i to g, and having label x_i^{-1}. A *path* in the graph is a sequence of edges or their inverses that constitutes a continuous path, i.e., the endpoint of the i-th edge equals the startpoint of the (i+1)th edge. The word associated with a path is the corresponding sequence of labels (consisting of generators or their inverses). Define the *distance* between two vertices g and g' in the Cayley graph to be the number of edges in a shortest path connecting them, which is equal to the length of the shortest word representing the group element $g^{-1}g'$. This metric is called the *word metric*.

If $\mathcal{P} = \langle x_1, \ldots, x_n \mid R_1, \ldots, R_m \rangle$ is a finite presentation for G, then $\Gamma(G, X)$ is equal to the 1-skeleton of the universal cover of the standard 2-complex $K_\mathcal{P}$ associated with the presentation. This can be seen as follows: If we attach 2-cells to all cyclic closed paths in $\Gamma(G, X)$ that read a defining relation, we obtain a 2-complex $K(G, X)$ which is obviously a covering of $K_\mathcal{P}$. To see that $K(G, X)$ is also simply connected (and hence is the universal covering $\widetilde{K_\mathcal{P}}$) one uses that every closed edge path p represents a word w which equals 1 in G, i.e., w is freely equivalent to a product of conjugates of defining relations or their inverses: $\prod_{j=1}^{n} u_j R_{i_j}^{\epsilon_j} u_j^{-1}$ ($\epsilon_j \in \{+1, -1\}$). This free equivalence produces a homotopy between p and a product of paths p_j that represent the factors $u_j R_{i_j}^{\epsilon_j} u_j^{-1}$. Clearly, the paths p_j are nullhomotopic in $K(G, X)$, hence, p itself is nullhomotopic.

Let F be the free group on the generators $x_1, \ldots, x_n$ of the presentation $\mathcal{P}$. We denote by $\mathbf{R}^{\pm 1}$ the set of all defining relations and their inverses, and by $\mathbf{R}^*$ the symmetrized set of defining relations.

Definition 3.2 *A finite, connected graph in the plane whose edges are oriented and labeled by generators of $\mathcal{P}$ is called a van Kampen diagram over $\mathcal{P}$ if the boundary path of every finite face reads a word in $\mathbf{R}^*$.*

Given a van Kampen diagram over $\mathcal{P}$. If we fill each finite face of this planar diagram with a 2-cell, we obtain a planar, connected, and simply connected 2-complex with oriented labeled edges. We adopt the convention that the label on a boundary path (of a 2-cell or of the entire planar diagram) is read clockwise starting at a given basepoint. Now let w be the label on the boundary path of the diagram with respect to a given basepoint. In each 2-cell we can mark a corner as a basepoint and label the 2-cell by a relator $r \in \mathbf{R}^{\pm 1}$ such that the label on the boundary path is r. This way we get a map from the diagram into $K_\mathcal{P}$ that represents a null-homotopy of the boundary path w (considered as a path in $K_\mathcal{P}$). This implies $w = 1$ in G. Therefore, a van Kampen diagram with boundary label w is also called a cancellation diagram for w.

For the converse: If w is a reduced word that represents the identity in G then there exists a van Kampen diagram over $\mathcal{P}$ whose boundary label is w. To show this we start by writing w as a product of conjugates of relations:

$$w = \prod_{j=1}^{k} u_j R_{i_j}^{\epsilon_j} u_j^{-1} \tag{1}$$

The right hand side of (1) which is a non-reduced word can be represented by a "bouquet of balloons"-diagram as in Figure 1. To get a diagram for the reduced word w, one has to realize a sequence of elementary free reductions geometrically (in the diagram). We do this one elementary reduction at a time, by identifying the pair of consecutive edges in the boundary path of the diagram that corresponds to the cancelling pair of letters. In certain cases such an identification may squeeze off a sphere that is attached to the rest of the diagram along an edge, and one preserves the planarity of the diagram by deleting the interior of this sphere (For details see [Jo80] pp. 215). This complication with spherical components can also be avoided by assuming that the number k of factors in (1) is minimal.

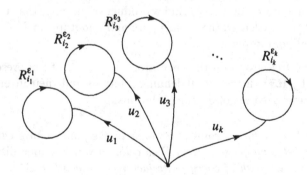

Figure VI.1. bouquet of balloons

The following notion of an isoperimetric function or an isoperimetric inequality originates from differential geometry and was introduced into geometric group theory by Gromov [Gr87]. However, a (quadratic) isoperimetric inequality was already used by Lyndon [Ly66] as principal tool in small-cancellation theory to solve the word problem for the non-metric cases. (Lyndon calls it the "area theorem")

Given a van Kampen diagram over the presentation $\mathcal{P}$ for the word w, the number of finite regions of the diagram is called its *combinatorial area*.

Definition 3.3 *A function $f: \mathbb{N} \to \mathbb{N}$ is called an* isoperimetric function *or* Dehn function *for $\mathcal{P}$ if it satisfies the following condition: for every reduced word w of length at most n representing 1 in G there exists a van Kampen diagram with combinatorial area at most $f(n)$.*

The condition in the definition is of course equivalent to the following: if $w = 1$ in G and $|w| \leq n$ then w can be written in F as a product of $k \leq f(n)$ conjugates of relations or their inverses: $w = \prod_{j=1}^{k} u_j R_{i_j}^{\epsilon_j} u_j^{-1}$. Some authors define the Dehn function to be the minimum of all isoperimetric functions, i.e., $f(n) = \max\{Area(w)|w = 1, |w| \leq n\}$, where $Area(w)$ is the minimal number of relators or their inverses needed in a van Kampen diagram for w.

In the context of discussing the complexity of the word problem, mainly the growth type of an isoperimetric function is important. Following Gersten [Ge90], we define: $f \preceq g$ if $\forall n \in \mathbb{N}, f(n) \leq Ag(Bn + C) + Dn + E$, where $A, \ldots, E$ are positive constants, and we define: $f \sim g$ if $f \preceq g$ and $g \preceq f$. It is not difficult to see that equivalence classes of isoperimetric functions are invariants of the group [Ge90]. They are even geometric invariants, i.e., invariants of the quasi isometry type of the group [A190].

Two metric spaces (X, d) and (X', d') are said to be *quasi isometric* if there exist maps $f: X \to X'$ and $f': X' \to X$ (not necessarily continuous) and constants $\lambda \geq 0, C \geq 0$ such that for all $x, y \in X$, and $x', y' \in X'$:

$$d'(f(x), f(y)) \leq \lambda d(x, y) + C \qquad d(f'(x'), f'(y')) \leq \lambda d'(x', y') + C$$

$$d(f'(f(x)), x) \leq C \qquad\qquad d'(f(f'(x')), x') \leq C$$

The first inequalities represent Lipschitz conditions, however, by the additive constant C, f and f' need not be continuous. The second inequalities express the failure of f and f' to be inverses by an error margin of C.

Two finitely generated groups are said to be quasi isometric if their Cayley graphs with respect to given finite sets of generators (and equipped with the word metric) are quasi isometric. It is not difficult to show that Cayley graphs of the same group with respect to different finite sets of generators are quasi isometric; therefore quasi isometry between groups is well defined. For a discussion of quasi isometry and geometric properties of groups see [Gh90], [GhHa90], [Ep92], and [GhHa91].

If G has an isoperimetric function of a certain type, such as linear, or quadratic etc., we say G *satisfies an isoperimetric linear (or quadratic) inequality.* The following result is stated in [Ge90] (also in [Ep92]) with parts of the proof sketched there.

Theorem 3.4 (Gersten)   *For any finitely presented group G the word problem for G is solvable, if and only if G has a recursive isoperimetric function.*□

Since this theorem is significant for the remainder of this chapter, we will discuss its proof in more detail. The insight that an isoperimetric function produces a solution of the word problem goes back to Lyndon's use of (quadratic) isoperimetric functions to solve the word problem for small-cancellation groups. The basic argument is this:

If $w = 1$ in G then there is a van Kampen diagram D for w with $k \le f(n)$ 2-cells (f an isoperimetric function for a given presentation of G, $n = |w|$). This implies: $w = \prod_{j=1}^{k} u_j R_{i_j}^{\epsilon_j} u_j^{-1}$ with $k \le f(n)$ factors. One can show that the lengths of the conjugating words u_j are also bounded, namely, by $(f(n) + 1)M + n/2$, where M is the maximal length of a defining relator. Hence, in order to decide if $w = 1$ in G or not, one has only to consider a finite list of products of conjugates of relations (which can be generated by a simple algorithm) and determine if one of these freely reduces to w. The entire process will be an algorithm for the word problem if the isoperimetric function f is algorithmically computable, i.e., recursive.

It remains to show that from a solution to the word problem for G one obtains a recursive isoperimetric function for G: Since $w = 1$ in G is defined by w being freely equivalent to a product of conjugates of defining relators or their inverses, such an algorithm for the word problem in G will produce for each reduced word w that is 1 in G a unique product of conjugates of relations: $\prod_{j=1}^{n} u_j R_{i_j}^{\epsilon_j} u_j^{-1}$ ("unique", since each step is uniquely determined because we have an algorithm). Now, let $f(n)$ be defined as

$$\max_{w=1 \text{ in } G, \ |w| \le n} \{k \in \mathbb{N} | k = \#\text{of factors in } w = \prod_{j=1}^{k} u_j R_{i_j}^{\epsilon_j} u_j^{-1}\},$$

where $\prod_{j=1}^{n} u_j R_{i_j}^{\epsilon_j} u_j^{-1}$ is the unique output of the algorithm on input w. Since the maximum is taken over a finite set, $f(n)$ is well defined and is an isoperimetric function for the given presentation of G. Since it is defined by an algorithm it is recursive.

If combinatorial area as geometric measure of a diagram is replaced by diameter, one obtains the concept of an isodiametric function. For a van Kampen diagram D with basepoint v_0 define $Diam_{v_0}(D) := \max_{v \in D^{(0)}} d(v_0, v)$, where d is the word metric of the 1-skeleton of the diagram. A function $f: \mathbb{N} \to \mathbb{N}$ is called an *isodiametric function* for $\mathcal{P}$ if for every word w of length at most n representing 1 in G there exists a van Kampen diagram D for w with basepoint v_0 and $Diam_{v_0}(D) \leq f(n)$. Equivalence classes (generated by " $\preceq$ ") of isodiametric functions are also geometric invariants of the group, and the property of having a recursive isodiametric function is again equivalent to G having solvable word problem. For a discussion of isodiametric functions and their relation to isoperimetric functions see [Co91], [Ge90], and [Ge91$_2$].

4   Word Hyperbolic Groups and Combings

Let G be a group with a finite set of generators $X = \{x_1, \dots, x_n\}$ and let $\Gamma = \Gamma(G, X)$ be the Cayley graph equipped with the word metric. The word metric is actually only defined on the set of vertices of Γ, although distance (in the wordmetric) is defined in terms of minimal length of paths connecting two vertices. A shortest path is called a *geodesic path*, a *geodesic triangle* consists of three geodesic paths that form a triangle (the paths may cross each other and one side may degenerate to a constant path).

The following definition of a hyperbolic group is due to Gromov [Gr87] who initiated the study of hyperbolic groups and of many other geometric properties of groups. There has been a general consensus to use the notation "word hyperbolic" or "negatively curved" for these groups to distinguish them from groups that are closer related to hyperbolic manifolds, and we will use the term "word hyperbolic group". Among the many equivalent definitions, we choose the one in terms of δ-thin triangles: Let δ be a non-negative real number. A geodesic triangle in Γ is called δ-*thin* if the distance from a point on one side to the union of the other two sides is bounded above by δ.

Definition 4.1 *We say the Cayley graph Γ is δ − hyperbolic if all geodesic triangles in Γ are δ-thin. The group G (with generating set X) is called* word hyperbolic *if $\Gamma(G, X)$ is δ-hyperbolic for some non-negative constant δ.*

The property of being word hyperbolic is independent of the specific finite generating set and even invariant under quasi isometry. However, the value of δ in the definition of δ-hyperbolic is sensitive to a change of generators.

Some group theoretic properties of word hyperbolic groups can be found in [Ly90]. Explicit proofs of the following equivalent characterizations of word hyperbolic, due to Gromov, can be found in [Sh91] and [CoDePa90]

Theorem 4.2 *For a finitely generated group G the following are equivalent:*

  (i) *G is word hyperbolic,*

  (ii) *G satisfies a linear isoperimetric inequality,*

  (iii) *there exists a Dehn presentation for G, i.e., a presentation satisfying Dehn's algorithm.* □

Another very influential recent development in combinatorial group theory concerns the application of automata theory to finitely generated groups in the study of automatic groups. We do not wish to go deep into automatic group theory; but concentrate instead on the geometric concept of a *combing* of a group which developed as a generalization of the notion of an automatic structure on a group. An automatic structure can be characterized ,using the geometry of the Cayley graph, as follows: it consists of a set of words in the generators and their inverses which represents all elements of G, forms a regular language (i.e., is the language accepted by a finite state automaton), and, viewed as a set of paths in the Cayley graph, has a characteristic geometric property, called the k-fellow-traveller property. If one extracts this characteristic geometric property from the concept of an automatic group, forgetting the requirement that the set of words representing G should be a regular language, one obtains the concept of a *combing*.

The notion of combing was introduced by Thurston and developed in [Sh90], [Ep92], [Ge92], [Al92], and [AlBr92]. Since the notation and definitions differ slightly in the different sources, we use our own. The ideas of the proofs in this section follow closely the account in [Sh90], [AlBr92] and [Al92].

Let G be a group with generating set X and let Γ be the corresponding Cayley graph. In the context of combings it is convenient to define a path in the Cayley graph to be a continuous map $p : [0, \infty[\to \Gamma$ which is at integer times at vertices (i.e., from $t = n$ to $t = n + 1$ the path either travels an edge between two vertices or pauses at a vertex). We will only consider finite paths in the sense that there exists a non-negative integer h such that $p|_{[h,\infty[} = \text{constant} = $ the endpoint of the path, and the minimum h with this property is called the length of the path: $|p|$, (this is the "time" it takes the path to reach its endpoint). Choosing $[0, \infty[$ as a universal domain allows us to consider paths of different lengths over the same parameter. Given the start point in Γ, a path is completely described by a sequence of elements in $X \cup X^{-1} \cup 1$ which provides the sequence of labels on the edges, the letter 1 representing a pause of length 1. If we delete the letters 1 in the sequence we get the "word associated with the path" which represents the group element $g^{-1}h$, where g is the startpoint and h the endpoint of the path; in particular if the startpoint is the identity element, as in the subsequent definition of "combing", the word represents the group element at the endpoint of the path.

Definition 4.3 *A* combing *of G with respect to X is a selection of a path $\sigma(g)$ from 1 to g in Γ for each $g \in G$, satisfying the following property: There exists an integer M such that, for each pair $g, g' \in G$ of distance 1 in Γ,*

$$d\left(\sigma(g)(t), \sigma(g')(t)\right) \leq M \quad \forall \text{ integers } t \in [0, \infty[. \tag{2}$$

Definition 4.4 *A* bicombing *of G with respect to X is a selection of a path $\sigma(g, h)$ from g to h in Γ for each pair $(g, h) \in G \times G$, satisfying the condition: There exists an integer M such that, for each two pairs $(g, h), (g', h')$ with $d(g, g') \leq 1$ and $d(h, h') \leq 1$,*

$$d((\sigma(g, h)(t), \sigma(g', h')(t)) \leq M \quad \forall \text{ integers } t \in [0, \infty[. \tag{3}$$

We call the paths $\sigma(g)$, $\sigma(g, h)$ the *combing lines, bicombing lines* respectively. Hence, condition (3) expresses that two points travelling simultaneously along bicombing lines that begin and end distance less or equal than 1 apart, will always remain within bounded distance M of each other, independent of the lengths of the paths. The formula (2) describes a similar behavior for combing lines. We call the distances in (2) and (3) which are bounded by M, the *geodesic differences* between adjacent (bi)combing lines. The following is an easy consequence of Definition 4.4:

Lemma 4.5 *If the bicombing σ satisfies (3) then for each pair of bicombing lines that begin and end distance less or equal than k apart (k a positive integer), the geodesic differences are bounded by kM.* □

A bicombing is said to be *equivariant* if $\sigma(g, h) = g \cdot \sigma(1, g^{-1}h) \quad \forall g, h \in G$, i.e., if the bicombing lines of translates of pairs (g, h) in Γ are translates of each other. "Translate" means translation under the operation of left-multiplication by a group element (which is an isometry of the Cayley graph). A group is said to be *(bi)combable* if it has a (bi)combing. Again, this notion is independent of the choice of finite generating set and also invariant under quasi isometry (see [Sh90]).

We say a combing is *linearly bounded, polynomially bounded, recursively bounded* if the length $|\sigma(g)| \leq f(d(1, g))$ with f linear, polynomial, recursive respectively; these terms generalize in the obvious way to bicombings.

The following theorem may be found in [Al92] and [Ep92].

Theorem 4.6   *(i) Every combable group is finitely presented.*

   (ii) If G admits a combing that is polynomially bounded of degree n then G satisfies a degree n + 1 polynomial isoperimetric inequality and a degree n polynomial isodiametric inequality.

   (iii) If G admits a recursively bounded combing then the word problem for G is solvable.

Proof: Let σ be a combing of G with respect to a finite generating set X and let $w = a_1 a_2 \ldots a_k$ be a reduced word in $X^{\pm 1}$ that equals 1 in G. Define $w_0 := 1$ and $w_i := a_1 \ldots a_i, (i = 1, \ldots, n)$.

Consider the closed path in Γ, based at the identity element, that reads w, then the group elements w_i are the vertices along this closed path. Let M be the constant associated with the combing in condition (2). If $|w| > 2M + 2$, consider the "fan" of combing lines $\sigma(w_i)$ that connect the basepoint 1 to the vertices w_i of the closed path. Now, connect the corresponding points $\sigma(w_i)(t)$, $\sigma(w_{i+1})(t)$ (t a positive integer $\leq \max\{|\sigma(w_i)|, |\sigma(w_{i+1})|\}$) of adjacent combing lines by geodesics (see Figure 2).

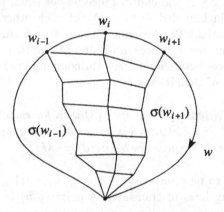

Figure VI.2. combing

By condition (2), the connecting geodesics between adjacent combing lines are not longer than M, therefore each triangle consisting of the combing lines $\sigma(w_i)$, $\sigma(w_{i+1})$, and the edge from w_i to w_{i+1} on the path w, breaks up into closed paths of lengths $\leq 2M + 2$ consisting of connecting geodesics and segments of length 1 along the bicombing lines (see Figure 2).

This way we have constructed a van Kampen diagram for w over the finite presentation $\mathcal{P}$ for G that uses the set X as generators and all relations of length $\leq 2M + 2$ as defining relations (if $|w| \leq 2M + 2$ then w itself is a defining relation of $\mathcal{P}$ and therefore has a van Kampen diagram with one region). The fact that each reduced word w that equals the identity in G has such a van Kampen diagram, proves that $\mathcal{P}$ is a finite presentation for G.

If the combing is bounded by a function f and $|w| = n$ then the number of regions in the van Kampen diagram for w (constructed above) will be bounded by the isoperimetric function: $f(n/2) \cdot n$; (here we use that $d(1, w_i) \leq n/2$ if $|w| = n$). This isoperimetric function will be polynomial of degree $n+1$ if f is polynomial of degree n, and recursive if f is recursive. It is also clear that the diameter of the diagram is bounded by $f(n/2) + M/2$, giving an isodiametric function of the same growth as f. $\qquad\qquad\square$

The condition 'recursively bounded' in (iii) is not necessary in order to get a solution to the word problem (see [Ge92]).

Theorem 4.7 ([Sh90]) *If G admits an equivariant, recursively bounded bicombing with respect to a finite generating set X then G has solvable conjugacy problem.*

Proof: Let σ be an equivariant bicombing of G with respect to X. We claim: this equivariant bicombing supplies for any pair of conjugate reduced words x, y a bound N, which only depends on the lengths $|x|$ and $|y|$, such that one can find a word of length $\leq N$ conjugating x to y. This means, in order to decide whether an arbitrary pair of reduced words is conjugate or not, we only have to consider the finite set of reduced words w of length $\leq N$ as possible conjugators. Since $\sigma|_{\{1\}\times G}$ defines a recursively bounded combing, the word problem is solvable for G and one can decide whether the set $\{wxw^{-1}y^{-1} | w \text{ reduced of length} \leq N\}$ contains a word representing the identity in G.

Although, so far we have been informal in using the same notation for a word representing a group element and the group element itself, in the following proof we distinguish the two objects by writing $\bar{w}$ for a group element represented by the word w. To prove the above claim, assume that $g \in G$ conjugates $\bar{x}$ to $\bar{y}$. Consider the bicombing line $\sigma(1, g)$ from 1 to g and its translate $\bar{y} \cdot \sigma(1, g) = \sigma(\bar{y}, \bar{y}g)$ from $\bar{y}$ to $\bar{y}g$. The distance of their start points 1 and $\bar{y}$ is $\leq |y|$, the distance of their endpoints g and $g\bar{x} = \bar{y}g$ is $\leq |x|$. Therefore they satisfy the hypothesis of Lemma 4.5 with $k = \max\{|x|, |y|\}$, and the geodesic differences between them are bounded by $M \cdot \max\{|x|, |y|\}$.

Now consider the closed rectangular path in Γ, based at the identity, that is depicted in Figure 3 $g\bar{x}g^{-1}\bar{y}^{-1} = 1$. The horizontal lines are the bicombing lines, the vertical lines are paths reading the words x and y respectively. For each positive integer $t \leq |\sigma(1,g)|$ we connect the points $\sigma(1,g)(t)$ and $\sigma(\bar{y},\bar{y}g)(t)$ by a geodesic path $\gamma(t)$. The lengths of each $\gamma(t)$ is $\leq M \cdot \max\{|x|,|y|\}$.

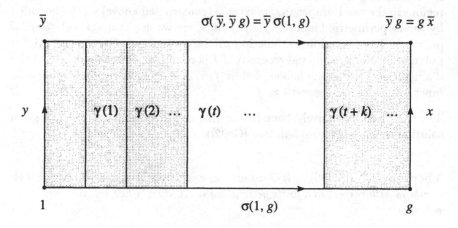

Figure VI.3. A bicombing

The trick for proving the claim now is the following: If the bicombing line $\sigma(1,g)$ is "too long" in the sense that the word associated with it has length greater than N which is the number of reduced words of length less or equal than $M \cdot \max\{|x|,|y|\}$, then at least two of the geodesic connecting paths, say $\gamma(t)$ and $\gamma(t+k)$ read the same word. In that case we can "cut out" the middle section of the rectangle in Figure 3 and glue the shaded right section onto the shaded left section along the word $\gamma(t) = \gamma(t+k)$. This is possible since the Cayley graph is "homogeneous", i.e., the two paths $\gamma(t)$ and $\gamma(t+k)$ whose labels are the same, correspond under an isometry of Γ that preserves all edge labels. The property that the horizontal paths are translates of each other (and, hence, read the same word) is preserved under this "surgery operation". (Here we need the equivariance of the bicombing.) By repeating this operation a sufficient number of times we arrive at a word w of length $\leq N$ such that $\bar{w}\bar{x}\bar{w}^{-1}\bar{y}^{-1} = 1$.   □

By a similar surgery argument as in the above proof, Gersten showed in [Ge92] that every combable group satisfies an exponential isoperimetric inequality. Meanwhile several generalized notions of combings and bicombings were introduced, (see [Bri92], [Me93] and [HuRo93]).

5 Curvature Tests

In this section, we would like to give criteria for presentations and 2-complexes, under which linear or quadratic isoperimetric inequalities are valid. The methods presented here, which are the hyperbolic weight test, the hyperbolic cycle test, and a uniform weight test (Theorem 5.9), may all be considered generalizations of small-cancellation-theory.

Consider a finite presentation $\mathcal{P} = < x_1, \ldots, x_n \mid R_1, \ldots, R_m >$ and its standard 2-complex $K_{\mathcal{P}}$. Then the *Whitehead graph* $W_{\mathcal{P}}$ (which was already defined in Chapter I, §3.1) is the boundary of a regular neighbourhood of the only vertex in $K_{\mathcal{P}}$. It consists of 2 vertices $+x_i$ and $-x_i$ for each generator x_i of $\mathcal{P}$ which correspond to the beginning and the end of the edge labeled x_i in $K_{\mathcal{P}}$. The edges of $W_{\mathcal{P}}$ are the *corners* of the 2-cells of the 2-complex. The *star graph* $S_{\mathcal{P}}$ was defined in the last chapter. It is the same as the Whitehead graph if no relator of $\mathcal{P}$ is a proper power. Let F denote the free group on the generators of $\mathcal{P}$.

If a relator R_i of $\mathcal{P}$ has the form $w_i^{k_i}$ with w_i not a proper power (we will call $|w_i|$ the period of R_i), then the star graph $S_{\mathcal{P}}$ is the Whitehead graph of the presentation $< x_1, \ldots, x_n \mid w_1, \ldots, w_m >$. We denote by $d(R_i)$ the length of R_i, that is the sum of the absolute values of the exponents of R_i.

Throughout this section we will use the stargraph S_p instead of the Whitehead graph, for the following reason: when dealing with algebraic questions such as the word problem, a rotation of a relation, which is a proper power, by any multiple of its period, is irrelevant (whereas, when dealing with homotopy invariants of the 2-complex, such rotations are quite significant, see Chapter V). Therefore we do not have to distinguish corners in such a relation that correspond under this periodicity; such corners are identified in the stargraph to one edge. By abuse of notation, from now on, we will often call an edge in the stargraph a corner.

A *cycle* is a non-constant closed path that is *cyclically reduced*, i.e., no oriented edge in the cyclic sequence is followed immediately by its inverse. Similarly we define non-closed reduced paths. A *simple cycle* is a cycle where no vertex is passed twice. A *weight function* is a real valued function on the edges of the graph, its values are called *weights*. If g is a weight function and z is a cycle we denote by $g(z)$ the sum of the weights that occur in the cycle where the weight of an edge that occurs several times in the cycle is counted with multiplicity. By $d(z)$ we denote the number of edges, that are traversed by z, again counting with multiplicity.

A *combinatorial map* (*in the strong sense*) is a cellular map that maps each open cell homeomorphically onto an open cell. A cell complex is said to be *combinatorial* if its attaching maps are combinatorial. We consider *diagrams* $f: M \to K_P$, where $M = S^2 - \cup_{i \leq \tau} D_i^2$ is a combinatorial cell decomposition of a 2-sphere minus $\tau > 0$ open 2-cells and f is a combinatorial map. For $\tau = 1$ this definition coincides with that of a van Kampen diagram given in §3. For the word problem or conjugacy problem respectively, τ will be one or two. Usually one works with diagrams M that are *reduced*, in the sense, that there are no two 2-cells $D_1 \neq D_2 \in M$, that have an arc t with label w in common with the following properties: D_1 is labeled $v_1 w$, D_2 is labeled $v_2 w^{-1}$ and $v_1 v_2$ freely reduces to 1. If M is not reduced, then we may perform reductions (i.e., deleting D_1, D_2) in M without changing the boundary label. If a diagram is reduced, then the link of every vertex of M gives rise to a cycle (or to a reduced path in the case of a boundary vertex) in S_P.

G. Huck made an observation which allows us to restrict ourselves to cycles and reduced paths in the star graph, where no edge is passed twice in different directions. Let z be a cycle or a reduced path in S_P, that uses an edge twice in different directions, corresponding to a vertex Q in a diagram M. Then there are two 2-cells $D_1 \neq D_2 \in M$ with Q in their boundary and with boundary labels $\delta D_i = v_i$ (start reading around D_i at Q according to a given orientation of M) such that $v_1 v_2$ freely reduces to 1. A reduction may be performed as in Figure 4. This notion of reduction includes the previously defined one.

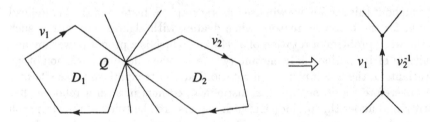

Figure VI.4. Reduction along a vertex

In the following discussion of the weight test and the cycle test, we only have to consider diagrams that are reduced in the stronger sense, that such pairs of 2-cells do not occur. Thus, when we speak of a cycle or a reduced path we assume that no edge is passed twice in different directions, and if we speak of a reduced diagram we assume that it is reduced in this stronger sense, i.e., the link of every vertex corresponds to a cycle or reduced path (that does not pass an edge twice in different directions).

At first, we want to define the hyperbolic weight test of Gersten [Ge87₁] and Pride [Pr88] together with a generalization of it, the hyperbolic cycletest of Huck/Rosebrock [HuRo92]. They are closely related to the weight test (which was presented in the last chapter) and the cycle test of [HuRo92] and differ from them by a slightly stricter hypothesis which yields that groups defined by presentations satisfying this hypothesis are word hyperbolic.

We say a presentation $\mathcal{P}$ satisfies the *hyperbolic weight test (WT)*, if there exists a weight function g for the star graph $S_\mathcal{P}$ satisfying the two conditions:

1. for all relators R of $\mathcal{P}$:

$$\sum_{\gamma \in R} g(\gamma) < d(R) - 2, \tag{4}$$

   where the sum ranges over all corners $\gamma \in R$,

2. for all simple cycles $z \in S_\mathcal{P}$

$$g(z) \geq 2. \tag{5}$$

Alternatively, we could have put a 'less or equal' sign in the first inequality and a 'strictly greater' sign in the second one. This would lead to the same statements and to very similar proofs in what follows.

In the original formulation of the (hyperbolic) weight test, condition (5) is required to hold for all cycles. Working with diagrams that are reduced (in the stronger sense) makes it possible to restrict (5) to simple cycles and it will be automatically satisfied for all cycles: It is easy to see that if all simple cycles satisfy (5) then so do cycles $z \in S_\mathcal{P}$ that pass each edge at most once. Such cycles will be called unsplittable, and we will see later that w.l.o.g. it suffices that all unsplittable cycles satisfy (5).

In order to define the hyperbolic cycle test, we need to consider sequences of cycles. A sequence of cycles describes the local incidence configuration of a 2-cell in a diagram: Consider a 2-cell D in a reduced diagram over $K_\mathcal{P}$ labeled by the relator R_i. Each vertex of the 2-cell in the diagram has either a disk neighborhood, described by a cycle in $S_\mathcal{P}$, or a half-disk neighborhood (if the vertex belongs to the boundary of the diagram), described by a reduced path in $S_\mathcal{P}$. When we list these cycles or reduced paths, in order, according to an orientation of the boundary of the 2-cell that follows the word R_i, we obtain what we call a "sequence of cycles for R_i" (which actually consists of cycles and reduced paths, unless the 2-cell is in the interior of the diagram). In addition, each cycle or reduced path in such a sequence of cycles has a preferred corner, namely the "inside corner" of D. This provides the geometric idea of the following definition.

Let R_i be a relator of $\mathcal{P}$ and D_i the corresponding 2-cell of $K_\mathcal{P}$. For each R_i there is a (in general infinite) set of *sequences of cycles* $\{Z_i^1, Z_i^2, \ldots\}$. Each *sequence of cycles* Z_i^j is an ordered set of $m = d(R_i)$ cycles or reduced paths $(z_1, \ldots, z_m)$ in S_P, together with a preferred edge β_t in each z_t (called the inside edge), satisfying the following three conditions:

1. No cycle or reduced path z_t of some Z_i^j passes an edge twice in different directions in S_P.

2. If we give the edges β_t the orientation induced by the paths z_t then the sequence of oriented edges $(\beta_1, \ldots, \beta_m)$ is the sequence of oriented corners read along the boundary of D_i in the direction of the word R_i.

3. Suppose $z_t = \gamma_1 \ldots \gamma_\nu$ and $z_{t+1} = \alpha_1 \ldots \alpha_\mu$ $(t + 1 = 1$ if $t = m)$ are consecutive cycles or reduced paths of the sequence of cycles Z_i^j and suppose $\beta_t = \gamma_l$ and $\beta_{t+1} = \alpha_k$ are the inside edges of z_t and z_{t+1} in Z_i^j. Then either a) γ_{l+1} and α_{k-1} are adjacent corners of the same relator R_s or b) γ_{l+1} and α_{k-1} do not exist (i.e., γ and α are reduced paths with γ_l the last edge of γ and α_k $(k = 1)$ the first edge of α). If, in either case, γ_l ends in a vertex $+a$ $(-a)$ of S_P then α_k starts in $-a$ $(+a)$ respectively, and, furthermore, in case a), the relators R_s and R_i have an edge a in common. This gives intuitively a local diagram (see Figure 5).

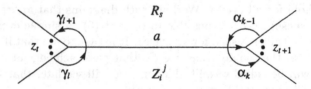

Figure VI.5. part of the relator R_i in M

We say $\mathcal{P}$ satisfies the *hyperbolic cycletest (CT)*, if for every sequence of cycles Z_i^j for every relator R_i there exists a weight function $g_i^j: \{\text{corners of } R_i\} \to \mathbb{R}$, that satisfies the following two inequalities:

1. for every sequence of cycles Z_i^j:
$$\sum_{\gamma \in R_i} g_i^j(\gamma) < d(R_i) - 2 \tag{6}$$

2. Let $z = \gamma_1 \ldots \gamma_\nu$ be a cycle with $\gamma_s \in R_{i_s}$, and for every corner γ_s let $Z_{i_s}^{j_s}$ be a sequence of cycles for R_{i_s} containing z such that the weight $g_{i_s}^{j_s}(\gamma_s)$ is defined. Then:
$$\sum_s g_{i_s}^{j_s}(\gamma_s) \geq 2 \tag{7}$$

From now on we will just speak of the *weight test* or *cycle test* and mean the hyperbolic weight test or hyperbolic cycle test. Note that for a given sequence of cycles Z_i^j for R_i the weight function g_i^j assigns weights only to the inside corners (i.e., the corners of R_i). Unlike the weight test, the cycletest does not assign unique weights to the corners of $S_{\mathcal{P}}$. However, if a corner occurs in a diagram then it belongs (as inside corner) to exactly one sequence of cycles (in the diagram) and, therefore has a well defined weight. If $\mathcal{P}$ satisfies the weight test, then it satisfies the cycletest, just by giving the corners of $S_{\mathcal{P}}$ fixed weights.

Our first goal is the next theorem. The proof in the case of the weight test is due to Pride [Pr88]; its generalization to the cycletest is from Huck/Rosebrock [HuRo92]:

Theorem 5.1 *If the presentation $\mathcal{P}$ satisfies the weight test or the cycle test then $\mathcal{P}$ has a linear isoperimetric inequality and, therefore, the corresponding group is word hyperbolic and has solvable word- and conjugacy problem.*

In order to prove this result we need some further notation: For any finite graph Γ call a reduced path or cycle w of Γ *splittable*, if it is the sum of a nontrivial reduced path or cycle z_0 and a nonempty set of cycles $z_1, \ldots, z_\mu$. By *sum* we mean: the number of occurrences of each edge in w is equal to the sum of the numbers of occurrences of each edge in the $z_i \in \{z_0, \ldots, z_\mu\}$.

Lemma 5.2 *In a finite graph there are only finitely many unsplittable reduced paths and cycles.*

Proof: Any path that uses an edge twice in the same direction is splittable which may be seen by the picture in Figure 6.

Figure VI.6. Splitting a path

There are only finitely many paths that use each edge at most once in every direction, because there are only finitely many words in the edges, that use each syllable only twice. □

In the case of the cycletest, we need a stronger definition of 'unsplittable'. For a fixed relator R_i, one cycle or reduced path z of a sequence of cycles Z_i^j must be compatible with its neighbors z', z'', in the sense that the two corners that are adjacent to the inside corner of z are compatible with the corresponding corners of z' and z'' as expressed in point 2. of the above definition (see Figure 5). So any splitting of cycles has to happen in such a way, that triples of adjacent corners of a cycle or reduced path are not separated. A cycle or reduced path z is called *splittable* (in the sense of the cycletest), if z traverses a subpath of length two in $S_{\mathcal{P}}$ consisting of edges a and b twice in the same direction. If one "switches tracks" at the vertex between the edges a and b in z (analogous to Figure 6), then one gets two reduced paths z_1 and z_2 that have z as their sum. It is easy to verify, that any triple of adjacent corners in z appears either in z_1 or in z_2. Iterating this process as often as possible produces a set of unsplittable paths (all of which, except possibly one, are cycles) whose sum is z.

For a given presentation $\mathcal{P}$ it suffices to find weight functions for sequences of cycles consisting of unsplittable cycles or reduced paths satisfying (6) and (7), in order to prove that $\mathcal{P}$ satisfies CT (see [HuRo92]): Choose for every cycle or reduced path a fixed decomposition into unsplittable components, and for every cycle or path in a sequence of cycles one of its unsplittable components so that the corners adjacent to the inside corner are preserved. This can be done in a unique way by an algorithm. W.l.o.g., the weights of an arbitrary sequence of cycles can then be chosen as the weights of the associated sequence of cycles with unsplittable paths (see the proof of 1.5 in [HuRo92]). In this case we say the weight functions are determined by sequences of cycles with unsplittable paths. As in the proof of Lemma 5.2 we can easily show that there are only finitely many sequences of cycles with unsplittable paths. This observation can be used to prove that the cycletest is decidable. It is also used in the following lemma (whose proof is left as an exercise) and in the proof of Theorem 5.1.

Lemma 5.3 *Let $\mathcal{P}$ satisfy the cycletest with weight functions g_i^j that are determined by sequences of cycles with unsplittable paths. Then there is a non-negative constant N such that every reduced path wthat occurs as link of a vertex in a diagram has weight $\geq -N$.*   □

Proof of Theorem 5.1:   Let $\mathcal{P}$ satisfy the cycle test with weight functions g_i^j. If $\mathcal{P}$ satisfies the weight test the proof is very similar and left as an exercise. By the above discussion we can assume w.l.o.g. that for every R_i there are only finitely many different weight functions g_i^j. Let ϵ be the smaller of $\min_{i,j}\{d(R_i)-\sum_{\gamma\in R}g_i^j(\gamma)-2\}$ and 1. Then (6) implies $\epsilon > 0$. Let $f: M \to K_{\mathcal{P}}$ be a reduced diagram where $M = S^2-\cup_{i\leq\tau}D_i^2$ as above. The weight functions g_i^j induce unique weights on the corners of M. Let $l(\delta M)$ be the sum of the lengths of each boundary component of M, i.e., the sum of the number of edges in each boundary path $\delta(D_i^2)$ (counting edges with multiplicity). Let F be the number of 2-cells of M. The following theorem is due to S. Pride [Pr88] and proves Theorem 5.1:

Theorem 5.4

$$F \leq \frac{3+N}{\epsilon} l(\delta M),$$

where N is the constant of Lemma 5.3.

Proof: We assume $F > 2$, otherwise the result is trivial. The weight functions g_i^j induce weights $g(\gamma)$ on the corners γ of all 2-cells of M. Let

$$s = \sum_{D^2 \in M} \sum_{\gamma \in D^2} g(\gamma).$$

Let c be the number of interior vertices of M, e the number of boundary vertices and V the total number of vertices of M. Then $c \geq V - l(\delta M)$ and $e \leq l(\delta M)$. The formula (7) and Lemma 5.3 imply $s \geq 2c - Ne$. All together we get

$$s \geq 2V - 2l(\delta M) - Nl(\delta M) \tag{8}$$

On the other hand $\sum_{\gamma \in D^2} g(\gamma) \leq d(D^2) - (2+\epsilon)$ and the definition of s gives: $s \leq \sum_{D^2 \in M}(d(D^2) - (2+\epsilon))$. If we denote by E the number of edges of M, the last inequality implies: $s \leq 2E - l(\delta M) - (2+\epsilon)F$. Together with (8) we get: $2V - 2l(\delta M) - Nl(\delta M) \leq 2E - l(\delta M) - (2+\epsilon)F$. This is equivalent to

$$2V - 2E + 2F \leq l(\delta M)(N+1) - \epsilon F$$

The Euler-characteristic of M gives $4 - 2\tau \leq l(\delta M)(N+1) - \epsilon F$. Since $\tau \leq l(\delta M)$ it follows $\epsilon F \leq l(\delta M)(N+3)$, which implies the theorem. $\square$

As an example to the weight- and cycletest we want to look at Wirtinger presentations.

A *Wirtinger presentation* is a finite presentation

$$\mathcal{P} = <x_1, \ldots, x_{n+1} \mid R_1, \ldots, R_n >,$$

such that each relator has the form $R_i : x_{i+1} = x_{\pi(i)}^{\epsilon_i} x_i x_{\pi(i)}^{-\epsilon_i}$ where $\epsilon_i = \pm 1$, $\pi(i) \in \{1, \ldots, n+1\}$. Such a presentation is sometimes called an *interval presentation*, because it may be represented as a labeled directed graph $I_{\mathcal{P}}$ (which is an interval) in the following way: For each generator x_i of $\mathcal{P}$ define a vertex labelled i and for each relator $R_i : x_{i+1} = x_{\pi(i)}^{\epsilon_i} x_i x_{\pi(i)}^{-\epsilon_i}$ define an edge from the vertex i to $i+1$ labelled $\pi(i)$. Orient this edge from $i+1$ to i if

$\epsilon_i = +1$ and from i to $i+1$ if $\epsilon_i = -1$. In the case of Wirtinger presentations the Whitehead graph and the star graph are the same, since there are no proper powers among the relators.

Consider a presentation $\mathcal{P}$ represented by the interval $I_{\mathcal{P}}$ of Figure 7 with any orientation on the edges. Assume $k \geq 2$ and $n \geq 3k$.

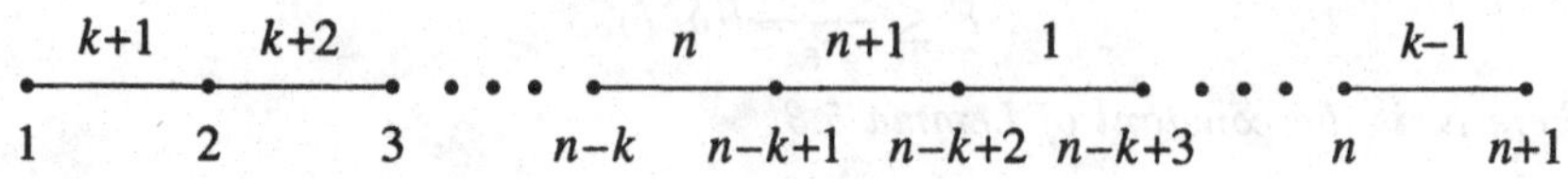

$k{+}1$	$k{+}2$		n	$n{+}1$	1		$k{-}1$
1	2	3	$n{-}k$	$n{-}k{+}1$ $n{-}k{+}2$ $n{-}k{+}3$		n	$n{+}1$

Figure VI.7. An interval presentation

Rosebrock has shown in [Ro91] that all cycles $z \in W_{\mathcal{P}}$ of these presentations have length at least four. This means that these presentations satisfy the small cancellation conditions C(4), T(4) (see [LySc77]). In [Ro91] there is also a careful analysis of all cycles of length four, which is necessary in order to prove weight- or cycletest conditions. In any interval presentation with at least two relators, there is for any $i \in \{2, \ldots, n\}$ a cycle of length four passing through the vertices $\epsilon_{i-1} x_{\pi(i-1)}$, $-x_i$, $-\epsilon_i x_{\pi(i)}$, $+x_i$ of the stargraph, in that order (see Figure 8). Call such a cycle an *ordinary cycle*.

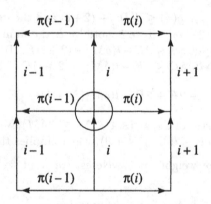

Figure VI.8. Ordinary cycle with $\epsilon_{i-1} = +1$, $\epsilon_i = -1$.

Theorem 5.5 *If $n \geq 4k$ and $k \geq 2$, then, for any orientation of the edges of $I_{\mathcal{P}}$, $\mathcal{P}$ satisfies the weight test.*

Proof: In [Ro91] it is shown that in this case all cycles of length four are ordinary. Then give weights according to Figure 9, where $\alpha_i = \nu/(n+2-i)$ and $\nu > 0$ is a fixed real number which will be chosen later.

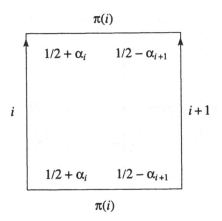

Figure VI.9. Weights for the relator R_i

Then

$$\sum_{\gamma \in R_i} g(\gamma) = 4 * 1/2 + 2\alpha_i - 2\alpha_{i+1} < 2,$$

since $\alpha_i < \alpha_{i+1}$. So (4) is satisfied. Now consider the ordinary cycle z of Figure 8. Its weight is 2, and, since all cycles of length four are ordinary, (5) is satisfied for cycles of length less than five. If z is a cycle of length $|z| \geq 5$ and we choose $\nu = 1/10$ then:

$$g(z) \geq |z|(1/2 - \alpha_{n+1}) \geq 5(1/2 - \nu/(n + 2 - (n + 1))) = 5/2 - 5\nu = 2$$

This is true since $1/2 - \alpha_{n+1}$ is the smallest weight defined. This proves (5).
□

It is not difficult to show that even in the case $n \geq 4k - 2, \quad k \geq 2$ these presentations satisfy the weight test. In this case there are some more cycles of length four that are not ordinary, but choosing weights with $\alpha_i = \nu/(5n-i)$ implies that they still satisfy (5).

Theorem 5.6 *If $n = 4k - 3$ and $k \geq 3$ then for any orientation of the edges of I_P, P satisfies the cycletest.*

Proof: We give 'in general' weights according to Figure 9, where $\alpha_i = \nu/(n + 2 - i)$ with $\nu > 0$ is a fixed real number which will be chosen later as above. Again there are no cycles of length smaller than four. The inequality (6) is satisfied because of $\alpha_i < \alpha_{i+1}$ and ordinary cycles fulfil (7) as in the previous proof. There are some more cycles of length four which cause problems like the cycle z around the vertex E in Figure 10.

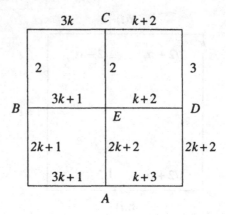

Figure VI.10. Non-ordinary cycle of length 4

If we would take the weights as suggested above, then $g(z) = 2 + \alpha_{k+2} + \alpha_2 - \alpha_{3k+1} - \alpha_{2k+2} < 2$. Observe that there is only some multiple λ of ν missing in order to get $g(z) \geq 2$, where λ is independent of ν. But in every sequence of cycles Z_i^j, where the vertex E occurs as in Figure 10, the cycles corresponding to the vertices A, B, C, D all have length at least five. So reduce the weights at every vertex of the set $\{A, B, C, D\}$ by $\lambda\nu/4$ and add this to the corresponding corner of the cycle z. Then $g(z) \geq 2$. Now each of the cycles w at the vertices A, B, C or D has weight $g(w) = 5/2 - c_w\nu$, where the constant c_w certainly depends on w. Now choose $\nu > 0$, such that $g(w) \geq 2$ for all possible cycles at all four vertices.

There are several more non-ordinary cycles of length four, but the method applied above for the cycle of Figure 10 works for all of these. Thereby ν has to be taken as the minimum of all the ν which appear in the corresponding cycles of length four. Cycles of length five or more are shown to satisfy (7) as in the previous proof.   □

The following generalization of small cancellation cases that yield a quadratic isoperimetric inequality is sketched in Gersten [Ge88₁] and goes back to Gromov [Gr87]. It extends the standard non-metric small cancellation conditions C(6), C(4)T(4), C(3)T(6) as well as the various non-homogeneous small cancellation cases considered by Juhasz [Ju86], [Ju89]. We give a detailed proof of this result following Gersten's outline. We start with some definitions:

A 2-complex is called *piecewise Euclidean* (PE) if each of its closed cells has the metric of a convex polygon in the Euclidean plane and these metrics agree on the overlaps. Let M be a finite PE 2-disc, i.e., a PE 2-complex that is homeomorphic to a 2-dimensional disk. Then it is possible to measure angles in the corners of 2-cells of M. Let μ be an inner vertex of M and let $W(\mu)$ be the sum of the angles that occur around μ. Now define the curvature at μ: $K(\mu) = 2\pi - W(\mu)$. Let $K(M) = \max K(\mu)$, where the maximum is taken over all inner vertices $\mu \in M$ (set $K(M) = 0$ if M has no inner vertices) and let l_M be the length of the boundary of M. The following theorem is due to Aleksandrov and Reshetnyak [Res61]:

Theorem 5.7 *Let M be a PE disc with $K(M) \leq 0$, then*

$$Area(M) \leq \frac{l_M^2}{4\pi} \qquad (9)$$

This result is certainly plausible. It gives as an upper estimate the area of an euclidean disk of the same boundary length.

Addendum:
Let M be a planar, connected, simply connected PE 2-complex (consisting of PE-disks and arcs where the disks are connected by common points or along arcs). Define the curvature $K(M)$ (as before) as the maximum of the curvatures of inner vertices of all the disks in M (or $= 0$ if M has no inner vertices). Then, by a simple induction over the number of disks in M, one can extend (9) to such a PE 2-complex M.

In the following discussion of the classes W and W^* of presentations we have to return to the original (weaker) definitions of a reduced path (or cycle) as a path (or non-constant closed path), respectively, in which no edge is followed immediately by its inverse. A diagram will be called reduced if the link of every vertex corresponds to a reduced path or cycle (in the above sense). Let K_P be the standard 2-complex of the presentation P with stargraph S_P. (We assume the relators of P are cyclically reduced.) We now define local small cancellation conditions on K_P. Let $f: M \to K_P$ be a reduced van Kampen diagram. Assume that M has no inner vertices of valence two. This may easily be achieved by removing these vertices and connecting the two arcs to one. Let $d(D)$ be the number of corners of the 2-cell $D \in M$ (not counting corners at deleted inner vertices of valence 2) and $d(v)$ the valence of the vertex v.

If P satisfies for every such diagram $f: M \to K_P$ and every inner 2-cell $D \in M$ one of the following conditions, then P (or K_P) is said to be *of type W*.

(i) $d(D) = 3$ $d(v) \geq 6$ for all vertices $v \in \delta D$

(ii) $d(D) = 4$

- $d(v) \geq 4$ for all $v \in \delta D$ or

- $d(v) \geq 5$ for two $v \in \delta D$ and $d(v) \geq 4$ for one more $v \in \delta D$ or

- $d(v) \geq 6$ for one $v \in \delta D$ and $d(v) \geq 4$ for another two $v \in \delta D$ or

- $d(v) \geq 6$ for at least two $v \in \delta D$

(iii) $d(D) = 5$

- $d(v) \geq 4$ for at least two $v \in \delta D$ or

- $d(v) \geq 6$ for one $v \in \delta D$

(iv) $d(D) \geq 6$

(vertices of M where several corners of the same 2-cell D occur count with multiplicity)

Certainly these conditions may be stated in terms of the presentation $\mathcal{P}$, like in small cancellation theory. Then we would have to speak of pieces and the C and T conditions (see [LySc77]). If only one of these conditions holds for the whole presentation then we are in the standard small cancellation theory. For example the condition (i) corresponds to the $C(3), T(6)$ case. All the other non-metric small cancellation conditions also occur in the list.

The above conditions defining the class W can be derived from a single curvature criterion. W is the class of standard 2-complexes for which every reduced van Kampen diagram (with degree 2 vertices deleted) is non-positively curved in the following sense: If the corners around an interior vertex v are given equal angles $2\pi/d(v)$, then every interior 2-cell D has the angle sum of a non-positively curved polygon ($\leq (d(D) - 2)\pi$). This class is called NPC (non-positive curvature) by Gromov [Gr87].

Before stating the main theorem of this part, we need a lemma from graph theory, whose proof is left as an exercise to the reader:

Lemma 5.8 *Let Γ be a finite graph where each vertex has valence at least two. Then each reduced path $w \in \Gamma$ may be completed by a reduced path $v \in \Gamma$ to a cycle $z = wv$.* $\square$

Theorem 5.9 *Let $\mathcal{P}$ be a finite presentation of type W, where each generator occurs at least twice in the set of relators and each relator is cyclically reduced. Let $f: M \to K_{\mathcal{P}}$ be a reduced van Kampen diagram. Then*

$$F \leq \frac{8l^2(\delta M)}{\sqrt{3}\pi} \tag{10}$$

where F is the number of 2-cells of M and $l(\delta M)$ is the number of boundary 1-cells of M (i.e., $\mathcal{P}$ satisfies a quadratic isoperimetric inequality and therefore has solvable word problem).

Proof: We first assume that M is a disc, where the boundary is reading off a cyclically reduced path in the free group on the generators of $\mathcal{P}$. If M consists of discs M_1, M_2 with exactly one boundary point in common (or connected by some path) and $F_i \leq \Phi(l(\delta M_i))$, $i = 1, 2$, then $F = F_1 + F_2 \leq \Phi(l(\delta M_1)) + \Phi(l(\delta M_2)) \leq \Phi(l(\delta M))$ for $\Phi(x) = 8x^2/(\sqrt{3}\pi)$. The result for arbitrary van Kampen diagrams follows by induction.

Now remove all interior vertices of valence two, and call the resulting diagram again M. There are no vertices of valence one because $\mathcal{P}$ is cyclically reduced. Then each interior vertex has valence greater or equal to three.

By the assumption that each generator occurs at least twice in the relators the vertices of $S_{\mathcal{P}}$ have valence ≥ 2. It is easy to show, by applying Lemma 5.8 to the links of the boundary vertices of the diagram M, that one can extend M to a reduced van Kampen diagram M' over $K_{\mathcal{P}}$ that contains M in its interior.

Now we define weights $g(\gamma)$ for the corners γ of M' that belong to the links of vertices of M, in the following way: If u is an inner vertex of M of valence $d(u) = k$, then each of the corners γ at u is assigned weight $g(\gamma) = 2/k$. If γ is a corner of M' incident to a boundary vertex v_i of M, define

$$g(\gamma) = \max\{1/3, 2/d(v_i)\}, \tag{11}$$

where $d(v_i)$ is the valence of v_i in M'. Then $g(c) \geq 2$ for any cycle $c \in M'$. The conditions of the definition of type W give for all $D \in M$

$$\sum_{\gamma \in D} g(\gamma) \leq d(D) - 2 \tag{12}$$

For example, if D is of type (ii) and $d(u_1) \geq 6$, $d(u_i) \geq 4$, $i \in \{2, 3\}$, then (12) may be seen as follows: $\sum_{\gamma \in D} g(\gamma) \leq 1 * 2/6 + 2 * 2/4 + 1 * 2/3 = 2 = d(D) - 2$. It is not hard to prove the following curvature formula by using the Euler-characteristic of M: (We write $\gamma \prec v_i$ if the corner γ occurs at the vertex v, and δM for the boundary of M)

$$\sum_{v \in \delta M} \sum_{\gamma \prec v, \ \gamma \text{ in } M} g(\gamma) \leq l(\delta M) - 2$$

Let τ be the number of inner edges of M that have at least one endpoint in δM, counting those edges twice that have both endpoints in δM. Then the curvature formula together with (11) implies

$$\tau \leq 2l(\delta M) - 6. \tag{13}$$

Now let M^* be the dual of M. M^* is a connected, simply connected planar diagram with boundary length $l(\delta M^*) = \tau$. We want to define a PE structure on M^* that realizes each 2-cell D^* in M^* as a regular $d(D^*)$-gon with each boundary edge having length one, and realizes each 1-cell that is not in the boundary of a 2-cell as a line segment of length one (such 1-cells occur in M^* if M contains 1-cells, not in δM, whose endpoints are in δM). This may not be possible directly, since closed 2-cells of M^* may have identifications on the boundary and, in that case, can not be isometric to a convex euclidean cell. However, after a suitable (stellar) subdivision, we can provide the subdivided diagram N^* with a PE structure that realizes each original 2-cell D^* by a PE subcomplex that corresponds to a subdivided regular $d(D^*)$-gon of edge length 1 with the necessary identifications on the boundary (and the curvature of vertices that are created by the subdivision is 0).

With this PE structure, N^* is a connected, simply connected planar PE 2-complex. We claim that $K(N^*) \leq 0$. We only have to show that if $u^* \in N^*$ is an interior vertex and c^* the corresponding cycle, then the sum of angles at c^*: $W(c^*)$ is $\geq 2\pi$. This is true if and only if it is true for any interior vertex of M^*. Let $D \in M$ be the 2-cell with c^* in its interior and boundary vertices $u_1, \ldots, u_t$, which are interior vertices of M. Interpret the weights above, times π, as angles. Then (12) implies:

$$W(c^*) = \sum_{\gamma \in D} (\pi - g(\gamma) * \pi) \geq d(D)\pi - \pi(d(D) - 2) = 2\pi$$

By the addendum to Theorem 5.7 we get $\tau^2/(4\pi) \geq Area(N^*)$. $Area(M^*) = Area(N^*)$ where the area of a subdivided 2-cell D^* is equal to the area of a regular $d(D^*)$-gon of edge length 1.

If F^* is the number of 2-cells of M^*, and $A_3 = \sqrt{3}/4$ the area of the equilateral triangle whose sides are of length 1, then

$$V - l(\delta M) = F^* \leq \tau^2/(4\pi A_3) = \tau^2/(\sqrt{3}\pi),$$

where V is the number of vertices of M. Let E be the number of edges of M. Each 2-cell of M has at least 3 boundary edges, so $2E \geq 3F + l(\delta M)$. The last two inequalities together with the Euler-characteristic of M yield: $F \leq \frac{2\tau^2}{\sqrt{3}\pi} + l(\delta M) - 2$. Now (13) gives the desired result. $\square$

There is a dual theorem to the one just stated which is easier to prove. We omit its proof, it will be published elsewhere. Let $\mathcal{P}$ be a presentation with cyclically reduced relators. Let $f \colon M \to K_{\mathcal{P}}$ be a reduced van Kampen diagram with inner vertices of valence two deleted. We also assume that $\mathcal{P}$ satisfies the small cancellation condition $C(3)$, i.e., in every such diagram every 2-cell has at least 3 corners. If $\mathcal{P}$ satisfies for every such diagram and every inner 0-cell $v \in M$ one of the following conditions, then $\mathcal{P}$ (or $K_{\mathcal{P}}$) is said to be *of type W^**.

  (i) $d(v) = 3$, $d(D) \geq 6$ for all 2-cells D with $v \in \delta D$.

  (ii) $d(v) = 4$

- $d(D) \geq 4$ for all 2-cells D with $v \in \delta D$ or
- $d(D) \geq 5$ for two 2-cells D and $d(D) \geq 4$ for another D with $v \in \delta D$ or
- $d(D) \geq 6$ for one 2-cell D and $d(D) \geq 4$ for two further 2-cells D with $v \in \delta D$ or
- $d(D) \geq 6$ for two 2-cells D with $v \in \delta D$

(iii) $d(v) = 5$

- $d(D) \geq 4$ for two 2-cells D with $v \in \delta D$ or
- $d(D) \geq 6$ for one 2-cell D with $v \in \delta D$

(iv) $d(v) \geq 6$

The following theorem is due to Huck/Rosebrock:

Theorem 5.10 *Let $\mathcal{P}$ be a finite $C(3)$ presentation of type W^*, where each relator is cyclically reduced. Let $f \colon M \to K_{\mathcal{P}}$ be a reduced van Kampen diagram with F 2-cells. Then*

$$F \leq \frac{l^2(\delta M)}{\sqrt{3}\pi} \tag{14}$$

At last, we want to give an example of a class of presentations, that satisfy the weight test in its weak form as defined for example in Chapter V (i.e., inequality (4) is not strict), but do not have a quadratic isoperimetric inequality. The weights will be certainly given in a non-uniform way, a generalization of the last two theorems will not work by leaving the uniform condition away.

Consider the presentation $B_{k,l} = < x, y \mid y x^k y^{-1} = x^l >$ of the Baumslag-Solitar group for $k * l \neq 0$ and $|k| \neq |l|$. Gersten shows in [Ge91$_1$], that this group has no polynomial isoperimetric inequality. But there are weights, such that the weight test of Chapter V is satisfied: If γ is an arc in $S_P = W_P$ that goes from a vertex $\pm x$ to a vertex $\pm y$ then define its weight $g(\gamma) = 1/2$. The remaining edges of W_P all go between $+x$ and $-x$ and get weight $g(\gamma) = 1$. It is easy to see that $B_{k,l}$ satisfies the non-hyperbolic weight test of Chapter V with these weights.

Chapter VII

Fox Ideals, $\mathcal{N}$-Torsion and Applications to Groups and 3-Manifolds

Martin Lustig[1]

In this chapter, we introduce Fox ideals $I_m(G)$ for each $m \geq 0$ by means of finitely generated free partial resolutions of the group G, and describe a practical method for computing $I_m(G)$ using representations of G. Fox ideals provide powerful tests for determining the rank of G (for $m = 1$), the deficiency of G ($m = 2$) and the homological dimension of G ($m \geq 2$). These are derived in §§1–2, and some (partly new) applications are given.

$\mathcal{N}$-torsion groups $\mathcal{N}^m(G)$, introduced in §3, are the direct analogues of the Whitehead group $Wh(G)$, with $\mathbb{Z}G$ replaced by $\mathbb{Z}G/I_m(G)$. In general, however, $\mathcal{N}^m(G)$ turns out to be much richer than $Wh(G)$, and a practical evaluation method is described (using again representations of G), which yields non-trivial values in a large variety of cases. For $m = 1$, they distinguish Nielsen equivalence classes of generating systems of G, and hence isotopy or homeomorphy classes of Heegaard splitting of 3-manifolds (see §4). For $m = 2$ (or even $m \geq 2$), $\mathcal{N}$-torsion provides a crucial tool for the distinction of (simple)-homotopy classes of m-dimensional cell complexes (see §5). Here Fox ideals and $\mathcal{N}$-torsion values are direct and natural generalizations of the bias modulus and the bias invariant respectively, as introduced in Chapter III (see 5.4 below).

---
[1]supported by a grant from the German-Israeli Foundation for Research and Development (G.I.F.)

Fox ideals and $\mathcal{N}$-torsion and their preliminary versions have been discovered in different contexts and with varying degree of generality by several authors; see, for example, [Dy85], [Ho-AnLaMe91], [Ho-An88], [LuMo91] or [Me90]. Our treatment here is closest to [Lu91$_2$], where the independence from particular resolutions of G was achieved through matrix representations, and $I_m(G)$ and $\mathcal{N}^m(G)$ have been established as group theoretic invariants.

1   Fox ideals

At first reading of this chapter, the reader may try to skip this section (except 1.1 and the statements of 1.4 and 1.7) and read directly §2, where concrete applications of Fox ideals in dimension $m = 1$ and $m = 2$ are given.

1.1   Fox ideals for modules and exact sequences

(a) Let C be a free $\mathbb{Z}G$-module with basis $s_1, \ldots, s_h$ and let T be a submodule of C. Define the *Fox ideal* $I_C(T)$ to be the two-sided ideal in $\mathbb{Z}G$ generated by the set

$$\{a_i \in \mathbb{Z}G \mid \text{ there exists } t \in T \text{ with } t = a_1 s_1 + \ldots + a_h s_h\}.$$

Since $I_C(T)$ is two-sided, it obviously does not depend on the choice of the basis for C. If T is finitely generated as $\mathbb{Z}G$-module then so is $I_C(T)$, since it suffices to consider those $t \in T$ which belong to a generating system.

(b) Consider any exact sequence

$$\mathbf{C}: \qquad C_m \xrightarrow{\partial_m} C_{m-1} \xrightarrow{\partial_{m-1}} \ldots \xrightarrow{\partial_1} C_0 \xrightarrow{\partial_0} \mathbb{Z}$$

of finitely generated free $\mathbb{Z}G$-modules C_i for all $0 \leq i \leq m$. Let $I_m(\mathbf{C})$ be the two-sided ideal $I_{C_m}(\ker(\partial_m)) \subset \mathbb{Z}G$. $I_m(\mathbf{C})$ is called the *m-th Fox ideal of* $\mathbf{C}$.

(c) For all $n \in \mathbb{Z}$, we define the two-sided ideal $I_{m,n}(G) \subset \mathbb{Z}G$ that is generated by the ideals $I_m(\mathbf{C})$ for all sequences $\mathbf{C}$ as in (b) which have *directed Euler characteristic* $(-1)^m \chi(\mathbf{C}) := \sum_{i=0}^{m} (-1)^{i+m} rk(\mathbf{C}_i)$ equal to n.

1.2 Comments on 1.1

1. For modules T and C as in 1.1 (a) and any $\mathbb{Z}G$-module $S \subset T$, one has $I_C(S) \subset I_C(T)$.

2. There is a close relationship between the Fox ideal and the trace ideal of the module T (see [Dy85]). Notice that $I_C(T)$ can alternatively be defined as the smallest ideal (= the intersection of all ideals) I such that the inclusion $T \xrightarrow{\ \subset\ } C$ induces the 0-map $\mathbb{Z}G/I \otimes T \to \mathbb{Z}G/I \otimes C$.

3. A *partial resolution* $\mathbf{C}$ of G, as considered in part (b), is given in particular by any (G, m)-complex K as defined in Chapter III, Definition 1.1, for $m \geq 2$. Thus, any such complex K also possesses an m-th Fox ideal.

4. The existence of such a partial resolution is a non-trivial condition on G, since all modules C_i are finitely generated.

5. At this point, the reader may think of 1.1 (c) as a cheap method of defining a Fox ideal that depends just on the group and not on a particular partial resolution or (G, m)-complex. However, it will be shown below that there is very little difference between the ideals $I_{m,n}(G)$ and $I_m(\mathbf{C})$, for any sequence $\mathbf{C}$ with $(-1)^m \chi(\mathbf{C}) = n$. Indeed, for finite groups they agree: see [Ho-AnLaMe91], Lemma 3.

1.3 Stabilization Lemma:

Consider any two exact sequences

$$\mathbf{C}: \qquad C_m \xrightarrow{\partial_m} C_{m-1} \xrightarrow{\partial_{m-1}} \ \ldots \ \xrightarrow{\partial_1} C_0 \xrightarrow{\partial_0} \mathbb{Z}$$

and

$$\mathbf{D}: \qquad D_m \xrightarrow{\delta_m} D_{m-1} \xrightarrow{\delta_{m-1}} \ \ldots \ \xrightarrow{\delta_1} D_0 \xrightarrow{\delta_0} \mathbb{Z}$$

as in 1.1 (b). There exist finitely generated free $\mathbb{Z}G$-modules $M_C, M_D, M_0, \ldots, M_{m-1}$, and exact sequences

$$\mathbf{C}^D: \qquad C_m \oplus M_C \xrightarrow{\partial'_m} M_{m-1} \xrightarrow{\partial'_{m-1}} \ \ldots \ \xrightarrow{\partial'_1} M_0 \xrightarrow{\partial'_0} \mathbb{Z}$$

and

$$\mathbf{D}^C: \qquad D_m \oplus M_D \xrightarrow{\delta'_m} M_{m-1} \xrightarrow{\delta'_{m-1}} \ \ldots \ \xrightarrow{\delta'_1} M_0 \xrightarrow{\delta'_0} \mathbb{Z},$$

with the following properties:

(a) the boundary operator $\partial_i' = \delta_i'$ agrees for all $i \leq m - 1$.

(b) $\mathbf{C}$ and $\mathbf{C}^D$ as well as $\mathbf{D}$ and $\mathbf{D}^C$ are pairs of stably isomorphic sequences (compare Chapter III, Lemma 1.2, or [Co73]). In particular one has

$$\chi(\mathbf{C}) = \chi(\mathbf{C}^D) \text{ and } \chi(\mathbf{D}) = \chi(\mathbf{D}^C),$$

and, for any choice of preferred bases in $\mathbf{C}$ and $\mathbf{D}$, there is a canonical set of preferred bases in $\mathbf{C}^D$ and $\mathbf{D}^C$. For any M_i, $i \leq m$, the bases coming from $\mathbf{C}^D$ and from $\mathbf{D}^C$ agree up to elementary matrix operations.

(c) One has $\ker(\partial_m') = \ker(\partial_m) \subset C_m, \ker(\delta_m') = \ker(\delta_m) \subset D_m$. In particular this gives $I_m(\mathbf{C}^D) = I_m(\mathbf{C}), I_m(\mathbf{D}^C) = I_m(\mathbf{D})$.

Proof: Let us assume by induction on k, $0 \leq k \leq m - 1$, that $C_i = D_i = M_i$ and $\partial_i = \delta_i$ for all $i \leq k - 1$.

Define $C_k' = C_k \oplus D_k, C_{k+1}' = C_{k+1} \oplus D_k$ and $C_i' = C_i$ for $i > k + 1$. Similarily define $D_k' = D_k \oplus C_k, D_{k+1}' = D_{k+1} \oplus C_k$ and $D_i' = D_i$ for $i > k + 1$. Define the boundary operators for the new sequences (as is standard for such a stabilization) to be the identity and the 0-map on the two new direct summands in dimension $k + 1$ and k respectively, and to agree with the old boundary operator everywhere else. Then change the preferred basis (if there is any) of $C_k' = C_k \oplus D_k$ by adding appropriate elements of C_k to the preferred basis elements of D_k such that the boundary operator ∂_k' on the resulting new basis elements coincides with the boundary map δ_k on the preferred basis of D_k. Do the analogue with D_k' and identify it with C_k' according to the newly obtained bases, thus defining M_k and $\partial_k' = \delta_k'$. This proves claims (a) and (b) by induction. Since the induction stops for $k = m - 1$, claim (c) follows from the observation that the above stabilization step preserves the kernel of the $(k + 1)$-th boundary operator.   □

1.4   Fox ideals and directed Euler characteristics for groups

Let G be a group of type FP_m, that is, G admits a finitely generated free m-th partial resolution $\mathbf{C}$ as in 1.1 (b).

(a) Then there is a lower bound to the directed Euler characteristic $(-1)^m \chi(\mathbf{C})$ for all such $\mathbf{C}$. The minimum will be denoted by $\chi_m(G) \in \mathbb{Z}$ and called the *m-th directed Euler characteristic of G*. (See 2.2 and 2.6 below for the group theoretic relevance of $\chi_m(G)$ for small m.)

(b) For all $n > \chi_m(G)$, one has $I_{m,n}(G) = \mathbb{Z}G$. We will call $I_{m,\chi_m(G)}$ the *m-th Fox ideal of G* and denote it by $I_m(G)$. (It is easy to see that $I_0(G)$ is just the augmentation ideal of G.)

Proof: (a) Let $\mathbf{C}$ and $\mathbf{D}$ be sequences as in 1.3 and let $\mathbf{C}^D$ and $\mathbf{D}^C$ be the corresponding sequences derived there. Consider the maps $C_m \oplus M_C \xrightarrow{\partial'_m} M_{m-1}$ and $D_m \oplus M_D \xrightarrow{\delta'_m} M_{m-1}$: From the construction in the proof of 1.3, we see that $im(\partial'_m) \cong im(\partial_m) \oplus M_C$. By considering the augmentation map $\mathbb{Z}G \to \mathbb{Z}$, we see that $im(\partial'_m)$ can not be generated by less than $\mathrm{rk}(M_C)$ elements. Hence,

$$(-1)^m \chi(\mathbf{D}) = (-1)^m \chi(\mathbf{D}^C) \geq (-1)^m \chi(\mathbf{C}^D) - \mathrm{rk}(C_m) = (-1)^m \chi(\mathbf{C}) - \mathrm{rk}(C_m).$$

(b) This is immediate from the definition: If

$$\mathbf{C}: \qquad C_m \xrightarrow{\partial_m} C_{m-1} \xrightarrow{\partial_{m-1}} \ \cdots \ \xrightarrow{\partial_1} C_0 \xrightarrow{\partial_0} \mathbb{Z}$$

is an exact sequence, then so is

$$\mathbf{C}': \qquad C'_m \xrightarrow{\partial'_m} C_{m-1} \xrightarrow{\partial_{m-1}} \ \cdots \ \xrightarrow{\partial_1} C_0 \xrightarrow{\partial_0} \mathbb{Z}$$

with $C'_m = C_m \oplus \mathbb{Z}G$, with $\partial'_m = \partial_m$ on the first summand and $\partial'_m = 0$ on the second. But then definition 1.1 (b) gives $I_m(\mathbf{C}') = \mathbb{Z}G$. Hence, it follows from definition 1.1 (c) that $I_{m,n}(G) = \mathbb{Z}G$ for any non-minimal n. $\square$

1.5 Open questions:

1. Do there exist $\mathbf{C}$ and $\mathbf{D}$, both with minimal directed Euler characteristic, that have distinct Fox ideals:

$$I_m(\mathbf{C}) \neq I_m(\mathbf{D}) \ ?$$

2. Does there exist a group G and some partial resolution $\mathbf{C}$ as above, with non-minimal directed Euler characteristic, $(-1)^m \chi(\mathbf{C}) > \chi_m(G)$, such that

$$I_m(\mathbf{C}) \neq \mathbb{Z}G \ ?$$

It will follow from 1.7 that examples of partial resolutions answering these questions in the affirmative will be very hard to verify, if they exist at all.

1.6 Lemma

Consider the following commutative diagram of $\mathbb{Z}G$-modules:

$$
\begin{array}{ccc}
 & C & \\
 & \searrow \partial & \\
f \downarrow\uparrow f' & & M \\
 & \nearrow \partial' & \\
 & D &
\end{array}
$$

Let C, D be free modules of finite rank with preferred bases, and denote by $[f], [f']$, etc. the corresponding $\mathbb{Z}G$-matrices with respect to the preferred bases. Then

(a) $[f' \cdot f] = 1$ modulo $I_C(\ker(\partial)) \subset \mathbb{Z}G$, and

(b) for any two-sided ideal $J \subset \mathbb{Z}G$, with $I_C(\ker(\partial)) \subset J$, one has:

$$[f \cdot f'] = 1 \text{ modulo } J \quad \Longleftrightarrow \quad I_D(\ker(\partial')) \subset J .$$

Proof: (a) From $\partial \cdot f' \cdot f = \partial' \cdot f = \partial$ we get $f'f(c) - c \in \ker(\partial)$ for all $c \in C$. But $\ker(\partial) = 0$ mod $I_C(\ker(\partial))$ follows directly from the definition of $I_C(\ker(\partial))$ in 1.1 (a).

(b) The implication $\Longleftarrow$ follows from (a) by the symmetry between C and D. For the converse implication, observe that $\partial \cdot f' = \partial'$ implies $f'(\ker(\partial')) \subset \ker(\partial)$ and hence $f'(\ker(\partial')) = 0$ mod J. By assumption, this implies $0 = ff'(\ker(\partial')) = \ker(\partial')$ mod J and hence $I_D(\ker(\partial')) \subset J$. $\square$

1.7 Computing $\chi_m(G)$ through representations

Let R be a commutative ring with $1 \neq 0$, let $\mathbf{C}$ be a sequence as in 1.1 (b), and let $\rho: \mathbb{Z}G \to \mathbb{M}_k(R)$ be a representation as $(k \times k)$-matrices over R with $\rho(1) = 1$ and $\rho(I_m(\mathbf{C})) = \{0\}$. Then $\rho(I_{m,n}(G)) = \{0\}$ for $n = (-1)^m\chi(\mathbf{C})$, and $\mathbf{C}$ realizes the minimal directed Euler characteristic:

$$\chi_m(G) = (-1)^m\chi(\mathbf{C}) \quad \text{(and hence } I_{m,n}(G) = I_m(G) \text{)}.$$

Proof: Let $\mathbf{D}$ be a second sequence as in 1.1 (b), with $(-1)^m\chi(\mathbf{D}) = n$, and let $\mathbf{C}^D$ and $\mathbf{D}^C$ be the corresponding sequences derived in 1.3. Since the last terms $C := C_m \oplus M_C$ and $D := D_m \oplus M_D$ of these sequences are both free, there exist maps f and f' as in 1.6 (for $M = M_{m-1}$). Part (a) of 1.6 gives $[f' \cdot f] = 1$ modulo $I_m(C)$ and hence $\rho([f]) \cdot \rho([f']) = 1$ (the change of order is due to the fact that we consider left modules). But in a matrix ring with commutative entries this implies $\rho([f']) \cdot \rho([f]) = 1$. Hence one has $[f \cdot f'] = 1$ modulo $\ker(\rho)$, which implies (by 1.6 (b)) $I_m(\mathbf{D}) \subset \ker(\rho)$. Since $\mathbf{D}$ was chosen arbitrarily with $(-1)^m \chi(\mathbf{D}) = n$, we obtain $I_{m,n}(G) \subset \ker(\rho)$. Thus $\chi_m(G) = (-1)^m \chi(\mathbf{C})$ follows directly from 1.4 (b), since the assumption $\rho(1) = 1$ implies $I_m(\mathbf{D}) \subset \ker(\rho) \neq \mathbb{Z}G$.   □

Notice that the statement of 1.7 is still valid if $\mathbb{M}_k(R)$ is replaced by any non-trivial ring with the property that left inverses are also right inverses.

2 Applications of Fox ideals: Tests for the rank, the deficiency and the homological dimension of a group

2.1 The classical Fox ideal

Every presentation $\mathcal{P} = \langle x_1, \ldots, x_n \mid R_1, R_2, \ldots \rangle$ of the group G by generators x_j and relators R_i gives rise to a canonical partial resolution $C_\mathcal{P}$ of G of length $m = 2$, which is the cellular chain complex of the universal cover $\tilde{K}$ of the standard 2-complex K associated to the presentation (see Chapter I, §1.4 (9)). In particular, the generating system $x = \{x_1, \ldots, x_n\}$ gives rise to a partial resolution $\mathbf{C}_x$ of G of length $m = 1$, with associated Fox ideal $I_x := I_1(\mathbf{C}_x)$. Indeed, this is the classical Fox ideal associated to the presentation $\mathcal{P}$ of G. I_x is more accessible than most of the more general Fox ideals discussed in the previous section, since $\mathcal{P}$ provides a canonical generating system: I_x is the two-sided ideal generated by the canonical images in $\mathbb{Z}G$ of the Fox derivatives $\partial R_i / \partial x_j$, see Chapter II, before Theorem 3.8.

2.2 The rank of G and $\chi_1(G)$

For any generating system of cardinality n the number $n - 1$ is an upper bound to the 1st directed Euler characteristic of G. Equivalently but more importantly, the value $\chi_1(G) + 1$ is a lower bound to the rank of G, by which we mean the minimal number of elements needed to generate G. This gives rise to the following non-trivial test for determining $\text{rank}(G)$, which is a direct consequence of 1.7.

2.3   A rank test ([Lu91$_2$])

Let G be a group presented by $\langle x_1, \ldots, x_n \mid R_1, R_2, \ldots \rangle$. Let $a_{i,j} \in \mathbb{Z}G$ denote the canonical images of the Fox derivatives $\partial R_i / \partial x_j \in F(x_1, \ldots, x_n)$. If for some $k \in \mathbb{N}$ and some commutative ring R with $1 \neq 0$, there exists a ring homomorphism $\rho : \mathbb{Z}G \to \mathbb{M}_k(R)$, with $\rho(1) = 1$ and with $\rho(a_{i,j}) = 0$ for all $a_{i,j}$, then

$$\text{rank } (G) = n. \qquad \qquad \square$$

2.4   Examples for the rank test

There are many large classes of groups where the rank test is applicable, and where furthermore the rank can not be determined by abelianization of G. The most fruitful (since technically easiest) method is given by considering the image of I_x in the abelianization of the group ring $\mathbb{Z}G$ (which is totally different from computing the rank of $G/[G,G]$). For example:

2.4.1 NEC-groups

Consider a non-Euclidean crystallographic group G with reflections:

$$G = \langle c_1, \ldots, c_n \mid c_1^2, \ldots, c_n^2, (c_1 c_2)^{h_1}, \ldots, (c_{n-1}, c_n)^{h_{n-1}}, (c_n c_1)^{h_n} \rangle.$$

The rank of such groups has been considered in [KaZi92], and it has been shown there that for certain values of the h_i the rank of G is bounded above by $2n/3$. However, there are many cases left over where the rank has not been fully determined. Some of them can be treated very easily by the rank test 2.3:

Proposition   *If all h_i have a common divisor $p \geq 2$, then the rank of G is n. (Notice that $\text{rank}(G/G') = \max(1, n - \#\{i \mid 1 \leq i \leq n, h_i \text{ odd }\})$.)*

Proof: The Fox derivatives $\partial R_i / \partial x_j$ as in 2.3 are computed as follows:

1. $\partial c_j^2 / \partial c_j = 1 + c_j$ and $\partial c_j^2 / \partial c_i = 0$ if $i \neq j$.

2. $\partial (c_j c_{j+1})^{h_j} / \partial c_i = (1 + (c_j c_{j+1}) + (c_j c_{j+1})^2 + \ldots + (c_j c_{j+1})^{h_j - 1})(\partial c_j c_{j+1} / \partial c_i)$
   (here j is understood modulo n).

Thus the map $\rho \colon \mathbb{Z}G \to \mathbb{Z}/p\mathbb{Z}, c_j \to -1$, sends all these Fox derivatives to 0, which proves $\text{rank}(G) = n$ by 2.3. $\qquad \square$

2.4.2 Coxeter groups

Let G be a Coxeter group, i.e., a group with presentation

$$G = \langle c_1, \ldots, c_n \mid c_1^2, \ldots, c_n^2, (c_i c_j)^{h_{ij}} \ (1 \le i, j \le n) \rangle,$$

where some of the h_{ij} may well be "infinite", i.e., there is no relation $(c_i c_j)^{h_{ij}} = 1$. The same proof as in 2.4.1 applies to give the following result, for which we do not know whether there exists another proof.

Proposition *If all finite h_{ij} have a common divisor $p \ge 2$, then the rank of G is n.* $\square$

2.5 The second Fox ideal $I_{\mathcal{P}}$ associated to a group presentation $\mathcal{P}$

Let us now assume that the presentation $\mathcal{P} = \langle x_1, \ldots, x_n \mid R_1, R_2, \ldots, R_q \rangle$ of the group G is finite. Then the associated partial resolution $\mathbf{C}_{\mathcal{P}}$ of G of length $m = 2$ is finitely generated as required in 1.1 (b), and hence determines a 2nd Fox ideal $I_{\mathcal{P}} := I_2(\mathbf{C}_{\mathcal{P}})$. In general, $I_{\mathcal{P}}$ will be much harder to compute than I_x in 2.1, since the kernel of the second chain map ∂_2 of $\mathbf{C}_{\mathcal{P}}$ is precisely the second homotopy group $\pi_2(K)$ of the standard 2-complex K associated to $\mathcal{P}$, and generators for π_2 are in general difficult to determine. However, in many concrete situations there are techniques available, see Chapter V: We will present below (see 2.7, 5.5 and 5.7) examples where $I_{\mathcal{P}}$ has been computed (or at least approximated) successfully.

2.6 The deficiency of G and $\chi_2(G)$

For any presentation $\mathcal{P}$, as in 2.5, we define the *deficiency* def$(\mathcal{P})$ of $\mathcal{P}$ to be the number $q - n$. Notice that some authors define the deficiency as $n - q$. We define the *deficiency* def(G) of G to be the minimum of the def$(\mathcal{P})$ for all presentations $\mathcal{P}$ of G. Notice that the computation of def(G) is in general a difficult task; see 2.8.

For any presentation P, the number $q - n + 1$ is an upper bound to the 2nd directed Euler characteristic of G. Conversely, $\chi_2(G)$ is a lower bound to the Euler characteristic of any 2-complex K with fundamental group $\pi_1 K = G$. In particular, the value $\chi_2(G) - 1$ is a lower bound to def $(\mathcal{P})$ for all $\mathcal{P}$ and hence to def (G). This gives rise to the following non-trivial test for determining def (G), as a direct consequence of 1.7.

2.7   A deficiency test ([Lu91$_2$])

Let G be a group presented by $\langle x_1, \ldots, x_n \mid R_1, \ldots, R_q \rangle$, and let K be the corresponding standard 2-complex. Let $\tilde{R}_1, \ldots, \tilde{R}_q$ denote the natural basis corresponding to the R_i for the second chain group $C_1(\tilde{K})$ of the universal covering of K. Let $T \subset C_2(\tilde{K})$ be a $\mathbb{Z}G$-module that contains $\pi_2 K = H_2(\tilde{K})$, and let

$$y_1 \; = \; a_{1,1}\tilde{R}_1 + \ldots + a_{1,q}\tilde{R}_q,$$

$$y_2 \; = \; a_{2,1}\tilde{R}_1 + \ldots + a_{2,q}\tilde{R}_q,$$

$$\ldots$$

be generators of T, with $a_{i,j} \in \mathbb{Z}G$. If there exists a ring homomorphism $\rho \colon \mathbb{Z}G \to \mathbb{M}_k(R)$ for $k \in \mathbb{N}$ and some commutative ring R with $1 \neq 0$, such that $\rho(1) = 1$ and $\rho(a_{i,j}) = 0$ for all $a_{i,j}$, then

$$\mathrm{def}\,(G) \; = \; q - n. \qquad \qquad \square$$

2.8   Efficiency

As stated already in 2.6, the deficiency of a group is not easy to determine: Of course, every finite presentation $\mathcal{P}$ gives an upper bound

$$\mathrm{def}(\mathcal{P}) \; \geq \; \mathrm{def}(G).$$

On the other hand, there is the well known homological lower bound for $\mathrm{def}(G)$,

$$\mathrm{def}(G) \; \geq \; s(H_2(G)) - rk(H_1(G))$$

(see Chapter IV, §4.2), where $s(\cdot)$ denotes the minimal number of $\mathbb{Z}$-module generators and $rk(\cdot)$ the dimension as $\mathbb{Q}$-vector space after tensoring with $\mathbb{Q}$. A presentation which meets this bound is called *efficient*, and a group G with such a presentation is called itself *efficient*. Until recently, the only known non-efficient groups were finite [Sw65]. Last year the author was asked by Pride whether the deficiency test 2.7 could be used to exhibit examples of non-efficient groups through second Fox ideals. The author suggested a construction principle for such groups as well as a concrete example. Later Bak-Pride [BaPr92] used the same ideas to give more examples of similar sort. The example below is close to the original one:

2.9 Non-efficient groups without torsion

Let $G = \langle x_1, x_2 \mid R \rangle$ be a 2-generator knot group with non-trivial Alexander polynomial $\Delta(t)$, e.g., the figure eight knot group (see [BuZi85]). We may assume that x_1 and x_2 are chosen so that the abelianization $q \colon G \to \langle t \mid - \rangle$ maps both x_1 and x_2 to t, which gives $\Delta(t) = \pm t^{k_i} q(\partial R/\partial x_i)$ for some $k_1, k_2 \in \mathbb{Z}$.

Proposition *The group $H = G \oplus \mathbb{Z}$ is non-efficient (and obviously torsion-free).*

Proof: The abelianization of H is $\mathbb{Z}^2$, and hence $rk(H_1(H)) = 2$. Knot complements are $K(\pi, 1)$-spaces, and without loss of generality we can assume that the above presentation of G is the standard presentation corresponding to a 2-complex K homotopy equivalent to the knot exterior. It follows that K is aspherical, and hence $L = K \times S^1$ is aspherical with $\pi_1 L = H$. Thus we can compute $H_2(H) = H_2(L) = H_2(K) \oplus (H_1(K) \otimes H_1(S^1))$ (e.g., see [GrHa81, 29.11.1]). But $H_1(K) = \mathbb{Z}$, which shows by the Euler characteristic of K that $H_2(K) = 0$. Thus $s(H_2(H)) = 1$.

On the other hand, we will use the deficiency test 2.7 in order to show that $\operatorname{def}(H) = 0$: Consider the presentation $\langle x_1, x_2, x_3 \mid R, S, T \rangle$ of H with $S = x_1 x_3 x_1^{-1} x_3^{-1}$ and $T = x_2 x_3 x_2^{-1} x_3^{-1}$, which is the standard 2-complex corresponding to the 2-skeleton L^2 of $L = K \times S^1$, where S^1 consists of a single 0-cell and a single 1-cell. From the asphericity of L we deduce that $\pi_2 L^2$ is generated by the boundary of the 3-cell of L, which is the element $(x_3 - 1)\tilde{R} + (\partial R/dx_1)\tilde{S} + (\partial R/dx_2)\tilde{T}$ (see also Chapter V, Theorem 3.2). Thus $I_2(C(\tilde{L}))$ is generated by $(x_3 - 1)$, $\partial R/dx_1$ and $\partial R/dx_2$. A representation as in 2.7 is now given by $\rho \colon \mathbb{Z}H \to \mathbb{Z}[t, t^{-1}]/(\Delta)$, with $\rho(x_1) = \rho(x_2) = t$ and $\rho(x_3) = 1$. $\qquad \square$

2.10 A Fox ideal test for the homological dimension of groups ([Lu91₂])

Let G be a group of type FL, and let

$$\mathbf{C} : \quad C_m \xrightarrow{\partial_m} C_{m-1} \xrightarrow{\partial_{m-1}} \ldots \xrightarrow{\partial_1} C_0 \xrightarrow{\partial_0} \mathbb{Z}$$

be an exact sequence of finitely generated free $\mathbb{Z}G$-modules. Assume $\ker(\partial_m) \neq 0$ (otherwise one has $\operatorname{hdim}(G) \leq m$). Let $s_1, \ldots, s_h$ be a basis of the m-th term C_m, and let $y_1, y_2, \ldots$ be $\mathbb{Z}G$-module generators of $\ker(\partial_m)$: For all y_i there exist expressions

$$y_i = a_{i,1} s_1 + \ldots + a_{i,m} s_m$$

with $a_{i,j} \in \mathbb{Z}G$. If for some $k \in \mathbb{N}$ and some commutative ring R with $1 \neq 0$, there exists a ring homomorphism $\rho: \mathbb{Z}G \to \mathbb{M}_k(R)$, with $\rho(1) = 1$ and with $\rho(a_{i,j}) = 0$ for all $a_{i,j}$, then

$$\mathrm{hdim}(G) \geq m + 1.$$

Proof: Observe first that $\mathrm{hdim}(G) < m + 1$ would imply (by [Br82], p. 203, Exercise 1) that there exists a resolution $\mathbf{D}$ of G of length $s < m + 1$ with all terms D_i free and finitely generated (since G was assumed to be of type FL). But the existence of any such resolution $\mathbf{D}$ gives immediately $I_t(G) = \{0\}$ for all $t \geq m \geq s$. However, the assumption $\ker(\partial_m) \neq \{0\}$ implies $I_m(\mathbf{C}) \neq \{0\}$ and hence (by 1.4 (b)) $(-1)^m \chi(\mathbf{C}) > \chi_m(G)$. This contradicts the existence of the assumed representation ρ, by 1.7. $\square$

3 $\mathcal{N}$-torsion: Basic theory

3.1   The $\mathcal{N}$-torsion groups $\mathcal{N}^m(G)$

For every group G of type FP_m, we define the *m-th $\mathcal{N}$-torsion group $\mathcal{N}^m(G)$* to be the (abelian) "Whitehead type group"

$$\mathcal{N}^m(G) := K_1(\mathbb{Z}G/I_m(G))/\pm G.$$

Here $K_1(\cdot)$ denotes the first K-group[2] from K-theory or simple-homotopy theory (see [Mi71] or [Co73]), with entries from the quotient ring $\mathbb{Z}G/I_m(G)$ of $\mathbb{Z}G$ modulo the m-th Fox ideal $I_m(G)$. Thus $\mathcal{N}^m(G)$ is the quotient of $K_1(\mathbb{Z}G/I_m(G))$ modulo all elements represented by diagonal matrices with diagonal entries $\pm g$ for some $g \in G$.

$\mathcal{N}^m(G)$ is called "degenerate", if $I_m(G) = \mathbb{Z}G$, and in this case we define formally $\mathcal{N}^m(G) = \{0\}$.

3.2   Whitehead torsion and $\mathcal{N}$-torsion

For all $m \in \mathbb{N}$, there is a canonical homomorphism

$$Wh(G) \to \mathcal{N}^m(G)$$

from the Whitehead group $Wh(G)$. This is an immediate consequence of the functorality of the group $K_1(R)$ for rings R and their homomorphisms.

---

[2]We actually use here a slight extension of the classical $K_1(\cdot)$ in order to include matrices which are only one-sided invertible.

3.3 The $\mathcal{N}$-torsion for a pair of partial resolutions with preferred bases

For every two exact sequences with preferred bases **C**, **D** as in 1.3 with $(-1)^m \chi(\mathbf{C}) = (-1)^m \chi(\mathbf{D}) = \chi_m(G)$, we define an element

$$\mathcal{N}^m(\mathbf{C}, \mathbf{D}) \in \mathcal{N}^m(G)$$

as follows: Let $\mathbf{C}^D, \mathbf{D}^C$ be the complexes with preferred bases derived from **C**, **D** as in the Stabilization Lemma 1.3, and let C, D denote their m-th term respectively. Consider any $\mathbb{Z}G$-homomorphism $f \colon C \to D$ as in 1.6 (with $M_{m-1} =: M$). Define $\mathcal{N}(\mathbf{C}, \mathbf{D})$ to be the element in $\mathcal{N}^m(G)$ represented by the image of the matrix $[f]$ (with respect to the preferred bases) under the quotient map $\mathbb{Z}G \to \mathbb{Z}G/I_m(G)$.

Notice that $\mathcal{N}^m(\mathbf{C}, \mathbf{D})$ does not depend on the choice of the map f, since any other such map differs from f only by elements that are equal to 0 modulo $I_m(G)$. As in standard simple-homotopy theory, it follows that the value of $\mathcal{N}^m(\mathbf{C}, \mathbf{D})$ is independent of the particular choice of stabilizations performed on **C** and **D** in order to define $\mathbf{C}^D$ and $\mathbf{D}^C$.

3.4 The cancellation rule

The group structure of $\mathcal{N}^m(G)$ yields for any sequences **C**, **D**, **E** as in 3.3:

$$\mathcal{N}^m(\mathbf{C}, \mathbf{D}) + \mathcal{N}^m(\mathbf{D}, \mathbf{E}) = \mathcal{N}^m(\mathbf{C}, \mathbf{E}).$$

In particular, if **C** and **D** are stably isomorphic with respect to their preferred bases, one has

$$\mathcal{N}^m(\mathbf{C}, \mathbf{D}) = 0 \quad \text{and} \quad \mathcal{N}^m(\mathbf{C}, \mathbf{E}) = \mathcal{N}^m(\mathbf{D}, \mathbf{E}).$$

These are direct consequences from the definitions and from the observation that a simultaneous stabilization for **C**, **D**, **E** can be achieved in three successive steps of pairwise stabilizations as in 1.3.   □

In the next sections, the relevance of $\mathcal{N}^m$-torsion for generating systems of G ($m = 1$) and for the (simple-)homotopy type of (G, m)-complexes ($m \geq 2$) will be explained and examples will be given. All of them use a particular technique for evaluating $\mathcal{N}^m(G)$ that we present now.

3.5 Evaluation of $\mathcal{N}^m(G)$ via the determinant

Let R be a commutative ring with unit $1 \neq 0$, let $k \in \mathbb{N}$, and let $\rho \colon \mathbb{Z}G \to \mathbb{M}_k(R)$ be a ring homomorphism with $\rho(1) = 1$ and $\rho(I_m) = \{0\} \subset \mathbb{M}_k(R)$.

By the functorality of K_1, the map ρ induces a homomorphism of abelian groups

$$K_1(\rho)\colon K_1(\mathbb{Z}G/I_m) \to K_1(\mathbb{M}_k(R)) \to K_1(R),$$

where the last map (an isomorphism !) is induced by the "forgetting the brackets" map. On $K_1(R)$, we have the determinant map det: $K_1(R) \to R^*$ into the multiplicative group of units R^* of R. Let τ_ρ denote the composite map

$$\tau_\rho\colon GL(\mathbb{Z}G/I_m) \xrightarrow{\text{definition}} K_1(\mathbb{Z}G/I_m) \xrightarrow{K_1(\rho)} K_1(R) \xrightarrow{\det} R^* .$$

We define the subgroup T_ρ of R^* as the τ_ρ-image of the subgroup of *trivial units* in $GL(\mathbb{Z}G/I_m)$, i.e., the elements represented by diagonal matrices with entries in $G \cup -G$. Let $\mathcal{N}^m(\rho)$ be the map $\mathcal{N}(G) \to R^*/T_\rho$ induced by τ_ρ.

3.6 $S\mathcal{N}^m(G)$, $\mathcal{N}^m(G)^*$ and $\mathcal{N}^m(G)_k$

Let $S\mathcal{N}^m(G) \subset \mathcal{N}^m(G)$ be the subgroup which consists of all elements $Z \in \mathcal{N}^m(G)$ with $\mathcal{N}(\rho)(Z) = 0$ for any R, k, and ρ as in 3.5, and denote by $\mathcal{N}^m(G)^*$ the quotient

$$\mathcal{N}^m(G)^* := \mathcal{N}^m(G)/S\mathcal{N}^m(G).$$

Similarly, we define, for any fixed $k \in \mathbb{N}$ and varying R and ρ, the groups $S\mathcal{N}^m(G)_k$ and $\mathcal{N}^m(G)_k = \mathcal{N}^m(G)/S\mathcal{N}^m(G)_k$. Notice that there is a canonical quotient $\mathcal{N}^m(G)^* \to \mathcal{N}^m(G)_k$.

All computations in the groups $\mathcal{N}^m(G)$ known to the author have been performed using some quotient $\mathcal{N}^m(G)_k$ of $\mathcal{N}^m(G)$. We know of no group G where the kernel $S\mathcal{N}^m(G)$ of the map $\mathcal{N}^m(G) \to \mathcal{N}^m(G)^*$ is proven to be non-trivial for some $m \in \mathbb{N}$ with $I_m(G) \neq 0$.

3.7   A direct sum formula for free products

We now state a formula for free products $G = G_1 * G_2$ that describes $\mathcal{N}^m(G)^*$ in terms of $\mathcal{N}^m(G_1)^*$ and $\mathcal{N}^m(G_2)^*$. It has been proved in [LuMo93$_2$] (stated there for $m = 1$, but the proof for $m \geq 2$ is exactly the same). The proof is based on a direct sum formula for "pure" rings due to Casson, but is too technical to be reproduced here.

The theorem below should be thought of as analogue to the celebrated free product formula for Whitehead groups:

$$Wh(G_1 * G_2) = Wh(G_1) \oplus Wh(G_2)$$

see [St65]. Unfortunately, the groups $\mathcal{N}^m(G)$ are less well behaved, as is indicated for $m = 2$ in [Si77], p.138 (partially reproduced in Chapter XII, §3, (17)). In [LuMo93$_2$], §3, an example is given with $\mathcal{N}^1(G_1 * G_2) = \{0\}$, while both $\mathcal{N}^1(G_1)$ and $\mathcal{N}^1(G_2)$ are non-trivial.

This difficulty is why we pass over to coefficients in a field $\mathfrak{K}$, thus obtaining torsion groups $\mathcal{N}^m(G; \mathfrak{K})$ and $\mathcal{N}^m(G; \mathfrak{K})^*$ defined in complete analogy to 3.1 and 3.6, with $\mathbb{Z}G$ replaced by $\mathfrak{K}[G]$.

Theorem   *Let $\mathfrak{K}$ be a field, $m \in \mathbb{N}$, and consider a free product of groups $G_1 * G_2$. If both $\mathcal{N}^m(G_1; \mathfrak{K})^*$ and $\mathcal{N}^m(G_2; \mathfrak{K})^*$ are non-degenerate, then the natural embeddings $G_1 \subset G_1 * G_2, G_2 \subset G_1 * G_2$ induce an isomorphism*

$$\mathcal{N}^m(G_1; \mathfrak{K})^* \oplus \mathcal{N}^m(G_2; \mathfrak{K})^* \overset{\cong}{\longrightarrow} \mathcal{N}^m(G_1 * G_2; \mathfrak{K})^* \ .$$

Using deep results of Waldhausen (see [Wa78]), it seems possible to derive the same statement for $\mathcal{N}^m(\cdot)$ rather than $\mathcal{N}^m(\cdot)^*$, but for all practical purposes this does not make a difference since the part $S\mathcal{N}^m(G)$ of $\mathcal{N}^m(G)$ is beyond the reach of our computational abilities.

Remark   Notice that the non-degeneracy of both $\mathcal{N}^m(G_1; \mathfrak{K})^*$ and $\mathcal{N}^m(G_2; \mathfrak{K})^*$ implies that the directed Euler characteristic of a free product is additive:

$$\chi_m(G_1 * G_2) = \chi_m(G_1) + \chi_m(G_2) - (-1)^m.$$

In general, one only has

$$\chi_m(G_1 * G_2) \leq \chi_m(G_1) + \chi_m(G_2) - (-1)^m.$$

Examples with strict inequality are given in [HoLuMe85] for $m = 2$.

4 $\mathcal{N}^1(G)$, Nielsen equivalence of generating systems and Heegaard splittings

4.1   Nielsen equivalence of generating systems

Let G be a finitely generated group with presentation

$$G = \langle x_1, \ldots, x_n \mid R_1, R_2, \ldots \rangle,$$

and let $y = \{y_1, \ldots, y_n\}$ be another generating system for G. Denote by $F(X)$ and $F(Y)$ the free groups on bases $X = \{X_1, \ldots, X_n\}$ and $Y = \{Y_1, \ldots, Y_n\}$

respectively, and let β_x, β_y be the canonical epimorphisms $F(X) \to G$, $F(Y) \to G$ given by $\beta_x(X_i) = x_i$ and $\beta_y(Y_j) = y_j$.

Definition   The systems x and y are said to be *Nielsen equivalent* if there is an isomorphism $\theta\colon F(Y) \to F(X)$ such that $\beta_x\theta = \beta_y$.

As in 2.1, we denote by $\partial/\partial X_i\colon \mathbb{Z}F(X) \to \mathbb{Z}F(X)$ the i-th Fox derivative of the integral group ring $\mathbb{Z}F(X)$. By a slight abuse of notation, we will denote any group homomorphism $F \to G$ and its $\mathbb{Z}$-linear extension to a ring homomorphism $\mathbb{Z}F \to \mathbb{Z}G$ by the same symbol. Similarly, we will not distinguish notationally between any ring homomorphism $A \to B$ and the induced homomorphism on the $(m \times m)$-matrix rings $\mathbb{M}_m(A) \to \mathbb{M}_m(B)$.

4.2   The torsion $\mathcal{N}(x, y)$ defined by a pair of generating systems

As described in 2.1, to every generating system $x = \{x_1, \dots, x_n\}$ of G there is associated the canonical partial resolution

$$\mathbf{C}_x\colon \qquad \mathbb{Z}G^{\{dx_1, \dots, dx_n\}} \xrightarrow{\partial_1} \mathbb{Z}G \to \mathbb{Z}$$

of G of length $m = 1$, with $\partial_1(dx_i) = x_i - 1$ for all $i = 1, \dots, n$. Thus, to every pair of generating systems x and y, there is defined in 3.3 a torsion value $\mathcal{N}(y, x) := \mathcal{N}^1(\mathbf{C}_y, \mathbf{C}_x) \in \mathcal{N}^1(G)$, provided that the directed Euler characteristic of $\mathbf{C}_x$ and of $\mathbf{C}_y$ is minimal. To any choice of expressions $y_j = w_j(x)$ of the y_j as words in the x_i corresponds the map

$$\mathbb{Z}G^{\{dy_1, \dots, dy_n\}} \to \mathbb{Z}G^{\{dx_1, \dots, dx_n\}}, \quad dy_j \to \sum_i \beta_x(\partial w_j(X)/\partial(X_i))dx_i \;,$$

which commutes with the first boundary operators (by the chain rule for Fox derivatives, see [BuZi85], Chapter 9 B). It follows directly from 3.3 that the element $\mathcal{N}(y, x)$ is represented by the matrix of Fox derivatives $\beta_x(\partial w_j(X)/\partial X_i)_{j,i}$.

4.3 $\mathcal{N}$-torsion as an invariant of Nielsen equivalence

The element $\mathcal{N}(y, x) \in \mathcal{N}^1(G)$ depends only on the Nielsen equivalence classes of x and y. In particular, $\mathcal{N}(y, x) \neq 0$ implies that x and y are not Nielsen equivalent.

This follows directly from the fact that distinct bases of a free group differ by finite sequences of elementary Nielsen operations, which in turn give rise to elementary $\mathbb{Z}G$-matrices. These are mapped to 0 in $\mathcal{N}(G)$, by definition of the functor $K_1(\cdot)$. For more details, see [LuMo91] or [LuMo93_2].

4.4 How to show $\mathcal{N}(x, y) \neq 0$ in practice

For this purpose one needs the following data:

1. A presentation $\langle x_1, \ldots, x_n \mid R_1, R_2, \ldots, R_q \rangle$ of G.

2. Expressions $y_j = w_j(x)$ of the y_j as words in the x_i.

One then computes the Fox derivatives $\partial R_h / \partial X_i$, $\partial Y_j / \partial X_i$ and their β_x-images in $\mathbb{Z}G$. Last one searches for commutative rings R with unit $1 \neq 0 \in R$ and representations $\rho \colon \mathbb{Z}G / I_m \to \mathbb{M}_k(R)$, with $\rho(1) = 1$ and $k \in \mathbb{N}$, such that the following conditions are satisfied:

(a) $\rho(\beta_x(\partial R_h / \partial X_i)) = 0 \in \mathbb{M}_k(R)$ for all i, h, and

(b) $\det \rho(\beta_x(\partial Y_j / \partial X_i))$ is not contained in the multiplicative subgroup of R^* generated by all $\det \rho(\pm x_i)$.

4.5 Examples

Applications of the method summarized in 4.4 can be found in [LuMo92, LuMo93$_1$, LuMo93$_2$, Lu91$_1$] for $k = 1$ (mostly 3-manifold groups), in [LuMo91, Lu91$_1$], for $k = 2$ (Fuchsian groups), in [Le90] (non-tameness of E. Stöhr's automorphism of $F/[F, F'']$) for $k = 3$ and in [MoSh93] for $k = 4$ (non-tame automorphisms of extensions of Burnside groups). We show below in detail some of the results:

4.5.1 Free products of cyclics (from [Lu91$_1$])

Let G be a free product of finite cyclic groups:

$$G = \langle s_1, \ldots, s_n \mid s_1^{\gamma_1}, \ldots, s_n^{\gamma_n} \rangle.$$

Then two generating systems $x = \{s_1^{u_1}, \ldots, s_n^{u_n}\}$ and $y = \{s_1^{v_1}, \ldots, s_n^{v_n}\}$ of G are Nielsen equivalent if and only if

$$u_i = \pm v_i \bmod \gamma_i$$

for all $i = 1, \ldots, n$.

(Note that by Grushko's theorem all generating systems of cardinality n for G are Nielsen equivalent to a system of the type given above for x and y).

Proof: For any generating system $x = \{x_1 = s_1^{u_1}, \ldots, x_n = s_n^{u_n}\}$, the group G admits a presentation $G = \langle x_1, \ldots, x_n \mid x_1^{\gamma_1}, \ldots, x_n^{\gamma_n} \rangle$. Thus we may assume without loss of generality that $u_i = 1$ for $i = 1, \ldots, n$. We then apply 4.4 and evaluate with respect to the quotient $q \colon \mathbb{Z}G \to \mathbb{Z}[G/G']$. We compute easily:

(1) The matrix $(q \circ \beta_x(\partial X_j^{v_j}/\partial X_i))_{1 \leq j, i \leq n}$ is a diagonal matrix with i-th entry
$$z_i := 1 + s_i + \ldots + s_i^{v_i - 1}.$$

(2) For the k-th relation $r_k = x_k^{\gamma_k}$, the image of the Fox derivative $q \circ \beta_x(\partial r_k/\partial X_i)$ is 0 $(i \neq k)$ or $S_k := 1 + s_k + \ldots + s_k^{\gamma_k - 1}$ $(i = k)$. We define
$$\rho \colon \mathbb{Z}G \to \mathbb{Z}[G/G']/(S_1, \ldots, S_n)$$
and observe that, for all $\zeta \in \mathbb{Z}G$, the equation $\rho(\zeta) = 0$ implies $q(\zeta) \cdot \prod_{k=1,\ldots,n}(s_k - 1) = 0$. In particular, $q(\zeta) \cdot \prod_{k=1,\ldots,n}(s_k - 1)$ determines $\rho(\zeta)$ for all $\zeta \in \mathbb{Z}G$.

From (2), we compute
$$\det q((\beta_x(\partial X_j^{v_j}/\partial X_i))_{1 \leq j, i \leq n}) = z_1 z_2 \ldots z_n,$$
and hence
$$\prod_{k=1,\ldots,n}(s_k - 1) \cdot \det \rho((\beta_x(\partial X_j^{v_j}/\partial X_i))_{1 \leq j, i \leq n}) = \prod_{k=1,\ldots,n}(s_k^{v_k - 1}).$$

On the other hand, for all $g \in G$ one has $q(\pm g) = \pm s_1^{m_1} \ldots s_n^{m_n}$ for some $m_k \in \mathbb{Z}/\gamma_k\mathbb{Z}$, and hence
$$\prod_{k=1,\ldots,n}(s_k - 1) \cdot q(\pm g) = \pm s_1^{m_1} \ldots s_n^{m_n} \cdot \prod_{k=1,\ldots,n}(s_k - 1).$$

The right side of this equation is a sum of elements of $q(G) \cup -q(G)$, where each s_k occurs only with exponents m_k or $m_k + 1$ modulo γ_k. If this equals $\prod_{k=1,\ldots,n}(s_k^{v_k-1})$, then $v_k \in \{1, 0, -1\} \subset \mathbb{Z}/\gamma_k\mathbb{Z}$ for all $k = 1, \ldots, n$. The case $v_k = 0$ is excluded by the hypothesis that y generates G.

The converse implication in the statement of 4.5.1 is obvious. $\square$

4.5.2 (Finite) quotients of free products of cyclics (from [Lu91$_1$])

Nielsen inequivalent generating system x and y, as given in 4.5.1, map to Nielsen inequivalent systems under the quotient maps

(a) $G \to G/G'' =: H$ (G'' is the second commutator subgroup $[G', G']$), and

(b) $G \to H/\{g^p \mid g \in H'\} =: J_p$ for any integer $p \geq 2$.

Proof: To prove (a), we modify the preceding proof by considering additional relators of type $r' := [g, h]$ ($= ghg^{-1}h^{-1}$), with $g, h \in G'$: The chain rule for Fox derivatives gives

$$\beta_x(\partial[v(X_1, \dots, X_n), w(X_1, \dots, X_n)]/\partial X_i)$$
$$= \beta_x((1 - w)\partial v(X_1, \dots, X_n)/\partial X_i + (v - 1)\partial w(X_1, \dots, X_n)\partial X_i) ,$$

but this mapped to 0 by $q \colon \mathbb{Z}G \to \mathbb{Z}[G/G']$ if $\beta_x(v)$, $\beta_x(w) \in G'$.

For (b), we also have to take into account relators of type $r'' = g^p$ for $g \in H'$: Replacing the quotient q by $q_p \colon \mathbb{Z}G \to (\mathbb{Z}/p\mathbb{Z})[G/G']$ doesn't affect the proof of 4.5.1, and this yields in addition that

$$\beta_x(\partial w(X_1, \dots, X_n)^p/\partial X_i) = \beta_x((1 + w + \dots + w^{p-1})\partial w(X_1, \dots, X_n)/\partial X_i)$$

is also mapped to 0 for all $\beta_x(w) \in G'$. □

Notice that the groups J_p in 4.5.2 (b) are finite [Lu91$_1$] and hence remarkably close to the groups

$$G/G' = \langle s_1 \mid s_1^{\gamma_1} \rangle \oplus \dots \oplus \langle s_n \mid s_n^{\gamma_n} \rangle .$$

In contrast to 4.5.2, one observes easily that the direct sum G/G' possesses only one Nielsen equivalence class of generating systems of cardinality n, unless the $\gamma_1, \dots, \gamma_n$ are multiples of a common prime.

4.5.3 Fuchsian groups (see [LuMo91, Lu91$_1$])

Let G be a Fuchsian group,

$$G = \langle s_1, \dots, s_m, a_1, b_1, \dots, a_g, b_g \mid s_1^{\gamma_1}, \dots, s_m^{\gamma_m}, s_1 s_2 \dots s_m [a_1, b_1] \dots [a_g, b_g] \rangle$$

with $m \geq 4$ or $g \geq 1$, and with all $\gamma_i \geq 3$. Assume further that for $m = 4$ and $g = 0$ at most one of the γ_i equals 3, 4 or 6. Two generating systems of G,

$$u = \{s_1^{u_1}, \dots, s_{j-1}^{u_{j-1}}, s_{j+1}^{u_{j+1}}, \dots, s_m^{u_m}, a_1, b_1, \dots, a_g, b_g\}$$

and

$$v = \{s_1^{v_1}, \dots, s_{k-1}^{v_{k-1}}, s_{k+1}^{v_{k+1}}, \dots, s_m^{v_m}, a_1, b_1, \dots, a_g, b_g\},$$

with $u_i, v_i \in \mathbb{Z}/\gamma_i\mathbb{Z}$ and gcd $(u_i, \gamma_i) = $ gcd $(v_i, \gamma_i) = 1$, are Nielsen equivalent if and only if

(1) $u_i = \pm v_i \bmod \gamma_i$ for $i \neq j, k$ and

(2) $v_j = \pm 1 \bmod \gamma_j$ and $u_k = \pm 1 \bmod \gamma_k$ if $j \neq k$.

Proof: We indicate here only the proof for the easier case (see [LuMo91]) with odd and pairwise relatively prime γ_i. The general case [Lu91₁] requires rather intricate evaluation considerations.

We first assume that all $u_i = 1$. From the last relation in the given presentation of G, we see that for all values of j the generating systems u are Nielsen equivalent, and hence we can assume $j = k$. A system of defining relators with respect to the generating system u are given by all $s_i^{\gamma_i}$ with $i \neq j$ and by $(s_1 s_2 \ldots s_{j-1} s_{j+1} \ldots s_m [a_1, b_1] \ldots [a_g, b_g])^{\gamma_j}$. Thus all Fox derivatives of the relators are multiples of the expressions $S_i := 1 + s_i + \ldots + s_i^{\gamma_i - 1}$. The matrix of Fox derivatives $[\partial v / \partial u]$ of the generating system v with respect to u is a diagonal matrix with $Z_i := 1 + s_i + \ldots + s_i^{v_i - 1}$ as i-th entry.

Every Fuchsian group has a faithful representation into $PSL_2(\mathbb{C})$, which lifts to a faithful representation $\rho: \mathbb{Z}G \to SL_2(\mathbb{C})$ if all γ_k are odd. Since every s_k is mapped to an elliptic element, $\rho(s_k)$ is conjugate to a diagonal matrix with diagonal entries ζ_i and ζ_i^{-1} for $\zeta_k = -e^{i\pi}/\gamma_k$. Thus one has immediately $\rho(S_k) = 0$ for all k. Since the determinant of every $\pm g \in G$ is 1, it follows that the determinant of $\rho([\partial u / \partial v])$ is equal to $\mathcal{N}(\rho)(v, u)$. One computes that

$$\mathcal{N}(\rho)\mathcal{N}(v, u) = \prod_{i \neq j} |1 - \zeta_i^{v_i}|^2 / |1 - \zeta_i|^2.$$

Hence, for generating systems u without the hypothesis $u_i = 1$ for all i we derive from the cancellation rule 3.4 that, defining formally $u_j = v_k = 1$,

$$\mathcal{N}(\rho)\mathcal{N}(v, u) = \prod_{i=1}^{m} |1 - \zeta_i^{v_i}|^2 / |1 - \zeta_i^{u_i}|^2.$$

The result follows from some number theoretic considerations (see Lemma 1.9 of [LuMo91]).   □

4.6   Applications to Heegaard splitting of 3-manifolds

The concept of a Heegaard splitting of a 3-manifold M^3 will be introduced in Chapter VIII, §1.1. It consists in describing M^3 as the union of two handlebodies along their boundaries. Any such handlebody of genus $g \geq 1$, or, more precisely, the isotopy class of its embedding into M^3, defines a surjection of a free group $F_g \to \pi_1 M^3$, and hence a Nielsen equivalence class of generating system of $\pi_1 M^3$. Thus it is natural to use $\mathcal{N}^1(G)$ for $G = \pi_1 M^3$ in order to distinguish isotopy classes of minimal Heegaard splittings of M^3. Note that this was the motivation why $\mathcal{N}$-torsion has been introduced in the first place (see [LuMo91]). Previously, non-isotopic (or non-homeomorphic) Heegaard splittings could be exhibited only for genus 2 (see [Zi88] for a survey of the state of the art at that time). We now describe some of the results on minimal Heegaard splitting to be found in the author's joint work with Moriah.

4.6.1 Seifert fibre spaces

Let M be a Seifert fibered 3-manifold that is orientable and has orientable base surface. Such a manifold is characterized by a system of invariants $\{g, e, (\alpha_1, \beta_1), \ldots, (\alpha_m, \beta_m)\}$, see [Se33]. A particular family of *vertical* Heegaard splittings $\Sigma\{i_0, \ldots, i_j\}$ of M, which goes back to work of Boileau and Zieschang, is described in [LuMo91] (recent work of Scharlemann and Schultens indicates that for almost all M every Heegaard splitting is isotopic to a vertical one). The first handlebody of $\Sigma\{i_0, \ldots, i_j\}$, $1 \leq j \leq m - 1$, is the regular neighborhood of the graph which consists of (1) the exceptional fibers with indices $i_0, \ldots, i_j$, (2) *horizontal* curves q_k (i.e., curves which inject into the base surface), where each q_k surrounds precisely one of the remaining exceptional fibers, except for one such fibre, and (3) horizontal arcs which connect the above curves; see Figure 1.

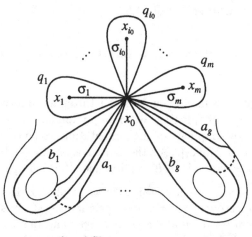

x_k = exceptional fibre σ_k = connecting arc

q_k = horizontal curve a_i, b_i = surface generators

Figure VII.1.

The second handlebody of $\Sigma\{i_0, \ldots, i_j\}$ is isotopic to the first handlebody of $\Sigma\{1, \ldots, m\}\backslash\{i_0, \ldots, i_j\}$, and in particular these two splittings are isotopic. Furthermore, if $\beta_i = \pm 1$ mod a_i, then the i-th exceptional fibre can be switched by an isotopy to the curve q_i, and conversely.

Theorem   (see [LuMo91], [Lu91$_1$])   *Let $m \geq 4$ or $g \geq 1$, and let $\alpha_i \geq 3$ for all $i = 1, \ldots, m$. Assume furthermore that for $m = 4$ and $g = 0$ at most one of the α_i equals 3, 4 or 6. Then $\Sigma\{i_0, \ldots, i_j\}$ is isotopic to $\Sigma\{k_0, \ldots, k_r\}$ if and only if $\{i_0, \ldots, i_j\}$ agrees with $\{k_0, \ldots, k_r\}$ or its complement $\{1, \ldots, m\} - \{k_0, \ldots, k_r\}$, up to those indices i with Seifert invariants $\beta_i = \pm 1$ mod α_i.*

Proof: (sketched, see [LuMo91]) The fundamental group of M quotients (modulo the conjugacy class represented by any regular fiber) to a Fuchsian group G as considered in 4.5.3, with $\gamma_i = \alpha_i$. Here the homotopy class of q_i maps to the element s_i, and that of the i-th exceptional fibre maps to $s_i^{v_i}$, with $v_i = \beta_i^{-1}$ mod γ_i. Hence the first handlebody of $\Sigma\{i_0, \ldots, i_j\}$ determines the Nielsen equivalence class of the generating system

$$s_{i_0}^{\beta_{i_0}}, s_{i_1}^{\beta_{i_1}}, \ldots, s_{i_j}^{\beta_{i_j}}, s_{l_1}, s_{l_2}, \ldots, s_{l_{m-j-2}},$$

where $\{i_0, \ldots, i_j\} \cup \{l_0, \ldots l_{m-j-2}\} = \{1, \ldots, m\}$. The second handlebody determines the generating system

$$s_{l_0}^{\beta_{l_0}}, s_{l_1}^{\beta_{l_1}}, \ldots, s_{l_{m-j-2}}^{\beta_{l_{m-j-2}}}, s_{i_1}, s_{i_2}, \ldots, s_{i_j}.$$

The claim follows now directly from 4.5.3. $\square$

4.6.2 A general criterion based on the Burau representation of braids

We will now consider 3-manifolds M given through surgery on a knot K. The results on Heegaard splittings of M imply analogous results on tunnel systems of K, but for simplicity of the presentation we ignore those here and refer the reader to [LuMo93$_1$]. For the same reason, we neglect the obvious generalization to the case where K is a link (see [LuMo93$_1$]).

Let K be any knot in $\mathbb{R}^3$, with a projection as $2n$-plat: K is obtained from a $2n$-braid B by connecting up the n pairs of neighboring strands on the top and those on the bottom by horizontal arcs, the *upper* and the *lower* bridges. Let $\mathbb{R}^2$ be any horizontal plane which intersects B (transversely), and compactify $\mathbb{R}^3$ to S^3 as well as $\mathbb{R}^2$ to S^2 by adding a point "at infinity". Then S^2 cuts $S^3 - int(N(K))$ in an upper and a lower part H_{top} and H_{bot}, and both are easily seen to be handlebodies, see Figure 2. They give rise to an *upper* and a *lower* Heegaard splitting $\Sigma_{top} = \{H'_{top}, H_{bot}\}$ and $\Sigma_{bot} = \{H_{top}, H'_{bot}\}$ of M

(obtained from surgery on K), since $H'_{top} := H_{top} \cup N$ and $H'_{bot} := H_{bot} \cup N$ are also handlebodies, for any solid torus N glued to $S^3 - int(N(K))$ along their boundaries. We show below that Σ_{top} and Σ_{bot} are non-isotopic and even non-homeomorphic (i.e., there is no homeomorphisms of M which maps Σ_{top} to Σ_{bot}) for a large variety of 3-manifolds M as above.

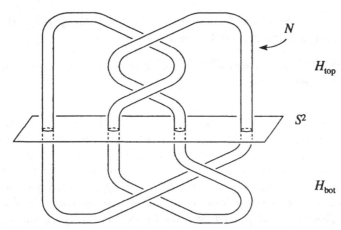

Figure VII.2.

Consider the Burau representation of B, i.e., the image under the map $\mathbb{Z}F(a_1,\ldots,a_{2n}) \to \mathbb{Z}[\langle t|-\rangle]$, $a_i \to t$, of the matrix of Fox derivatives $(\partial b_j/\partial a_i)_{j,i}$, where $\{a_1,\ldots,a_{2n}\}$ and $\{b_1,\ldots,b_{2n}\}$ are generating systems of the free group $\pi_1(\mathbb{R}^3 \to \hat{B})$. Here $\hat{B}$ is obtained from B by adding infinite trivial braids to B at the top and the bottom, and the a_i correspond to these bottom strands, whereas the b_i correspond to the top stands (alternatively, see [BuZi85], pp.158). Let $\hat{\rho}(B)$ denote the image of the Burau representation of B, when t is evaluated at -1. Let α be the gcd of the entries of the Matrix $L \circ \hat{\rho}(B) \circ M$, and let β be the determinant of the matrix $N \circ \hat{\rho}(B) \circ M$, for L, M and N determined as follows: Let L be the $(n \times 2n)$-matrix (a_{ij}) with

$$a_{ij} = \begin{cases} 1 & \text{for } j = 2i-1 \\ -1 & \text{for } j = 2i \\ 0 & \text{otherwise} \end{cases},$$

let M be the $(2n \times n)$-matrix (a_{ij}) with

$$a_{ij} = \begin{cases} 1 & \text{for } i = 2j-1 \\ 1 & \text{for } i = 2j \\ 0 & \text{otherwise} \end{cases},$$

and let N be the $(n \times 2n)$-matrix $(a_{i,j})$ with

$$a_{ij} = \begin{cases} 1 & \text{for } j = 2i-1 \\ 0 & \text{otherwise} \end{cases}.$$

Theorem   ([LuMo93$_1$])   *Let M be a closed 3-manifold obtained by p/q-surgery on K, with p even.*

(1) *Then $\alpha \neq 1$ implies that $rank(\pi_1(M))$ and hence the Heegaard genus of M (i.e., the minimal genus of any Heegaard surface of M) equal n.*

(2) *If $\beta \neq \pm 1$ mod α and $2^{n-1} \neq \pm 1$ mod α, then the genus n Heegaard splittings Σ_{top} and Σ_{bot} of M are non-isotopic.*

(3) *If $\beta \neq \pm \beta^{-1}$ mod α and $2^{2(n-1)} \neq \pm 1$ mod α, then Σ_{top} and Σ_{bot} of M are non-homeomorphic.*

Proof: We consider the Wirtinger presentation of the knot group $\hat{G} = \pi_1(S^3 - int(N(K)))$ associated to the $2n$-plat projection of K. To any of the generators x_i $(i = 1, \ldots, n)$ corresponding to the lower bridges, we formally introduce the inverse generator x_i'. We can then find a new presentation based on the $2n$ generators $x_1, \ldots, x_n, x_1', \ldots, x_n'$ by successively eliminating all other Wirtinger generators from the bottom to the top, using the Wirtinger relations. This inductive procedure gives words $w_1, \ldots, w_{2n}$ in the x_i, x_i' for the top bridge generators and their inverses $y_1, y_1^{-1}, y_2, y_2^{-1}, \ldots, y_n, y_n^{-1}$. Replacing x_i' by x_i^{-1} hence gives a presentation for the knot group

$$\hat{G} = \langle x_1, \ldots, x_n \mid w_1^{-1} = w_2, \ldots, w_{2n-1}^{-1} = w_{2n} \rangle.$$

A similar presentation can be found if one starts with the top generators and their inverses and eliminates all the other generators successively from the top to the bottom.

The entries in the evaluated Burau matrix $\hat{\rho}(B)$ are now precisely the images of the Fox derivatives $\partial w_j^{\epsilon_j}/\partial x_i^{\epsilon_i}, \epsilon_k = (-1)^k$, under the evaluated abelianization map $q: \mathbb{Z}\hat{G} \to \mathbb{Z}[\langle t \mid - \rangle]/(t+1) = \mathbb{Z}$. A consequence of mapping t to -1 is that the exponents ϵ_k can be suppressed. From the definition of L, M and N it follows directly that $L \circ \hat{\rho}(B) \circ M$ is precisely the q-image of the matrix of Fox derivatives $\partial w_{2j-1} w_{2j}/\partial x_i$, and that the matrix $N \circ \hat{\rho}(B) \circ M$ is the q-image of the matrix of Fox derivatives $\partial y_k/\partial x_i$. Thus, α generates an ideal in $\mathbb{Z}$ which contains $q(I_1(\hat{G}))$. This proves part (1) of the Theorem, since $G = \pi_1(M)$ is obtained from $\hat{G}$ by adding the relator $x_1^p \lambda^q$, where λ is a certain element in the second commutator subgroup of $\hat{G}$, and hence the Fox derivatives of $x_1^p \lambda^q$ have zero q-image for even p (see Remark 5.1 of [LuMo93$_1$] or compare 4.5.2 (a) above).

The hypothesis $\beta \neq \pm \beta^{-1}$ implies that the generating systems $x = \{x_1, \ldots, x_n\}$ and $y = \{y_1, \ldots, y_n\}$ are not Nielsen equivalent, since β gives precisely the image of $\mathcal{N}(y, x)$ in $(\mathbb{Z}/\alpha\mathbb{Z})^*/(\pm 1)$ under the evaluation map induced by q.

But x and y are generating systems corresponding to H_{bot} and H_{top} respectively. Thus there is no isotopy of M which maps H_{bot} to H_{top}. Hence, for part (2), it suffices to show that H_{bot} can not be isotoped to H'_{bot}. An ingenious computation entirely due to my coauthor ([LuMo93$_1$], Proposition 5.2) shows that, for a suitable generating system z in the Nielsen equivalence class determined by H'_{bot}, the image of $\mathcal{N}(z,x)$ in $(\mathbb{Z}/\alpha\mathbb{Z})^*/(\pm 1)$ under the evaluation map induced by q is equal to 2^{1-n} mod α.

The arguments for part (3) are slightly more complicated, but coincide with those in the proof of Theorem 0.3 of [LuMo93$_1$] and are omitted here.   □

4.6.3 Applications to 2-bridge knots and Montesinos knots

Let $K = K(\alpha, \beta)$ be a 2-bridge knot in standard projection as 4-plat, see Figure 3. The invariants α and β agree precisely with the numbers derived in 4.6.2 from the plat projection (see [LuMo93$_1$], §3).

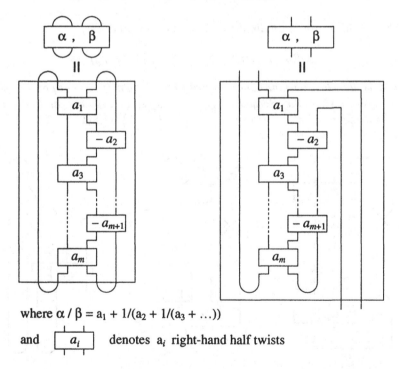

where $\alpha / \beta = a_1 + 1/(a_2 + 1/(a_3 + ...))$

and $\boxed{a_i}$ denotes a_i right-hand half twists

Figure VII.3.

Thus we obtain as direct consequence of 4.6.2 a new proof of the following:

Corollary   *Let M be obtained from the 2-bridge knot $K(\alpha,\beta)$ by p/q-surgery with even p. If $\beta \neq \pm 1$ mod α and $2^{n-1} \neq \pm 1$ mod α, then M admits at least two non-isotopic Heegaard splitting Σ_{top} and Σ_{bot}. If furthermore $\beta \neq \pm\beta^{-1}$ mod α and $2^{2(n-1)} \neq \pm 1$ mod α, then Σ_{top} and Σ_{bot} are even non-homeomorphic.*   □

Next, we consider Montesinos knots $K = \mathcal{M}(e; (\alpha_1,\beta_1),\ldots,(\alpha_g,\beta_g))$, see Figure 4 (a) and [BuZi85], Chapter 12. It has been shown in [BoZi85] that K has a projection as $2g$-plat, see Figure 4 (b). In [LuMo92] for each g-tuple $(\epsilon_1,\ldots,\epsilon_g) \in \{1,-1\}^g \backslash \{\pm(1,\ldots,1)\}$ such a projection $P(\epsilon_1,\ldots,\epsilon_g)$ has been exhibited, with corresponding Heegaard splittings $\Sigma_{top}(\epsilon_1,\ldots,\epsilon_g)$ and $\Sigma_{bot}(\epsilon_1,\ldots,\epsilon_g)$. As a consequence of the Theorem in 4.6.2, one can compute that many of them are non-homeomorphic, see Theorem D of [LuMo92]. In particular, this yields:

Theorem   ([LuMo92])   *Let $K \subset S^3$ be the Montesinos knot*

$$\mathcal{M}(e; (\alpha_1,\beta_1),\ldots,(\alpha_g,\beta_g)),$$

such that all β_i are mutually distinct odd primes, and let

$$gcd(\alpha_1,\ldots,\alpha_g) \;>\; 2^{2g-1}(\beta_1 \cdot \ldots \cdot \beta_g)^2 \;.$$

Let M be the closed 3-manifold obtained from S^3 by m/n-surgery on K, where m is even. Then the Heegaard splittings $\Sigma_y(\epsilon_1,\ldots,\epsilon_g)$ are mutually non-homeomorphic for all g-tuples $(\epsilon_1,\ldots,\epsilon_g) \in \{1,-1\}^g \backslash \{\pm(1,\ldots,1)\}$.   □

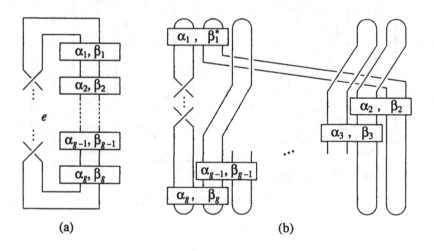

(a)   (b)

Figure VII.4.

Corollary   (see [LuMo92])   *For each $g \geq 3$, there exist infinitely many hyperbolic 3-manifolds of genus g with at least $2^g - 2$ pairwise non-homeomorphic minimal Heegaard splittings.*   □

A similar statement for Seifert fibered rather than hyperbolic 3-manifolds can be derived as in 4.6.1 (see [LuMo91], §0, after Theorem 2).

5 $\mathcal{N}$-torsion as generalization of the bias and (simple)-homotopy of (G, m)-complexes

5.1   The topological problem

Question   Let K, L be two (finite) m-dimensional cell complexes with isomorphic fundamental groups, equal Euler characteristics and vanishing homotopy groups $\pi_i(K) = \pi_i(L) = 0$ for $i = 2, \ldots, m - 1$. Are K and L homotopy equivalent or even simple-homotopy equivalent?

This topological question translates directly into the algebraic setting of section 3: The complexes K, L are (G, m)-complexes for some identification isomorphisms $i_K \colon \pi_1 K \to G$, $i_L \colon \pi_1 L \to G$, and hence the augmented chain complexes $\mathbf{C} := \mathbf{C}(\tilde{K}), \mathbf{D} := \mathbf{C}(\tilde{L})$ associated to the universal coverings of K, L respectively are partial resolutions of G with preferred bases as in 3.3. In particular, one obtains a torsion value $\mathcal{N}^m(\mathbf{C}, \mathbf{D}) \in \mathcal{N}^m(G)$ associated to K and L, which is crucial for the above simple-homotopy question (see 5.2 below).

Warning The identifications $i_k \colon \pi_1 K \to G$ and $i_L \colon \pi_1 L \to G$ are by no means canonical. Different identifications will, in general, give distinct values $\mathcal{N}^m(\mathbf{C}, \mathbf{D}) \in \mathcal{N}^m(G)$!!!

As a consequence one has to consider for the above problem all different possible identifications i_K, i_L.

We also want to draw the reader's attention immediately to the second major difficulty in evaluating $\mathcal{N}^m(\mathbf{C}, \mathbf{D})$; unlike the case $m = 1$ treated in section 4, the data given for $m \geq 2$ in general do not include a generating system of $\pi_m(K) = H_m(\mathbf{C})$ or $\pi_m(L) = H_m(\mathbf{D})$. Fortunately, in the case $m = 2$, which is of greatest interest to us, there are a variety of techniques available which deal with that problem for important classes of group presentations; see Chapter V. An alternative approach is described in 5.4 below.

5.2 $\mathcal{N}$-torsion test for simple-homotopy equivalence

Let K, L be finite CW-complexes as in 5.1. If

$$\mathcal{N}^m(\mathbf{C}, \mathbf{D}) \neq 0 \in \mathcal{N}^m(G),$$

then there exists no simple-homotopy equivalence h: $K \to L$ with induced map $h_* = i_L^{-1} i_K$ on the fundamental groups of K, L.

Proof: By construction, there exist chain isomorphisms with trivial Whitehead torsion between $\mathbf{C}$ and $\mathbf{C}^D$ as well as between $\mathbf{D}$ and $\mathbf{D}^C$. So any simple-homotopy equivalence h: $K \to L$ would give rise to a simple-homotopy equivalence h': $\mathbf{C}^D \to \mathbf{D}^C$. By the exactness of $\mathbf{D}^C$ the map h' then were chain homotopic (see [Co73], pp. 72) to a $\mathbb{Z}G$-chain map g: $\mathbf{C}^D \to \mathbf{D}^C$ which is the identity on all M_i (for $0 \leq i \leq m - 1$), whereas on the m-th term it determines a map f as in 3.3. Here $[f]$ would represent the trivial element in $Wh(G)$, with $\mathcal{N}^m(\mathbf{C}, \mathbf{D}) = 0 \in \mathcal{N}^m(G)$ as an immediate consequence of 3.2.
□

5.3   A homotopy equivalence test

Let $\mathcal{D}^m(G)$ denote the quotient of $\mathcal{N}^m(G)$ modulo the canonical image of $Wh(G)$ as in 3.2. Such a group was first considered by Dyer in [Dy85].

Corollary   *If, in the situation of 5.2, the image of $\mathcal{N}^m(\mathbf{C}, \mathbf{D})$ is non-zero in $\mathcal{D}^m(G)$, then there exists no homotopy equivalence h: $K \to L$ with induced map $h_* = i_L^{-1} i_K$ on the fundamental groups.*   □

5.4   Relation between $\mathcal{N}$-torsion and bias

As described in 3.5, the torsion group $\mathcal{N}^m(G)$ is usually evaluated through representations ρ: $\mathbb{Z}G \to \mathbb{M}_k(R)$ with $\rho(1) = 1$ and $\rho(I_m) = \{0\} \subset \mathbb{M}_k(R)$. A particular such evaluation is obtained by considering the image $(q) \subset \mathbb{Z}$ of the Fox ideal $I_m(G)$ under the augmentation homomorphism $\mathbb{Z}G \to \mathbb{Z}$ and defining ρ to be the resulting map $\mathbb{Z}G \to \mathbb{Z}/q\mathbb{Z}$ ($= \mathbb{M}_1(\mathbb{Z}/q\mathbb{Z})$). If $q \neq 1$, it follows directly from the definitions that the induced map $\mathcal{N}(\rho)$ as in 3.5 maps $\mathcal{N}^m(\mathbf{C}, \mathbf{D})$ precisely to the bias $b(\alpha) \in (\mathbb{Z}/q\mathbb{Z})^*/\{\pm 1\}$ as defined in Chapter III, Definition 1.7.

This observation suggests a viewpoint which has been particularly advocated in [Ho-AnLaMe91]: Rather than working with $\mathcal{N}^m(G)$, which corresponds geometrically to the universal covering, or with the classical bias, which corresponds to the base complex, it might be useful to consider intermediate (regular) coverings. Of course, these correspond to quotients $\mathbb{Z}G \to \mathbb{Z}[G/N]$ and hence can be viewed as the first step of an evaluation of $\mathcal{N}^m(G)$ (followed for example by a matrix representation ρ of $\mathbb{Z}[G/N]$ as above). However, these particular quotients of $\mathbb{Z}G$ have the advantage over arbitrary algebraic quotients that they come with a canonical *bias ideal* [3] $I_m(G, N)$ in $\mathbb{Z}G$ which always contains the Fox ideal $I_m(G)$, but may be in practice easier to compute (e.g., see [Me90]). Here $I_m(G, N)$ is defined as the Fox ideal (in the sense of 1.1 (a)) associated to the m-th homology group of that covering, understood as submodule of the m-th chain group. The advantage of this approach is that module generators for H_m (and hence ideal generators for $I_m(G, N)$) may be found for suitable intermediate coverings by geometric arguments.

5.5   Remainder (see [Co73])

For $m \geq 3$, there is a 1-1 relation between the topology and the algebra in 5.1. More precisely:

(a) Complexes K and L as in 5.1 are homotopy equivalent if and only if there are identifications i_K, i_L and $\mathbb{Z}G$-chain maps $f\colon \mathbf{C} \to \mathbf{D}$, $g\colon \mathbf{D} \to \mathbf{C}$ such that the compositions are chain homotopic to the identity. K and L are simple-homotopy equivalent if and only if there exist such f, g with $\tau(f) = \tau(g) = 0 \in Wh(G)$.

(b) To any partial resolution $\mathbf{C}$ with preferred bases as in 3.3, there exists a (G, m)-complex K with augmented chain complex $\mathbf{C} := \mathbf{C}(\tilde{K})$. In particular, every torsion value $\tau \in Wh(G)$ is realizable by some homotopy equivalence of (G, m)-complexes $h\colon K \to L$.

For $m = 2$, part (a) is correct, but (b) fails in general. The reason is that attaching 2-cells has a direct influence on the fundamental group and can hence not be done freely for the purpose of creating distinct simple-homotopy types. Indeed, it had been open until recently, whether homotopy type and simple-homotopy type agree for 2-dimensional complexes. The first counterexamples were given in [Lu91₁] and [Me90].

---

[3]Notice that bias ideals have been defined in [Ho-AnLaMe91] in a somewhat more restrictive way.

5.6   2-Complexes that are homotopy equivalent but not simple-homotopy equivalent

The simplest known examples of this kind (see [Lu91₁]) are the standard 2-complexes for the presentations

$$K: \quad \langle x, y, z \mid y^3, yx^{10}y^{-1}x^{-5}, [x^7, z] \rangle$$

$$L: \quad \langle x, y, z \mid y^3, yx^{10}y^{-1}x^{-5}, x^{14}zx^{14}z^{-1}x^{-7}zx^{-21}z^{-1} \rangle.$$

A homotopy equivalence is easy to find. In order to show that there exists no simple-homotopy equivalence, Theorem 5.2 was used in [Lu91₁] (where the invariant $\mathcal{N}^2(G)$ is yet in a slightly less formalized form). The evaluation of $\mathcal{N}^2(\mathbf{C}, \mathbf{D})$ was performed by factoring the group $\mathcal{N}^2(G)$ via the representation

$$\mathbb{Z}G \to \mathbb{Z}/7\mathbb{Z}[G/\langle x^7, y, z \rangle] = \mathbb{Z}/7\mathbb{Z}[\langle \beta \mid \beta^5 \rangle],$$

which maps $\mathcal{N}^2(G)$ to $Wh(\mathbb{Z}/7\mathbb{Z}[\langle \beta \mid \beta^5 \rangle])$ and $\mathcal{N}^2(\mathbf{C}, \mathbf{D})$ to the well known non-zero "Kaplansky unit" $1 + \beta - \beta^3$. For more detail, see [Lu91₁], §2.

In [Me90] a more general, rather elaborate technique is described how to produce for any given torsion value $\tau \in Wh(G)$ (G finitely presented) pairs of finite homotopy equivalent 2-complexes K, L with

$$\pi_1 K = \pi_1 L = G * (\mathbb{Z}/2\mathbb{Z} \oplus \mathbb{Z}/4\mathbb{Z}) * \ldots * (\mathbb{Z}/2\mathbb{Z} \oplus \mathbb{Z}/4\mathbb{Z}),$$

and a homotopy equivalence $f \colon K \to L$ which realizes τ. In certain cases τ cannot be realized by any self-equivalence $K \to K$, and hence K and L are not simple-homotopy equivalent (see also Chapter XII, §3.2).

5.7   Simple-homotopy for free products

We now use the direct sum formula 3.7 to compose by wedge product pairs of homotopy equivalent 2-complexes K, L with the property that for some field $\mathfrak{K}$[4]

(*)   all possible choices for i_K and i_L, as in 5.1, yield a non-zero torsion value in $\mathcal{N}^2(G; \mathfrak{K})^*$.

In particular, K and L are not simple-homotopy equivalent (by 5.2).

---

[4]Compare also to Chapter XII, §3.

Theorem   Let K_i, L_i $(i = 1, \ldots, n)$ be 2-complexes such that, for some permutation $\pi: \{1, \ldots, n\} \to \{1, \ldots, n\}$ all pairs, K_i and $L_{\pi(i)}$, are homotopy equivalent, and that, for every such π, there is at least one pair, K_i and $L_{\pi(i)}$, with property (*). Assume further that the fundamental groups $G_i = \pi_1 K_i$ are freely indecomposable and not isomorphic to $\mathbb{Z}$. Then the pair of wedge products, $K = K_1 \vee \ldots \vee K_n$ and $L = L_1 \vee \ldots \vee L_n$, also has property (*).

Proof: From the assumption that every G_i is freely indecomposable and different from $\mathbb{Z}$, it follows that all possible identifications

$$\pi_1(K_1 \vee \ldots \vee K_n) = \pi(L_1 \vee \ldots \vee L_n) =: G$$

come from identifications of the factors through some permutation π. The claim hence follows directly from the assumption that there is always a pair, K_i and $L_{\pi(i)}$, with property (*), by 3.7 and 5.2.   □

5.8   Large families of homotopy equivalent but pairwise simple-homotopy inequivalent 2-complexes

It follows directly from 5.7 that any wedge product between n copies of the complex K and m copies of the complex L as defined in 5.6 is simple-homotopy equivalent to another such wedge product if and only if the numbers for n and m coincide.

This makes it possible to construct for any $q \in \mathbb{N}$ compact 2-complexes $K_1, \ldots, K_q$ which are all homotopy equivalent, but no two of them are simple-homotopy equivalent. The problem of finding an infinite such family seems to be still open.

Chapter VIII

(Singular) 3-Manifolds

Cynthia Hog–Angeloni and Allan J. Sieradski

While 2–complexes warrant study for their own complexities, the unresolved conjectures for 2–complexes described in Chapters X–XII arose from considerations of 2–dimensional spines of manifolds. This chapter investigates 2–complexes that are spines of closed orientable 3–manifolds and special 2–polyhedra that are spines of singular 3–manifolds.

1   3–Manifolds

This first section presents connections between combinatorial group theory and closed orientable 3–manifolds. A homotopy classification of these manifolds and an investigation of their spines are presented from the viewpoint of combinatorial squashings for them.

1.1   Representing 3–manifolds

By a 3–manifold, we mean a connected (second countable Hausdorff) space in which each point has a 3–ball neighborhood. This view of 3–manifolds as collections of overlapping 3–balls is too unstructured to contribute to their classification. To better understand 3–manifolds, especially closed orientable ones, numerous methods of representing them have been developed.

One can represent 3–manifolds as assemblages of basic bounded 3–manifolds that are glued along their boundary components in regulated manners.

Triangulations   The simplest assemblage system is a *triangulation*, that is, a finite collection of tetrahedra whose 2–faces are identified in pairs via simplicial isomorphisms. The triangulability of closed 3–manifolds was established by E. Moise [Mo51]. This inherent structure of closed 3–manifolds contributes indirectly to their classification by offering a starting point for most other representations of 3–manifolds.

Heegaard Splittings   In a closed orientable 3–manifold there are regular neighborhoods of the 1–skeletons of a triangulation and its dual that are solid handlebodies sharing the same boundary surface. Any such decomposition of a 3–manifold M into two genus g handlebodies identified along their boundaries surfaces is called a *Heegaard decomposition* or *splitting* of M of genus g. In [Zi88], H. Zieschang offers a survey of results on Heegaard splittings; Chapter VII describes later work of Lustig and Moriah. Any attempt to classify 3–manifolds via Heegaard splittings leads to the study of the mapping class group (of isotopy classes of homeomorphisms) of closed orientable surfaces. See J. S. Birman's work [Bi75₃] and Zieschang's volume [Zi81].

Or one can represent 3–manifolds as special modifications of a basic one.

Surgery   A. H. Wallace has shown [Wa60] that each closed orientable 3–manifold is produced by surgery on a framed link in the 3–sphere S^3. The surgery alters each framed link component according to a rational surgery coefficient b/a, replacing meridional discs by discs whose attaching loop travels the tubular neighborhood a times longitudinally and b times meridionally. W. B. R. Lickorish has a proof [Li62] that one can always use a link whose components are trivial knots and whose framing or surgery coefficient is ± 1. M. Takahashi [Ta91] has shown that +1-surgery on each component suffices.

Branched Coverings   According to J. W. Alexander [Al20], every closed orientable 3–manifold can be constructed as a branched covering of S^3, branched over a link. It is possible to make the branch set (of points that are not evenly covered) a knot and the associated unbranched covering of the complement a 3–fold (irregular) covering, according to independent refinements of H.M. Hilden [Hi74] and J.M. Montesinos [Mo74]. Montesinos' proof is based upon a surgery description of the 3–manifold. Hilden's proof [Hi76] is based upon a Heegaard splitting of the 3–manifold.

A general weakness of these schemes of representing 3–manifolds is non-uniqueness. There are more recent results that reduce the variables involved in the representations:

W. P. Thurston established the existence of *universal links* $\mathcal{L} \subset S^3$, which are defined by the property that every closed orientable 3–manifold is a branched cover over $\mathcal{L} \subset S^3$. H.M. Hilden, M.T. Lozano, and J.M. Montesinos estab-

lished the universality of the Whitehead link and the Borromean rings as well
as that of the knot 9_{46} [HiLoMo83] and the figure-eight knot [HiLoMo85$_1$].
Each universal link provides the class of all closed oriented 3–manifolds the
structure of a lattice corresponding to the finite index subgroups of the fundamental group of that universal link, e.g., see Hempel's article [He90].

D. Cooper and W. P. Thurston [CoTh88] proved that every closed orientable
3–manifold can be triangulated so that the link of each vertex is one of five
possibilities: ∂(octahedron), $(\partial(\text{tetrahedron}))'$, $(\Sigma(\partial(\text{triangle})))'$,
$(\Sigma(\partial(\text{square})))'$, or $(\Sigma(\partial(\text{pentagon}))'$. (The prime ($'$) represents barycentric
subdivision and the sigma (Σ) represents suspension.) Their proof utilizes a
special paving of the 3–manifold by cubes. Their constructing of the paving
is based upon the universality of the Borromean rings and a straight-forward
paving of S^3 that respects the Borromean rings.

In further investigations of universality, H. M. Hilden, M. T. Lozano, and J.
M. Montesinos [HiLoMo85$_2$] deduced that any closed orientable 3–manifold
can by *pentagulated*, that is, obtained from a finite set of dodecahedra by
gluing along their pentagonal faces in pairs. In [Ba91], P. Bandieri proved
that every 3–manifold admits a decomposition into octahedra.

Patterns of Contact   So any closed 3–manifold can be viewed as an assemblage of either tetrahedra, cubes, octahedra, or dodecahedra whose faces are
glued together in pairs. (The question of icosahedra appears to be open.)

A suitable partial gluing of the units in any such assemblage yields a single
more involved polyhedral 3–ball with triangular, square, or pentagonal faces
which, when identified in pairs, produces the 3–manifold. The partial gluing
involves the faces penetrated by a maximal tree in the dual complex for the
assemblage. So one can construct any closed 3–manifold M using a single
polyhedral 3–ball P whose finitely many boundary faces are glued together
in pairs. The interior of P becomes an open 3–ball whose boundary meets
itself in the manifold M along an embedded 2–complex K, which is a *spine* of
M in the sense of Chapter I. But here the focus is on viewing the spine as the
boundary points of contact that arise when a 3–ball is fully inflated, filling
all available space in the closed 3–manifold M. The manifold M is orientable
exactly when the paired faces of P are oppositely oriented in its boundary.

The face identification procedure is a very traditional method of constructing
3–manifolds. Poincaré's original construction [Po04] of a homology 3–sphere
is that of a solid dodecahedron in which opposite faces are identified with a
$\pi/5$ twist (see Figure 1a). The Weber-Seifert manifold [WeSe33] is obtained
from the dodecahedron by using a $3\pi/5$ twist (Figure 1b). The vertex labels and edge labels in Figure 1 show the face identifications that give these
distinct dodecahedral spaces.

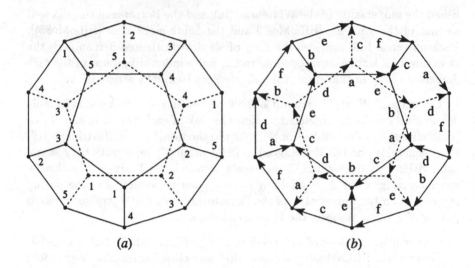

(a)   (b)

Figure VIII.1. Spherical and hyperbolic dodecahedral spaces

Of course, not every pairing of oppositely oriented boundary faces of a poly-hedral 3–ball P yields an oriented 3–manifold. The only troublesome points in the resulting quotient space $M = P/\sim$ are the 0–cells of $K = (\partial P)/\sim$ that arise from the vertices of P. They have small neighborhoods that are cones over surfaces S_i constructed from hemispherical caps in P about its vertices. The classical argument [SeTh45, 60, Satz 1] expresses the manifold property in terms of Euler-characteristic:

(1) M is a closed orientable 3–manifold iff each surface S_i is a 2–sphere iff $\chi(\cup S_i)$ is twice the number of 0–cells of M iff $\chi(K) = 1$ iff $\chi(M) = 0$.

Geometric Structures   The description of a 3–manifold M as a *geometric* polyhedron P with face identifications via geometric isometries can provide the view that one has in $M = P/\sim$ when one sights along geodesics.

The *triple torus* $S^1 \times S^1 \times S^1$ provides a simple Euclidean example, as it results when opposite faces of a unit cube are identified under translation. Since the cube serves as a fundamental region for the group $\mathbb{Z}^3$ generated by three mutually orthogonal translations of Euclidean space $\mathbb{E}^3$, this construction also gives $S^1 \times S^1 \times S^1$ the Euclidean structure of the orbit space $\mathbb{E}^3/\mathbb{Z}^3$. Lines in $\mathbb{E}^3$ descend to geodesics in the orbit space; so the view along geodesics from any point in $S^1 \times S^1 \times S^1$ is that of the $\mathbb{Z}^3$ lattice of unit cubes in the universal covering $\mathbb{E}^3$. Spectators in $S^1 \times S^1 \times S^1$ see $\mathbb{Z}^3$ copies of themselves.

Poincaré's spherical dodecahedral space is the orbit space of the 3–sphere under action of the binary icosahedral group $< 5, 3, 2 >$ as a group of 120 spherical isometries. There is a dodecahedral fundamental region whose translates under the group $< 5, 3, 2 >$ tessellate S^3 into two solid torii of 60 dodecahedra each, with exactly 3 dodecahedra sharing each edge and 4 meeting at each vertex. This tessellation by regular spherical dodecahedra with 120° dihedral angles gives the view that one has in spherical dodecahedral space.

When a polyhedron P appears in hyperbolic 3–space H^3 and its planar faces are paired via isometries that satisfy certain cyclic conditions, Poincaré's polyhedron theorem (see [Ma71]) shows that the polyhedron is a fundamental region for the discontinuous group of isometries G generated by the identifications of the sides. In this case, the manifold $M = P/\sim$ acquires a hyperbolic structure as the orbit manifold H^3/G. The tessellation of hyperbolic 3–space via translated copies of the polyhedron gives the hyperbolic world view that one has in that orbit 3–manifold. For the Weber-Seifert hyperbolic dodecahedral space, this view shows a tesselation of hyperbolic 3–space by regular hyperbolic dodecahedra with 72° dihedral angles, with exactly 5 dodecahedra sharing each edge and 20 meeting at each vertex. This is an infinitely richer view than the one in spherical dodecahedral space.

Open manifolds can often be handled by considering polyhedra with deleted vertices, such as, an *ideal* hyperbolic polyhedron with vertices at infinity. In W. Thurston's notes, *The Geometry and Topology of 3–Manifolds*, the complement of the figure-eight knot, Whitehead link, and Borromean rings are given complete hyperbolic structure via expression as ideal hyperbolic polyhedra with face identifications. In the same manner, A. Hatcher [Ha81] established the hyperbolic structure of arithmetic type of certain other link complements. W. W. Menasco [Me83] gives a constructive algorithm for representing any link complement as an ideal polyhedron (not necessarily hyperbolic) with face identifications and works out the structure for some that are hyperbolic. The software *Snappea* distributed by J. Weeks implements this algorithm and determines the hyperbolic status of the polyhedron.

1.2   Squashing maps for closed 3–manifolds

The method of representing a closed orientable 3–manifold as a single polyhedron with face identifications is one whose tie to combinatorial group theory is most immediate. So, in keeping with the theme of this volume, we devote the rest of this section to this method. We show, for example, how to capture the homotopy type of these manifolds and how to recognize 2–dimensional spines of these manifolds from this viewpoint.

Combinatorial and fibered 2-cells   We first generalize the face identification process by defining squashing maps as in [Si86]. As in Chapter II, any 2-cell c^2 in a combinatorial complex L has a combinatorial characteristic map $\psi : (P_k, \dot{P}_k) \to (L^2, L^1)$, where $(P_k, \dot{P}_k)$ is the regular k-gon combinatorial complex on (B^2, S^1) for some $k > 1$, by which c^2 and each boundary 1-cell c^1 receive a linear parametrization. Two such 2-cells are *identifiable* if they are modeled on the same polygonal complex P_k. A *product 2-cell* (f^2, ϕ) in L has a combinatorial characteristic map $\phi : B^1 \times B^1 \to L$; it may be viewed as *fibered* by the projection $B^1 \times B^1 \to B^1$ onto the first factor, with *fibered side cells* $f_{\pm}^d = \phi(\{\pm 1\} \times B^1)$ of dimension $d = 0$ or 1, and ordinary *end 1-cells* $e_{\pm}^1 = \phi(B^1 \times \{\pm 1\})$, and finally *sections* $s_t = \phi(B^1 \times \{t\})$ $(-1 \leq t \leq +1)$. In any combinatorial 2-complex L, a collection of product 2-cells may be designated as fibered, provided that the fibered side cells bound only the sides of other fibered 2-cells. Examples of fibered 2-cells appear in Figure 2, with sections traced out; other examples have $e_+^1 = e_-^1$ and/or $f_+^d = f_-^d$.

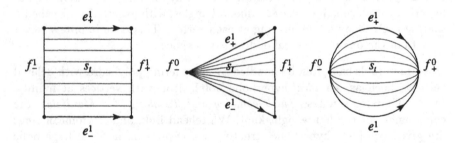

Figure VIII.2. Some fibered 2-cells

Squashing Maps   A *squashable complex* C on the 2-sphere S^2 is a combinatorial complex, together with a grouping of its non-fibered 2-cells in oppositely oriented, identifiable pairs $(c_{\pm}^2, \psi_{\pm})$. An identification of the paired 2-cells $(c_{\pm}^2, \psi_{\pm})$ in C according to their combinatorial characteristic maps $\psi_{\pm} : P_k \to S^2$, as well as the identification of all sections $\phi(B^1 \times \{t\})$ $(-1 \leq t \leq +1)$ of each fibered 2-cell (f^2, ϕ) in C, *squashes* the ball-sphere pair (B^3, S^2) into a combinatorial complex pair $(M, K) = (B^3//C, S^2//C)$. The resulting combinatorial quotient map

$$q = q(C) : (B^3, S^2) \to (B^3//C, S^2//C) = (M, K)$$

is called the *squashing map* associated with C. The 2-skeleton K involves 2-cells $(d^2, \theta) = q(c_{\pm}^2, \psi_{\pm})$; and $q : (B^3, S^2) \to (M, K)$ is a combinatorial characteristic map for the single 3-cell $d^3 = q(\mathring{B}^3)$ of M. When M is a 3-manifold, K serves as a spine, as $B^3 - \{0\} \searrow S^2$ implies $M - \{q(0)\} \searrow K$.

An easy extension of the classical argument [SeTh45, 60, Satz 1] shows:

Lemma 1.1 *A squashing map* $q : (B^3, S^2) \to (M, K)$ *yields a closed oriented 3–manifold* M *with spine* K *iff* $\chi(K) = 1$.

The technique, described above, of partially assembling a triangulation, coupled with the technique of collapsing a maximal tree in a simplicial spine, show:

Lemma 1.2 *Every closed oriented 3–manifold* M *has a presentation via a squashing map* $q : (B^3, S^2) \to (M, K)$. *One can ensure that either there are no fibered 2–cells in* S^2 *or that the spine* K *has a single 0–cell, but it is not always possible to have both of these features.*

Vacuum sealing a known spine   To illustrate Lemma 1.2 and the impossibility, in general, of achieving both features mentioned, we produce squashing maps for two very simple closed oriented 3–manifolds, S^3 and $S^1 \times S^2$. The technique amounts to *vacuum sealing* a 2–sphere around a spine K, using the cells of K to imprint a squashable complex C on S^2. The details can be best expressed using a *handle decomposition* of a *thickening* or regular neighborhood of the spine.

Let K be a combinatorial 2–complex, with its k–cells denoted by d^k ($k = 0, 1, 2$). Let M be a closed orientable 3–manifold that has K as a spine. A regular neighborhood $N(K)$ of K in M can be assembled from k–*handles* $H(d^k) \equiv B^{3-k} \times B^k$ whose cores $\kappa H(d^k) \equiv \{0\} \times B^k$ lie on the k–cells $d^k \in K$ ($k = 0, 1, 2$). The 1–handles $H(d^1) \equiv B^2 \times B^1$ are solid cylinders that attach by their *end discs* $\epsilon H(d^1) \equiv B^2 \times \partial B^1$ to disjoint discs in the boundary of the 0–handles $H(d^0) \equiv B^3$ at the ends of the 1–cells d^1. The 2–handles $H(d^2) \equiv B^1 \times B^2$ are solid cylinders that attach by their *edge annuli* $\epsilon H(d^2) \equiv B^1 \times \partial B^2$ to annuli in the boundary of the union of the 0–handles and 1–handles. The handles have disjoint interiors.

Because K is a spine of M, the boundary $\partial N(K)$ is a topological 2–sphere S^2 bounding a complementary 3–ball B^3 in M. The handles can be chosen so that their attachments cover the boundaries of the 0–handles and 1–handles, save for some product 2–cells that lie on the *side cylinders* $\sigma H(d^1) \equiv \partial B^2 \times B^1$ of the 1–handles and that have cross-sections paralleling the 1–cells $d^1 \in K$. Then the spherical boundary $\partial N(K)$ is completely tiled by those product 2–cells, together with the paired *side discs* $\sigma H(d^2) \equiv \partial B^1 \times B^2$ of the 2–handles These cells define a squashable complex C on $S^2 = \partial N(K)$ as the non-paired 2–cells of C are product cells that meet side-to-side. A suitable regular neighborhood collapse $\partial N(K) \searrow K$ produces the squashing map $q(C) : \partial N(K) = S^2 \to K$. This is the *vacuum sealing* procedure.

Three sphere   The squashable complex C on S^2 in Figure 3a consists of two sliced hemispherical 2-cells that are displayed in clamshell fashion, hinged open at $1 \subset S^1$. This complex C arises from *vacuum sealing* a 2-sphere around the *dunce hat* $K = d^0 \cup d_x^1 \cup_{xx^{-1}x} d^2$ (modeled on the presentation $\langle x \mid xx^{-1}x \rangle$) embedded as a spine in the 3-sphere S^3 just as in Figure I.12. Try it for yourself. So the squashing map $q(C) : (B^3, S^2) \to (S^3, K)$ determined by C yields the 3-sphere S^3, with the contractible dunce hat K as spine.

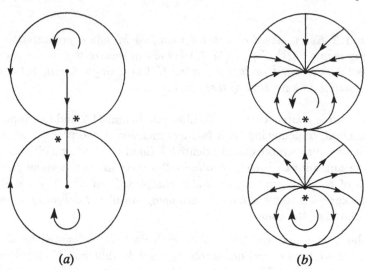

(a)   (b)

Figure VIII.3. S^3 and $S^1 \times S^2$

The product $S^1 \times S^2$ The squashable complex C in Figure 3b arises from *vacuum sealing* a 2-sphere around the *pin-cushion* $K = d^0 \cup d_x^1 \cup_{xx^{-1}} d^2$ (modeled on the presentation $\langle x \mid xx^{-1} \rangle$) embedded as a spine in $S^1 \times S^2$. It contains two fibered 2-cells, in addition to paired 2-cells that form d^2. Figure 4 shows how this squashing arises from a handle decomposition of a regular neighborhood $N(K)$ in $S^1 \times S^2$. The two fibered 2-cells meet side to side at the (vertical) meridian of the 0-handle $H(d^0)$ and end at the attachment annulus $\epsilon H(d^2)$ for the 2-handle. An application of [Si86, Theorem 3] provides a surgery description showing that the manifold $M = B^3//C$ really is $S^1 \times S^2$. The complex C is the unique one yielding K; so $S^1 \times S^2$ is the unique closed orientable 3-manifold with this spine. Finally, any squashing q producing $M = S^1 \times S^2$ and a spine with a single 0-cell d^0 must contain fibered 2-cells. For, in the universal covering complex $\tilde{M} = R^3 - \{0\}$, the 3-cells above the single 3-cell $d^3 = q(\overset{\circ}{B}{}^3)$ are *permuted* by a radial expansion $\rho : R^3 \to R^3$ to tile $\tilde{M} = R^3 - \{0\}$. Their attaching maps must span consecutive 0-cells $\tilde{d}^0$ and $\rho(\tilde{d}^0)$ and perform some collapsing at these bottlenecks.

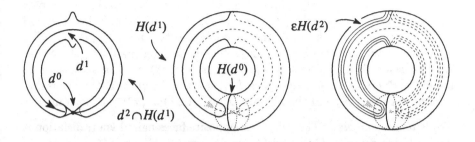

$$H(d^1)$$

$$d^1$$

$$d^0$$

$$d^2 \cap H(d^1)$$

$$\varepsilon H(d^2)$$

$$H(d^0)$$

Figure VIII.4.

When the fibered 2–cells of a squashable complex C do not meet to contain a circular fiber, they may be preliminarily squashed from C, stretching the paired 2–cells accordingly. The equator of Figure 3b is just such a fibered circle; it spans a disc in B^3 that forms a 2–sphere which doesn't bound a 3–ball in $S^1 \times S^2$. So the phenomenon of essential fibered 2–cells in a squashable complex relates to the irreducibility (see Subsection 1.5) of the 3–manifold.

Taut identities   When a squashed 2–sphere $K = S^2//C$ has a single 0–cell, it is the model $K_\mathcal{P}$ of a group presentation $\mathcal{P} = \langle \mathbf{x} \mid \mathbf{r} \rangle$ for $\pi_1(K)$. Then the squashing map $q : S^2 \to K$ can be interpreted as a realization of an identity sequence $\omega = ((w_1, r_1)^{\epsilon_1}, \ldots, (w_m, r_m)^{\epsilon_m})$ for $\mathcal{P}$ as in Chapter II, Section 2. The identity sequence is revealed when one dissects the spherical complex C into an array of polygonal balloons and strings as in Figure II.6. The string assignments into $K_\mathcal{P}$ are words $\{w_i \in F(\mathbf{x})\}$ and the polygon assignments are signed characteristic maps $\{\phi_{r_i}^{\epsilon_i}\}$ of the 2–cells c_r^2 ($r \in \mathbf{r}$).

For example, the squashing map for Figure 3a represents the trivial identity $(1, xx^{-1}x)(1, xx^{-1}x)^{-1}$ for the dunce hat presentation $\langle x \mid xx^{-1}x \rangle$. The squashing map for Figure 3b represents the identity $(1, xx^{-1})(x^{-1}, xx^{-1})^{-1}$ for the pin-cushion presentation $\langle x \mid xx^{-1} \rangle$.

In [Si86], an identity for a presentation $\mathcal{P}$ is called *taut* if it can be realized as above by a squashing map $q : S^2 \to K$ that produces the model $K = K_\mathcal{P}$ of the presentation. So you may think of a taut identity as one that determines its presentation. Because the model $K_\mathcal{P}$ has $\chi(K) = 1$ if and only if the presentation $\mathcal{P}$ is *balanced* (i.e., has an equal number of generators and relators), the two preceding lemmas yield:

Theorem 1.3 ([Si86], (7)) *A group π is the fundamental group of a closed orientable 3–manifold if and only if there is a taut identity ω for a balanced presentation $\mathcal{P}$ for π. Also, any squashing map $q_\omega : S^2 \to K$ representing a taut identity ω forms a 3–manifold $M = K \cup_{q_\omega} B^3$ with spine $K = K_\mathcal{P}$.*

Lens spaces   For relatively prime integers $1 \le q \le p$, the squashable complex $C_{p,q}$ comprised of two regular p-gon complexes $P_p^{\pm}$, with boundaries identified under a $2\pi q/p$-twist, produces the pseudo-projective plane $K_p = S^2//C_{p,q}$ modeled on the presentation $\langle x \mid x^p \rangle$. The squashing map $q(C_{p,q}) : S^2 \to K_p$ represents the taut identity $\omega_q = (1, x^p)(x^q, x^p)^{-1}$. The associated 3-manifold $B^3//C_{p,q}$ is the lens space $L_{p,q}$. (When $(p, q) \ne 1$, the squashing does not produce K_p and the identity $\omega_q = (1, x^p)(x^q, x^p)^{-1}$ is not taut.)

Triple torus example   The cube with opposite faces paired via translation is a squashable complex C that yields the 2-complex $K = S^2//C$ modeled on the presentation $\langle x, y, z \mid [x, y], [y, z], [z, x] \rangle$ of $\mathbb{Z}^3$. The associated 3-manifold $M = B^3//C$ is the triple torus $S^1 \times S^1 \times S^1$. The partial dissection in Figure 5 shows that the squashing map $q(C) : S^2 \to K$ realizes the taut identity

$$(1, [x, y])(y, [z, x])^{-1}(1, [y, z])(z, [x, y])^{-1}(1, [z, x])(x, [y, z])^{-1}.$$

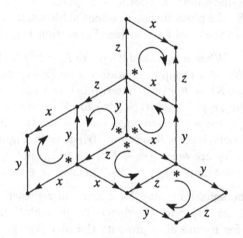

Figure VIII.5. $S^1 \times S^1 \times S^1$

Hyperbolic dodecahedral space   The squashing of the dodecahedron via the identification of opposite faces under $3\pi/5$ twists produces hyperbolic dodecahedral space and a spine modeled on a six-generator six-relator presentation. The reader may derive the taut identity realized by this squashing map.

Spherical dodecahedral space   The squashing of the dodecahedron via the identification of opposite faces under $\pi/5$ twists produces spherical dodecahedral space and a spine that has five 0-cells, and so it doesn't immediately yield a presentation of the fundamental group. In [Si86, Section 5], it is shown that the squashable complex C_6 on the suspension of the 6-gon, displayed in

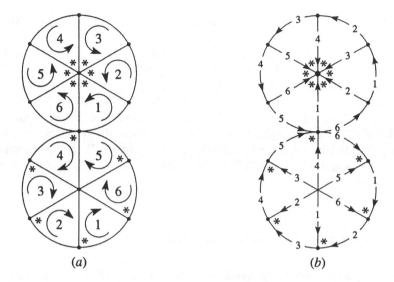

(a)   (b)

Figure VIII.6. Spherical dodecahedral space

open clamshell fashion in Figure 6a with paired 2–cells co-labeled, produces spherical dodecahedral space and a spine modeled on the presentation

$$\mathcal{P}_6 = \langle x_i (i \in \mathbb{Z}_6) \mid x_i^{-1} x_{i-1} x_{i+1} (i \in \mathbb{Z}_6) \rangle.$$

To verify this, one first checks that the identification of the paired 2–cells, matching up their starred reference points and their orientations, causes the 0–cells to coalesce into a single 0–cell and the 1–cells to coalesce into six 1–cells as labeled in Figure 6b. So the boundaries of the six 2–cells read the six relators $r_i = x_i^{-1} x_{i-1} x_{i+1} (i \in \mathbb{Z}_6)$ of the presentation $\mathcal{P}_6$. Second, one identifies the resulting manifold as spherical dodecahedral space by means of a surgery description derived using [Si86, Theorem 3]. A dissection of C_6, placing the six lower 2–cells in a pin-wheel arrangement around the six upper 2–cells, shows that the squashing map realizes the taut identity $\Pi_{i \in \mathbb{Z}_6} (1, r_i)(x_{i-1}^{-1}, r_{i+4})^{-1}$.

The squashing construction implies that a taut identity ω for a balanced presentation $\mathcal{P} = \langle \mathbf{x} \mid \mathbf{r} \rangle$ is a *paired identity*: every relator $r \in \mathbf{r}$ appears in exactly two entries of ω and they have opposite exponents, say $(w_{r,+}, r)^{+1}$ and $(w_{r,-}, r)^{-1}$. The following observation is used below to establish a homotopy classification (Theorem 1.8) for closed oriented 3–manifolds.

Theorem 1.4 *The quotients $w_{r,+} w_{r,-}^{-1} \in F(\mathbf{x})$ associated with the paired entries in a taut identity $\omega = ((w_1, r_1)^{\epsilon_1}, \cdots, (w_m, r_m)^{\epsilon_m})$ for a balanced presentation $\mathcal{P} = \langle \mathbf{x} \mid \mathbf{r} \rangle$ represent a set of generators for the group presented.*

Proof: Let $q(C) : S^2 \to K$ be a squashing map realizing the taut identity ω. For all paired 2-cells $(c_{r,\pm}^2, \psi_{r,\pm})$ of C, we attach to B^3 a polygonal post $P_k \times B^1$, gluing its ends $P_{k(r)} \times \{\pm 1\}$ to $c_{r,\pm}^2$ via $\psi_{r,\pm}$ (here, $k(r) = length(r)$ for $r \in \mathbf{r}$). By [Si86, (15)], this forms a handlebody $H(C)$ that is one half of a Heegaard splitting of the manifold $M = B^3//C$. (The other half of the handle body is constructed from a collection of *plugs* $B^1 \times B^2$, one for each generator $x \in \mathbf{x}$ glued by its *tread* $B^1 \times S^1$ along the annulus of fibered boundary 2-cells of $H(C)$ that project to the 1-cell c_x^1.) Since the quotients $w_{r,+}w_{r,-}^{-1}$ are represented by closed curves in the handlebody $H(C)$ that generate its free fundamental group $\pi_1(H(C))$, the theorem follows from the Heegaard splitting of M.   □

1.3   Detecting 3-manifold spines

Reinterpreting Neuwirth's Algorithm in the light of §1.2   As indicated in Chapter I, §3.1, L. Neuwirth's algorithm [Ne68] can decide whether a standard complex K is the spine of a closed orientable 3-manifold. When the algorithm detects K as such a spine, the proof, in effect, yields a squashable complex on the 2-sphere boundary of the thickening of K (see the vacuum sealing technique following Lemma 1.2). So applying the algorithm is equivalent to assembling a squashing map $q : S^2 \to K$ from pairs of oppositely oriented copies of the combinatorial characteristic maps $\psi_r : P_{k(r)} \to K$ for the 2-cells c_r^2 of K, which are indexed by the relators $r \in \mathbf{r}$ of an associated balanced group presentation[1] $\mathcal{P} = \langle \mathbf{x} \mid \mathbf{r} \rangle$. ($P_{k(r)}$ denotes the regular $k(r)$-gon, where $k(r)$ is the length of the relator $r \in \mathbf{r}$.) It is *insufficient* to just build a combinatorial map q into K from these pairs; one must check that squashing the paired and fibered 2-cells creates K, i.e., identifies all the 0-cells as well as the 1-cells according to their labels $x \in \mathbf{x}$. Then $M = K \cup_q B^3$ is a closed oriented 3-manifold by Lemma 1.1.

As a more visual alternative to Neuwirth's algorithm, one can construct all such spherical diagrams for the presentation $\mathcal{P}$ and apply the check for squashing maps to determine all appearances of the model $K_{\mathcal{P}}$ for the presentation as a spine of a closed oriented 3-manifold. Neuwirth's algorithm works well for short lengths of the relators. But, when the 2-complex really is a 3-manifold spine, it is usually quicker to construct a squashing map directly than to apply Neuwirth's algorithm.

---

[1]Throughout this section, a *presentation* is an alphabet $\mathbf{x}$ together with a collection of not necessarily distinct elements from the free semigroup of $\mathbf{x}$ and its formal inverse $\mathbf{x}^{-1}$. Two presentations are considered the same if the 2-complexes determined by them are isomorphic.

For example, Neuwirth's presentation [Ne68]

$$\mathcal{P}_n = \langle\, x_i \ (i \in \mathbb{Z}_n) \mid x_1 \cdots x_{i-1} x_i^{-1} x_{i+1} \cdots x_n (i \in \mathbb{Z}_n) \rangle$$

presents the spine of a closed oriented 3–manifold because there is the squashable complex $C(\mathcal{P}_n)$ depicted in Figure 7.

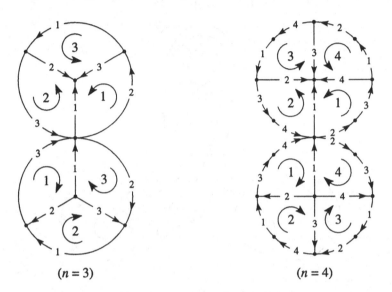

$$(n = 3) \qquad\qquad\qquad (n = 4)$$

Figure VIII.7. Neuwirth's example

We will turn back to squashings in §1.4 but for the moment carry the discussion of embeddability of 2–complexes into 3–manifolds into a new direction.

Railroad-systems   R.P. Osborne and R.S. Stevens (see [OsSt74], [OsSt77]) have their own approach for detecting spines of (closed) 3–manifolds, based upon an observation which was stated in Chapter I, (51). They derive relator conditions by considering the neighborhoods of the edges, the handles of the thickening. The advantage of Osborne and Stevens' approach is that necessary algebraic conditions on the shape of the relators are derived which do not result from Neuwirth's algorithm (see Theorems 1.7 and 1.8 below).

Below, we give a different development of Osborne-Stevens' *railroad-systems*.

First, rather than considering the original problem (whether a given group presentation fits on a handlebody as in Chapter I, (51)), we consider a structurally enriched problem: whether a group presentation *with given syllable decomposition of the relators* fits on a handlebody (see Lemma 1.5 and Proposition 1.6). Second, we shall discuss how an answer of the enriched problem (can) lead to an answer of the original problem (see Theorem 1.7).

For motivation, consider the boundary curve k of a handlebody V in Figure 8. Corresponding to the arcs $\lambda_1, \lambda_2, \lambda_3$ of intersection of k with the handles H_x, H_y, and $H_{z_{,,}}$ we read xxy^0 and it is not possible to isotope k so that its intersection with the H_i has fewer components.

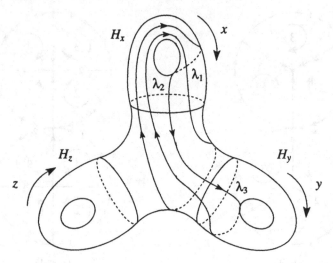

Figure VIII.8. Handlebody figure

If a presentation $\mathcal{P} = \langle x_1, \ldots, x_i, \ldots, x_n \mid r_1, \ldots, r_j, \ldots, r_m \rangle$ fits on a handlebody V with handles H_i, the arcs[2] λ_q of intersection of the disjoint relator curves k_j with H_i can be assumed to be proper, simple and nontrivial on H_i. (A proper arc λ in a surface F is *trivial* if it can be homotoped relative $\partial\lambda$ into ∂F.) Each arc $\lambda_q \subset H_i$ corresponds to a *syllable* $x_i^{e_q}$ in some relator, possibly $e_q = 0$ despite λ_q being nontrivial, as the syllable only reads the longitudes of H_i (e. g. λ_3 in Figure 8). Note that two choices are involved in the definition of the handles H_i of V: First, a system of meridian discs for V, second a system of connecting arcs (longitudes) on the sphere obtained by cutting ∂V along the chosen meridians.

Thus, when $\mathcal{P}$ fits on V, each relator r_j acquires from V a *syllable-decomposition* $r_j = \Pi x_{j_k}^{e_{j_k}}$, with $e_{j_k} \in \mathbb{Z}$. Of course, every r_j has a unique *natural* syllable-decomposition for which $e_{j_k} \neq 0$ and $j_k = j_{k+1}$ implies that e_{j_k} and $e_{j_{k+1}}$ have different signs, but we the one we read off V need not be natural. As a further example, there is the *trivial* syllable-decomposition where $e_{j_k} \in \{\pm 1\}$.

---

[2] If some k_j runs entirely on an H_i, isotope a little arc onto $\overline{V - \cup H_i}$

Lemma 1.5 *If $\mathcal{P}$ with given syllable-decomposition fits on a handlebody, then the absolute values $|\,e_{j_k}\,|$ of the x_i-exponents of the syllable-decomposition of the relators take on at most 3 distinct values m_i, p_i, and $m_i + p_i$, where $(m_i, p_i) = 1$.*

Proof: Two disjoint proper simple nontrivial arcs λ_1, λ_2 on a handle H_i are isotopic iff $\partial \lambda_1$ doesn't separate $\partial \lambda_2$ on the boundary circle of H_i. To see this, close non-separating λ_1 and λ_2 into disjoint simple curves h_1, h_2 using a disc filling the puncture of H_i. Removing a regular neighborhood of h_1 from the resulting torus T leaves an annulus in which the only nontrivial simple closed curve is parallel to the boundary. Hence, h_1 and h_2 are isotopic, whence also λ_1 and λ_2. Conversely, if $\partial \lambda_1$ and $\partial \lambda_2$ do separate, the analogous construction yields generators h_1 and h_2 for $\pi_1(T)$ which implies that λ_1 and λ_2 aren't isotopic.

For each isotopy class of arcs λ_q on H_i, choose a representative $\tilde{\lambda}_q$. By the above reasoning, their boundaries $\partial \tilde{\lambda}_q$ separate each other in pairs so that we can assume that each $\partial \tilde{\lambda}_q$ is a pair of antipodal points on ∂H_i. Passing again to T, we can close the $\tilde{\lambda}_q$ to nontrivial simple closed curves h_q, where $\cap h_q$ is a singleton. Assume that there are at least two of them, h_1 and h_2. In terms of canonical generators a and b of $\pi_1(T)$, represented by the inclusions $\alpha : S^1 \times \{1\} \to S^1 \times S^1 = T$ and $\beta : \{1\} \times S^1 \to S^1 \times S^1 = T$, let $[h_1] = a^p b^q$ and $[h_2] = a^m b^s$ wrt some orientation of the involved curves[3]. A pair $a^p b^q$, $a^m b^s$ of nontrivial elements of $\pi_1(S^1 \times S^1)$ is represented by a pair h_1, h_2 of simple closed curves that meet transversally in a single point iff $ps - mq = \pm 1$. (The point is that the two curves h_1, h_2 cut T into a square, making it possible to define a homeomorphism $H : T \to T$ such that $H \circ \alpha = h_1$ and $H \circ \beta = h_2$. The induced automorphism $H_\# = \left(\begin{smallmatrix} p & q \\ m & s \end{smallmatrix} \right)$ of the free abelian group $\pi_1(T)$ necessarily has determinant ± 1. See [He76, Lemma 2.9]).

Consequently: If the presentation $\mathcal{P}$, with its given syllable-decomposition, fits on a handlebody, then the absolute values of the exponents of syllables x_i^p and x_i^m associated with the arcs $\tilde{\lambda}_1$ and $\tilde{\lambda}_2$ are relatively prime. If $m = 0$ (resp. $p = 0$), then $|\,p\,| = 1$, (resp. $|\,m\,| = 1$).

If there is a third isotopy class h_3, it must cut diagonally across the square cut out of T by h_1 and h_2 and so leaves no further room for a fourth isotopy class h_4 to exist. So $[H \circ \delta] = [h_3]$ where δ is one of the four directed diagonals of $S^1 \times S^1$. Because $[\delta] = [\alpha]^{\pm 1}[\beta]^{\pm 1}$, then $[h_3] = [h_1]^{\pm 1}[h_2]^{\pm 1}$. Thus the absolute value of the exponent of the syllable associated with $\tilde{\lambda}_3$ equals the absolute value of the sum or the difference of m and p. In the

---

[3]Up to this point, only loops have been considered; and, strictly speaking, they become elements of π_1 only after equipped with an orientation.

latter case, exchanging the roles of $\tilde{\lambda}_2$ and $\tilde{\lambda}_3$ gives rise to the transition $\{m, p, m - p =: m'\} \rightarrow \{p + m', p, m'\}$ □

Osborne and Stevens [OsSt77] use this observation to devise a new picture of M, called *railroad-system* (abbrev., *RR-system*) as follows. In the plane $\mathbb{E}^2$, draw disjoint regular hexagons labelled by the generators x_i. The plane stands for the handlebody surface $\overline{\partial V - \cup H_i}$, the handles H_i have to be imagined as sitting on the hexagons. The edges of each hexagon are labelled clockwise (according to an orientation of $\mathbb{E}^2$) by m_i, $m_i + p_i$, p_i, $-m_i$, $-(m_i + p_i)$, $-p_i$. The (pairwise disjoint simple sub-)arcs λ_q of the relation curves k_j running on $\overline{\partial V - \cup H_i}$ connect the hexagons. Their endpoints avoid the corners of the hexagons and we get from λ_q to the next arc λ_{q+1} of k_j by proceeding along a line segment in the hexagon orthogonal to two of its edges. (See Figure 9.)

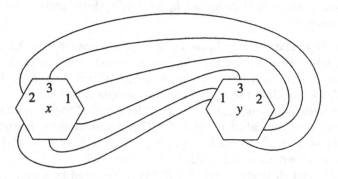

Figure VIII.9. Relators read x^3yxy and x^2y^3

Adding to the hexagons handles H_i inscribed with the arcs λ_q achieving the values m_i, $m_i + p_i$, or p_i returns us to a thickening of K^1. Because [He76, Lemma 2.9] has a valid converse, we have:

Proposition 1.6 $\mathcal{P}$ *with its given syllable-decomposition fits on a handlebody if and only if there is an RR-system for $\mathcal{P}$.*   □

While an RR-system gives a good view of the manifold, the ambiguity of syllable decompositions still prevents derivation of relator restrictions. The conclusion of Lemma 1.5 is always fulfilled for the monosyllables and the values $m_i = 1$, $p_i = 0$ $\forall i$. But it needn't hold for the natural syllable decomposition, as in the following example, due to Osborne-Stevens: For $\langle x, y \mid x^4y, x^2y \rangle$, the set 2, 4 doesn't equal $m, m + p, p$ as required for Lemma 1.5; whereas the syllable decomposition $x^2y^0x^2y$ of the first relator gives rise to the RR-system of Figure 10.

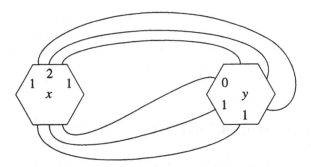

Figure VIII.10. RR-system for $\langle x, y \mid x^2 y^0 x^2 y, x^2 y \rangle$

On the other hand, for a restricted class of presentations (see Theorem 1.7 below), 0-labels as well as *recurrent* arcs having both endpoints on the same hexagon without constituting an entire relator can be eliminated from the RR-system, changing only the syllable decomposition of the relators. In order to switch back to the non-enriched problem (see the discussion ahead of Figure 8), we assume the presentation $\mathcal{P}$ to be *syllable-reduced*, i.e., the relators are reduced and it is impossible to decrease the total number of syllables of (the natural syllable decomposition of) $\mathcal{P}$ by conjugating some of the x_i, $i \neq i_0$, with a fixed x_{i_0}-power and then reducing. Note that any presentation can be transformed to a syllable-reduced one by a finite number of elementary steps. Note further that reduced two-generator presentations are always syllable–reduced.

The case $n = 2$ of the next theorem was given in [OsSt74, Theorem 1], for arbitrary n, see [Ho-An92].

Theorem 1.7 *Let* $\mathcal{P} = \langle x_1, \ldots, x_n \mid r_1, \ldots, r_m \rangle$ *be syllable-reduced and assume that for each i there is an x_i-syllable in some r_j with exponent $e_{j_k} \neq \pm 1$. If $\mathcal{P}$ is a presentation for an orientable 3-manifold, there is an RR-system for the natural syllable-decomposition of the r_j. In particular, the absolute values of the x_i-exponents in the r_j take on at most three distinct values m_i, p_i, and $m_i + p_i$, where $(m_i, p_i) = 1$.*

For presentations satisfying the assumptions of this theorem, one can amalgamate Neuwirth's and Osborne-Stevens' ideas by rendering Neuwirth's definitions into permutations of a set of certain points on the edges of the hexagons, thus constructing an algorithm which in many cases is shorter than Neuwirth's.

Osborne offers in [Os78] the simplest closed 3–manifolds, catalogued according to group presentations for their spines. Of course, a spine may not uniquely determine a 3–manifold, a fact illustrated by the pseudo-projective spine $K_{(x|x^5)}$ shared by the non-homeomorphic lens spaces $L_{5,1}$ and $L_{5,2}$. The problem of classifying 3–manifolds with a specific spine has been considered by R. S. Stevens [St75] and A. Cavicchioli and F. Spaggiari [CaSp92].

1.4   Homotopy classification

As indicated in Chapter I, prior to Theorem 3.1, a closed oriented 3–manifold M is not, in general, determined by a spine K. However, we show in Theorem 1.9 below that M is determined up to oriented homotopy type by the homotopy class $[q] \in \pi_2(K)$ of a squashing map $q : (B^3, S^2) \to (M, K)$, in fact, by a related element $\zeta_M \in H_3(\pi_1(M))$ that is independent of the spine K. More precisely, we use squashing maps to prove G.A. Swarup's homotopy classification [Sw74] of closed, oriented, 3–manifolds M by their fundamental class $\zeta_M \in H_3(\pi_1(M))$. Swarup's work refines an earlier homotopy classification result by C.B. Thomas [Th67].

Fundamental class   Associated with each closed oriented 3–manifold M is an element $\zeta_M \in H_3(\pi_1(M))$, called the *fundamental class* of M; namely, ζ_M is the image $\zeta_M = i_*(z_M) \in H_3(J) \equiv H_3(\pi_1(M))$ of the traditional fundamental class $z_M \in H_3(M) = \mathbb{Z}$ under the homomorphism induced by the inclusion $i : M \subset J$, where J is an aspherical CW complex with the same 2–skeleton $J^2 = K$ as M. There is a way to express the fundamental class ζ_M using any combinatorial presentation of M by a squashing map $q : (B^3, S^2) \to (M, K)$. The spherical attaching map $q : S^2 \to K$ is homologically trivial because it identifies the non-fibered 2–cells of S^2 in oppositely oriented pairs. So its homotopy class $[q] \in \pi_2(K)$ determines a coset

$$\zeta_q \in \Gamma(K) = Ker(h : \pi_2(K) \to H_2(K))/(I(\pi_1(K)) \cdot \pi_2(K)),$$

where h is the Hurewicz homomorphism and $I(\pi_1(K))$ is the augmentation ideal of the integral group ring $\mathbb{Z}\pi_1(K)$.

The following summarizes Propositions 1 and 2 in [Si86].

Lemma 1.8 *For a closed oriented 3–manifold M and spine K, there is a natural isomorphism $H_3(\pi_1(M)) \approx \Gamma(K)$, under which $\zeta_M \in H_3(\pi_1(M))$ corresponds to $\zeta_q \in \Gamma(K)$ for any squashing map $q : (B^3, S^2) \to (M, K)$.*

Proof: For any group G and free resolution (C_*, ∂_*) of the trivial module $\mathbb{Z}$ over the integral group ring $\mathbb{Z}G$, the derived groups $H_*(\mathbb{Z} \otimes_{\mathbb{Z}G} C_*)$ are taken as the integral homology groups $H_*(G)$ of G. The connecting homomorphism associated with the short exact sequence $0 \to I(G) \cdot C_* \to C_* \to \mathbb{Z} \otimes_{\mathbb{Z}G} C_* \to 0$ of chain complexes over $\mathbb{Z}G$, where $I(G)$ is the augmentation ideal of G, are isomorphisms (C_* being free)

$$H_{n+1}(G) \equiv H_{n+1}(\mathbb{Z} \otimes_{\mathbb{Z}G} C_*) \approx H_n(I(G) \cdot C_*) \qquad (n \geq 1).$$

In particular, $H_3(G)$ can be calculated entirely in terms of the second boundary operator $\partial_2 : C_2 \to C_1$ as the quotient

$$H_2(I(G) \cdot C_*) = (Ker\ \partial_2 \cap I(G))/I(G) \cdot Ker\ \partial_2.$$

When $J \supset M$ is an aspherical complex with $J^2 = K$, the cellular chain complex $C_*(\tilde{J})$ of the universal covering $\tilde{J} \supset \tilde{M}$, with the $\mathbb{Z}\pi_1(M)$-module structure given by the action of $\pi_1(M)$ as covering transformations, is a free resolution of $\mathbb{Z}$ over $\mathbb{Z}\pi_1(M)$. Using this resolution and the identification $C_*(J) \equiv \mathbb{Z} \otimes_{\mathbb{Z}G} C_*(\tilde{J})$, the connecting isomorphism yields

$$H_3(J) \equiv H_3(\pi_1(M)) \approx Ker(h : \pi_2(K) \to H_2(K))/(I(\pi_1(K)) \cdot \pi_2(K)).$$

In the previous display, we have used the identification

$$\pi_2(K) \equiv H_2(\tilde{K}) \equiv Ker\ \partial_2(\tilde{J}) \leq C_2(\tilde{J})$$

of Lemma 3.3, Chapter II. This isomorphism $H_3(J) \approx \Gamma(K)$ calculates the homotopy boundary in K of the cellular homology 3–cycles in J, which yields the claim of the lemma.   □

Here is Swarup's classification theorem:

Theorem 1.9 ([Sw74, Theorem B]) *Let $\alpha : \pi_1(M) \to \pi_1(N)$ be an isomorphism for closed oriented 3–manifolds M and N. There is an oriented (i.e., degree one) homotopy equivalence $F : M \to N$ with $F_\# = \alpha$ if and only if $\alpha_* : H_3(\pi_1(M)) \to H_3(\pi_1(N))$ preserves the fundamental class: $\alpha_*(\zeta_M) = \zeta_N$.*

Proof: A degree one map $F : M \to N$ induces $F_* : H_3(M) \to H_3(N)$ satisfying $F_*(z_M) = z_N$. Under the correspondence of Lemma 1.8, this implies that $\alpha_*(\zeta_M) = \zeta_N$.

For the converse, consider squashing maps

$$q_M : (B^3, S^2) \to (M, K) \quad \text{and} \quad q_N : (B^3, S^2) \to (N, L)$$

realizing taut identities $\omega = \prod_i (w_i, r_i)^{\epsilon_i}$ and $\nu = \prod_j (z_j, s_j)^{\delta_j}$ for presentations $\mathcal{P} = \langle \mathbf{x} \mid \mathbf{r} \rangle$ and $\mathcal{Q} = \langle \mathbf{y} \mid \mathbf{s} \rangle$ on which the spines K and L are modeled.

Let $(w_{r,\pm}, r)^{\pm 1}$, $r \in \mathbf{r}$, and $(z_{s,\pm}, s)^{\pm 1}$, $s \in \mathbf{s}$, denote the paired entries of these taut identities. Using the correspondence in Lemma 1.8, the condition $\alpha_*(\zeta_M) = \zeta_N$ translates to say that any map $F : K \to L$ inducing α on π_1 carries $[q_M] \in \pi_2(K)$ to $[q_N] \in \pi_2(L)$, up to an element of $I(\pi_1(L)) \cdot \pi_2(L)$.

Equivalently, the lifting $\tilde{F} : \tilde{K} \to \tilde{L}$ induces a chain homomorphism $C_2(\tilde{F})$:

$$\pi_2(K) \overset{p_{K\#}}{\leftarrow} \pi_2(\tilde{K}) \overset{h_3}{\to} H_2(\tilde{K}) = Ker\, \partial_2(\tilde{K}) \le C_2(\tilde{K})$$

$$\downarrow F_\# \qquad\quad \downarrow \tilde{F}_\# \qquad\qquad\quad \downarrow \tilde{F}_* \qquad\qquad\qquad\quad \downarrow C_2(\tilde{F})$$

$$\pi_2(L) \overset{p_{L\#}}{\leftarrow} \pi_2(\tilde{L}) \overset{h_3}{\to} H_2(\tilde{L}) = Ker\, \partial_2(\tilde{L}) \le C_2(\tilde{L})$$

that carries

$$[q_M] = \sum_{r \in \mathbf{r}} (w_{r,+} - w_{r,-})\, \tilde{c}_r^2 \in C_2(\tilde{K}) \quad \text{to} \quad [q_N] = \sum_{s \in \mathbf{s}} (z_{s,+} - z_{s,-})\, \tilde{c}_s^2 \in C_2(\tilde{L}),$$

up to an element of $\delta \in I(\pi_1(L)) \cdot \pi_2(L)$, i.e., $C_2(\tilde{K})([q_M]) = [q_N] - \delta$.

As in Chapter II, Lemma 3.5, the modification $F^\gamma : K \to L$ of F by any chosen $\mathbb{Z}\pi_1(K)$-cochain $\gamma : C_2(\tilde{K}) \to {}_\alpha\pi_2(L)$ induces

$$C_2(\tilde{F}^\gamma) = C_2(\tilde{F}) + \gamma : C_2(\tilde{K}) \to {}_\alpha C_2(\tilde{L}).$$

Now (1) α is an isomorphism, (2) the quotients $w_{r,+} w_{r,-}^{-1}$ $(r \in \mathbf{r})$ associated with the paired entries in the taut identity $\omega = \prod_i (w_i, r_i)^{\epsilon_i}$ generate the fundamental group $\pi_1(K)$ (Theorem 1.4), and (3) the augmentation ideal $I(\pi_1(L))$ of $\mathbb{Z}\pi_1(L)$ is generated by elements of the form $(y - 1)$. It follows that, by suitable choices of the spherical elements $\gamma(\tilde{c}_r^2) \in \pi_2(L)$, the image

$$\gamma([q_M]) = \sum_{r \in \mathbf{r}} (\alpha(w_{r,+}) - \alpha(w_{r,-}))\, \gamma(\tilde{c}_r^2)$$

in $\pi_2(L) \le C_2(\tilde{L})$ can be made any prescribed element $\delta \in I(\pi_1(L)) \cdot \pi_2(L)$. So some modification $F^\gamma : K \to L$ carries $[q_M] \in \pi_2(K)$ exactly to $[q_N] \in \pi_2(L)$. This F^γ extends to a degree one map $G : M = K \cup_{q_M} B^3 \to L \cup_{q_N} B^3 = N$ of these 3–manifolds, mapping the 3–cell in M along a homotopy $F^\gamma \circ q_M \simeq q_N$ in L and over the 3–cell in N. Such a map G is necessarily a homotopy equivalence by [Sw74, Lemma 1.1].   □

Homotopy Type Invariant   So the pair $(\pi_1(M), \zeta_M \in H_3(\pi_1(M)))$ is a complete invariant of homotopy type for closed oriented 3–manifolds M. The preceding proof avoids Swarup's invocation (see, e.g., [Sw73, Theorem 1.0]) of Stallings' Splitting Theorem [St65$_1$, St59], which establishes Kneser's conjecture that a free product factorization of the fundamental group is realizable by a connected sum factorization of a closed 3–manifold. But a fair amount of 3–manifold topology is lurking in Swarup's proof of the cited lemma.

Restrictions on the invariant ζ_M and the classifying group $H_3(\pi_1(M))$ are easily derived. The 3–manifold group $\pi_1(M)$ has some free product factorization

$$(G_1 * \cdots * G_l) * (F_1 * \cdots * F_m) * (T_1 * \cdots * T_n)$$

into $l \geq 0$ infinite non-cyclic indecomposable factors G_i $(1 \leq i \leq l)$, $m \geq 0$ finite factors F_j with orders d_j $(1 \leq j \leq m)$, and $n \geq 0$ infinite cyclic factors T_k with generators $t_k(1 \leq k \leq n)$. Stallings' Splitting Theorem implies:

Theorem 1.10 ([Si86, Theorem 2]) *Associated with the free product factorization into indecomposables*

$$\pi_1(M) = (G_1 * \cdots * G_l) * (F_1 * \cdots * F_m) * (T_1 * \cdots * T_n)$$

is a direct sum decomposition

$$H_3(\pi_1(M)) = (\mathbb{Z} \oplus \cdots \oplus \mathbb{Z}) \oplus (\mathbb{Z}_{d_1} \oplus \cdots \oplus \mathbb{Z}_{d_m})$$

of l infinite cyclic and m finite cyclic factors, under which $\zeta_M \in H_3(\pi_1(M))$ corresponds to an $(l + m)$-tuple of generators of the summands. □

So there are these obvious questions:

(1) For the fundamental group G of a closed orientable 3–manifold, which tuples of generators of the summands of $H_3(G)$ correspond to fundamental classes of closed orientable 3–manifolds M with $\pi_1(M) = G$?

(2) What spines are required to represent all such classes?

(3) What are the orbits of the 3–manifold fundamental classes $\zeta_M \in H_3(G)$ under the action of $\alpha \in Aut\ G$?

Answers to these questions would constitute a homotopy classification of the closed oriented 3–manifolds with fundamental group $\pi_1(M) = G$. For example, Whitehead's homotopy classification [Wh41$_2$, Theorem 10] of the 3–dimensional lens spaces ($L_{p,q} \simeq L_{p,s}$ if and only if $qs \equiv \pm m^2 (mod\ p)$ for some m) can be derived in this way.

Implications for Spines There are homotopy features of the 2–dimensional spines that follow from the 3–manifold topology. The proof of Theorem 1.10 is modeled on Swarup's derivation [Sw73, Theorem 3.2] of a presentation of the $\pi_2(M)$ for a closed orientable 3–manifold M. An intermediate stage in the proof of Theorem 1.10 involves the following description of $\pi_2(K)$ for the spine K of a closed oriented 3–manifold.

Theorem 1.11 ([Si86, Proposition 3]) *Let* $q : S^2 \to K$ *be a squashing map, where* $\chi(K) = 1$, *and let* $\pi_1(K)$ *have a free product factorization into indecomposables as in Theorem 1.10. Then there is a* $\mathbb{Z}\pi_1(K)$-*module presentation* $\langle x_i, y_j, v_k \mid N(F_j) \cdot y_j = 0 \rangle$ *for* $\pi_2(K)$, *where* $N(F_j)$ *is the norm element in* $\mathbb{Z}\pi_1(K)$ *for the finite factor* F_j *of* $\pi_1(K)$. *Further,* $[q]$ *represents the sum* $\Sigma_i\, x_i + \Sigma_j\, y_j + \Sigma_k\, (1 - t_k)v_k$. □

As observed in [Si86], it follows that a squashing map $q : S^2 \to K$ with $\chi(K) = 1$ represents a $\mathbb{Z}\pi_1(K)$-module generator for $\pi_2(K)$ if and only if $\pi_1(K)$ is indecomposable under free products and is not infinite cyclic. Furthermore, $\pi_2(K)$ equals $\mathbb{Z}\pi_1(K)$ or $\mathbb{Z}\pi_1(K)/N(\pi_1(K))$ as $\pi_1(K)$ is infinite or finite. These are the cases of spines of closed aspherical 3–manifolds and (homotopy) spherical space forms, respectively.

In these cases, do copies of the squashing map $q : S^2 \to K$ always *tile* the universal covering $p : \tilde{K} \to K$, as for the squashings of the triple torus and the lens spaces? In the first case, where $\pi_2(K) = \mathbb{Z}\pi_1(K)$ and is generated by $[q]$, the spine K curiously determines the squashing map $q : S^2 \to K$ up to homotopy and a unit of $\mathbb{Z}\pi_1(K)$.

1.5   Topological Classification

Topological and geometrical classifications of closed oriented 3–manifolds begin with connected sum decompositions via embedded separating 2–spheres. A 3–manifold is *prime* if every embedded separating 2–sphere bounds a ball, equivalently, in each connected sum decomposition one of the factors is the 3–sphere. A 3–manifold is *irreducible* if every embedded 2–sphere bounds a ball. The only orientable prime 3–manifold that is not irreducible is $S^2 \times S^1$.

Prime Factorizations   In [Kn29], H. Kneser proved that any compact 3–manifold can be expressed as a finite connected sum of primes, and J. Milnor has shown [Mi62] that the prime factors are unique when M is orientable. The prime factorization of a closed oriented 3–manifold M gives rise to a free product factorization of $\pi_1(M)$ whose entries are the fundamental groups of the prime factors. These free product factors are indecomposable, as in Theorem 1.10, by Stallings' Splitting Theorem [St65$_1$, St59]. Below we describe some of what is known about the topological classification of these three types of closed, orientable, prime factors.

Space Form Factors   A closed, orientable, prime 3–manifold with finite fundamental group is the orbit space of a finite group action on a homotopy 3–sphere. So the classification of prime factors with finite fundamental group requires resolution of the Poincaré Conjecture and an analysis of all free actions of finite groups on S^3. C. B. Thomas [Th79] has completed the classification begun by J. W. Milnor [Mi57] of groups that can act freely on S^3; any such group must be isomorphic with a standard orthogonal group (i.e., a subgroup of $SO(4)$). It follows that the only possible exotic 3–manifolds with universal cover S^3 must arise from nonstandard actions of standard orthogonal groups on S^3 and that if any closed 3–manifold with finite non-orthogonal fundamental group exists, its universal cover is a counterexample to the 3–dimensional Poincaré Conjecture. (It is known [Th78] that all free actions on S^3 of certain orthogonal groups are conjugate to orthogonal actions.) The topological classification of lens spaces ($L_{p,q} \cong L_{p,s}$ if and only if $s \equiv \pm q^{\pm 1}(mod\ p)$) shows that the homotopy and topological classification differ for closed, orientable, prime 3–manifolds with finite fundamental group. For other evidence of this difference, see, for example, the work of Evans and Maxwell [EvMa77].

Aspherical Factors   By the Sphere Theorem ([He76, Theorem 4.3]), a closed, orientable, prime 3–manifold with infinite non-cyclic fundamental group (equivalently, closed, orientable, irreducible 3–manifold with infinite fundamental group) is aspherical, a $K(\pi, 1)$ manifold. If, in addition, it is sufficiently large in the sense of containing a 2–sided incompressible surface, it is called a *Haken* manifold. By the closed case of the work of Haken and Waldhausen (see, [Wa68]), a homotopy equivalence between two such closed Haken 3–manifolds is homotopic to a homeomorphism. Since homotopy equivalences of aspherical spaces correspond to isomorphisms of their fundamental groups, the closed Haken 3–manifolds are topologically classified by their fundamental group. All non-Haken closed, orientable, irreducible 3–manifolds with infinite fundamental group appear to be Seifert fiber spaces or admit a hyperbolic structure, but their analysis is far from complete. For an algorithm to decide whether a 3–manifold is a Haken manifold, see [JaOr84]. For closed 3–manifolds that do admit a complete hyperbolic structure, any homotopic equivalence is homotopic to an isometry by Mostow's Rigidity Theorem. So in fact the geometric structure of these manifolds is even determined by their fundamental group.

Remaining Factor   Finally, the only closed, orientable, prime 3–manifold with infinite cyclic fundamental group is the non-irreducible $S^2 \times S^1$.

Topological Invariant   V. G. Turaev [Tu88] has paired Swarup's homotopy type invariant $(\pi_1(M), \zeta_M \in H_3(\pi_1(M)))$ with a torsion element $\theta(M)$ in the ring of fractions of the rational group ring $\mathbf{Q}(H_1 M)$ to form a complete topological invariant for *geometric* 3–manifolds, by which he means the closed, oriented 3–manifolds each of whose prime factors is a hyperbolic, Seifert fibered, or Haken 3–manifold.

2   Singular 3–Manifolds

2.1   Presentation classes of singular 3-manifolds

In this section, we discuss (singular) 3–manifold thickenings of special polyhedra. For the terminology, see Chapter I, §3.1. While not every special
polyhedron embeds into some 3–manifold, it has been shown in Chapter I,
Theorem 3.2, that any one K^2 of them admits a canonical (up to homeomor-
phism) *singular* 3-manifold $\check{M}^3(K^2) \supset K^2$ such that $\check{M}^3 \searrow K^2$. This singular
3–manifold–thickening procedure induces a *bijection between 3–deformation
types of special polyhedra and equivalence classes of singular 3–manifolds* under certain elementary surgery operations defined below, see Theorem 2.2.

Definition (see Chapter I, §3.1): A *singular 3-manifold* $\check{M}^3$ is a compact
connected polyhedron in which the link of each point is either D^2 (*boundary
point*), S^2 (*inner point*) or the projective plane P^2 (*singular point*). The set
of boundary points is assumed to be nonempty.

Every singular 3–manifold $\check{M}^3$ has a *special spine* $K^2(\check{M}^3)$, i. e. a spine that
is a special polyhedron: In the absence of singular points, this is Theorem
3.1a of Chapter I; in the general case, one mimics the proof of Theorem 3.1a
considering additionally "pineapples" that are cones cP^2 on the projective
plane (Exercise). This special spine is not unique but any two choices K^2
and L^2 3-deform into each other via $\check{M}^3 : K^2 \nearrow \check{M}^3 \searrow L^2$.

It is clear that if $\check{M}^3$ arises as the thickening $\check{M}^3(K^2)$ of a special polyhedron
K^2, one choice of special spine is K^2 itself. Conversely, starting with a given
singular 3-manifold $\check{M}^3$, choose a spine $K^2(\check{M}^3)$ that contains all singular
points of $\check{M}^3$ in its intrinsic 2–skeleton and essentially[4] (Existence is left as
an exercise. Hint: find a collapse of cP^2 to a special polyhedron analogously
to Bing's house.) Then the thickening construction returns $\check{M}^3$.

Thus, thickening and collapsing as above are inverse operations for appropriate choices of $K^2(\check{M}^3)$.

We are now going to analyze how the thickening changes when applying 3-
deformations to a special polyhedron. These changes will define the desired
equivalence relation $\sim$ on singular 3–manifolds used in Theorem 2.2.

First, we introduce the elementary surgery operations $T_i^{\pm 1}$ ($i \in \{1,2,3,5\}$)
on special polyhedra (compare Chapter XI, §5.1). Each consists of a removal
of a certain local model from the special polyhedron K^2 (if possible) and

---

[4] *Essential* means that the link of the singular point with respect to the disc is a generator
of π_1 of the projective plane, which is the link with respect to $\check{M}^3$. We shall tacitly assume
the analogous situation henceforth.

replacement of it by another one that is glued in along the same graph, as indicated in Figure 11. Each substitution $T_i^{\pm 1}$ can be achieved by a 3-deformation of K^2 (Exercise, use Lemma 2.1 of Chapter I).

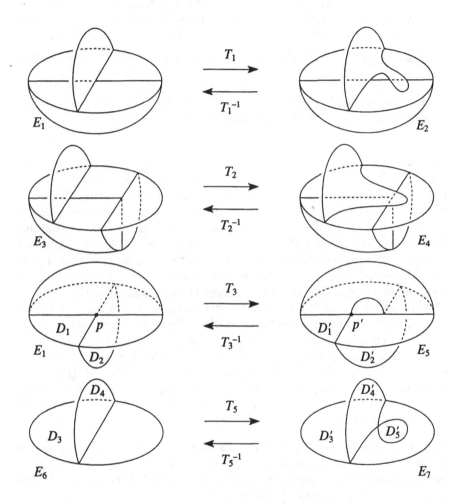

Figure VIII.11.

By [Ma87$_1$, Ma87$_2$, Pi88] (see Chapter XI, Theorem 9), we have

- A special polyhedron K^2 3-deforms to L^2 if and only if some sequence of operations T_1, T_2, T_3 and their inverses transforms K^2 to L^2.

Furthermore, by [Ma87$_2$] (see Chapter XI, Theorem 13),

- T_3^{-1} is dispensable provided L^2 cannot be embedded into a 3–manifold.

Or, T_3^{-1} is always dispensable if T_5^{-1} is added instead: If L'^2 is obtained from L^2 by an operation T_5, it cannot be embedded into a 3–manifold (see Chapter I, Remark 2 to Theorem 3.1). Then, by the above remarks, there is a sequence $K^2 \xrightarrow{T_1^{\pm 1}, T_2^{\pm 1}, T_3} L'^2$. Combining with $L'^2 \xrightarrow{T_5^{-1}} L^2$ we see that

- A special polyhedron K^2 3-deforms to L^2 if and only if some sequence of operations[5] $T_1^{\pm 1}, T_2^{\pm 1}, T_5^{\pm 1}$ and T_3 transforms K^2 to L^2.

Compare thickenings $\check{M}^3(K^2)$ and $\check{M}^3(L^2)$ before and after a T_1^{-1} operation. Note that both local models E_1 and E_2 are embeddable into $\mathbb{R}^3$. Choosing them small enough not to contain any singular point of the thickening shows that $\check{M}^3(K^2)$ and $\check{M}^3(L^2)$ are homeomorphic: The homeomorphism is a modification of the identity given by isotopically straightening the gluing band of the upper thickened disc. Similarly, the moves T_1, T_2, and T_2^{-1} don't change the (homeomorphism type of the) resulting singular 3–manifold. The situation is rather different for the remaining moves $T_i^{\pm 1}$ ($i = 3, 5$): Consider for example the discs D_1 and D_1' in Figure 11 before and after the move T_3. The thickening of the boundary curve ∂D_1 changes orientation (solid torus resp. Klein bottle) because the local model at p forces D_2 to run "downwards" before the move and "upwards" after the move. Hence, either D_1 or D_1' will contain a singular point of the thickening. In general, the distribution of singular points on $K^2 \subset \check{M}^3(K^2)$ determines the one on $L^2 \subset \check{M}^3(L^2)$. If we know which of the (six) discs of K^2 involved in T_3 carry singular points we can decide the locus of singular points after the move, but this would require considering 2^6 cases.

To bypass this multiplicity, we introduce an elementary surgery operation on singular 3–manifolds that allows the addition or deletion of two singular points on discs of K^2 minus a regular neighbourhood of its intrinsic 1– and 0–skeleton (see Quinn [Qu81], Metzler [Me85]):

Definition((0,2)-move)[6]: If $\check{M}^3$ is locally a thickening of a 2–disc D^2 without singular points, then replace this piece by a thickening of a 2–disc with two singular points in its interior.

---

[5]In fact, T_5 could be avoided but we will see below that doing so (other than T_3^{-1}!) doesn't give any advantage.

[6]Following Quinn, we denote the moves on singular 3–manifolds by the number of singular points which are involved in the local situation before and after. For example, the inverse $(0,2)^{-1}$ of a $(0,2)$–move is also written $(2,0)$.

In other words, take out a piece of the form $D^2 \times I$ that is glued into $\check{M}^3$ along $\partial D^2 \times I$ and substitute it by a piece $(cP^2 \#_\partial cP^2)$ (The symbol $\#_\partial$ denotes the boundary connected sum of singular 3–manifolds, i. e. $\check{M}_1^3 \#_\partial \check{M}_2^3 = \check{M}_1^3 \cup_{D^2} \check{M}_2^3$ where $\partial \check{M}_1^3 \supset D^2 \subset \partial \check{M}_2^3$). The latter is glued into $\check{M}^3$ along $S^1 \times I$, where S^1 is an orienting curve cutting both crosscaps of $\partial(cP^2 \#_\partial cP^2)$ exactly once. This process is described by the formula $(\check{N} \cup_{S^1 \times I} D^2 \times I) \sim (\check{N} \cup_{S^1 \times I} (cP^2 \#_\partial cP^2))$. Note the analogy to the definition of the operations $T_i^{\pm 1}$ on special polyhedra given before Fig. 11: Here, a certain local model is removed from the singular 3–manifold and replaced by another one that is glued in along the same 2–manifold.

Lemma 2.1 *The $(0,2)$–move (and its inverse) can be realized by a 3–deformation of singular 3–manifolds.*

Proof: The collapse

$$(cP^2, \text{Moebius strip}) \searrow \text{Moebius strip union a disc glued on its waist}$$

(described in the proof of Theorem 3.2 of Chapter I) applied twice yields

$(cP^2 \#_\partial cP^2, S^1 \times I) \searrow$ Two Moebius strips union discs on their waists, glued together along a band $\searrow$ (Twisted) annulus with a disc on its middle line,

which in turn expands to $(D^2 \times I, S^1 \times I)$.   □

Definition ((4,3)–move): If $\check{M}^3$ is locally a thickening of E_1 with singular points[7] as in Figure 12a, then replace it by a thickening of E_5 with singular points as in Figure 12b.

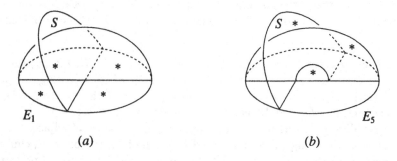

(a)   (b)

Figure VIII.12.

[7]The intersection of the "sheet" S with the $T \times I$–part is contained in the boundary of S. In Figure 12a, all singular points are in the horizontal part of $T \times I$, in Figure 12b none.

In other words, take a piece of the form $(cP^2\#_\partial cP^2\#_\partial cP^2\#_\partial cP^n)$ out of $\check{M}^3$ leaving some, say $\check{N}^3$, and substitute it by a piece $(cP^2\#_\partial cP^2\#_\partial cP^2)$. These pieces are glued to $\check{N}^3$ along a regular neighbourhood on $\partial\check{N}^3$ of the boundary graph of E_1 resp. E_5 (the complete graph on 4 points) which is an annulus with two bands attached to it (see again Fig. 12) resulting in a torus with two holes.

Let $\check{M}_1^3$ result from $\check{M}^3$ by an application of the $(4,3)$–move. Then

(Exercise) $\check{M}^3 \searrow \check{N}^3 \cup E_1 \nearrow^3 \check{N}^3 \cup E_5 \nearrow \check{M}_1^3$ *rel* $\check{N}^3$

where $E_1 \cap \check{N}^3 = \{$free faces of $E_1\} \subset \partial\check{N}^3$ has a torus with two holes as a regular neighbourhood on $\partial\check{N}^3$ and the same property holds for E_5 instead of E_1.

Investigating the definitions of the elementary surgery moves on singular 3–manifolds, we can state

Theorem 2.2 *There is a bijection between 3–deformation types of special polyhedra and equivalence classes of singular 3–manifolds under the elementary surgery moves $(0,2)$, $(2,0)$ and $(4,3)$.*   □

(compare [Qu81], also compare Chapter I, Theorem 2.4).

Proof: It is now straightforward that thickenings of special polyhedra K^2 and L^2, where L^2 is the result of an application of T_3 to K^2, are related by a sequence of these moves: By $(0,2)$–moves, create singular points on discs which contain horizontal discs of $E_1 \subset K^2$, inside $\check{M}^3(K^2)$. We may assume that E_1 captures exactly the singular points prescribed by Figure 12; then perform the corresponding $(4,3)$–move on $\check{M}^3(K^2)$. Eliminating pairs of singular points on discs of L^2 - where possible - by $(2,0)$–moves leads to $\check{M}^3(L^2)$.

Turn your attention to a situation $K^2 \xrightarrow{T_5} L^2$. Again by Chapter I, Remark 2 to Theorem 3.1, in Figure 11 some inner point of D_5' is a singular point of $\check{M}^3(L^2)$. Up to moves $(0,2)$ and choice of size of the local models, we can assume that D_3' and D_5' carry one singular point of $\check{M}_2^3 \sim \check{M}^3(L^2)$ each, the other discs of E_7 none. Comparing how permutations of adjacent sheets are transported along edges as in the discussion of T_3, the piece E_6 could be assumed to carry no singular point, but it is convenient here to perform a $(0,2)$-move in order to have two of them on D_3 inside $\check{M}_1 \sim \check{M}^3(K^2)$.

Recall that a small disc around a singular point thickens to a cP^2 that intersects the surrounding singular 3–manifold in a Moebius strip. The boundary line of this Moebius strip can thus be found on $\partial \breve{M}_1^3$, (see Figure 13a) and bounds the disc $D^2 = \overline{P^2 - Moebiusstrip}$ there. Isotoping the gluing band of the thickened disc D_4 from Figure 11 across D^2, we obtain exactly $\breve{M}_2^3$ (see Figure 13b): $\breve{M}_1^3$ and $\breve{M}_2^3$ are in fact homeomorphic. Hence, $\breve{M}^3(K^2)$ and $\breve{M}^3(L^2)$ are related by a sequence of $(0,2)^{\pm 1}$–moves.

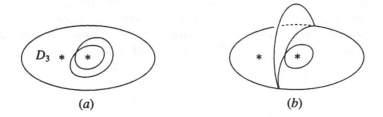

(a) (b)

Figure VIII.13.

Summarizing, we conclude for special polyhedra K^2 and L^2

> If K^2 transforms to L^2 by a sequence of moves $T_1^{\pm 1}, T_2^{\pm 1}, T_5^{\pm 1}$ and T_3, then $\breve{M}^3(K^2)$ transforms into $\breve{M}^3(L^2)$ by a sequence of moves $(0,2)$, $(2,0)$ and $(4,3)$.

In other words, the map $K^2 \to \breve{M}^3(K^2)$ is well defined on equivalence classes. Because thickening and collapsing are themselves 3–deformations, Lemma 2.1 and the Exercise after the definition of the $(4,3)$–move imply that the inverse map $\breve{M}^3(K^2) \to K^2$ is well defined on equivalence classes either. This completes the proof of Theorem 2.2.   □

2.2   Examples and discussion

Quinn (see [Qu81], [Ho83]) gives a definition of the singular 3–manifold–moves in §2.1 in terms of regular neighbourhoods of embedded 1– and 2–dimensional manifolds as follows:

> $(0,2)$–move: Replace the regular neighbourhood of a properly embedded arc with the regular neighbourhood of a properly embedded disc containing two singular points

and the $(4,3)$–move as

(4,3)–move: Replace the regular neighbourhood of a properly embedded disc containing four singular points with the regular neighbourhood of a properly embedded disc containing three singular points together with an embedded arc from the disc to the boundary.

Here, and for the remainder of this section, all embeddings are assumed to be *in general position* to singular points, i. e. discs are allowed to contain singular points only essentially (compare footnote 4) and arcs don't meet them at all. We refer to [Ho83] for a proof that regular neighbourhoods of properly embedded discs D^2 in singular 3–manifolds are homeomorphic to $cP^2 \#_\partial \ldots \#_\partial cP^2$ (as many summands as singular points hit) where the regular neighbourhood of ∂D^2 is an orienting curve on $\partial(cP^2 \#_\partial \ldots \#_\partial cP^2)$ as well as for the equivalence of the above definitions to those in §2.1.

The $(cP^2 \#_\partial \ldots \#_\partial cP^2)$ constitute the simplest examples for singular 3–manifolds. They are all collapsible and thus by Theorem 2.2 equivalent to the 3–ball. Conversely, all *collapsible* singular 3–manifolds are of this form. The proof of this generalization of Chapter I, (46), is left as an exercise.

In order to visualize some noncollapsible example, remove a regular neighbourhood of a knotted arc from a ball (obtaining a knot complement) and plug the top end of the hole with a slightly thickened meridian disc (compare Figure 253 in [St80]), first without singular points, then with two singular points on it. In the latter case you have obtained an $\check{M}_0^3$ which by construction is the result of one (0,2)–move on (the former case) B^3. We claim that $\check{M}_0^3$ is not homeomorphic to $(cP^2 \#_\partial cP^2)$. To see this, remove from $\check{M}_0^3$ an open regular neighbourhood around the singular points thus obtaining an ordinary manifold $\overset{\circ}{M}_0^3 \subset \check{M}_0^3$, called $\check{M}_0^3$-*bored*. A homeomorphism between singular 3–manifolds induces one between the bored manifolds. But $(cP^2 \#_\partial cP^2)$-bored has fundamental group $\mathbb{Z}_2 * \mathbb{Z}_2$, whereas by the Seifert-van Kampen Theorem, $\pi_1(\overset{\circ}{M}_0^3)$ is the free product of $\mathbb{Z}_2 * \mathbb{Z}_2$ and a knot group, amalgamated over Z. In particular, the knot group injects into $\pi_1(\overset{\circ}{M}_0^3)$, whereas Kurosh's Theorem shows that this is not the case for $\mathbb{Z}_2 * \mathbb{Z}_2$.

For a further discussion of singular 3–manifolds see Chapter XII, §2.3.

Chapter IX

Cancellation Results for 2-Complexes and 4-Manifolds and Some Applications

Ian Hambleton and Matthias Kreck

This is a survey chapter. The idea is to summarize some recent work which illustrates in one way or another the connection between topology in dimension 2 and the study of 4-dimensional manifolds. There are almost no new results and no result is proved completely in the paper. Instead, in each section we collect together some related statements and motivation, and give a sketch of some typical or important steps in the proofs.

1 A Cancellation Theorem for 2-Complexes

Any two finite 2-complexes K, K' with isomorphic fundamental groups become *simple* homotopy equivalent after wedging with a sufficiently large (finite) number of S^2's (see chapter I, (40)). Furthermore, if $\alpha : \pi_1(K, x_0) \to \pi_1(K', x'_0)$ is a given isomorphism and K, K' have the same Euler characteristic, then there is a simple-homotopy equivalence $f : K \vee rS^2 \to K' \vee rS^2$ inducing α on the fundamental groups. For a given group π, the minimal number r with the property above for all finite 2-complexes with this fundamental group is called the *stable range*.

It is known that for finite fundamental groups the stable range is always ≤ 2 ([Dy81], Theorem 3). The main result of this section is the following.

Theorem 1.1 ([HaKr92$_1$]) *Let K and K' be finite 2-complexes with the same Euler characteristic and finite fundamental group. Let $\alpha : \pi_1(K, x_0) \to \pi_1(K', x_0')$ be a given isomorphism and suppose that $K \simeq K_0 \vee S^2$. Then there is a simple-homotopy equivalence $f : K \to K'$ inducing α on the fundamental groups.*

The analogous result for "homotopy type" instead of "simple-homotopy type" was proved by W. Browning ([Br78], 5.4; see also Chapter III, §2).

This is the best possible result in general ([Me76]; see also Chapter III, §2 and this chapter, §4); but for special fundamental groups like cyclic groups [Me76], [DySi73] or more generally finite subgroups of $SO(3)$ ([HaKr92$_1$]; see also [La91] for the groups $D(4n)$) it can sometimes be improved (see Theorem 1.3).

Proof: Let $h : K \vee rS^2 \to K' \vee rS^2$ be a simple-homotopy equivalence as above, inducing a given isomorphism α on the fundamental groups. We will prove the theorem inductively and thus we may assume that $r = 1$. Our strategy is to construct a simple self-equivalence of K such that, after composing with this, we obtain $h' : K \vee S^2 \to K' \vee S^2$ which fixes the element p_1 of π_2 represented by the S^2 factor. Then the composition of h' with the inclusion and projection gives a homotopy equivalence $f : K \to K'$, which by the additivity formula for the Whitehead torsion is simple.

To construct such a simple self-equivalence of K, one naturally first considers the corresponding algebraic problem of constructing an automorphism of π_2 preserving π_1 and then realizing it by a simple self-equivalence.

We fix some notation. Let $A = \mathbb{Z}[\pi_1(K)]$, $L = \pi_2(K_0)$ and let $P = P_0 \oplus P_1$ be the A-submodule of $\pi_2(K_0 \vee S^2 \vee S^2)$ generated by $\pi_2(S^2 \vee S^2)$. We note that the A-module L has $(A, \mathbb{Z})$–free rank ≥ 1 at all primes p not dividing the order of $\pi_1(K)$. This notion was introduced in [HaKr92$_1$] and means that there exists an integer r such that $(\mathbb{Z}^r \oplus L)_p$ has free rank ≥ 1 over A_p, where we consider $\mathbb{Z}$ as A-module via the augmentation map. In this case, the reason that $\pi_2(K_0)$ has $(A, \mathbb{Z})$-free rank ≥ 1 is that L fits into an exact sequence

$$(1) \qquad 0 \to L \to C_2 \to C_1 \to C_0 \to \mathbb{Z} \to 0$$

with the modules $C_i = C_i(\tilde{K})$ finitely generated free A-modules.

More generally, any lattice L with a resolution (1) by finitely generated projective A-modules C_i is unique up to direct sum with projectives. The stable class is denoted $\Omega^3\mathbb{Z}$. Such lattices with minimal Z-rank need not contain any projective direct summands over $A = Z\pi$, but rationally contain all the representations of π except perhaps the trivial one. Then L has $(A, \mathbb{Z})$-free rank ≥ 1 at all primes not dividing the order of π.

We need the following notation. If $M = M_1 \oplus M_2$ is a direct sum splitting of an A-module, then $E(M_1, M_2)$ denotes the subgroup of $GL(M)$ generated by the elementary automorphisms ([Ba68], p.182). This is the group generated by automorphisms of the form $1 + f$ and $1 + g$, where $f: M_1 \to M_2$ and $g: M_2 \to M_1$ are arbitrary A-homomorphisms. An element of an A-module is called unimodular if there is a homomorphism to A mapping it to 1. The main algebraic ingredient of our proof is the following result whose proof we will sketch at the end of this section.

Theorem 1.2 [HaKr92$_1$], Corollary 1.12 and Lemma 1.16) *Let $M = P \oplus L$ be an A-lattice, where $P = p_0 A \oplus p_1 A = P_0 \oplus P_1$ and L has $(A, \mathbb{Z})$-free rank ≥ 1. Then the group $G = \langle E(P_0, L \oplus P_1), E(P_1, L \oplus P_0) \rangle$ acts transitively on unimodular elements in $L \oplus P$.*

To finish our proof, we have to realize elements in G by simple-homotopy self equivalences of $K_0 \vee 2S^2 = K \vee S^2$ inducing the identity on π_1. It is enough to do this for $E(P_1, L \oplus P_0)$. This group is generated by automorphisms of the form $1 + f$ and $1 + g$, where $f: L \oplus P_0 \to P_1$ and $g: P_1 \to L \oplus P_0$ are arbitrary A-homomorphisms. Recall that $P_1 = p_1 A$ and $L \oplus P_0 = \pi_2(K)$. Consider the map $Id \vee u: K \vee S^2 \to K \vee S^2$, where $u = (g(p_1), p_1) \in \pi_2(K \vee S^2) = \pi_2(K) \oplus p_1 A$. It realizes $1 + g$ and its restriction to K is the identity and it also induces the identity on $(K \vee S^2)/K = S^2$. Thus the additivity formula for the Whitehead torsion implies that the torsion of $Id \vee u$ vanishes.

To realize $1 + f$, we note that $f : L \oplus P_1 = \pi_2(K) = H_2(K; A) \to P_1 = A$ factors through $H_2(K, K^1; A)$, with K^1 the 1-skeleton. The reason for this is that we have an exact sequence

$$\text{Hom}_A(H_2(K, K^1; A), A) \to \text{Hom}_A(H_2(K; A), A) \to \text{Ext}^1_A(H_1(K^1; A), A)$$

and the last group vanishes since $H_1(K^1; A)$ is $\mathbb{Z}$-torsion free. Choose a factorization map $\bar{f}: H_2(K, K^1; A) \to A$, where $H_2(K, K^1; A)$ is a free A-module generated by the 2-cells of K (appropriately connected to the base point). Denote this basis by $e_1, .., e_k$. Now write $K = K^1 \cup D^2 \cup ... \cup D^2$. Pinch off the 2-cells to obtain $K \vee rS^2$ and denote the projection map by $p: K \to K \vee kS^2$. Consider the composition map $\beta: K \to K \vee kS^2 \to K \vee S^2$, where the second map is $Id \vee \bar{f}(e_1) \vee ... \vee \bar{f}(e_k)$. By construction the induced map on π_2 is $1 \oplus f$ and the composition $K \to K \vee S^2 \to K$ is homotopic to Id. Finally, consider $\beta \vee Id: K \vee S^2 \to K \vee S^2$ realizing $1 + f$. Its restriction to

S^2 and the induced map on K are homotopic to the identity, implying from the additivity of the Whitehead torsion that $\beta \vee Id$ has trivial torsion. □

Without proof, we state the full classification result for 2-complexes with fundamental group a finite subgroup of $SO(3)$. Recall that the finite subgroups G of $SO(3)$ are cyclic, dihedral, A_4, S_4, and A_5.

Theorem 1.3 *Let π be a finite subgroup of $SO(3)$. If K and K' are finite 2-complexes with fundamental group π and the same Euler characteristic, and if $\alpha : \pi_1(K, x_0) \to \pi_1(K', x_0')$ is a given isomorphism, then there is a simple-homotopy equivalence $f : K \to K'$ inducing α on the fundamental groups.*

The proof runs along the same lines as above but needs several additional steps. For π cyclic or $\pi = \mathbb{Z}/2 \times \mathbb{Z}/2$, this was proved in [Me76], [DySi73]. The result for $\pi = D(4n)$, the dihedral group of order $4n$, has recently been obtained by P. Latiolais [La91]. Our methods give a new proof in these cases.

Now, we give a sketch of the proof of Theorem 1.2.

Proof of Theorem 1.2: Recall that a *lattice* is a finitely generated right A-module that is torsion free over $\mathbb{Z}$. Our proof is based on an improvement of the Bass transitivity theorem ([Ba68], pp.178-184), which assumed that M has free rank $\geq d + 2$ where d is the Krull dimension of the ring A. In our special case of lattices over group rings of finite groups (Krull dimension $= 1$), we are able to obtain a transitivity theorem assuming only free rank ≥ 2 when the lattice M contains a summand L which has $(A, \mathbb{Z})$-free rank ≥ 1. Thus, the improvement here is that a particular type of non-free modules which occurs geometrically as $\pi_2(K)$ can play the role of a free module in producing algebraic transitivity.

We denote the augmentation map by $\epsilon : A \to B = \mathbb{Z}$. This is a surjective ring homomorphism. If M is an A-lattice we get an induced homomorphism

$$\epsilon_* : M \to M \otimes_A \mathbb{Z}.$$

Recall that for an element $x \in M$, $O_M(x)$ is the left ideal in A generated by

$$\{f(x) \mid f \in \mathrm{Hom}_A(M, A)\}.$$

If $O_M(x) = A$, we say that x is unimodular. We need two easy facts whose proofs are omitted.

Proposition 1.4 *Let M be an A-lattice and $A' = A/A\mathfrak{t}$ for an ideal $\mathfrak{t} \in \mathbb{Z}$ such that the localized order $A_\mathfrak{t}$ is maximal. Then the induced map*

$$\mathrm{Hom}_A(M, A) \to \mathrm{Hom}_{A'}(M', A')$$

is surjective, where $M' = M/M\mathfrak{t}$.

Proposition 1.5 ([Ba73], (2.5.2), p.225) *If C is a semisimple algebra, then for each a, $b \in C$ there exists $r \in C$ such that $C(a + rb) = Ca + Cb$.*

Now let $x = p_0 a + p_1 b + v \in M$ be a unimodular element, with $p = p_0 a + p_1 b \in P$ and $v \in L$, so that $O(x) = Aa + Ab + O(v)$. Since the elementary matrices $E_n(\mathbb{Z})$ act transitively on unimodular elements in $\mathbb{Z}^n$ for $n \geq 2$, we may assume that $\epsilon_*(x) = \epsilon_*(p_0)$. In the proof, we use the stability assumption on L to move x so that its component in $p_0 A \oplus L$ is unimodular. Then we move x to p_0 to prove the statement about unimodular elements in M. At each step, we use only elements σ of G fixing $\epsilon_*(p_0)$.

Lemma 1.6 *Let S be a finite set of (non-zero) primes in $\mathbb{Z}$, and $\bar{A} = A/\mathfrak{g}A$ where $\mathfrak{g}$ is the product of all the primes $\mathfrak{p} \in S$. Then after applying an element $\tau \in E(P_1 \oplus L, P_0)$ to x, $O(\bar{x}) = \bar{A}\bar{a} = \bar{A}$ and $\epsilon_*(x) = \epsilon_*(p_0)$.*

Proof: The semi-simple quotient ring $\bar{A}/\text{Rad}\,\bar{A} = \bar{C} \times \bar{C}'$, where $\bar{C} = \bar{B}/\text{Rad}\bar{B}$ and C' is a complementary direct factor. Here "Rad" denotes the Jacobson radical [CuRe62]. Since $\epsilon_*(x) = \epsilon_*(p_0)$, a projects to 1 in the $\bar{C}$ component of the semisimple quotient. Since $Aa + O(p_1 b + v) = A$, there exists $c \in O(p_1 b + v)$ such that $Aa + c$ contains 1, and c projects to zero in $\bar{B}$. By Proposition 1.5, there exists $z \in A$ with $A(a + zc) = A(\text{mod } g)$ and a map $g: P_1 \oplus L \to p_0 A \subseteq M$ with $g(p_1 b + v) = p_0 zc$. Extend g to a map from M to M by zero on the complement. Then $\tau = 1 + g$ is an element of $E(P_1 \oplus L, P_0)$ and $\tau(x)$ has the desired properties. $\square$

We apply Lemma 1.6 to the set S of primes $\mathfrak{p} \in \mathbb{Z}$ at which A is not maximal, or L does not have (A, B)-free rank ≥ 1.

Lemma 1.7 *If $x = p_0 a + p_1 b + v \in M$ is a unimodular element for which $Aa + \mathfrak{g}A = A$, then after applying an element $\tau \in E(P_1, L)$ we have $x = p_0 a + p_1 b + v$ with $p_0 a + v$ unimodular and $\epsilon_*(x) = \epsilon_*(p_0)$.*

Proof: Let $\mathfrak{t} \subseteq \mathbb{Z}$ denote the ideal which is maximal among those such that $A\mathfrak{t} \subseteq Aa$. It is not hard to see that $\mathfrak{g}$ is relatively prime to $\mathfrak{t}$, and so $A_{\mathfrak{t}}$ is a maximal order.

Now we project to the semilocal ring $A' = A/A\mathfrak{t}$. This is a finite quotient ring of the maximal order $A_{\mathfrak{t}}$, and so the projection $\epsilon': A' \to B'$ splits and $A' = B' \times C'$. Since over the B' factor a projects to 1, we have $(Aa)' = A'$. Over the complementary factor C' we use a suitable $\tau \in E(p_1' C', L')$, so that after applying τ we achieve the condition

$$A'a' + O(v') = A'$$

over both factors of A'. This is an application of Proposition 1.5 to the component of x in $L' \oplus p_1' C'$ using the fact that $C' \subseteq L'$. The necessary homomorphism $g \in \mathrm{Hom}_{A'}(P_1', L')$, which is the identity over B', can be lifted to $\mathrm{Hom}_A(P_1, L)$ since P_1 is projective and extended to M by zero on $p_0 A \oplus L$.

We now lift the relation above to A using Proposition 1.4 and obtain

$$Aa + O(v) + At = A.$$

But $At \subseteq Aa$ so $v + p_0 a$ is unimodular. $\square$

We now complete the proof of Theorem 1.2 by the following:

Lemma 1.8 Let $x = p_0 a + p_1 b + v$ and $\epsilon_*(x) = p_0$. Suppose that $z = p_0 a + v$ is unimodular, and write $L \oplus P_0 = zA \oplus L_0$. Then there exist elementary automorphisms $\tau_1 \in E(zA, P_1)$, $\tau_2 \in E(P_1, P_0)$, $\tau_3 \in E(P_0, P_1)$ and $\tau_4 \in E(P_0, L)$ such that $\tau_4 \tau_2^{-1} \tau_3 \tau_2 \tau_1(x) = p_0$ and the product fixes $\epsilon_*(p_0)$.

Proof: This is the argument of [Ba68, pp. 183-184]. Let $g_1(z) = p_1(1 - a - b)$, with $g_1(L_0) = 0$. Define $g_2(p_1) = p_0$, $g_3(p_0) = p_1(a - 1)$, and $g_4(p_0) = -v$, where the homomorphisms are extended to the obvious complements by zero. If $\tau_i = 1 + g_i$, then

$$\tau_4 \tau_2^{-1} \tau_3 \tau_2 \tau_1 (x) = p_0.$$

The product fixes $\epsilon_*(p_0)$ and lies in $E(P_1, P_0 \oplus L)$. $\square$

This finishes the proof of Theorem 1.2.

2   Stable Classification of 4-Manifolds

There is a close analogy between the stable classification of homotopy types of 2-complexes and homeomorphism types of 4-manifolds. To indicate this analogy, consider the thickening functor from finite 2-complexes to closed 4-manifolds obtained by embedding a 2-complex X as polyhedron in $\mathbb{R}^5$ and taking the boundary of a smooth regular neighborhood (compare Chapter I, §3). If two 2-complexes are simple-homotopy equivalent, the corresponding 4-manifolds are s-cobordant (implying homeomorphic, if the fundamental groups are poly-(finite or cyclic) [Fr84]) and we denote the corresponding s-cobordism class by $M(X)$. If we replace the 2-complex by its 1-point unification with S^2, the corresponding 4-manifold changes by connected sum with $S^2 \times S^2$. This indicates the analogy of stable equivalence classes of 2-complexes with the following notation for 4-manifolds.

Definition 2.1 *Two smooth (topological) closed 4-manifolds M_0 and M_1 are stably diffeomorphic (homeomorphic) if the connected sums $M_0 \# r(S^2 \times S^2)$ and $M_1 \# r(S^2 \times S^2)$ are diffeomorphic (homeomorphic) for some integer r.*

Since the smooth stable s-cobordism theorem (implying that two s-cobordant 4-manifolds are stably diffeomorphic) holds [Qu83], the stable diffeomorphism class of $M(X)$ is determined by the stable simple-homotopy class of X and so (see §1) by $\pi_1(X)$.

Compared to the 2-complexes, it is not true that for 4-manifolds the stable classification needs only the fundamental group and the Euler characteristic as invariants. At least one has to control basic properties like orientability and existence of a spin-structure and in addition for oriented manifolds the signature.

The following definition turns out to be very useful for coding the fundamental group together with orientability and spin-structure information. Let M be a topological 4-manifold. Abbreviate $\pi_1(M) = \pi$. Let $u : M \to K(\pi, 1)$ be a classifying map of the universal covering $\tilde{M}$. Then we have an isomorphism $u^* : H^1(\pi; \mathbb{Z}/2) \to H^1(M; \mathbb{Z}/2)$ and an exact sequence $0 \to H^2(\pi; \mathbb{Z}/2) \to H^2(M; \mathbb{Z}/2) \to H^2(\tilde{M}; \mathbb{Z}/2)$ [Br82]. Thus we can pull back $w_1(M)$ by u from a class denoted $w_1 \in H^1(\pi; \mathbb{Z}/2)$ and, if $w_2(\tilde{M}) = 0$, $w_2(M)$ from a class denoted $w_2 \in H^2(\pi; \mathbb{Z}/2)$. If $w_2(\tilde{M}) \neq 0$, we define $w_2 = \infty$. There is an obvious notion of isomorphism classes of the triple (π, w_1, w_2) and we denote the isomorphism class by $[\pi, w_1, w_2]$.

Definition 2.2 *For a topological 4-manifold M, we call the isomorphism class $[\pi, w_1, w_2]$ the algebraic normal 1-type.*

The algebraic normal 1-type determines the geometric normal 1-type , called the normal 1-type , as follows. We begin with the smooth case. Let M be a smooth manifold. If $w_2 = \infty$ (corresponding to $w_2(\tilde{M}) \neq 0$), then we define the normal 1-type as follows. Consider the real line bundle $E \to K(\pi, 1)$ with $w_1(E) = w_1$ and the composition

$$K(\pi, 1) \times BSO \xrightarrow{E \times p} BO \times BO \longrightarrow \oplus BO,$$

where $E : K(\pi, 1) \to BO$ is the classifying map of the stable bundle given by E and $\oplus$ is the H-space structure on BO given by the Whitney sum. We denote the corresponding fibration by $B[\pi, w_1, \infty]$. If $w_2 \neq \infty$, we define the normal 1-type as the fibration $p : B(\pi, w_1, w_2) \longrightarrow BO$ given by the following pullback square

$$B(\pi, w_1, w_2) \qquad\longrightarrow\qquad K(\pi, 1)$$

$$p \downarrow \qquad\qquad\qquad\qquad \downarrow w_1 \times w_2$$

$$BO \xrightarrow{\ w_1(EO) \times w_2(EO)\ } K(\mathbb{Z}/2, 1) \times K(\mathbb{Z}/2, 2).$$

where $w_i(EO)$ are the Stiefel-Whitney classes of the universal bundle and we interpret w_i as maps to $K(\mathbb{Z}/2, i)$. The fibre homotopy type of

$$p : B(\pi, w_1, w_2) \longrightarrow BO$$

is determined by the isomorphism class of (π, w_1, w_2) and is denoted by $B[\pi, w_1, w_2]$, the normal 1-type.

If $w_1 = 0$, $B[\pi, 0, w_2]$ factorizes over BSO and we choose one of the possible lifts. This way we consider $B[\pi, 0, w_2]$ as fibrations over BSO. To deal the oriented case ($w_1 = 0$) and the non-oriented case simultaneously we write $p : B(\pi, w_1, w_2) \longrightarrow B(S)O$.

For topological manifolds, one can make the obvious changes (replace the linear normal bundle by the topological normal bundle given by a map $\nu : M \to B(S)Top$) to obtain from the algebraic normal 1-type the normal 1-type $p : B(\pi, w_1, w_2) \longrightarrow B(S)Top$.

The following theorem plays a central role in the stable classification of 4-manifolds. Given a fibration $B \to B(S)O$, abbreviated for short as B, we consider the B-bordism group $\Omega_n(B)$ consisting of bordism classes of closed smooth n-manifolds, which are oriented, if the fibration is over BSO, together with a lift $\bar\nu$ over B of the normal Gauss-map $\nu : M \to B(S)O$ [St68]. Such a lift is called a *normal 1-smoothing* if $\bar\nu$ is a 2-equivalence. It is easy to check that, if the algebraic normal 1-type of M is $[\pi, w_1, w_2]$, by construction of $B[\pi, 0, w_2]$, M admits a normal 1-smoothing in $B[\pi, w_1, w_2]$. Similarly, for topological manifolds, one starts with a fibration $B \to B(S)Top$, abbreviated for short as B^{Top}, and introduces the analogous bordism group of topological manifolds denoted $\Omega_n(B^{Top})$.

Theorem 2.3 ([Kr85]) *Two smooth (topological) 4-manifolds M_0 and M_1 with the same algebraic normal 1-type $[\pi, w_1, w_2]$ are stably diffeomorphic (homeomorphic), if and only if they have the same Euler characteristic and if they admit normal 1-smoothings $\bar\nu_0$ and $\bar\nu_1$ respectively such that $(M_0, \bar\nu_0)$ and $(M_1, \bar\nu_1)$ represent the same bordism class in $\Omega_4(B[\pi, w_1, w_2])$ (in $\Omega_4(B^{Top}[\pi, w_1, w_2])$).*

If one wants to apply this theorem, one has to compute the bordism group $\Omega_4(B[\pi, w_1, w_2])$ or $\Omega_4(B^{Top}[\pi, w_1, w_2])$. In general, this is not easy; but, under some assumptions, it follows from the Atiyah-Hirzebruch bordism spectral sequence ([CoF164]). For example, if $w_2 = \infty$ (i.e., $w_2(\tilde{M}) \neq 0$) and $w_1 = 0$, then $\Omega_4(B[\pi, w_1, w_2]) = \Omega_4(K(\pi, 1))$ and $\Omega_4(B^{Top}[\pi, w_1, w_2]) = \Omega_4^{(Top)}(K(\pi, 1))$, where the right side is the oriented smooth (topological) singular bordism group in $K(\pi, 1)$. In this situation the choice of a normal 1-smoothing is equivalent to the choice of a map $u : M \to K(\pi, 1)$ inducing an isomorphism on π or equivalently a representative of a classifying map of the universal covering. The different choices are obtained by composing with an automorphism of π acting as self equivalences of $K(\pi, 1)$. Now, $\Omega_i = 0$ for $1 \leq i \leq 3$ and $\Omega_0 \cong \Omega_4 \cong \mathbb{Z}$, where in the last case the isomorphism is given by the signature (cf [MiSt74]). In the topological case, one has an additional term $\mathbb{Z}/2$ detected by the Kirby-Siebenmann obstruction KS [KiSi77]. Thus, from the Atiyah-Hirzebruch spectral sequence, one has $\Omega_4(K(\pi, 1)) \cong \mathbb{Z} \oplus H_4(K(\pi, 1); \mathbb{Z})$ in the smooth case, and $\Omega_4^{(Top)}(K(\pi, 1)) \cong \mathbb{Z} \oplus H_4(K(\pi, 1); \mathbb{Z}) \oplus \mathbb{Z}/2$ in the topological case. The isomorphism is given by the signature of M, the image of the fundamental class $u_*([M])$ in $H_4(K(\pi, 1); \mathbb{Z})$, and, in the topological case, in addition the KS-invariant. This proves the first part of the following theorem.

Theorem 2.4 *Two oriented smooth (topological) 4-manifolds M_0 and M_1 with the same fundamental group and with $w_2(\tilde{M}_i) \neq 0$ are stably diffeomorphic (homeomorphic), if and only if they have the same Euler characteristic and signature, if $u_*(M_0) = u_*(M_1) \in H_4(K(\pi, 1); \mathbb{Z})/Out(\pi)$ and, in the topological case, $KS(M_0) = KS(M_1)$.*

Arbitrary values of the signature and the class in $H_4(K(\pi, 1); \mathbb{Z})/Out(\pi)$ and, in the topological case, of $KS \in \mathbb{Z}/2$ can be realized.

Proof: We are left with the realization statement. This follows since by surgery any element in the corresponding bordism group can be realized by a manifold, such that u induces an isomorphism of π and with $w_2(\tilde{M}) \neq 0$. $\square$

One can use the same surgery method to say much more about the stable classification of 4-manifolds. For instance, if the manifolds M_i are equipped with spin-structures, they are stably diffeomorphic (homeomorphic) if and only if they have the same Euler characteristic and (M_0, u_0) and (M_1, u_1) represent the same element in the singular smooth (topological) bordism group of spin-manifolds together with maps to $K(\pi, 1)$. But this bordism group is much harder to compute and a general answer is not known. In the next theorems, we list some results for manifolds with special fundamental groups which can easily be obtained along these lines of arguments.

Theorem 2.5 *Let M_0 and M_1 be smooth (topological), oriented 4-manifolds with $w_2(\tilde{M}_i) = 0$ and $\pi_1(M_i) = \pi$. If $H_i(\pi, \mathbb{Z}/2) = 0$ for $1 \leq i \leq 3$, then M_0 and M_1 are stably diffeomorphic (homeomorphic) if and only if they have the same Euler characteristic, signature, $u_*(M_0) = u_*(M_1) \in H_4(K(\pi,1); \mathbb{Z})/Out(\pi)$ and, in the topological case, if $KS(M_0) = KS(M_1)$.*

If M is smooth, then the signature, abbreviated by σ, is by Rohlin's Theorem divisible by 16; and arbitrary values divisible by 16 of the signature and the class in $H_4(K(\pi,1); \mathbb{Z})/Out(\pi)$ and, in the topological case, of $KS \in \mathbb{Z}/2$ can be realized.

Theorem 2.5 in particular covers all finite fundamental groups of odd order.

Theorem 2.6 *Let M_0 and M_1 be smooth (topological), oriented 4-manifolds with $w_2(M_i) = 0$ and cyclic fundamental group $\pi_1(M_i) = \pi$. Then M_0 and M_1 are stably diffeomorphic (homeomorphic) if and only if both admit a spin structure or both do not admit a spin structure and they have same Euler characteristic, signature and, in the topological case, if $KS(M_0) = KS(M_1)$.*

The signature is always divisible by 8 and in the smooth case, if $w_2(M) \neq 0$, every integer divisible by 8 can be realized and, if $w_2(M) = 0$, all integers divisible by 16 can be realized. In the topological case, every integer divisible by 8 can be realized and, if $w_2(M) \neq 0$, one can prescribe $KS \in \mathbb{Z}/2$ arbitrarily, whereas, if $w_2(M) = 0$, $KS = \sigma(M)/8 \bmod 2$.

3   A Cancellation Theorem for Topological 4-Manifolds

In this section we prove a cancellation theorem for topological 4-manifolds which is analogous to Theorem 1.1.

Theorem 3.1 ([HaKr92$_2$], Theorem B) *Let X and Y be closed oriented topological 4-manifolds with finite fundamental group. Suppose that for some r the connected sum $X \# r(S^2 \times S^2)$ is homeomorphic to $Y \# r(S^2 \times S^2)$. If $X = X_0 \#(S^2 \times S^2)$, then X is homeomorphic to Y.*

Note that the assumption that X splits off one $S^2 \times S^2$ cannot be omitted, in general. There are, for example, even simply–connected closed topological 4-manifolds that are stably homeomorphic but not homeomorphic because they have non–isometric intersection forms. Examples of distinct but stably homeomorphic manifolds with finite fundamental group and the same equivariant intersection form were constructed in [KrSc84]. We will discuss these examples in the next section.

Before we prove this theorem, we formulate the following immediate corollary to it and Theorems 2.4, 2.5 and 2.6.

Corollary 3.2 *Let M_0 and M_1 be closed oriented topological manifolds with finite fundamental group π, such that one of the three conditions are fulfilled: i) $w_2(\tilde{M}_i) \neq 0$, ii) $w_2(\tilde{M}_i) = 0$ and π cyclic, iii) $H_i(\pi, \mathbb{Z}/2) = 0$ for $1 \leq i \leq 3$. Suppose that $M_0 = X\#(S^2 \times S^2)$. Then M_0 is homeomorphic to M_1 if and only if both admit a spin structure or both do not admit a spin structure and they have same Euler characteristic, signature, Kirby-Siebenmann obstruction and $u_*(M_0) = u_*(M_1) \in H_4(K(\pi,1); \mathbb{Z})/Out(\pi)$.*

As in the proof of Theorem 1.1, there is an algebraic and a geometric part in the proof of Theorem 3.1. We begin by stating the algebraic input. As in the last section, we set $A = \mathbb{Z}[\pi]$ and we equip A with the anti-involution $a \mapsto \bar{a}$ mapping an element in π to its inverse. As common in algebra, we consider right A-modules but note that with the help of the anti-involution one can pass from right to left modules and vice versa. Thus, whenever the module comes naturally with a left action, we pass to the corresponding right action. In particular, we do this for the dual of a right module V , which we denote by $\bar{V}$. A quadratic A-module V is an A-module together with a hermitian form $\langle -, - \rangle$ and a quadratic refinement q in the sense of ([Wa70], Chapter 5) with values in $A/\{a - \bar{a}\}$. It has $(A, \mathbb{Z})$-*hyperbolic rank* ≥ 1 at a prime $p \in \mathbb{Z}$ if there exists an integer r such that $(H(\mathbb{Z}^r) \oplus V)_p$ has free hyperbolic rank ≥ 1 over A_p. Here the hyperbolic form $H(W)$ of an A-module W is the form on $W \oplus \bar{W}$ which is trivial on W and $\bar{W}$ and evaluation between W and $\bar{W}$, and where the quadratic refinement vanishes on W and $\bar{W}$. The hyperbolic rank is $\geq s$ if the quadratic form splits off $H(A^s)$.

We need various subgroups of the isometries on a quadratic module. If $P = p_0 A \oplus p_1 A$ is A-free of rank 2, we denote by $E(P)$ the group generated by elementary triangular matrices having 1 on the diagonal and by $H(E(P))$ the induced isometries on the hyperbolic space $H(P)$. A *transvection* ([Ba73], p.91) of V is a unitary automorphism $\sigma = \sigma_{u,a,v} : V \to V$ given by

$$\sigma(x) = x + u\langle v, x \rangle - v\langle u, x \rangle - ua\langle u, x \rangle,$$

where $u, v \in V$ and $a \in A$ satisfy the conditions

$$q(u) = 0 \in A/\{a - \bar{a}\}, \langle u, v \rangle = 0, q(v) = a \in A/\{a - \bar{a}\}.$$

For any submodule $L \subseteq V$,

$$L^{\perp} = \{\imath \in V | \langle x, y \rangle = 0 \text{ for all } y \subset L\}.$$

If $V = V' \perp V''$ is an orthogonal direct sum, with $L' \subseteq V'$ a *totally isotropic* submodule (*i.e.* $h(x, y) = 0 \pmod{\{a - \bar{a}\}}$ for all $x, y \in L'$), and $L'' \subseteq V''$, then we define

$$EU(V', L'; L'') = \langle \sigma_{u,a,v} | u \in L' \text{ and } v \in L'' \rangle$$

and in the special case $V = P \perp \bar{P}$

$$EU(H(P)) = EU(P, P, \bar{P}).$$

A hyperbolic plane is a quadratic module isomorphic to $H(A)$. A hyperbolic pair consists of two vectors u and v with $q(u) = q(v) = 0$ and $< u, v >= 1$.

Theorem 3.3 ([HaKr92$_2$], Theorem 1.20 and Lemma 1.21) *Let V be a quadratic module which has $(A, \mathbb{Z})$-hyperbolic rank ≥ 1 at all but finitely many primes, and put $M = V \perp H(P)$, where $P = p_0 A \oplus p_1 A$ is A-free of rank 2. Then*

$$G = \langle EU(H(P), Q; V), H(E(P)) \cdot EU(H(P)) \rangle$$

where $Q = P$ or $\bar{P}$, acts transitively on the set of q-unimodular elements of a fixed length, and the set of hyperbolic pairs and hyperbolic planes in M.

Here an element $x \in M$ is q-unimodular if there exists $y \in M$ such that $\langle x, y \rangle = 1$.

This theorem is the quadratic analogue of Theorem 1.2 and the proof uses the same philosophy. We will apply this algebraic cancellation theorem to prove Theorem 3.1. We need some preparations.

Proposition 3.4 *Let X be a closed oriented topological 4–manifold with finite fundamental group, and let $A = \mathbb{Z}[\pi_1(X)]$. There is an A–submodule V of $\pi_2(X)$ which supports a quadratic refinement of the intersection form on X. In addition, V has $(A, \mathbb{Z})$–hyperbolic rank ≥ 1 at all but finitely many primes.*

Proof: Since our algebraic result uses quadratic modules and the intersection form on $\pi_2(X)$ does in general not admit a quadratic refinement, we take the submodule $V = \ker (\langle w_2, - \rangle : \pi_2(X) \to \mathbb{Z}/2)$ on which the intersection form

S_X has a quadratic refinement $q : V \to A/\{\nu - \bar{\nu}\}$ defined as in [Wa70, Chapter 5].

Next we check that V has $(A, \mathbb{Z})$–hyperbolic rank ≥ 1 at all odd primes not dividing the order of $\pi_1(X)$. Since X is a closed manifold, the components of the multi–signature of S_X are all equal (compare [Le77]). On the other hand, from [HaKr88, 2.4] we know that $\pi_2(X)_{(p)}$ is isomorphic to the localization of $I \oplus I^* \oplus A^\ell$, where I denotes the augmentation ideal of A. It follows that the components of S_X are indefinite at all non–trivial characters of $\pi_1(X)$. Since S_X is unimodular when restricted to V_p, for p as above, we conclude that V has $(A, \mathbb{Z})$ hyperbolic rank ≥ 1 at all odd primes not dividing the order of $\pi_1(X)$. □

We need the following result of Cappell–Shaneson. In the statement a standard basis for the summand $H_2(S^2 \times S^2, \mathbb{Z})$ of $H_2(X\#(S^2 \times S^2), \mathbb{Z})$ is denoted by $\{p_0, q_0\}$.

Theorem 3.5 ([CaSh71],1.5) *Let X be a compact, connected smooth (topological) manifold of dimension four, and suppose $X = X_0\#(S^2 \times S^2)$ for some manifold X_0. Let $\omega \in H_2(X; A) \cong \pi_2(X)$ with $w_2(\omega) = 0$ and let $a \in A = \mathbb{Z}[\pi_1(X)]$ be any element such that $q(\omega) \equiv a(\mathrm{mod}\{a - \bar{a}\})$. Then there is a base point preserving diffeomorphism (homeomorphism) ϕ of $X\#(S^2 \times S^2)$ with itself which preserves local orientations and induces the identity on $\pi_1(X\#(S^2 \times S^2))$, so that $\phi_*(p_0) = p_0$, $\phi_*(q_0) = q_0 + \omega - p_0 a$, and $\phi_*(\xi) = \xi - (\xi \cdot \omega)p_0$ for $\xi \in H_2(X; A)$.*

In order to prove Theorem 3.1, we need to realize transvections by homeomorphisms of $X\#r(S^2 \times S^2)$. For the rest of this section, we fix the notation

$$K\pi_2(X) = \ker\left(\langle w_2, -\rangle : \pi_2(X) \to \mathbb{Z}/2\right)$$

for the submodule of the intersection form on $H_2(X; A)$ on which a quadratic refinement is defined. We denote by $H(P_0)$, where $P_0 = p_0 A$, the summand of $H_2(X\#(S^2 \times S^2); A)$ given by $H_2(S^2 \times S^2; A)$. As further copies of $S^2 \times S^2$ are added to X by connected sum, we denote all these hyperbolic factors of the intersection form by $H(P)$. Note that Theorem 3.5 allows us to realize the transvections $\sigma_{p_0, a, v}$ by self-homeomorphisms of $X\#(S^2 \times S^2)$ for any $v \in K\pi_2(X_0)$, in the case when $X = X_0\#(S^2 \times S^2)$. Cappell and Shaneson use this to realize many isometries (see [CaSh71, Theorem A2]), but the conclusions given are not in the exact form we need.

Corollary 3.6 *Suppose that $K\pi_2(X) = V_0 \perp V_1$ with V_0 non-singular under the intersection form S_X. Then, for any transvection $\sigma_{p,a,v}$ on $K\pi_2(X) \perp H(P_0)$ with $p \in V_0 \perp P_0$ and $v \in K\pi_2(X)$, the stabilized isometry $\sigma_{p,a,v} \oplus Id_{2(S^2 \times S^2)}$ can be realized by a self-homeomorphism of $X\#3(S^2 \times S^2)$.*

Proof: First, we consider a unimodular isotropic element $p \in V_0 \perp P_0$. Since $V_0 \perp H(P_0)$ is non-singular, p is automatically a hyperbolic element and thus by Freedman [Fr84] we can re-split $X\#(S^2 \times S^2) = X'\#(S^2 \times S^2)$ such that p is represented by $S^2 \times *$. Thus $\sigma_{p,a,v} \oplus Id_{S^2 \times S^2}$ can be realized by a self-homeomorphism on $(X'\#(S^2 \times S^2))\#(S^2 \times S^2)$ for all $v \in K\pi_2(X)$ with $\langle v, p \rangle = 0$.

Next, we consider the transvection $\sigma_{p,0,v}$ for an arbitrary $p \in V_0 \perp P_0$, but assume that $v \in K\pi_2(X)$ is isotropic. Then we write $p = \sum p_i$ with $p_i \in V_0 \perp P_0$ unimodular and $\langle v, p_i \rangle = 0$. This uses the fact that $A = \mathbb{Z}[\pi_1(X)]$ and $P_0 \cong A$. We obtain: $\sigma_{p,0,v} = \sigma_{v,0,-p} = \sigma_{v,0,-\sum p_i} = \prod \sigma_{p_i,0,v}$. Thus $\sigma_{p,0,v} \oplus Id_{S^2 \times S^2}$ is realizable by a self-homeomorphism on $(X\#(S^2 \times S^2))\#(S^2 \times S^2)$, since $\sigma_{p_i,0,v}\#Id_{S^2 \times S^2}$ is realizable.

Finally, we realize an arbitrary transvection $\sigma_{p,a,v}\#Id_{2(S^2 \times S^2)}$, of the form required, by a homeomorphism on $(X\#(S^2 \times S^2))\#(S^2 \times S^2)$. We use the fact that v can be expressed as $v = \sum v_i$ with $v_i \in K\pi_2(X) \perp H_2(S^2 \times S^2; A)$ isotropic and $\langle v_i, p \rangle = 0$. Thus $\sigma_{p,a,v} \oplus Id_{2(S^2 \times S^2)} = \prod \sigma_{p,0,v_i} \oplus Id_{S^2 \times S^2}$ which by the considerations above is realizable.   □

Corollary 3.7 *Let X_0 be a topological 4-manifold, $V = K\pi_2(X_0)$ and consider an element $\varphi \in EU(H(P), Q; V)$, for $Q = P, \bar{P}$, as an isometry of the intersection form of $X_0\#2(S^2 \times S^2)$. Then the stabilized isometry $\varphi \oplus Id_{2(S^2 \times S^2)}$ can be realized by a self-homeomorphism of $X_0\#4(S^2 \times S^2)$.*

Proof: By definition, the group $EU(H(P), Q; V)$ is generated by transvections $\sigma p, a, v$ with $p \in P$ or $\bar{P}$ and $v \in V$ fulfilling the conditions of a transvection. It is enough to consider the case $p \in P$. Now Corollary 3.6 applies with the splitting $K\pi_2(X) = V \perp H(A)$ with $H(A)$ the first summand of $H(P)$. This shows that for each $\varphi \in EU(H(P), Q; V)$, the isometry $\varphi \oplus Id_{2(S^2 \times S^2)}$ can be realized by a self-homeomorphism on $(X_0\#2(S^2 \times S^2))\#2(S^2 \times S^2)$.   □

Proof of Theorem 3.1: By induction, it is enough to consider the case $r = 1$. Let $f : X\#(S^2 \times S^2) \to Y\#(S^2 \times S^2)$ be a homeomorphism. We will apply Theorem 3.3 and Corollary 3.6 to show that there is a self-homeomorphism g of $X\#3(S^2 \times S^2)$ such that $(f\#Id) \cdot g$ induces the identity on the hyperbolic form corresponding to $\# 3(S^2 \times S^2)$ in $H_2(X\#3(S^2 \times S^2); A)$. Then it follows that X and Y are s-cobordant ([Kr85], Theorem 3.1). By Freedman [Fr84], X and Y are homeomorphic.

To begin, we apply Theorem 3.3 to

$$V \oplus H(P) \subseteq H_2(X_0\#2(S^2 \times S^2); A),$$

where $P = A \oplus A$ and $V = K\pi_2(X_0)$. This gives an isometry

$$\varphi \in G = \langle EU(H(P), Q; V), H(E(P)) \cdot EU(H(P)) \rangle,$$

where $Q = P$ or $\bar{P}$, such that $f_* \cdot \varphi$ induces the identity on $H_2(2(S^2 \times S^2); A) \subseteq H_2(X_0 \# 2(S^2 \times S^2); A)$. We finish the proof by showing that for each $\varphi \in G$, $\varphi \oplus Id$ can be realized by a self-homeomorphism on $X_0 \# 4(S^2 \times S^2)$. Note that by definition $G \subseteq Aut(H_2(X_0 \# 2(S^2 \times S^2); A))$.

The elements of $EU(H(P), Q; V)$ are handled by Corollary 3.7. In addition, we have to realize an arbitrary element in $H(E(P)) \cdot EU(H(P))$, stabilized by the identity, by a self-homeomorphism of $(X_0 \# 4(S^2 \times S^2))$. This follows again from Corollary 3.6 and the considerations above since this group is generated by transvections $\sigma_{p,a,x}$ with $p \in P_0$ or P_1 ([Ba73], p.142-143).   □

As in the case of 2-complexes we want to finish this section by stating without proofs two classification results for oriented 4-manifolds with special fundamental groups which follow from more refined cancellation results and Theorems 2.5 and 2.6.

We begin with the complete classification for finite cyclic fundamental groups. The following notation is useful for encoding the different possibilities of the vanishing of the second Stiefel-Whitney class. The w_2-type is I, if $w_2(\tilde{M}) \neq 0$, II, if $w_2(M) = 0$, or III, if $w_2(\tilde{M}) = 0$ and $w_2(M) \neq 0$.

Theorem 3.8 ([Fr84], 1-connected case; [HaKr92₃], general case) *Let M be a closed, oriented 4-manifold with finite cyclic fundamental group. Then M is classified up to homeomorphism by the fundamental group, the intersection form on $H_2(M, \mathbb{Z})/Tors$, the w_2-type, and the Kirby-Siebenmann invariant. Moreover, any isometry of the intersection form can be realized by a homeomorphism. All invariants can be realized except in the case of w_2-type II, where KS is determined by the intersection form.*

Next we give an explicit bound for the difference between the Euler characteristic e and the absolute value of the signature σ for odd order fundamental groups guaranteeing cancellation. Combined with Theorem 2.5, this gives a homeomorphism classification under these stability assumptions. For any finite group π, let $d(\pi)$ denote the minimal $\mathbb{Z}$-rank for the abelian group $\Omega^3 \mathbb{Z} \otimes_{\mathbb{Z}\pi} \mathbb{Z}$. Here we minimize over all representatives of $\Omega^3 \mathbb{Z}$, obtained from a free resolution of length three (see section 1) of $\mathbb{Z}$ over the ring $\mathbb{Z}\pi$. Let $b_2(M)$ denote the rank of $H_2(M; \mathbb{Z})$.

Theorem 3.9 [HaKr92₃]) *Let M be a closed oriented manifold of dimension four, and let $\pi_1(M) = \pi$ be a finite group of odd order. When $w_2(\tilde{M}) =$*

0 *(resp.* $w_2(\tilde{M}) \neq 0$*), assume that* $b_2(M) - |\sigma(M)| > 2d(\pi)$*, (resp.* $>$ $2d(\pi) + 2$*). Then* M *is classified up to homeomorphism by the signature, Euler characteristic, type, Kirby-Siebenmann invariant, and fundamental class in* $H_4(\pi, \mathbb{Z})/Out(\pi)$.

The *type* is the parity (even or odd) of the intersection form on M.

4   A Homotopy Non-Cancellation Theorem for Smooth 4-Manifolds

In the case of 2-complexes, it was not easy to give non-cancellation examples, e.g., of 2-complexes X and Y such that $X \vee S^2$ is (simple-) homotopy equivalent to $Y \vee S^2$ but X not (simple-) homotopy equivalent to Y. The first examples were only published in 1976 (see references in Chapter I, following (40)).

In the case of topological 4-manifolds, the existence of closed topological 4-manifolds X and Y such that $X\#(S^2 \times S^2)$ is homeomorphic to $Y\#(S^2 \times S^2)$ but X not homeomorphic or equivalently not homotopy-equivalent to Y follows easily from Freedman's classification of 1-connected 4-manifolds (see Theorem 3.8). There are for instance 1-connected topological 4-manifolds X with intersection form $E_8 \oplus E_8$ and Y with intersection form E_{16}, where E_8 and E_{16} are the indecomposable even negative definite unimodular forms over $\mathbb{Z}$ with signature 8 and 16, respectively. These forms become isometric after adding a hyperbolic plane [Se73] and thus by Theorem 3.8, $X\#(S^2 \times S^2)$ is homeomorphic to $Y\#(S^2 \times S^2)$ but X is not homeomorphic to Y.

In the case of smooth 4-manifolds with finite fundamental group, it is not so easy to find non-cancellation examples, which here means manifolds X and Y such that $X\#(S^2 \times S^2)$ is diffeomorphic to $Y\#(S^2 \times S^2)$ but X is not diffeomorphic to Y. The method used above in the topological category finding manifolds with non-isometric definite intersection form which are stably homeomorphic cannot work in the smooth category since by Donaldson's Theorem [Do83] the only definite forms realized as intersection forms of smooth 4-manifolds are up to sign the standard Euclidean forms.

In this situation it is natural to try to make use of the non-cancellation examples of 2-complexes by applying the thickening construction (see the beginning of §2). This was carried out in [KrSc84] and we summarize these examples.

Here is the main result. Recall that we denote the boundary of a thickening of a 2-complex X in $\mathbb{R}^5$ by $M(X)$.

Theorem 4.1 ([KrSc84], Theorem III.3) *Suppose $G = (\mathbb{Z}/p)^s$ is elementary abelian where p is a prime congruent to 1 mod 4 and $s > 1$ is odd. Then there exist finite 2-dimensional CW complexes X and Y such that $M(X)$ and $M(Y)$ are not homotopy equivalent but $M(X)\#r(S^2 \times S^2)$ is diffeomorphic to $M(Y)\#r(S^2 \times S^2)$ for $r > 0$.*

Remark: The homotopy type of 4-manifolds with odd order fundamental group is determined by the quadratic 2-type consisting of the quadruple (π_1, π_2, k, s), where π_2 has to be considered as module over π_1, $k \in H^3(\pi_1; \pi_2)$ is the first k-invariant and s is the equivariant intersection form on π_2 [HaKr88], [Ba88]. In the examples that we will describe in the following, the triple (π_1, π_2, s) is isomorphic for $M(X)$ and $M(Y)$ ([KrSc84], p.21) and thus the manifolds are distinguished by the k-invariant, but this is not the way we prove our result.

The simplest examples for our theorem are derived from Metzler's theorem, a special case of which is stated below (see also Chapter III, §§1 and 2). Note, that a presentation of a group defines a 2-complex with this group as fundamental group by attaching to a wedge of r circles, r the number of generators, 2-cells according to the relations.

Theorem 4.2 ([Me76]) *For $s \geq 2$ and $(q, p) = 1$, the presentations*

$$< a_1, ..., a_s; a_i^p = 1, [a_1^q, a_2] = 1, [a_i, a_j] = 1, 1 \leq i < j \leq s, (i,j) \neq (1,2) >$$

of $(\mathbb{Z}/p)^s$ determine 2-complexes $X(q)$. $X(q)$ and $X(q')$ are not homotopy-equivalent, if $q \neq \pm k^{s-1}q'$ mod p for all k.

If one considers the boundary $M(X(q))$ of a thickening of $X(q)$ one gets examples of non-cancellation examples of smooth 4-manifolds, if Metzler's invariant or some weakening of it survives as invariant of the thickening. We don't know, if the full invariant survives but some partial information does. Theorem 4.1 is a consequence of the following Proposition.

Proposition 4.3 *Let $X(q)$ be as in Theorem 4.2. Then, if $s > 1$ is odd and p is a prime congruent to 1 mod 4, $M(X(q))$ and $M(X(q'))$ are not homotopy equivalent if qq'^{-1} is not a square mod p.*

Since $M(X(q))$ and $M(X(q'))$ are stably diffeomorphic, Theorem 4.1 follows.

Proof. In the following, we give a sketch of the proof of Proposition 4.0. For the details see [KrSc84].

Since $\pi_1(X(q)) = \pi_1(M(X(q))$ and $\pi_1(X(q')) = \pi_1(M(X(q'))$ are isomorphic we choose an isomorphism, a polarization, between them and denote the group by π.

Denote the cellular chain complex over $A = \mathbb{Z}[\pi]$ of the universal covering of $M(X(q))$ and $M(X(q'))$ by C and C'. Then it is easy to show by standard homological algebra that there is a chain map $h : C \to C'$ inducing the identity on $H_0(..; \mathbb{Z})$ and $H_4(..; \mathbb{Z})$. Denote the 0th Tate cohomology of an A-module M by $\hat{H}^0(M) = M^\pi/N(M)$, where M^π is the fixed point set and $N(M)$ consists of the norm elements. If f is an A-module homomorphism we denote the induced map between the Tate cohomologies by $\hat{H}^0(f)$. If h is a chain map as above, then $\hat{H}^0(h_*)$ is an isomorphism.

Consider the equivariant intersection form on the middle homology of the universal covering. This induces an equivariant symmetric bilinear form on $H_2(\tilde{X}(q))^\pi = H_2(C)^\pi$. Any orientation preserving homotopy equivalence which induces the given isomorphism on π_1 induces a map respecting this bilinear form.

Thus, if $M(X(q))$ and $M(X(q'))$ are orientation preserving homotopy equivalent inducing the given isomorphism on π_1, then $\hat{H}^0(h_*)$ is induced by an **isometry** from $H_2(C)^\pi$ to $H_2(C')^\pi$.

Thus we get an invariant of polarized oriented homotopy types by the set of all isomorphisms $\hat{H}^0(h_*)$ modulo those induced by isometries from $H_2(C)^\pi$ to $H_2(C')^\pi$. Dividing out the different choices of polarizations equivalently of automorphisms of π and using the fact that $M(X)$ always admits an orientation reversing diffeomorphism ($M(X)$ can be described as a double of a 4-dimensional thickening and interchanging the two halves gives the orientation reversing diffeomorphism), one gets a homotopy invariant.

The main work of [KrSc84] is to show that this invariant is non-trivial if qq'^{-1} is not a square mod p. For this one can rather easily compute a representative of this invariant but it is not so easy to decide when it is non-trivial. We get our result by weakening the invariant, namely we pass to an L-theoretic invariant. More precisely, it is not difficult to show that the restriction of the intersection form to $H_2(C)^\pi$ is up to scaling by a constant a hyperbolic form over $\mathbb{Z}$. It induces the hyperbolic form over $\mathbb{Z}/p$ on $\hat{H}^0(H_2(C))$. After appropriately identifying $H_2(C)$ with $H_2(C')$ our invariant given by $\hat{H}^0(h_*)$ gives an automorphism of determinant 1 of $\hat{H}^0(H_2(C))$, which turns out to be an isometry. Stable equivalence classes of isometries of determinant 1 represent elements in the Wall group $L_1^0(\mathbb{Z}/p)$ [Wa70]. If $M(X(q))$ and $M(X(q'))$ are homotopy equivalent, this element in $L_1^0(\mathbb{Z}/p)$ is induced from an isometry of $H_2(C)$. Since an isometry of a scaled hyperbolic form over $\mathbb{Z}$ is an isometry of the hyperbolic form itself it is in the image of the reduction map from $L_1^0(\mathbb{Z})$ to $L_1^0(\mathbb{Z}/p)$.

For the manifolds $M(X(q))$ and $M(X(q'))$, $\hat{H}^0(H_2(C))$ is isometric to the hyperbolic form on $\mathbb{Z}/p$ and the invariant in $L_1^0(\mathbb{Z}/p)$ is represented by a diagonal matrix of rank 2 over $\mathbb{Z}/p$ with entries (qq'^{-1}) and $(qq'^{-1})^{-1}$. The different choices of a polarization of the fundamental groups correspond to an action of $Aut(\pi)$. It turns out that $Aut(\pi)$ acts on $\hat{H}^0(H_2(C))$ by diagonal matrices of rank 2 over $\mathbb{Z}/p$ with entries r^{s-1}, r^{1-s} for some r prime to p. Thus, if s is odd, the action is trivial.

To finish the proof, we need the following information from [Wa76]. The Wall group $L_1^0(\mathbb{Z})$ is isomorphic to $\mathbb{Z}/2$ generated by the diagonal matrix of rank 2 with entries $(-1, -1)$. The Wall group $L_1^0(\mathbb{Z}/p)$ is isomorphic to $\mathbb{Z}/2$ generated by a diagonal matrix of rank 2 with entries (r, r^{-1}), where r is a non-square mod p. Thus, if p is congruent to 1 *mod* 4, the reduction map is trivial finishing the argument.   □

Remark: Comparing Theorem 4.2 and Proposition 4.3 one sees that the invariant used there is considerably weaker than Metzler's. It would be interesting to know if one actually is losing information by passing from 2-complexes to boundaries of 5-dimensional thickenings.

5   A Non-Cancellation Example for Simple-Homotopy Equivalent Topological 4-Manifolds

The non-cancellation examples in Section 4 were non-homotopy equivalent but stably diffeomorphic smooth 4-manifolds. As mentioned before one can get other examples from exotic structures on closed smooth oriented 4-manifolds. They are homeomorphic, not diffeomorphic but stably diffeomorphic. Such examples are much more complicated than the ones described in Section 4 since the only known way to distinguishing them is by Donaldson invariants. We will describe many exotic structures in Section 6.

The most delicate question one can ask in the topological category in connection with non-cancellation examples is whether there are simple- homotopy equivalent non-homeomorphic but stably homeomorphic topological closed 4-manifolds. Recently, in joint work with Peter Teichner, we found the first examples of this type. We will describe them here. The examples constitute another link between 2-dimensional topology and 4-manifolds since they are distinguished by a codimension 2 invariant.

We begin with a notation. According to Freedman there exists a unique non-smoothable 4-manifold which is homotopy-equivalent to $\mathbb{CP}^2$, the Chern manifold denoted $\mathbb{CH}$. We will see that there is a similar manifold corresponding to $\mathbb{RP}^4$, a unique non-smoothable 4-manifold homotopy equivalent to $\mathbb{RP}^4$, denoted $\mathbb{RH}$.

Theorem 5.1 ([HaKrTe92]) *The simple-homotopy equivalent closed 4-manifolds $\mathbb{RP}^4\#\mathbb{CP}^2$ and $\mathbb{RH}\#\mathbb{CH}$ are not homeomorphic but homeomorphic after connected sum with r copies of $S^2 \times S^2$.*

Remark: In [HaKrTe92] it is actually shown that $r = 1$ works, but we don't need this to get our non-cancellation examples. We don't know whether $\mathbb{RH}\#\mathbb{CH}$ admits a smooth structure. The only known obstruction, the Kirby-Siebenmann obstruction vanishes, since it is non-trivial on both summands and is additive under connected sum. If a smooth structure exists, then one gets examples of stably diffeomorphic simple-homotopy equivalent smooth 4-manifolds that are not homeomorphic.

Proof: We begin with the construction of $\mathbb{RH}$. According to Freedman [Fr84], there exists a unique simply connected topological 4-manifold with intersection form isomorphic to E_8, the unique negative definite form with signature -8. We denote this manifold by $M(E_8)$. The Kirby-Siebenmann obstruction of $M(E_8)$ is $KS(M(E_8)) = 1$. This follows since the Kirby Siebenmann obstruction of a $TopSpin$-manifold (i.e. w_1 and w_2 vanish) is equal to $1/8\sigma(M)$ mod 2. Consider $\mathbb{RP}^4\#M(E_8)$. The quadratic intersection form of this manifold is $E_8 \otimes_{\mathbb{Z}} A$, where $A = \mathbb{Z}[\mathbb{Z}/2]$ equipped with the anti-involution which here in the non-oriented case maps the nontrivial element τ in $\mathbb{Z}/2$ to $-\tau$. This form is stably (i.e. after adding a hyperbolic form) isomorphic to a hyperbolic form. This follows for instance from the fact that the map of Wall groups $L_0(\mathbb{Z}) \to L_0(\mathbb{Z}[\mathbb{Z}/2])$ is trivial [Wa70]. By Freedman [Fr84], one can decompose the manifold as the connected sum of some topological manifold M' and $\# \, r(S^2 \times S^2)$, if the quadratic intersection form of a manifold M splits off a hyperbolic form of rank $2r$. Applying this to $\mathbb{RP}^4\#M(E_8)\#r(S^2 \times S^2)$ one can decompose this as the connected sum of $(r + 8)(S^2 \times S^2)$ and some manifold which we will denote by $\mathbb{RH}$. By construction this manifold has fundamental group $\mathbb{Z}/2$ and Euler characteristic 1. Thus the manifold is homotopy equivalent to $\mathbb{RP}^4$. One can prove that this manifold is unique up to homeomorphism but for our context we don't need this and call any manifold constructed this way by the same name. Since

$KS(M(E_8)) = 1$ and the Kirby-Siebenmann obstruction is additive under connected sum, $KS(\mathbb{RH}) = 1$.

Next we show that $\mathbb{RP}^4 \# \mathbb{CP}^2$ and $\mathbb{RH} \# \mathbb{CH}$ are stably homeomorphic. For this we apply Theorem 2.3. Obviously, both manifolds have the same normal 1-type: $[\mathbb{Z}/2, x, \infty]$, where x generates $H^1(\mathbb{Z}/2; \mathbb{Z}/2)$. The geometric normal 1-type is the trivial fibration $Id : BO \to BO$. Thus the relevant bordism group is the non-oriented topological bordism group $\mathfrak{N}_4^{Top}$, which is isomorphic to $\mathbb{Z}/2 \oplus \mathbb{Z}/2 \oplus \mathbb{Z}/2$, detected by w_1^4, w_4 and KS. This follows, since, if $KS = 0$, the manifold is bordant to a smooth manifold ([Fr84], [FrQu90]) and the smooth non-oriented bordism group is detected by w_1^4 and w_4 [Th54]. By construction, all these invariants agree for $\mathbb{RP}^4 \# \mathbb{CP}^2$ and $\mathbb{RH} \# \mathbb{CH}$. Thus, by Theorem 2.3, they are stably homeomorphic.

To finish, we have to show that they are not homeomorphic. This will follow from the construction and computation of an invariant which roughly speaking is defined as follows. Let M be one of the manifolds we want to distinguish. $H^2(M; \mathbb{Z}) \cong \mathbb{Z} \oplus \mathbb{Z}/2$. Let $c \in H^2(M; \mathbb{Z})$ be a class which reduces to $w_2\nu(M)$ and which generates $H^2(M; \mathbb{Z})/Tors$. Such a class is unique up to sign. Now, represent c by a map to $\mathbb{CP}^N$ for some large N. After making this map transversal to $\mathbb{CP}^{N-1}$, the inverse image of $\mathbb{CP}^{N-1}$ is a surface Σ in M (transversality holds in the topological category, see e.g. [FrQu90]) and it inherits from M a so called normal Pin^+-structure, which is unique up to sign in the corresponding bordism group (for details see [HaKrTe92], §2). Here Pin^+ is the central extension

$$0 \longrightarrow \mathbb{Z}/2 \longrightarrow Pin^{\pm}(n) \longrightarrow O \longrightarrow 0$$

classified by $w_2 + w_1^2$. We obtain a fibration

$$p : BPin^+ \longrightarrow BO.$$

A normal Pin^+-structure is a lift of the normal Gauss map to Pin^+. According to Brown [Br72], a Pin^+-structure on a surface Σ determines a quadratic refinement with values in $\mathbb{Z}/4$ of the intersection form on $H^2(\Sigma; \mathbb{Z}/2)$. The Witt group of such forms is isomorphic to $\mathbb{Z}/8$ and the corresponding element represented by the quadratic refinement on Σ is denoted by $\pm \mathrm{arf}\ (M) \in \mathbb{Z}/8$.

This is our invariant and it is obviously a homeomorphism invariant. Note that one can define the same sort of invariant on $M \# r(S^2 \times S^2)$ after choosing a cohomology class c reducing to w_2. But, if $r > 0$, this invariant depends on the choice of c (not only up to sign) and loses all its information (to indicate the dependence on c we denote the invariant by $\mathrm{arf}\ (M, c)$). But it turns out that it takes different values for $\mathbb{RP}^4 \# \mathbb{CP}^2$ and $\mathbb{RH} \# \mathbb{CH}$.

The reason for this is the following. The class c is the sum $c_1 + c_2$ corresponding to the connected sum decomposition of our manifolds. The arf-invariant is additive under connected sum. For oriented manifolds, one has the following formula ([KiTa89], Cor.9.3):

$$2 \cdot \operatorname{arf}(M, c) = c \circ c - \sigma(M) + 8 \cdot KS(M) \bmod 16,$$

where $\sigma(M)$ is the signature of M. Thus, arf $(\mathbb{CP}^2, c_2) = 0$ and arf $(\mathbb{CH}) = 4$ (mod)8. By construction of $\mathbb{RH}$ we see that $(\mathbb{RH}, c_1) \# (4(S^2 \times S^2), 0) = (\mathbb{RP}^4, c_1) \# (E_8, 0)$. Thus, from the formula above, $\pm \operatorname{arf}(\mathbb{RP}^4, c_1) = \pm \operatorname{arf}(\mathbb{RH}, c_1)$. □

6　Application of Cancellation to Exotic Structures on 4-Manifolds

In this section, we study the existence of exotic structures on many algebraic surfaces with finite fundamental group. From the point of view of cancellation problems for 4-manifolds the construction of exotic structures on oriented closed 4-manifolds is equivalent to the construction of homeomorphic smooth manifolds which are stably diffeomorphic but not diffeomorphic. The reason for this is that homeomorphic oriented smooth closed 4-manifolds are automatically stably diffeomorphic, a result that can rather easily be derived from Theorem 2.3 by comparing the topological and the smooth bordism group of the corresponding normal 1-type ([Kr84₁], for another proof see [Go84]). To distinguish stably diffeomorphic smooth oriented closed 4-manifolds, one has to find rather delicate invariants. These are provided by the Donaldson polynomials [Do90], which are defined for closed oriented smooth 4-manifolds with some additional restrictions. For instance, these restrictions are fulfilled for all 1-connected algebraic surfaces. We will base our examples of exotic structures on the following result of Donaldson.

Theorem 6.1 ([Do90]) *Let X be a 1-connected compact algebraic surface without singularities. Then X is not diffeomorphic to a connected sum $M_1 \# M_2$ unless M_1 or M_2 have negative definite intersection form.*

To apply this theorem to the construction of exotic structures on closed 4-manifolds, it is sufficient to find an algebraic surface with finite fundamental group X and a smooth 4-manifold M, such that X and M are homeomorphic but the universal covering $\tilde{M}$ is diffeomorphic to a connected sum $M_1 \# M_2$ where M_1 and M_2 do not have negative definite intersection form.

The following result is an application of this method showing the existence of an exotic structure on surfaces where the sum of the signature σ and the Euler characteristic e is sufficiently large.

Theorem 6.2 ([HaKr90]) *Let π be a finite group. Then there is a constant $c(\pi)$ such that a compact non-singular algebraic surface X with $\pi_1(X) \cong \pi$ and $\sigma(X) + e(X) \geq c(\pi)$ has at least two smooth structures.*

Note that by a construction of Shafarevic ([Sh74, p. 402 ff]) for each finite group π there are algebraic surfaces with fundamental group π and arbitrarily large $\sigma(X) + e(X)$ (compare, [HaKr90, p. 109] and the following remark).

Remark: In [HaKr90], we used instead of $\sigma(X) + e(X) \geq c(\pi)$ the condition $c_1^2(X) \geq 0$ and $e(X)$ sufficiently large. We thank Stefan Bauer for pointing out that our proof works under this slightly better condition.

Proof: The first ingredient in the proof is the following Proposition. We say that two closed topological 4-manifolds M_0 and M_1 are *weakly stably homeomorphic* if there exists a natural number r and and integers s_0 and s_1 such that $M_0 \# r(S^2 \times S^2) \# s_0 K$ is homeomorphic to $M_1 \# r(S^2 \times S^2) \# s_1 K$. Here K is the Kummer surface (K_3-surface), the quartic in $\mathbb{CP}^3$, and for s negative we mean by sK the connected sum of $-s$ copies of K with its negative orientation. Recall that K is a 1-connected 4-manifold with signature -16 and Euler characteristic 24.

Proposition 6.3 *Let π be a finite group. Then the set of weakly stable homeomorphism classes of closed smooth oriented 4-manifolds with fundamental group π is finite.*

With this proposition we proceed as follows. For each weakly stable homeomorphism class α, choose a representative M_α with $e(M_\alpha)$ minimal and $-8 \leq \sigma(M_\alpha) < 8$ and suppose $M_\alpha \cong M_\alpha' \# S^2 \times S^2$, if π is trivial. Then, for each closed oriented smooth 4-manifold X with fundamental group isomorphic to π, there exist α and s such that X is stably homeomorphic to $M_\alpha \# sK$. If $e(X) > e(M_\alpha \# sK)$, then Theorem 3.1 implies that X is homeomorphic to $Y = M_\alpha \# r(S^2 \times S^2) \# sK$ for some $r > 0$. Now, Donaldson's Theorem 6.1 implies that, if X is an algebraic surface, then X and Y are not diffeomorphic, since $\tilde{X}$ is again a compact algebraic surface and for π non-trivial $\tilde{Y} = Y' \# \tilde{S^2} \times S^2$ decomposes as the connected sum of two smooth manifolds with indefinite intersection forms, and for π trivial we assumed that the same holds for Y. Now the proof of Theorem 6.2 is finished if we can find a number $c(\pi)$ such that $e(X) \geq e(M_\alpha \# sK)$ for any algebraic surface X with fundamental group π and $\sigma(X) + e(X) \geq c(\pi)$. It is actually enough to do this for minimal surfaces X since the condition $\sigma(X) + e(X) \geq c(\pi)$ is invariant under blow ups and also $X \# k \cdot \overline{\mathbb{CP}}^2$ and $Y \# k \cdot \overline{\mathbb{CP}}^2$ remain non-diffeomorphic by Donaldson's Theorem.

To compare for a minimal surface X, $e(X)$ with $e(M_\alpha \# s \cdot K)$, we express $e(M_\alpha \# s \cdot K)$ in terms of $e(M_\alpha)$, $\sigma(M_\alpha)$ and $\sigma(X)$:

$$e(M_\alpha \# s \cdot K) = e(M_\alpha) + 22 \cdot \mid s \mid$$

and

$$\sigma(X) = \sigma(M_\alpha \# s \cdot K) = \sigma(M_\alpha) - 16s.$$

This implies

$$e(M_\alpha \# s \cdot K) = e(M_\alpha) + (11/8) \mid \sigma(X) - \sigma(M_\alpha) \mid .$$

We have the following inequality for algebraic surfaces:

$$e(X) - (11/8) \mid \sigma(X) \mid \geq (1/12)e(X).$$

If $\sigma(X) < 0$, this is an immediate consequence of the signature theorem $(\sigma(X) = \frac{c_1^2(X) - 2e(X)}{3})$ and the fact that a minimal surface has $c_1^2 \geq 0$. If $\sigma(X) \geq 0$, the signature theorem implies

$$e(X) - (11/8) \mid \sigma(X) \mid \geq (1/12)e(X) + (11/24)(4e(X) - c_1^2(X)).$$

Thus we are finished for surfaces fulfilling $(4e - c_1^2) \geq 0$. For surfaces of general type, this is a consequence of the inequality of Miyaoka-Yau ([BaPeVa84], p. 212). The only minimal surfaces with finite fundamental group and $\sigma(X) \geq 0$ are diffeomorphic to $\mathbb{C}P^2$, $S^2 \times S^2$ or $\mathbb{C}P^2 \# \overline{\mathbb{C}P}^2$ (this follows from the Enriques-Kodaira classification ([BaPeVa84], p. 187ff)), for which the inequality holds.

Using this inequality together with the formula for $e(M_\alpha \# s \cdot K)$ above (note that $\sigma(X) = \sigma(M_\alpha)$ mod 16 and $-8 \leq \sigma(M_\alpha) < 8$) we get:

$$
\begin{aligned}
e(X) - e(M_\alpha \# s \cdot K) &= e(X) - (11/8) \mid \sigma(X) \mid -e(M_\alpha) \pm (11/8)\sigma(M_\alpha) \\
&\geq (1/12)e(X) - e(M_\alpha) - 11.
\end{aligned}
$$

Since $2e(X) > \sigma(X) + e(X)$, we see that if $\sigma(X) + e(X) \geq 24(e(M_\alpha) + 11)$ we have $e(X) > e(M_\alpha \# s \cdot K)$.

As there are only finitely many M_α's, we can define

$$c(\pi) := 24 \cdot \max\{e(M_\alpha) + 11\},$$

finishing the proof of our theorem. □

Proof: (of Proposition 6.3) The proof is an application of Theorem 2.3. First, we note that for a fixed algebraic normal 1-type $[\pi, w_1, w_2]$, the bordism group $\Omega_4(B^{Top}[\pi, w_1, w_2]) \otimes \mathbb{Q}$ is isomorphic to $\mathbb{Q}$, the isomorphism is given by the signature. This is an easy consequence of the Atiyah-Hirzebruch spectral sequence. K is a 1-connected spin-manifold and thus the connected sum with K does not change the algebraic normal 1-type. Since $\sigma(K) = -16$, the set of weakly stable homeomorphism classes of manifolds with fixed algebraic normal 1-types is finite. But, if we fix π, the set of algebraic normal 1-types is finite since $H^1(\pi; \mathbb{Z}/2)$ and $H^2(\pi; \mathbb{Z}/2)$ are finite.   □

This result and stronger results for special fundamental groups led us to the following conjecture.

Conjecture: *A compact non-singular algebraic surface with finite fundamental group has at least two smooth structures.*

We note that a minimal surface with finite fundamental group has $c_1^2 \geq 0$ (this follows from the classification, e.g. [BaPeVa84], p. 188). But if $c_1^2 \geq 0$ and $\sigma(X) + e(X) < c(\pi)$, the Euler characteristic can only take finitely many values. On the other hand, there are only finitely many homeomorphism types of closed oriented 4-manifolds with prescribed finite fundamental group π and fixed Euler characteristic ([HaKr88], Corollary 1.5). Thus we obtain:

Corollary 6.4 *Let π be a finite group. Then the conjecture holds for all but perhaps a finite number of homeomorphism types of minimal algebraic surfaces X with fundamental group π.*

Based on similar arguments as above and some more delicate computations of Donaldson invariants one gets the following result, which we state without proof.

Theorem 6.5 ([HaKr92₃]) *(i) The conjecture holds for all algebraic surfaces with finite non-trivial cyclic fundamental group.*

(ii) The conjecture holds for all elliptic surfaces X with finite fundamental group except perhaps if X has geometric genus 0, where the statement holds after blowing up once replacing X by $X \# \overline{\mathbb{CP}}^2$.

7   Topological Embeddings of 2-Spheres into 1-Connected 4-Manifolds and Pseudo-free Group Actions

We finish this paper with two further applications of cancellation to 4-dimensional topology. The first is again a link between 2- and 4-dimensional topology and concerns the existence and uniqueness of locally flat simple embeddings of 2-spheres in a 1-connected 4-manifold N. These problems were substantially settled in [LeWi90] for homology classes of odd divisibility. Let $x \in H_2(N; \mathbb{Z})$. Then $x = dy$ with y primitive and d is called the *divisibility* of x. Such embeddings are called *simple* if the fundamental group of the complement is abelian (and hence isomorphic to $G = \mathbb{Z}/d$). Denote $y \cdot y$ by m, and let $b_2(N)$ and $\sigma(N)$ denote the rank and signature of the intersection form on $H_2(N; \mathbb{Z})$. A homology class x is called *characteristic*, if its reduction mod 2 is dual to w_2.

Theorem 7.1 ([HaKr92$_2$]) *Let N be a closed 1-connected topological 4-manifold.*

i) Let $x \in H_2(N; \mathbb{Z})$ be a homology class of divisibility $d \neq 0$. Then x can be represented by a simple locally flat embedded 2-sphere in N if and only if

$$KS(N) = (1/8)(\sigma(N) - x \cdot x) \bmod 2$$

when x is a characteristic class, and if

$$b_2(N) \geq \max_{0 \leq j < d} |\sigma(N) - 2j(d - j)(1/d^2)x \cdot x|.$$

ii) Any two locally flat simple embeddings of S^2 in N representing the homology class x are ambiently isotopic if $b_2(N) > |\sigma(N)| + 2$ and

$$b_2(N) > \max_{0 \leq j < d} |\sigma(N) - 2j(d - j)(1/d^2)x \cdot x|.$$

The proof will be based on the original idea of V. Rochlin (as in [LeWi90]). The embedding problem will be studied via an associated semi-free cyclic group action which is the same as a branched covering: if $f : S^2 \to N$ is an embedding representing a homology class of divisibility d, then there is a d-fold branched cyclic covering (M, G) over N, branched along $f(S^2)$.

This correspondence connects the embedding problem with the second topic of this section. It is the classification of actions of finite cyclic groups on 1–connected 4–manifolds, where we assume that the group action has a singular set consisting of isolated points. We also assume that the singular set of the action is *non-empty*: free actions, or equivalently 4-manifolds with finite cyclic fundamental group, were classified in Theorem 3.8. For earlier work in this direction compare [EdEw90], [Wi90]. The following result is a slight generalization of ([HaKr92$_2$], Corollary 4.1).

Theorem 7.2 (compare [HaKr92$_2$], Corollary 4.1) *Let M be a closed, oriented, simply-connected topological 4–manifold. Let G be a finite cyclic group* *acting locally linearly on M, preserving the orientation, with non-empty finite singular set. Let M_0 denote the complement of a set of disjoint open G-invariant 4-disks around the singular set, and assume that $X = M_0/G = W \# (S^2 \times S^2)$, where $\partial W = \partial(M_0/G)$. Then the action (M, G) is classified up to equivariant homeomorphism by the w_2-type, the local singular data, the signature and Euler characteristic of M and the Kirby–Siebenmann invariant of M_0/G.*

The "w_2-$type$" is I, II or III, if $w_2(M) \neq 0$, if $w_2(X) = 0$ or if $w_2(M) = 0$ and $w_2(X) \neq 0$ resp. The "local singular data" is the equivalence class of pairs consisting of the tangential G-representations at the singular set together with, when M is spin and $|G|$ is even, a preferred set of spin structures on the lens spaces bounding $X = M_0/G$. To describe this preferred set note that the w_2-type determines the normal 1-type of X. If M is spin and $|G|$ is even, then a normal 1-smoothing on X determines a spin-structure on $\nu(X) - L$, where L is a complex line bundle with $w_2(L) = w_2(X)$ and both possible spin-structures occur. Now, consider the boundary components $\partial_i X$. If the map from $H^2(X; \mathbb{Z}/2)$ to $H^2(\partial_i X; \mathbb{Z}/2)$ is non-trivial for some i then it is an isomorphism and, since $\partial_i X$ is spin, X is spin. In this case we choose L the trivial bundle and the preferred set of spin structures is the restriction of any spin structure on X to ∂X. If the map from $H^2(X; \mathbb{Z}/2)$ to $H^2(\partial_i X; \mathbb{Z}/2)$ is trivial for all i, then the restriction of L to $\partial_i X$ is stably trivial for all i and a normal 1-structure on X determines a spin structure on ∂X. Any of these gives the preferred set in this second case.

We also remark that $KS(M_0/G) = KS(M_0) = KS(M)$ when G has odd order, since connected sum with the Chern manifold changes the $\mathbb{Z}/2$-valued Kirby-Siebenmann invariant.

The proof of both theorems is similar in spirit but the proof of Theorem 7.1 is rather lengthy. We will prove Theorem 7.2 in detail and only give a sketch for Theorem 7.1 and refer to [HaKr92$_2$] for the details. Let G act on M with fixed point set either a 2-sphere and semi-free action (Theorem 7.1, ii)) or with finite singular set with prescribed fixed point data (Theorem 7.2). Then we denote by M_0 the complement of an open equivariant tubular neighborhood around the fixed point set resp. singular set. Given another action choose a homeomorphism between the boundaries of $X = M_0/G$. We have to show that the homeomorphism type of M_0/G rel. boundary is determined by the data. For this one first proves that the homeomorphism extends stably. This is an application of a relative version of Theorem 2.3. This relative version says that a homeomorphism between two compact topological 4-manifolds M_0 and M_1 with the same algebraic normal 1-type $[\pi, w_1, w_2]$ extends to a stable homeomorphism, if and only if they have the same Euler characteristic and if they admit normal 1-smoothings $\bar{\nu}_0$ and $\bar{\nu}_1$ resp., which are compatible with the homeomorphism between the boundaries and such that the union of $(M_0, \bar{\nu}_0)$ and $(M_1, \bar{\nu}_1)$ via the homeomorphism along the boundaries represent zero in $\Omega_4(B^{Top}[\pi, w_1, w_2])$ ([Kr85], Theorem 2.1). By assumption there exist compatible normal 1-smoothings. In our situation, the Atiyah-Hirzebruch spectral sequence implies that this bordism group is determined by the signature and Kirby-Siebenmann obstruction. Then one uses a relative version of Theorem 3.1 to cancel ([HaKr92$_2$], Corollary 3.6). For this one has to show that one can split off $S^2 \times S^2$ from M_0/G, something which is assumed in Theorem 7.2, and which follows from the inequalities in Theorem 7.1, ii) and the existence result in Theorem 19 i). This finishes the proof of Theorem 7.2. For Theorem 7.1, ii) one has to show that the resulting homeomorphism of N mapping the two embedded 2-spheres into each other is isotopic to Id. For this one carries the program above out with more care to control the induced map on homology which has to be the identity. Then one applies a Theorem from [Kr79] which says that a self-homeomorphism on a 1-connected 4-manifold inducing Id on homology is pseudo-isotopic to the identity. By a theorem of Perron [Pe86], this implies the existence of an isotopy.

To prove Theorem 7.1 i) one uses again a stabilization argument. The point will be to construct an embedding of S^2 into $N' = N \# r(S^2 \times S^2)$ for some r representing $x + 0$. Now, consider the ramified covering M' over N', ramified over the embedded 2-sphere. One has to carry out the construction of N' and the embedding in such a way that $H_2(M'; \mathbb{Z})$, considered as module over $\mathbb{Z}[G]$ with equivariant intersection form splits off a hyperbolic summand of rank r, such that the fixed point set under the G-action on this orthogonal complement is isomorphic to $H_2(N; \mathbb{Z})$ and the homology class represented by the embedded 2-sphere is x. This will follow from some purely algebraic arguments. Then it is not difficult to cancel the hyperbolic summand geometrically using Freedman's techniques, to realize the homology class x by an embedded 2-sphere in the original manifold N.

Chapter X

J. H. C. Whitehead's Asphericity Question

William A. Bogley

1   Introduction

This article is concerned with a question posed by Whitehead in 1941 [Wh41₁]:

> "Is any subcomplex of an aspherical, 2-dimensional complex itself aspherical?"

By a 2-dimensional complex is meant a CW complex in which each cell has dimension less than three; in short, a 2-complex. By definition, a connected complex X is aspherical if its universal covering complex $\tilde{X}$ is contractible. For a connected 2-complex X, this is equivalent to $\pi_2(X) = 0$ by use of Whitehead's Theorem (II.2.12) for $\tilde{X}$.

This question remains unanswered despite considerable expense of effort; a wide variety of results is scattered throughout the literature. The present intent is to survey these efforts and to present both a summary of the published results and an overview of the methods that have been used in the study of the problem.

The context of Whitehead's question is discussed in Section 2. In Section 3, hypotheses on the cellular structure of a subcomplex of an aspherical 2-complex are considered under which that subcomplex must be aspherical. Section 4 begins with a result of Howie [Ho83₂] that essentially bifurcates the study of Whitehead's question into a 'finite case' and an 'infinite case'. Each of these cases leads to open questions in combinatorial group theory. The finite case affirms a relation between Whitehead's question and the asphericity of knot complements, and includes a connection to the Andrews-Curtis conjecture. (See Chapters I and XII of this volume.) The discussion in Sections 5-7 is algebraic, in fact group theoretic; the work treated in these sections relates the asphericity of K to the subgroup structure of $\ker(\pi_1 K \to \pi_1 L)$, where K is a connected subcomplex of an aspherical 2-complex L. Asphericity of K is related to homological properties of $\pi_1 K$ in Section 8 via a theorem of Kaplansky [Ka72, Mo69]. Topological considerations are revisited in Section 9, where null-homotopies of spherical maps are considered directly by means of framed links. Throughout the article, the advantage of hindsight is freely employed. Little heed is paid to chronology and some results have been somewhat paraphrased. The author is solely responsible for any omissions or misrepresentations. The concluding Section 10 contains questions for future work, and is co-authored with J. Howie.

2   The Context of Whitehead's Question

Whitehead's work in [Wh41$_1$] concerns the relation between $\pi_n X$ and $\pi_n Y$ where Y is obtained from the path connected space X by attaching n-cells ($n \geq 2$). Among the results one finds an exact sequence

$$\pi_n X \to \pi_n Y \to R_n \to \pi_{n-1} X,$$

where R_n is an explicitly given group upon which $\pi_1 X$ acts on the left [Wh41$_1$, Theorems 3 and 4]. Subsequent papers [Wh46, Wh49$_2$] reveal R_n as the relative homotopy group $\pi_n(Y, X)$ and the above as a portion of the long exact homotopy sequence for the pair (Y, X). When $n \geq 3$, R_n is a free $\mathbb{Z}\pi_1 X$-module with basis in one-to-one correspondence with the n-cells of $Y - X$. (Still with $n \geq 3$, Whitehead also gives an explicit description of $\mathbb{Z}\pi_1 X$-module generators for $\ker(\pi_n X \to \pi_n Y)$ [Wh41$_1$, Lemma 4] involving a certain product

$$\pi_{n-1} X \otimes \pi_2 X \to \pi_n X$$

[Wh41$_1$, pages 411-413].)

The situation is radically different when $n = 2$, principally because $\pi_2(Y, X)$ is non-abelian, in general. In [Wh41$_1$] the group R_2 (there denoted h_Γ) is defined in terms of generators and relations. The isomorphism $R_2 \cong \pi_2(Y, X)$ is recognized in [Wh46]. In [Wh49$_2$] the terminology is applied that $\pi_2(Y, X) \to \pi_1 X$ is a free crossed $\pi_1 X$-module when Y is obtained from X by attaching 2-cells. (See Chapters II and IV of this volume). Subsequent expositions of this result appear in [Br80, Co51, Fe83, Ra80, Si80]. See [Br84, BrHi78, BrHu82, Dy87$_1$, Pe49, GuRa81] for applications and further developments.

The description of relative π_2 in terms of generators and relators prompted Whitehead to pose his question, as it is thus seen to reduce to a problem in algebra. For suppose that an aspherical 2-complex L is a union of subcomplexes K and K' where $K \cap K'$ is the 1-skeleton $L^{(1)}$. Comparing the exact homotopy sequence of $(K, L^{(1)})$ with that of $(L, L^{(1)})$, it is easy to see that K is aspherical $\Leftrightarrow \pi_2 K = 0 \Leftrightarrow \pi_2(K, L^{(1)}) \to \pi_2(L, L^{(1)})$ is injective. The point is that Whitehead's description of relative π_2 shows that $\pi_2(L, L^{(1)})$ is an explicitly given quotient of the free product $\pi_2(K, L^{(1)}) * \pi_2(K', L^{(1)})$, where each of the factors is explicitly given in terms of generators and relators. The question of whether $\pi_2(K, L^{(1)}) \to \pi_2(L, L^{(1)})$ is injective is therefore a (difficult) problem in combinatorial group theory; it has been treated algebraically in [GiHi89, Gi92]. Further reformulations of Whitehead's question based on the free crossed module structure appear in [Pa63, (14.1)] and [Si80, (4)].

Interest in Whitehead's question can be motivated by the fact that the complement of any tame knot in the 3-sphere has the homotopy type of a 2-complex that can be embedded in a finite contractible 2-complex. A positive solution to Whitehead's question therefore holds the promise of a new proof of the asphericity of knot complements. A footnote included in the midst of Whitehead's original question [Wh41$_1$, Footnote 30] suggests that this prospect may have been uppermost in Whitehead's mind at the time. See Section 4 for further discussion of this and related issues.

From a more general perspective, computation of the second homotopy group of 2-complexes, including the asphericity of 2-complexes, is a difficult problem that lies at the heart of the general problem of computing the homotopy groups of CW complexes. Using results from [Wh39], an effective algorithm for computing all higher homotopy modules of CW complexes in terms of generators and relations would be available if such existed for the nth homotopy module of n-complexes ($n \geq 2$). Whitehead's work on crossed modules provides an abstract algebraic description of the second homotopy group of a 2-complex [Wh41$_1$, page 427]. Abelianizing, one obtains the homological description of π_2 in terms of Reidemeister chains [Re34, Re50, Wh46]. (See also Chapter II, Theorem 3.8, in this volume.) However, as Whitehead himself observes [Wh41$_1$, page 409] [Wh49$_2$, page 495], neither of these descriptions leads to effective general calculations of π_2. Nor do they shed any practical light on Whitehead's question on the heredity of asphericity. Work on this question may therefore by viewed as fundamental to the further understanding of asphericity of 2-complexes and of the mechanisms that create the algebraic structure of π_2.

Much of the work that has been done on Whitehead's question has a strong group-theoretic flavor. Indeed, in the decades since it was posed, the question has played direct or indirect roles in the development of several new themes in group theory. The homological characterization of locally indicable groups achieved in [HoSc83] is a logical continuation of work of Adams [Ad55] on Whitehead's question. (See Section 6 below.) The study of equations over groups is certainly related to asphericity in some way (see e.g. Section 5); however the exact nature of this relation is not fully understood.

3   Structural Results

Let K be a connected subcomplex of an aspherical 2-complex L. This section discusses hypotheses on the cell structure of K which guarantee that K is aspherical. Of course, any 1-dimensional CW complex is aspherical. Cockcroft [Co54] used Lyndon's Simple Identity Theorem [Ly50] to show that if K has just a single 2-cell, then K is aspherical. This implies, for example, that if L is an aspherical 2-complex with at most two 2-cells, then every subcomplex of L is aspherical. Here we will use [Ho82] in place of the Simple Identity Theorem to considerably extend Cockcroft's result.

A fundamental observation of Cockcroft [Co54] is that the Hurewicz homomorphism $\pi_2 K \to H_2 K$ is trivial. This follows from the fact that the composite $\pi_2 K \to H_2 K \to H_2 L$ factors through $\pi_2 L = 0$ and that $H_2 K \to H_2 L$ is injective, as L is 2-dimensional. In other words, every spherical map $S^2 \to K$ is homologically trivial. 2-complexes with this latter property are therefore said to be *Cockcroft*. The Cockcroft property has a notable consequence for the attaching maps of 2-cells as seen below in 3.1 and 3.2.

An element g in a group G is a *proper power in G* if there exists an integer $n > 1$ and an element $a \in G$ such that $a^n = g$. Note that the identity element is thus a proper power. A loop $f : S^1 \to X$ in a path connected space X is a *proper power in X* if, after connecting f to the basepoint of X by a path, the resulting element of $\pi_1 X$ is a proper power in $\pi_1 X$. This is well-defined since varying choices of connecting path or basepoint vary the resulting elements of $\pi_1 X$ only up to conjugacy in $\pi_1 X$. Finally, if a 2-cell e is attached to a space X, then e is *attached by a proper power in X* if an attaching map $S^1 \to X$ for e is a proper power in X. As an example, let Y be the 2-complex modeled on $(x, y, t : xyx^{-1}y^{-1}, txytyx)$ and let X be the subcomplex modeled on $(x, y, t : xyx^{-1}y^{-1})$. The 2-cell of $Y - X$ is not attached by a proper power in the 1-skeleton $Y^{(1)}$, but it is attached by a proper power in X. This is because $xy = yx$ in $\pi_1 X$, and so $txytyx = (txy)^2$ in $\pi_1 X$. The following two lemmas are closely related to [Co54, pages 383-384], [Hu79, Proposition 1] and [Hu81, Proposition 5].

Lemma 3.1 *If Y is a connected Cockcroft 2-complex with torsion-free fundamental group and e is a 2-cell of Y, then e is not attached by a proper power in $Y - e$.*

Proof: Let $F : B^2 \to Y$ be a based version of a characteristic map for e, obtained by connecting the basepoint of Y to a point in the boundary of e. The restriction $f = F|_{S^1} : S^1 \to Y - e$ is then a based version of an attaching map for e. Just suppose that $a \in \pi_1(Y - e)$ and that n is an integer such that $a^n = [f]$ in $\pi_1(Y - e)$. First note that $n \neq 0$; for otherwise Y could be 3-deformed to $(Y - e) \vee S^2$, which is not Cockcroft. It suffices to show that $n = \pm 1$. The following commutative diagram has exact rows.

$$
\begin{array}{ccccccc}
\pi_2 Y & \xrightarrow{j} & \pi_2(Y, Y - e) & \xrightarrow{\partial} & \pi_1(Y - e) & \xrightarrow{i} & \pi_1 Y \\
\downarrow{\scriptstyle 0} & & \downarrow{\scriptstyle h} & & & & \\
H_2 Y & \to & H_2(Y, Y - e) & & & &
\end{array}
$$

Since $i([f]) = 1$ and $\pi_1 Y$ is torsion-free, it follows that $i(a) = 1$. Thus there exists an element $A \in \pi_2(Y, Y - e)$ such that $\partial(A) = a$. Now, $\partial([F]A^{-n}) = [f]a^{-n} = 1$, so $[F]A^{-n} \in \operatorname{im} j$. On the other hand, Y is Cockcroft (i.e., $\pi_2 Y \xrightarrow{0} H_2 Y$), and so $h([F]A^{-n}) = h([F]) - nh(A) = 0$. Since $H_2(Y, Y - e)$ is infinite cyclic, generated by $h([F])$, it follows that $n = \pm 1$.   □

Lemma 3.2 *If K is a connected subcomplex of an aspherical 2-complex L and e is a 2-cell of K, then e is not attached by a proper power in $K - e$.* □

Proof: Since L is Cockcroft and has torsion-free fundamental group (indeed, $\pi_1 L$ has cohomological dimension at most two), e is not attached by a proper power in $L - e$ by 3.1. It follows immediately that e is not attached by a proper power in $K - e$.   □

Following [Ho82], a 2-complex X is *reducible* if for each finite subcomplex Y of X, either $Y \subseteq X^{(1)}$ or else there exists a subcomplex Z of Y such that Y is obtained from Z by attaching a single 1-cell e^1, and at most a single 2-cell e^2; the 2-cell, if it exists, is required to *strictly involve* the 1-cell e^1, in the sense that its attaching map $S^1 \to Z \cup e^1$ is not freely homotopic in $Z \cup e^1$ to a loop in Z. A typical reducible 2-complex is built in the following way. Begin with a 1-complex and attach a single new 1-cell and a single new 2-cell, where the attaching map for the new 2-cell strictly involves the new 1-cell. Repeat this process as often as you like. The notion of reducible 2-complexes is a natural generalization of *staggered* 2-complexes [Ge87$_1$, Ho87] and of staggered group presentations [Ma30, Ly50]. Howie proves [Ho82, Corollary 5.4]:

If X is a reducible 2-complex such that for each 2-cell e of X, e is not attached by a proper power in X − e, then X is aspherical.

Removing the proper powers hypothesis, no general computation of π_2 is known for reducible 2-complexes. Nevertheless, with 3.2 one has the following result, which does not seem to have been observed in the literature.

Theorem 3.3 *Connected reducible subcomplexes of aspherical 2-complexes are aspherical.*   □

This result includes the fact that a staggered subcomplex of an aspherical 2-complex is aspherical [Ly50, Hu81]. Finally, Cockcroft's result is a consequence of 3.3.

Corollary 3.4 ([Co54]) *If a connected subcomplex of an aspherical 2-complex has just a single 2-cell, then that subcomplex is aspherical.*   □

4   Reductions, Evidence and Test Cases

Suppose that K is a connected subcomplex of an aspherical 2-complex L. As is well known, there is no loss of generality in the study of Whitehead's question if one assumes either of the following:

- L is obtained from K by attaching 2-cells, or

- K is finite and L is contractible.

The first reduction is available because $\pi_2 K = 0 \Leftrightarrow \pi_2(K \cup L^{(1)}) = 0$. The second reduction is available by compact supports, since the universal covering complex $\tilde{L}$ is contractible. Unless otherwise stated, neither of these assumptions will be in force in the sequel. A deeper reduction of the problem appears in a theorem of Howie.

Theorem 4.1 ([Ho83$_2$]) *If the answer to Whitehead's question is NO, then there exists a connected 2-complex L such that either*

1. *L is finite and contractible and L − e is not aspherical for some open 2-cell e of L, or*

2. *L is the union of an infinite ascending chain of finite connected non-aspherical subcomplexes $K_0 \subset K_1 \subset \ldots \subset K_{i-1} \subset K_i \subset \ldots$, where each inclusion $K_{i-1} \subset K_i$ is nullhomotopic.* □

As Howie points out, a weaker form of this result, obtained by replacing the word 'contractible' in 4.1.1 by 'aspherical,' is elementary. A proof of this weaker form can be extracted from [Ho83$_2$] as follows. Given that the answer to Whitehead's question is NO, assume first that there exists a *finite* aspherical 2-complex Y with a connected non-aspherical subcomplex X. Suppose that $Y - X$ has 2-cells $e_1, \ldots, e_n (n \geq 1)$, and let m be the minimum among all i such that $X \cup Y^{(1)} \cup e_1 \cup \ldots \cup e_i$ is aspherical. Then $1 \leq m \leq n$, so taking $L = X \cup Y^{(1)} \cup e_1 \cup \ldots \cup e_m$ and $e = e_m$ produces an example as in the weaker form of 4.1.1. Assume next that the answer to Whitehead's question is NO, but that connected subcomplexes of finite aspherical 2-complexes are aspherical. Select an aspherical 2-complex Y with a connected non-aspherical subcomplex X. Let $p : \tilde{Y} \to Y$ be the universal covering, and let $\bar{X}$ be a connected component of $p^{-1}(X)$. Then $\bar{X}$ is not aspherical and so, by compact supports, has a finite connected non-aspherical subcomplex K_0. Since $\tilde{Y}$ is contractible, the inclusion $K_0 \subset \tilde{Y}$ is nullhomotopic, and so extends over the cone CK_0 to a map $CK_0 \to \tilde{Y}$. Since CK_0 is finite, this extension has its image in a finite connected subcomplex K_1 of $\tilde{Y}$. The inclusion $K_0 \subset K_1$ is nullhomotopic and K_1 is not aspherical by hypothesis. Repeat the argument with K_0 replaced by K_1 and iterate. This produces the desired chain $K_0 \subset K_1 \ldots$; the result is completed by setting $L = \bigcup_i K_i$.

The proof of Theorem 4.1 employs *towers* of 2-complexes, a technique that Howie adapted from 3-manifold theory. For example, Papakyriakopoulos relied on tower constructions for his proof of the Sphere Theorem and the consequent asphericity of knot complements [Pa57]. Towers of 2-complexes have proved useful in a variety of contexts. See [Ho79, Ho81$_2$, Ho82, Ho83$_2$, Ge83]. Their use in [Ho79] will be outlined in Section 7 below.

Each of the two cases in 4.1 can be reduced to a problem in combinatorial group theory. The possibility of constructing an example as in the 'infinite case' 4.1.2 has been considered by Dyer [Dy91$_1$], and leads naturally to the study of increasingly subtle versions of the Cockcroft property. Note that if $K_0 \subset K_1 \subset \ldots$ is a chain as in 4.1.2, then each inclusion $K_{i-1} \subset K_i$ induces the trivial map on π_2. Following [BrDy81], let X be a connected 2-complex and let $N \leq \pi_1 X$. The 2-complex X is *N-Cockcroft* if the lifted Hurewicz map $\pi_2 X \to H_2 X_N$ is trivial, where $X_N \to X$ is the covering corresponding to N. Note that if X is N-Cockcroft and $N' \leq \pi_1 X$ contains some $\pi_1 X$-conjugate of

N, then X is N'-Cockcroft. Also, X is Cockcroft $\Leftrightarrow$ X is $\pi_1 X$-Cockcroft, while X is aspherical $\Leftrightarrow$ X is $\{1\}$-Cockcroft. Suppose now that X is a subcomplex of a connected 2-complex Y.

Lemma 4.2 *The inclusion-induced map* $\pi_2 X \to \pi_2 Y$ *is trivial if and only if* X *is* $\ker i_\#$-*Cockcroft, where* $i_\# : \pi_1 X \to \pi_1 Y$ *is the inclusion-induced homomorphism of fundamental groups.*

Proof: Let $p : \tilde{Y} \to Y$ be the universal covering and let $\bar{X}$ be a connected component of $p^{-1}(X)$; the restriction of p then determines the covering $\bar{X} \to X$ corresponding to $\ker i_\# \leq \pi_1 X$. Since $\pi_2 Y \to H_2 \tilde{Y}$ and $H_2 \bar{X} \to H_2 \tilde{Y}$ are both injective, it readily follows that $\pi_2 X \overset{0}{\to} \pi_2 Y \Leftrightarrow \pi_2 X \overset{0}{\to} H_2 \bar{X}$.   □

Suppose that a connected 2-complex L is given as an ascending union $K_0 \subset K_1 \subset \ldots \subset \bigcup_i K_i = L$ as in 4.1.2. Replacing each K_i by $K_i \cup L^{(1)}$, we have that for each $i \geq 1$, K_i is obtained from K_{i-1} by attaching 2-cells, and the inclusion-induced map $\pi_2 K_{i-1} \to \pi_2 K_i$ is trivial. Lemma 4.2 reveals the main difficulty in attempting to construct such an example: Having constructed K_{i-1}, one must add 2-cells in such a way that the resulting adjunction space K_i has a suitable Cockcroft property. Perhaps even more problematic is the fact that one must be able to do this infinitely many times. The requirements are detailed in the following theorem, which is a slight modification of a theorem of Dyer [Dy91₁].

Theorem 4.3 *If there exists an example as in 4.1.2, then there exists a finite connected non-aspherical 2-complex* K *and an infinite ascending chain* $\{1\} = N_0 < N_1 < N_2 < \ldots < \pi_1 K$ *of normal subgroups of* $\pi_1 K$ *such that the following three properties hold.*

1. *K is N_1-Cockcroft.*

2. *For each $i \geq 1$ there is a subset* $\mathbf{r}_i \subset \pi_1 K$ *such that*

    (a) *$\{r_i N_{i-1} : r_i \in \mathbf{r}_i\}$ normally generates N_i/N_{i-1} in $\pi_1 K/N_{i-1}$, and*

    (b) *$\{r_i N_{i-1}[N_{i+1}, N_{i-1}] : r_i \in \mathbf{r}_i\}$ is a free basis for the $\mathbb{Z}(\pi_1 K/N_{i+1})$-module*

    $$\mathbb{Z} \otimes_{\mathbb{Z} N_{i+1}} H_1(N_i/N_{i-1}) \cong N_i/(N_{i-1}[N_i, N_{i+1}]).$$

3. *For each $i \geq 1$ the natural map $H_2(N_{i+1}/N_{i-1}) \to H_2(N_{i+1}/N_i)$ is injective.*

*Furthermore, if such a 2-complex K exists, then the answer to Whitehead's
question is NO.* $\square$

Given K, N_i and $\mathbf{r}_i$ as in 4.3, set $K = K_0$ and for each $i \geq 1$ let K_i be obtained
from K_{i-1} by attaching 2-cells along based loops representing the elements
$r_i \in \mathbf{r}_i$. One then has exact sequences

$$1 \to N_i/N_{i-1} \to \pi_1 K/N_{i-1} \to \pi_1 K/N_i \to 1$$

where $\pi_1 K_i \cong \pi_1 K/N_i$. In particular, $N_{i+1}/N_i \leq \pi_1 K/N_i$ and $H_1(N_i/N_{i-1}) \cong$
$N_i/(N_{i-1}[N_i, N_i])$ has $\pi_1 K/N_i$-action induced by conjugation in $\pi_1 K/N_{i-1}$ (or
rather in $\pi_1 K$). Killing the action of N_{i+1}, the group $N_i/(N_{i-1}[N_i, N_{i+1}])$
becomes a $\mathbb{Z}(\pi_1 K/N_{i+1})$-module with the elements listed in 2(b) as module
generators. The condition 1 gives that $\pi_2 K_0 \overset{0}{\to} \pi_2 K_1$ by 4.2. In addition,
K_0 is N_2-Cockcroft. Given that K_{i-1} is N_{i+1}/N_{i-1}-Cockcroft, the conditions
2 and 3 together provide necessary and sufficient conditions for K_i to be
N_{i+1}/N_i-Cockcroft. By 4.2 and compact supports, the 2-complex $L = \bigcup_i K_i$
is aspherical.

Consider next the 'finite case' in 4.1.1. Work in this area has led to the study
of knots and ribbons, and of groups of the form $F/[R, S]$ where R, S are
normal subgroups of a finitely generated free group F. The latter considerations arise as follows. A *normal factorization* of a group G is an expression
$G = R_1 \ldots R_n$ of G as a product of a finite number of normal subgroups of
G. A normal factorization $F = R_1 \ldots R_n$ of a finitely generated free group F
is *efficient* if for $i = 1, \ldots, n$ there exist normal generating sets $\mathbf{r}_i$ for R_i in F
such that $|\mathbf{r}_1| + \ldots + |\mathbf{r}_n| = \text{rank} F$.

If A and B are subgroups of a group G, then $[A, B]$ denotes the subgroup of
G generated by all commutators $[a, b] = aba^{-1}b^{-1}$ $(a \in A, b \in B)$. If A and B
are normal in G, then so is $[A, B]$, and $[A, B] \subseteq A \cap B$. The nth term of the
lower central series of G is denoted G_n, and is defined inductively by $G_1 = G$
and $G_{n+1} = [G, G_n]$.

Lemma 4.4 [Bo91]) *The following two statements are logically equivalent.*

1. *Connected subcomplexes of finite contractible 2-complexes are aspherical.*

2. *If R and S are distinct factors from an efficient normal factorization
   of a finitely generated free group, then $R \cap S \subseteq [R, S]$.*

Proof: $2 \Rightarrow 1$: Let K be a connected subcomplex of a finite contractible 2-complex L. It suffices to show that $\pi_2 K = 0$. Replacing K by $K \cup L^{(1)}$, one may assume that L is obtained from K by attaching 2-cells. If K has just a single 2-cell, then K is aspherical by 3.4. Suppose then that K is a union of subcomplexes $K = K_r \cup K_s$ where each of K_r, K_s has at least one 2-cell and $K_r \cap K_s = L^{(1)}$. Now, $L = K \cup K_t$ where K_t is obtained from $L^{(1)}$ by attaching the 2-cells of $L - K$. Let $F = \pi_1 L^{(1)}$, a finitely generated free group, and let R (resp. S, T) denote the kernel of the homomorphism of fundamental groups induced by the inclusion of $L^{(1)}$ in K_r (resp. K_s, K_t). Then $F = RST$ since $\pi_1 L$ is trivial; this normal factorization of F is efficient because $H_2 L = 0$. As normal generators of R (resp. S, T), take the based homotopy classes of the attaching maps for the 2-cells of K_r (resp. K_s, K_t). By induction, each of K_r, K_s is aspherical, and by [GuRa81, Theorem 1], $\pi_2 K \cong (R \cap S)/[R, S]$. The conclusion 1 follows immediately from 2.

$1 \Rightarrow 2$: Suppose that R and S are distinct factors from an efficient normal factorization of a finitely generated free group F. Upon multiplication of the complementary factors there is a an efficient normal factorization $F = RST$ of F. There are sets $\mathbf{r}, \mathbf{s}, \mathbf{t}$ of normal generators for R, S, T in F such that $|\mathbf{r}| + |\mathbf{s}| + |\mathbf{t}| = \operatorname{rank} F$. Let X be a one-point union of circles, with one circle for each element of a basis of F; thus $\pi_1 X \cong F$. Let K be the 2-complex obtained by attaching 2-cells to X along based loops representing the elements of $\mathbf{r} \cup \mathbf{s}$, and let L be obtained from K by further attaching 2-cells along based loops representing the elements of $\mathbf{t}$. Then L is contractible since $F = RST$ is efficient. The hypothesis 1 implies that K is aspherical. By [GuRa81, Theorem 1], there is an epimorphism $\pi_2 K \to (R \cap S)/[R, S]$, and so $R \cap S = [R, S]$. □

Suppose that $F = RST$ is an efficient normal factorization of a finitely generated free group F, where $R \neq S$, and let $Q = F/[R, S]$. Using [GuRa81, Theorem 1], it is not difficult to show that $R \cap S = [R, ST] \cap [S, RT] \subseteq F_2$ [Bo91, Lemma 7], and so there is a natural homomorphism $(R \cap S)/[R, S] \to Q_2/Q_3 = F_2/[R, S]F_3$ which is induced by the inclusion of $R \cap S$ into F_2. Compare [BoGu92, Section 6]—for other choices of $R, S \trianglelefteq F$, this homomorphism is often nontrivial. In the present situation however, it turns out that $R \cap S \subseteq [R, S]F_3$, and so this homomorphism is trivial. As such, there is an inclusion-induced homomorphism $(R \cap S)/[R, S] \to Q_3/Q_4 = [R, S]F_3/[R, S]F_4$. One might hope to be able to use such a homomorphism to eventually detect an element of $(R \cap S)/[R, S]$ for suitably chosen R, S, F, and thus give a negative answer to Whitehead's question. The following result dashes such hopes.

Theorem 4.5 ([Bo91]) *If R and S are distinct factors from an efficient normal factorization of a finitely generated free group F, then*

$$R \cap S \subseteq \bigcap_{n \geq 1} [R, S] F_n.$$

□

It follows that the quotient $(R \cap S)/[R, S]$ embeds naturally in $Q_\omega = \bigcap_{n \geq 1} Q_n$. The proof of 4.5 essentially amounts to a determination of the structure of the Lie algebra that is built out of the lower central series of Q [Bo91, Theorem 2]. In particular it is shown that Q_n/Q_{n+1} is finitely generated and free abelian for all $n \geq 1$. It follows from a result attributed to Hall-Jennings in [Gu87, p. 75] that $Q_n = Q \cap (IQ)^n$ for all $n \geq 1$, where IQ denotes the augmentation ideal in the integral group ring $\mathbb{Z}Q$. Little seems to be known about Q_ω however. A group G is *residually nilpotent* if $G_\omega = 1$. We remark that residual nilpotence of groups of the form $F/[R, R]$ has been studied extensively. See [Gu87, Chapter II.4] and the references cited there.

There are many test cases to consider in the setting of 4.1.1. The model of any finite balanced presentation for the trivial group is a finite contractible 2-complex. (Conversely, any finite contractible 2-complex can be 3-deformed to such a model.) The Andrews-Curtis conjecture predicts that any finite contractible 2-complex can be 3-deformed to a point. (See Chapters I and XII in this volume.) For the Whitehead question, one is interested to know whether an aspherical subcomplex results upon deletion of a single 2-cell from a finite contractible 2-complex L. This is unknown even if one assumes that L can be 3-deformed to a point. Indeed, a large and concrete family of test cases for the Whitehead question arises in this setting. For if e is an open 2-cell of L, an analysis given in [Ho83$_2$] details the effect on $L - e$ of a 3-deformation of L.

Let Γ be a graph with vertices $V\Gamma$ and (geometric) edges $E\Gamma$. Assume that each edge of Γ is oriented and is labeled by a vertex of Γ. (Thus Γ is a *labeled oriented graph*.) Associated to Γ is a group presentation

$$\mathcal{P}(\Gamma) = (V\Gamma : i(e)\lambda(e)t(e)^{-1}\lambda(e)^{-1}(e \in E\Gamma)).$$

Here, $i(e), \lambda(e)$ and $t(e)$ denote the initial vertex, label and terminal vertex of the edge $e \in E\Gamma$, respectively. Let $K(\Gamma)$ denote the 2-complex modeled on $\mathcal{P}(\Gamma)$. If Γ is a finite tree, then upon adding a single relation of the form $v = 1, (v \in V\Gamma)$, there results a finite balanced presentation of the trivial group whose model contains $K(\Gamma)$ as a subcomplex, and which can be 3-deformed to a point (exercise).

Theorem 4.6 ([Ho83$_2$]) *If a finite 2-complex L can be 3-deformed to a point and e is an open 2-cell of L, then there exists a finite labeled oriented tree Γ such that $L - e$ can be 3-deformed to $K(\Gamma)$.* □

In particular, if the Andrews-Curtis conjecture is true (i.e., if each finite contractible 2-complex can be 3-deformed to a point), then the complexes $K(\Gamma)$ constitute *all* test cases for the Whitehead question in the finite case 4.1.1.

The 2-complexes $K(\Gamma)$ carry a large body of evidence to support the hypothesis that the answer to Whitehead's question might be YES. For if k is any tame knot in the 3-sphere S^3, then there is a finite labeled oriented tree Γ such that $K(\Gamma)$ embeds in the complement $S^3 - k$ as a spine. (That is, $K(\Gamma)$ is a strong deformation retract of $S^3 - k$.) Indeed, a suitable Γ arises via a Wirtinger presentation taken from a planar projection of k which contains only double points. (See e.g. [BuZi85].) Thus, for all Γ that arise in this way, $K(\Gamma)$ is aspherical, since simplicial knot complements are aspherical [Pa57]. The bewildering variety of knots therefore provides that many of the $K(\Gamma)$ are aspherical. However, not all of the $K(\Gamma)$ arise as spines of knot complements [Ho85, Ro91] and it is not known at present exactly which of the $K(\Gamma)$ do arise in this way. On the other hand, each $K(\Gamma)$ does appear as a spine of a certain four manifold, namely a ribbon disc complement, which is the complement in the four-ball of a properly embedded 2-disc. Of course, it is unknown at present whether all ribbon disc complements are aspherical. See [Ho83$_2$] for a discussion (including a survey of some errors that have appeared in the literature) and [Ho85] for partial results.

5　On the π_1-Kernel

Let K be a connected subcomplex of an aspherical 2-complex L, and let G be the kernel of the inclusion-induced homomorphism $\pi_1 K \to \pi_1 L$ of fundamental groups. Sections 5-7 concentrate on group-theoretic hypotheses on G under which K is guaranteed to be aspherical.

Let $p : \tilde{L} \to L$ be the universal cover and let $\bar{K}$ be a connected component of $p^{-1}(K)$. The restriction of p to $\bar{K}$ is a regular covering of K; indeed, $p_\#(\pi_1 \bar{K}) = G \trianglelefteq \pi_1 K$. Attention is focused on G by the following lemma, which was first observed by Cockcroft [Co54, page 376].

Lemma 5.1 *If $G = \ker(\pi_1 K \to \pi_1 L)$ is the trivial group, then K is aspherical.*

Proof: When $\pi_1 K \to \pi_1 L$ is injective, $\bar{K} \to K$ is the universal cover of K. Thus, $\pi_2 K \cong H_2 \bar{K} \leq H_2 \tilde{L} \cong \pi_2 L = 0$. □

This leads to a consideration the following problem:

If X is a connected subcomplex of a connected 2-complex Y, under what circumstances does the inclusion of X into Y induce a monomorphism of fundamental groups?

This in turn is equivalent to the study of equations over groups, a longstanding problem in combinatorial group theory with roots in work of B. H. Neumann [Ne43]. There is a rich literature on this problem. See for example [BrS80, EdHo91, Ge83, Ge87$_1$, GeRo62, Ho81$_2$, Ho83$_1$, Kl92, Kr85, Ne43, Rot76, Sh81]. For an algebraic formulation of this problem, consult [LySc77, page 49] or [Ho81$_2$]. The cited topological interpretation appears in [Ho81$_2$]. The following result is due to Gerstenhaber and Rothaus [GeRo62, Rot76].

If X is a connected subcomplex of a connected 2-complex Y, $H_2(Y, X) = 0$ and $\pi_1 X$ is locally residually finite, then the inclusion-induced homomorphism $\pi_1 X \to \pi_1 Y$ is injective.

A group H is *locally residually finite* if for each finitely generated subgroup F of H and for each nontrivial element $1 \neq f \in F$, there exists a normal subgroup N of F such that N has finite index in F and $f \notin N$. The proof of the Gerstenhaber-Rothaus theorem appeals to the homotopy theory of maps between Lie groups, first showing that the conclusion holds whenever $H_2(Y, X) = 0$ and $\pi_1 X$ is a subgroup of a compact Lie group, and to the representation theory of finite groups. The result was applied to Whitehead's question by Beckmann in his thesis. A group is *locally finite* if each of its finitely generated subgroups is finite.

Theorem 5.2 ([Be80$_1$]) *Let K be a connected subcomplex of an aspherical 2-complex L. If $G = \ker(\pi_1 K \to \pi_1 L)$ is locally finite, then G is trivial, and so K is aspherical.*

Proof: The restricted covering $\bar{K} \to K$ has $\pi_1 \bar{K} \cong G$. Since $\tilde{L}$ is contractible, the homology sequence for the pair $(\tilde{L}, \bar{K})$ reveals that $H_1 \bar{K} \cong H_2(\tilde{L}, \bar{K})$ is free abelian. Since G is locally finite, $H_1 \bar{K}$ is a torsion group, and so $H_2(\tilde{L}, \bar{K}) = 0$. Since locally finite groups are locally residually finite, the Gerstenhaber-Rothaus result implies that $\pi_1 \bar{K} \to \pi_1 \tilde{L}$ is injective, and so $G = 1$. Asphericity of K follows from 5.1. □

Beckmann's result is a descendant of a theorem of Cockcroft [Co54], in which it is assumed that $\pi_1 K$ is finite and that L is obtained from K by attaching 2-cells. The argument in 5.2 does not extend to the case where G is locally residually finite, since the abelianization of a locally residually finite group need not be a torsion group. For example, free groups are locally residually finite ([IIa49] [LySc77, page 195]).

6   Acyclic Coverings

Results on the π_1-kernel in the setting of Whitehead's question were first given by Cockcroft [Co54]. These results were recovered and generalized by Adams [Ad55] using very different methods. Adams's argument shows that any connected subcomplex of an aspherical 2-complex must have an acyclic regular covering complex. (A 2-complex X is *acyclic* if $H_1X = H_2X = 0$.)

Let G be a group and let $\bar{A}$ be an abelian group. The group G is *conservative over* A if whenever $\bar{X} \to X$ is a regular covering of 2-complexes whose group of covering automorphisms is isomorphic to G and $H_2(X; A) = 0$, then $H_2(\bar{X}; A) = 0$. The group G is *conservative* if it is conservative over all abelian groups A. Moderately surprising is the fact [HoSc83, Theorems 1 and 2] that if G is conservative over $\mathbb{Z}$, then G is conservative. The notion of conservativity is due to Adams [Ad55], who established a number of closure properties of the class of conservative groups, and proved the fundamental fact that the infinite cyclic group is conservative. Homological and group-theoretic characterizations of conservativity have since appeared, as in the following theorem of Howie and Schneebeli [HoSc83].

The following are equivalent for a group G:

1. *G is conservative;*

2. *If R is any commutative ring with 1 and $\phi : M \to N$ is a homomorphism of projective left RG-modules such that $1 \otimes \phi : R \otimes_{RG} M \to R \otimes_{RG} N$ is injective, then ϕ is injective;*

3. *Each nontrivial finitely generated subgroup of G has an infinite cyclic homomorphic image.*

The equivalence of conservativity with the homological condition 2 was (essentially) observed by Adams [Ad55] (see also [HoSc83, Lemma 3.2]). Strebel [St74] proved that $2 \Rightarrow 3$. Groups satisfying 3 are said to be *locally indicable*; this group-theoretic property was first considered by Higman [Hi40] in his study of units and zero divisors in group rings. The implication $3 \Rightarrow 2$ for $R = \mathbb{Z}$ or $\mathbb{Z}_p$ was proved independently by Gersten [Ge83] and by Howie and Schneebeli [HoSc83]. The stated result appears in [HoSc83].

Locally indicable groups have also been studied in connection with equations over groups. See [BrS80, Ge83, Ho81$_2$, Kr85, Sh81]. Groups which are known to be locally indicable (i.e., conservative) include free groups, torsion-free nilpotent groups [St74], torsion free one-relator groups [BrS80, Ho82], knot groups [Ho82], certain ribbon disc groups [Ho85] and also the fundamental group of any reducible 2-complex X such that for each 2-cell e of X, e is not attached by a proper power in $X - e$ [Ho82].

The following universal coefficients lemma will be useful. An elementary proof is to be found in [Ad55, page 487].

Lemma 6.1 *For a 2-complex X, $H_2(X; \mathbf{Z}/n\mathbf{Z}) = 0$ for all $n \in \mathbf{Z}$ if and only if $H_2 X = 0$ and $H_1 X$ is torsion-free.* □

Following Adams [Ad55], define the *conservative radical* (or *Adams radical*) of a group G to be

$$\mathcal{A}(G) = \bigcap \{N \trianglelefteq G : G/N \text{ is conservative}\}.$$

This is a characteristic subgroup in the sense that if $f : G \to H$ is a group homomorphism, then $f(\mathcal{A}(G)) \subseteq \mathcal{A}(H)$. In particular, $\mathcal{A}(G) \trianglelefteq G$. Further, as Adams points out, $G/\mathcal{A}(G)$ is conservative. These facts are not difficult to verify using the equivalence of conservativity and local indicability. For example, $G/\mathcal{A}(G)$ embeds in $\prod\{G/N : N \trianglelefteq G \text{ and } G/N \text{ is conservative}\}$; local indicability is hereditary and is preserved under direct products of groups. Finally, G is locally indicable if and only if $\mathcal{A}(G) = 1$. The following statement of Adams' theorem includes an observation due to Cohen [Co78, page 103].

Theorem 6.2 ([Ad55]) *Let K be a connected subcomplex of an aspherical 2-complex L and let $G = \ker(\pi_1 K \to \pi_1 L)$. Then, the regular covering complex of K corresponding to $\mathcal{A}(G) \trianglelefteq \pi_1 K$ is acyclic.*

Proof: Let $p : \tilde{L} \to L$ be the universal covering and let $\bar{K}$ be a connected component of $p^{-1}(K)$. Let $q : \hat{K} \to \bar{K}$ be the covering of $\bar{K}$ corresponding to $\mathcal{A}(G) \trianglelefteq G$. (That is, $q_\#(\pi_1 \hat{K}) = \mathcal{A}(G)$.) Then, $p|_{\bar{K}} \circ q : \hat{K} \to K$ is the covering of K corresponding to $\mathcal{A}(G) \leq \pi_1 K$. This covering of K is regular since $\mathcal{A}(G)$ is characteristic in the normal subgroup G of $\pi_1 K$, and so $\mathcal{A}(G) \trianglelefteq \pi_1 K$. It suffices to show that $H_1 \hat{K} = H_2 \hat{K} = 0$.

As in the proof of 5.2, the homology sequence for $(\tilde{L}, \bar{K})$ reveals that $H_2 \bar{K} = 0$ and that $H_1 \bar{K}$ is free abelian. By 6.1, $H_2(\bar{K}; \mathbf{Z}/n\mathbf{Z}) = 0$ for all $n \in \mathbf{Z}$. Since the automorphism group $G/\mathcal{A}(G)$ of q is conservative, $H_2(\hat{K}; \mathbb{Z}/n\mathbb{Z}) = 0$ for all $n \in \mathbb{Z}$. In particular, $H_2\hat{K} = 0$. Further, by 6.1, one has that $H_1\hat{K} \cong \mathcal{A}(G)/[\mathcal{A}(G), \mathcal{A}(G)]$ is torsion-free. The exact sequence

$$1 \to \mathcal{A}(G)/[\mathcal{A}(G), \mathcal{A}(G)] \to G/[\mathcal{A}(G), \mathcal{A}(G)] \to G/\mathcal{A}(G) \to 1$$

now shows that $G/[\mathcal{A}(G), \mathcal{A}(G)]$ is an extension of a locally indicable group by a locally indicable group, and so is locally indicable. Thus it follows that $\mathcal{A}(G) \subseteq [\mathcal{A}(G), \mathcal{A}(G)]$. The other containment being trivial, one has that $H_1\hat{K} \cong \mathcal{A}(G)/[\mathcal{A}(G), \mathcal{A}(G)] = 0$. $\square$

For a connected 2-complex X, let $\mathbf{ra}(X)$ denote the set of all normal subgroups P of $\pi_1 X$ such that X_P is acyclic, where $X_P \to X$ is the covering corresponding to P. (Here, $\mathbf{ra}$ stands for 'regular acyclic'.) Adams' theorem 6.2 says that if K is a connected subcomplex of an aspherical 2-complex L and $G = \ker(\pi_1 K \to \pi_1 L)$, then $\mathcal{A}(G) \in \mathbf{ra}(K) \neq \emptyset$. A group G is *perfect* if $H_1 G = 0$ (i.e., $G = [G, G]$), and is *superperfect* if $H_1 G = H_2 G = 0$.

Theorem 6.3 *Let X be a connected 2-complex.*

1. *X is aspherical $\Leftrightarrow$ $\{1\} \in \mathbf{ra}(X)$.*

2. *For $P \trianglelefteq \pi_1 X$, $P \in \mathbf{ra}(X)$ $\Leftrightarrow$ P is superperfect and X is P-Cockcroft.*

3. *[BrDy81, HoSc83] If $P \in \mathbf{ra}(X)$, then*

   (a) *$\operatorname{cd} \pi_1 X/P \leq 2$ and*

   (b) *For all $c \in \pi_1 X$ and $n \in \mathbb{Z}$, $c^n \in P \Rightarrow c \in P$.*

4. *If $\mathbf{ra}(X) \neq \emptyset$, then*

   (a) *[Dy91$_1$] if A is a ring with trivial $\pi_1 X$-action and $i = 1, 2$, then the cup product*

   $$H^i(\pi_1 X, A) \otimes H^2(\pi_1 X, A) \overset{\cup}{\to} H^{i+2}(\pi_1 X, A)$$

   is trivial, and

   (b) *[Mo90] all Massey products with codomain $H_3\pi_1 X$ vanish.*

Proof: The claim 1 is clear because a simply connected CW complex is contractible if and only if it is acyclic. Let $P \trianglelefteq \pi_1 X$ and let $p : X_P \to X$ be the regular covering corresponding to P (i.e. $p_\#(\pi_1 X_P) = P$). Then $H_1 P \cong H_1 X_P$ and there is the Hopf sequence

$$\pi_2 X_P \to H_2 X_P \to H_2 P \to 0.$$

It follows that if P is superperfect and X_P is P-Cockcroft, then X_P is acyclic, and conversely, proving 2. If X_P is acyclic, then the augmented cellular chain complex $C_*(X_P) \to \mathbb{Z} \to 0$ provides a free $\mathbb{Z}\pi_1 X/P$-resolution of $\mathbb{Z}$ of length two. This proves 3a; the conclusion 3b follows immediately from the fact that groups of finite cohomological dimension are torsion-free. The proof of 4 is omitted.   □

If X is aspherical, then $\mathrm{cd}\,\pi_1 X \le 2$, and so one way to show that X is *not* aspherical is to display some non-vanishing (co)homology element in a dimension greater than two. (For that matter, it would suffice to show that $\pi_1 X$ has nontrivial torsion.) If K is a connected subcomplex of an aspherical 2-complex, then $\mathbf{ra}(K) \ne \emptyset$ and the results in 6.3 show that $\pi_1 K$ is to a large extent 'homologically invisible'. Compare 4.5 and see [Dy91$_1$] for further results of this type.

Let K be a connected subcomplex of an aspherical 2-complex L and let $G = \ker(\pi_1 K \to \pi_1 L)$. By 6.2, $\mathcal{A}(G)$ is perfect. From this, Adams recovered two results of Cockcroft. Namely, if G (or $\pi_1 K$) is free or abelian, then G has no nontrivial perfect subgroups, and so K is aspherical by 6.3.1. From a more general perspective, while the abelianization of the conservative radical is always a torsion group [Ge83, Proposition 2.4], the conservative radical of an arbitrary group need not be perfect; any finite group is equal to its own conservative radical.

The *perfect radical* of a group G is the subgroup $\mathcal{P}(G)$ of G generated by the family of all perfect subgroups of G. (This is also referred to as the *maximal perfect subgroup* of G.) As with the conservative radical, $\mathcal{P}(G)$ is a characteristic, and hence normal, subgroup of G. This is also the intersection of all terms in the transfinite derived series of G. In the setting of 6.2, it follows that $\mathcal{A}(G) \subseteq \mathcal{P}(G)$. Even in this setting, this inclusion can be strict. For suppose that G is the fundamental group of the complement of a nontrivial knot in the 3-sphere with trivial Alexander polynomial [BuZi85, page 120]. Then $\mathcal{A}(G)$ is trivial, since G is locally indicable [Ho82]. Meanwhile, $\mathcal{P}(G)$ is the derived group of G, which is nontrivial since the knot is nontrivial. Adams [Ad55, page 487] gives an example with the same properties; in this case G is a torsion-free one-relator group. All of these examples occur within the Whitehead setting. On the other hand, the containment $\mathcal{A}(G) \subseteq \mathcal{P}(G)$ is a special feature of the Whitehead setting. For example, a finite abelian group is equal to its own conservative radical, while its perfect radical is trivial.

Theorem 6.4 ([BrDy81, BrDySt83]) *If K is a connected subcomplex of an aspherical 2-complex L and $G = \ker(\pi_1 K \to \pi_1 L)$, then $\mathcal{P}(G) \in \mathbf{ra}(K)$.*   □

The proof of 6.4 relies on Strebel's theory of $\mathbf{E}$-groups [St74]. The key step lies in the fact [St74, proof of Proposition 2.4] that $G/\mathcal{P}(G)$ is conservative in this case. In closing this section, there is the following theorem of Dyer.

Theorem 6.5 ([Dy93]) *Let K be a connected subcomplex of an aspherical 2-complex L and assume that $\pi_1 K$ is finitely presented. If $\mathbf{ra}(K)$ contains a finite group, then that group is trivial, and so K is aspherical.*   □

It should be noted that Beckmann [Be80$_2$] and Dunwoody [Du80] have both exhibited non-aspherical finite connected acyclic 2-complexes with torsion-free fundamental groups, thus answering a question of Cohen [Co78, page 104].

7   Finitely Generated Perfect Subgroups

As always, let K be a connected subcomplex of an aspherical 2-complex L and let $G = \ker(\pi_1 K \to \pi_1 L)$. Using the fact that locally indicable groups are conservative, Adams's result 6.3 implies that if G is locally indicable, then K is aspherical. (An alternate proof of this follows quickly from [Ho82, Theorem 5.2].) The following theorem of Howie shows that a weaker local hypothesis on G also implies the asphericity of K. The proof employs the towers mentioned in Section 4.

Theorem 7.1 ([Ho79]) *If G is locally nonperfect (i.e., has no nontrivial finitely generated perfect subgroups), then K is aspherical.*

Proof: As in Chapter II of this volume, $\pi_2 K$ is generated under the homotopy action of the edge-path groupoid $\pi_1(K^{(1)}, K^{(0)})$ by the homotopy classes of *spherical diagrams* $f : C \to K$, where C is a cell decomposition of the 2-sphere (see [Ge87$_1$]). Thus, f carries open cells of C homeomorphically onto open cells of K. In particular, the image of a spherical diagram is a subcomplex of the codomain complex. Let f be such, representing $[f] \in \pi_2 K$ (for suitable choice of basepoint). It suffices to show that $[f] = 0$. The following tower strategy shows how to factor f through a finite connected acyclic 2-complex.

Let $p : \tilde{L} \to L$ be the universal cover, and let $\bar{K}$ be a connected component of $p^{-1}(K)$. Since S^2 is simply connected, f lifts to a spherical diagram $f_0 :$

$C \to \bar{K}$. Then $f_0(C) = K_0$ is a finite connected subcomplex of $\bar{K}$, and $f_0 : C \to K_0$ is surjective. As K_0 is a subcomplex of the contractible 2-complex $\tilde{L}$, $H_2(K_0; \mathbb{Z}/n\mathbb{Z}) = 0$ for all $n \in \mathbb{Z}$. By 6.1, the finitely generated abelian group $H_1 K_0$ is torsion-free, and hence free abelian.

If $\pi_1 K_0$ is not perfect, then there exists an epimorphism $\pi_1 K_0 \to \mathbb{Z}$, and hence a regular covering $p_1 : \bar{K}_1 \to K_0$ with automorphism group $\mathbb{Z}$. (In short, p_1 is a $\mathbb{Z}$-*covering*.) Further, f_0 lifts to a surjective spherical diagram $f_1 : C \to K_1$, where K_1 is a finite connected subcomplex of $\bar{K}_1$. The crucial fact is this:

$$K_1 \text{ has more 0-cells than does } K_0.$$

For otherwise p_1 would restrict to a bijection $K_1^{(0)} \to K_0^{(0)}$ of 0-skeleta, and to a combinatorial surjection $K_1^{(1)} \to K_0^{(1)}$ of 1-skeleta. From this it would follow that the map $\pi_1 K_1 \to \pi_1 K_0$ is surjective, and hence that $p_{1\#} : \pi_1 \bar{K}_1 \to \pi_1 K_0$ is surjective, a contradiction.

Since $\mathbb{Z}$ is conservative [Ad55, Proposition 1], it follows that $H_2(\bar{K}_1; \mathbb{Z}/n\mathbb{Z}) = 0$ for all $n \in \mathbb{Z}$. Since K_1 is a subcomplex of the 2-complex $\bar{K}_1$, one further has that $H_2(K_1; \mathbb{Z}/n\mathbb{Z}) = 0$ for all $n \in \mathbb{Z}$. Applying 6.1, either $\pi_1 K_1$ is perfect or else the lifting process can be repeated. As successive factorizations of f through finite connected complexes K_i are obtained, the number of 0-cells in the K_i strictly increases. However, the number of 0-cells in each K_i is bounded above by the number of 0-cells in the finite complex C. Therefore, the process must stop at some stage: After some finite number of steps, say m, one finds that K_m has perfect fundamental group. The net result is a factorization $f_0 = g \circ f_m : C \to K_m \to \bar{K}$ where f_m is surjective, K_m is acyclic and g is a composite of inclusions and $\mathbb{Z}$-coverings. (N.B. This factorization is referred to as a *maximal tower lifting* of f_0; the existence of such is the heart of the towers method. See [Ho81$_2$, Lemma 3.1].)

The proof is completed as follows. Since $\pi_1 \bar{K} \cong G$ is locally nonperfect, the homomorphism $g_\# : \pi_1 K_m \to \pi_1 \bar{K}$ is trivial, and so g lifts through the universal covering $u : \tilde{K} \to \bar{K}$ to a map $\tilde{g} : K_m \to \tilde{K}$. Now, $H_2 K_m = 0$ and the Hurewicz homomorphism $\pi_2 \tilde{K} \to H_2 \tilde{K}$ is an isomorphism. This implies that $g_\# : \pi_2 K_m \to \pi_2 \bar{K}$ is trivial. It follows that $[f_0] = 0$ in $\pi_2 \bar{K}$ and hence that $[f] = 0$ in $\pi_2 K$. □

With a further application of the tower method, Howie proves the following local version of 6.3.3b.

Theorem 7.2 ([Ho81$_1$]) *Let K be a connected subcomplex of an aspherical 2-complex L and let $G = \ker(\pi_1 K \to \pi_1 L)$. If c is an element of finite order in $\pi_1 K$, then there exists a finitely generated perfect subgroup P of G such that $c \in P$.* □

8 Kaplansky's Theorem

Among the results from [Co54] is the fact that if K is a finite connected subcomplex of an aspherical 2-complex and $\pi_1 K$ is a free group, then K is aspherical. As noted above, this result was recovered by Adams, even without the assumption that K is finite. Nevertheless, Cockcroft's approach merits note.

Let K be a finite connected subcomplex of an aspherical 2-complex, where $\pi_1 K$ is a free group. Cockcroft's proof that K is aspherical requires these facts:

1. K is Cockcroft;

2. There exists a finite aspherical 2-complex Y such that $\pi_1 Y \cong \pi_1 K$;

3. Any $\mathbb{Z}\pi_1 K$-module epimorphism between free $\mathbb{Z}\pi_1 K$-modules of the same finite rank is an isomorphism.

The fact 1 is a feature of any subcomplex of an aspherical 2-complex, as has been noted in Section 3. For fact 2, take Y to be an appropriate one-point union of circles. As for 3, Cockcroft appealed to the fact [Ne49, page 213] that if $\pi_1 K$ is a free group, then $\mathbb{Z}\pi_1 K$ can be embedded in a division ring. Cockcroft used 1 and 2 to show that there exists a finitely generated free $\mathbb{Z}\pi_1 K$-module F such that $\pi_2 K \oplus F \cong F$. The fact 3 then implies that $\pi_2 K = 0$. For if $\eta : F \to \pi_2 K \oplus F$ is an isomorphism, then composing with the projection $p : \pi_2 K \oplus F \to F$ yields an epimorphism $p \circ \eta : F \to F$, which by 3 is an isomorphism. Since $\eta^{-1}(\pi_2 K) \subseteq \ker(p \circ \eta) = 0$, it follows that $\pi_2 K = 0$.

The condition 2 is restrictive, depending heavily on the isomorphism type of $\pi_1 K$. However, it turns out that the condition 3 is not a restriction, as seen in the following theorem of Kaplansky [Ka72, Mo69]:

Let G be a group and let A and B be $n \times n$ matrices over the complex group ring $\mathbb{C}G$, where n is a positive integer. If $AB = 1$, then $BA = 1$.

It follows that the condition 3 holds regardless of the structure of $\pi_1 K$. The proof of Kaplansky's result is analytic. In search of an algebraic proof, Kaplansky asks [Ka72, page 122] whether the same result holds if $\mathbb{C}$ is replaced by $\mathbb{Z}/p\mathbb{Z}$. Topological applications of Kaplansky's theorem are given in [CoSw61]. Here is a compilation of subsequent applications to Whitehead's question.

Theorem 8.1 ([BrDy81, Co78, GuRa81]) *The following are equivalent for a finite connected Cockcroft 2-complex X:*

1. X *is aspherical;*

2. *There exists a finite aspherical 2-complex Y such that $\pi_1 X \cong \pi_1 Y$;*

3. *cd $\pi_1 X \leq 2$ and $\pi_1 X$ is of type* FL*;*

4. *cd $\pi_1 X \leq 3$, $H_3 \pi_1 X = 0$ and $\pi_1 X$ is of type* FL*.*

A group G is of *type* FL if the trivial module $\mathbb{Z}$ admits a resolution of finite length by finitely generated free $\mathbb{Z}G$-modules. The group G has *cohomological dimension* less than or equal to n (cd $G \leq n$) if there is a projective $\mathbb{Z}G$-resolution of $\mathbb{Z}$ of length n. The implication $2 \Rightarrow 1$ is a consequence of [BrDy81, Lemma 1.4]. (See also [BrDy81, page 432] and [CoSw61].) The equivalence $1 \Leftrightarrow 3$ appears in [GuRa81]. The implication $4 \Rightarrow 1$ is from [Co78].

Proof: The implications $1 \Rightarrow 2 \Rightarrow 3 \Rightarrow 4$ are trivial. Set $G = \pi_1 X$. Since cd $G \leq 3$ and G is of type FL, there is a free resolution of $\mathbb{Z}$

$$0 \to F_3 \to F_2 \to F_1 \to F_0 \to \mathbb{Z} \to 0$$

by finitely generated free $\mathbb{Z}G$-modules [Br82, XIII.6, Exercise 1]. There is also the augmented cellular chain complex

$$0 \to \pi_2 X \to \tilde{C}_2 \to \tilde{C}_1 \to \tilde{C}_0 \to \mathbb{Z} \to 0$$

of the universal cover $\tilde{X}$. By Schanuel's lemma [Br82, VIII.4.4], there is an isomorphism

$$\pi_2 X \oplus F_2 \oplus \tilde{C}_1 \oplus F_0 \cong F_3 \oplus \tilde{C}_2 \oplus F_1 \oplus \tilde{C}_0$$

of $\mathbb{Z}G$-modules. Let $\text{rk}_{\mathbb{Z}G}$ (resp. $\text{rk}_{\mathbb{Z}}$) denote the rank of a finitely generated free $\mathbb{Z}G$-module (resp. the torsion-free rank of a finitely generated abelian group). Now,

$$\sum_{i=0}^{3}(-1)^i \text{rk}_{\mathbb{Z}G} F_i = \sum_{i=0}^{3}(-1)^i \text{rk}_{\mathbb{Z}} (\mathbb{Z} \otimes_{\mathbb{Z}G} F_i) = \sum_{i=0}^{3}(-1)^i \text{rk}_{\mathbb{Z}} H_i G$$

since, for instance, the alternating sum of the torsion-free ranks of the homology groups of a finite chain complex of finitely generated abelian groups is equal to the corresponding alternating sum of the torsion-free ranks of the chain groups themselves. More to the point, the hypotheses give that $H_3G = 0$ and that X is Cockcroft, whence $H_2G \cong H_2X$. As such,

$$\sum_{i=0}^{3}(-1)^i \mathrm{rk}_{\mathbf{Z}} \, H_iG = \sum_{i=0}^{2}(-1)^i \mathrm{rk}_{\mathbf{Z}} \, H_iX = \sum_{i=0}^{2}(-1)^i \mathrm{rk}_{\mathbf{Z}} \, C_i = \sum_{i=0}^{2}(-1)^i \mathrm{rk}_{\mathbf{Z}G} \, \tilde{C}_i$$

where C_i denotes the ith cellular chain group of X. Upon equating ranks of finitely generated free $\mathbf{Z}G$-modules, it follows that there is an isomorphism

$$F = F_2 \oplus \tilde{C}_1 \oplus F_0 \cong F_3 \oplus \tilde{C}_2 \oplus F_1 \oplus \tilde{C}_0$$

of $\mathbf{Z}G$-modules. Thus, one finds that $\pi_2X \oplus F \cong F$. By Kaplansky's theorem, $\pi_2X = 0$.   □

It is perhaps worth noting that in each of [BrDy81, GuRa81, Co78], the aforementioned portions of 8.1 were proved under the stronger hypothesis that X is a finite connected subcomplex of an aspherical 2-complex, in which case X is certainly Cockcroft. On the other hand, there are lots of finite connected Cockcroft 2-complexes X which cannot appear as subcomplexes of aspherical 2-complexes. For example, take X to be the (non-aspherical) 2-complex modeled on either the presentation $(x, y : xyx^{-1}y, yxy^{-1}x)$ for the quaternion group of order eight, or the presentation $(a, b, c : [a, b], [b, c], [c, a])$ for the free abelian group of rank three.

If X is a connected 2-complex, then $\mathrm{cd}\,\pi_1X \leq 3$ if and only if π_2X is projective as a $\mathbf{Z}\pi_1X$-module [Br82]. When X is Cockcroft, $H_3\pi_1X = 0$ if and only if $\mathbf{Z} \otimes_{\mathbf{Z}\pi_1X} \pi_2X = 0$ (i.e., π_2X is *perfect* as a $\mathbf{Z}\pi_1X$-module). In [HoSc83, Corollary 3.4], conditions on π_1X are given under which there are no nonzero (finitely generated) perfect projective $\mathbf{Z}\pi_1X$-modules. When such an X is Cockcroft, $\pi_2X = 0$ if and only if $\mathrm{cd}\,\pi_1X \leq 3$ and $H_3\pi_1X = 0$.

9   Framed Links

Perhaps the most direct approach to Whitehead's question, indeed to asphericity in general, is to ask what a null-homotopy of a spherical map looks like. This leads to a consideration of linkages in the 3-ball.

Let K be a subcomplex of a 2-complex L, and let $f : S^2 \to K$ be a spherical map. The map f is nullhomotopic in L if and only if it extends to a map $H : B^3 \to L$ of the 3-ball into L. Simplicial techniques (e.g. [Br80, Si80, Wh41₁]) and Chapter II, Lemma 2.4 of this volume) provide that after a homotopy of maps of pairs $(B^3, S^2) \to (L, K)$, the closure of the inverse image of the open 2-cells of L under such a null-homotopy is a *framed link* in B^3. This means that for each open 2-cell $c^2 \subset L$, each connected component of the closure of $H^{-1}(c^2)$ is either an embedded solid torus $D^2 \times S^1$ in the interior of B^3, or else an embedded solid cylinder $D^2 \times B^1$ which meets the boundary sphere S^2 in just its end discs $D^2 \times \pm 1$. In either case, the null-homotopy H maps each cross-sectional disc $D^2 \times \{z\}$ characteristically onto the closed 2-cell $\bar{c}^2$. The solid tori and cylinders may be taken to be pairwise disjoint; together they comprise the framed link of the null-homotopy H.

Stefan [St82] showed that if L is assumed to be aspherical, then there are restrictions on what sorts of framed links can arise from null-homotopies of spherical maps into L. These results have been extended by Wolf [Wo91]. A rough outline of these arguments goes as follows.

Let Λ be the framed link of a null-homotopy $H : B^3 \to L$. Deleting the interior of Λ, the restriction of H determines a homomorphism h of the link complement group into the free group $\pi_1 L^{(1)}$. Taking a planar projection of the link that intersects itself only in double points, the link complement group admits a (Wirtinger) presentation of the form $\mathcal{P}(\Gamma)$, where Γ is a labeled oriented graph in the sense of Section 4 (Toroidal components in Λ give rise to simple circuits in Γ.) The generators are realized topologically by certain meridional loops around the boundary of Λ that are connected to a base point by arcs in B^3. Thus, generators in this Wirtinger presentation are mapped by h to certain $\pi_1 L^{(1)}$-conjugates of the homotopy classes of suitably based versions of the attaching maps for the 2-cells of L. (Remark: Whitehead [Wh41$_1$] used the Wirtinger relations to show that the homomorphism h lifts to a homomorphism of the link complement group into $\pi_2(L, L^{(1)})$. This alone is a good heuristic explanation for the form of the so-called 'Peiffer relations' for relative π_2—see [Br80] and Chapter II in this volume.)

Following Wolf [Wo91], consider a loop in the link complement that travels around a longitude in the boundary of a toroidal component of Λ in such a way that it does not link with a core of that embedded toroidal component (a 'zero-parallel' in Wolf's terminology). Connecting this loop to the base-point in B^3 and applying h, the resulting element $A \in \pi_1 L^{(1)}$ is a power of a conjugate of an attaching map for a 2-cell of L. (The power arises from any twisting that may be present in the embedding of the toroidal component.) On the other hand, this based loop represents an element in the link comple-ment group, and so under h determines a product B of $\pi_1 L^{(1)}$-conjugates of attaching maps. The elements A, B lie in the kernel of the inclusion-induced homomorphism $\pi_1 L^{(1)} \to \pi_1 L$, and so may be lifted to elements α, β in the relative group $\pi_2(L, L^{(1)})$. Further, $\alpha\beta^{-1}$ lies in the kernel of the boundary map $\pi_2(L, L^{(1)}) \to \pi_1 L^{(1)}$. When L is aspherical, it follows that $\alpha\beta^{-1} = 1$. Whitehead's description of $\pi_2(L, L^{(1)})$ as a free crossed $\pi_1 L^{(1)}$-module then allows one to make conclusions about the nature of the elements α, β, A, B, and eventually about the link Λ and the null-homotopy H. Key ingredients in these conclusions are Papkyriakopoulos' characterization of Peiffer identities [Pa63] (see Chapter II of this volume), and the fact that the centralizer of any element in a free group (such as $\pi_1 L^{(1)}$) is cyclic.

It is interesting to note that many of the results that Stefan and Wolf obtained by these methods are actually true if L is assumed only to be Cockcroft. A notable exception is [Wo91, Theorem 5].

Theorem 9.1 ([Wo91]) *If the cellular model of the presentation* $(\mathbf{x} : r, s, \mathbf{t})$ *is aspherical, then* r *and* s *do not commute in the group presented by* $(\mathbf{x} : \mathbf{t})$.

□

The point is that if r and s commute modulo $\mathbf{t}$, then there is an identity among these relations that gives rise to a spherical map of a certain form. Wolf uses framed links to argue that such a map cannot arise in an aspherical 2-complex.

Returning to Whitehead's question, suppose that $H : (B^3, S^2) \to (L, K)$ is a null-homotopy for a spherical map into K, where L is aspherical. It is then natural to consider only that portion of the framed link that corresponds to the inverse image of the 2-cells of $L - K$. (This consists entirely of toroidal components.) The same considerations as above then apply, with $\pi_1 L^{(1)}$ replaced by $\pi_1(K \cup L^{(1)})$. In particular, one still has a free crossed module description of $\pi_2(L, K \cup L^{(1)})$. What is lacking here is that one knows little about the centralizers of elements in $\pi_1 K$ in general.

Sieradski showed in [Si80] that there are spherical maps into aspherical 2-complexes for which no null-homotopy has a framed link that can be *geometrically split* in a sense made precise in [Si80, Section 4]. This resolved in the negative the question of whether an aspherical 2-complex is necessarily *diagrammatically aspherical*, a result also obtained by Chiswell [ChCoHu81]. (See [CoHu82, pp. 178-179] and [Hu81, Proposition 8] for further discussions.) Sieradski showed [Si80, Theorem 1] that an aspherical 2-complex L is diagrammatically aspherical if and only if each spherical map into L admits a null-homotopy whose framed link is geometrically split. On the other hand, diagrammatic asphericity is inherited by subcomplexes. This indicates that the heart of Whitehead's question is intrinsically tied to the complications inherent in (unsplittable) links, a daunting prospect.

A reasonable alternative might be to try to use the complexities of links to show that the answer to Whitehead's question is NO. A construction due to Metzler provides one possible approach. Let Γ be a labeled oriented graph, and consider a spherical diagram $f : C \to K(\Gamma)$. Thus, C is a tessellation of

S^2 and f carries open cells in C homeomorphically onto open cells of $K(\Gamma)$. Each 2-cell c of C has four oriented boundary 1-cells, labeled in sequence by a word of the form $i(e)\lambda(e)t(e)^{-1}\lambda(e)^{-1}$, where e is an edge of Γ. Thicken S^2 to a product $S^2 \times D^1$. Place two arcs in c joining midpoints of opposite boundary 1-cells of c so that the arc joining the two 1-cells labeled by $\lambda(e)$ overcrosses the arc joining $i(e), t(e)$. Having done this for each 2-cell of C, there results a link $\mathcal{L}$ in $S^2 \times D^1$. (This is the reverse of the process that produces a Wirtinger presentation for a link complement group.) The link $\mathcal{L}$ comes naturally equipped with a planar projection; each segment between undercrossings of this projection is labeled by a vertex of Γ, and each component of $\mathcal{L}$ can be consistently oriented. Conversely, a planar link projection that is suitably labeled and oriented determines a spherical diagram in $K(\Gamma)$. This construction was used in [Ro90] to produce examples of finite labeled oriented trees Γ for which $K(\Gamma)$ is aspherical, but is not diagrammatically aspherical, thus answering a question of Gersten [Ge87$_1$]. Applications to the Whitehead question of some combination of this link construction and the geometry of framed links await further developments.

10   Open Questions

(with J. Howie)

1. *Labeled oriented trees:* Are ribbon disc complements aspherical? That is, if Γ is a finite labeled oriented tree, is $\pi_2 K(\Gamma) = 0$? Howie has shown [Ho85] that if Γ has diameter not more than three, then $K(\Gamma)$ is aspherical and $\pi_1 K(\Gamma)$ is locally indicable. It appears that direct attempts to exploit the combinatorial structure of labeled oriented trees are bound to be complicated.

2. *Residual nilpotence:* Let R, S be distinct factors from an efficient normal factorization of a finitely generated free group F, where the factorization involves at least three distinct terms. Is $R \cap S$ contained in $[R, S]$? By 4.5 it would suffice to show that the group $Q = F/[R, S]$ is residually nilpotent. An example from [Bo91] suggests that this may not be the case in general. Failing that, can one prove an analogue of 4.5 that 'pushes $R \cap S$ down the derived series of Q'? Namely, is $(R \cap S)/[R, S]$ contained in $\mathcal{P}(Q)$?

3. *Acyclic coverings:* For a connected 2-complex X, let $\mathbf{c}(X)$ denote the set of all subgroups N of $\pi_1 X$ such that X is N-Cockcroft. Clearly, $\mathbf{ra}(X) \subseteq \mathbf{c}(X)$. Both of these classes of subgroups are partially ordered by inclusion. Properties of $\mathbf{c}(X)$ and related classes are discussed in [DyHa92, GiHo$_1$, GiHo$_2$, Pr92$_2$]; in particular[DyHa92, GiHo$_1$], the poset $\mathbf{c}(X)$ has minimal elements. (See also [HoSc83, Proposition 3.1] and Chapter IV of this volume.) Does

$\mathbf{ra}(X)$ have minimal elements? Is the intersection of perfect groups a perfect group?

Suppose that K is a connected subcomplex of an aspherical 2-complex L and $G = \ker(\pi_1 K \to \pi_1 L)$. As discussed in Section 6, $\mathcal{A}(G), \mathcal{P}(G) \in \mathbf{ra}(K)$ with $\mathcal{A}(G) \subseteq \mathcal{P}(G)$. Is $\mathcal{A}(G)$ minimal in $\mathbf{c}(K)$? If this were known to be the case, then Whitehead's question would become: Is G locally indicable?

4. *Equations over groups and the Kervaire problem:* If X is a connected subcomplex of a connected 2-complex Y, under what circumstances does the inclusion of X into Y induce a monomorphism of fundamental groups? As a special case, suppose Y is contractible, and X is connected and acyclic. The question is then equivalent to: is X contractible? If this question has a positive solution, then there are no counterexamples to the Whitehead conjecture with tower-height 0 (in the sense of the proof of 7.1).

Is $\pi_1 X \to \pi_1 Y$ injective if $\pi_1 X$ is locally nonperfect and $H_2(Y, X : \mathbb{Z}/p\mathbb{Z}) = 0$ for all primes p? (See [Kr85].)

Suppose that K is a connected subcomplex of an aspherical 2-complex L and $G = \ker(\pi_1 K \to \pi_1 L)$. Is K aspherical if G is assumed to be (locally) residually finite?

5. *Homology and group theory:* Suppose that K is a connected subcomplex of an aspherical 2-complex. Is $\pi_1 K$ torsion-free? Is $\operatorname{cd} \pi_1 K \leq 3$? Is $H_3 \pi_1 K = 0$? Is $\pi_2 K$ finitely generated as a $\mathbb{Z}\pi_1 K$-module? Can anything be said about the centralizer of an element in $\pi_1 K$?

Chapter XI

Zeeman's Collapsing Conjecture

Sergei Matveev[1] and Dale Rolfsen[2]

1 Introduction

The subject of this chapter is the rather audacious conjecture put forward in 1964 by E. C. Zeeman.

Zeeman's Conjecture (Z) *If P^2 is a contractible 2-dimensional polyhedron, then P^2 is 1-collapsible, that is, $P^2 \times I$ collapses to a point.*

As already pointed out in Chapter I, §4.2, (Z) implies both the Poincaré conjecture (P) and the Andrews-Curtis conjecture (AC). It is an affirmation of the subtlety of low-dimensional topology that these old basic conjectures are still unsolved, despite strenuous efforts of generations of topologists. The attempts to solve (Z) have led mathematicians to discover novel ideas and powerful methods in low-dimensional topology, and to a deeper understanding of the strange and mysterious world of 2-dimensional complexes.

Although there are many candidates for counterexamples, (Z) has not been refuted (if it is false!) because, for the present, we have no methods of detecting non-collapsibility for contractible 3-dimensional polyhedra of the form $P^2 \times I$. As a result, the main achievements in investigation of (Z) consist in

[1]This author was partially supported by GB-6 from Chelyabinsk State University and by a grant from the Canadian Natural Sciences and Engineering Research Council

[2]This author was partially supported by by a grant from the Canadian Natural Sciences and Engineering Research Council.

(1) proving it for different special types of P^2;

(2) proving of weakened and disproving of strengthened versions of (Z).

The first contributions to (1) were made by P. Dierker, W. B. R. Lickorish, D. Gillman [Di68, Li70, Gi86] and may be called *collapsing by adding a cell.* They proved the 1-collapsibility for each 2-dimensional polyhedron P^2 satisfying the following: there exists a cell B^k of dimension $k \leq 3$ such that $P^2 \cap B^k$ is a $(k-1)$-cell in the boundary of B^k and $P^2 \cup B^k \searrow \{*\}$ (see §3.1). Note that if $P_0^2 = P^2$ 3-deforms to a point, then for an integer $n > 0$ there exist a sequence

$$P_0^2 \subset P_0^2 \cup B_0^{k_0} \searrow P_1^2 \subset P_1^2 \cup B_1^{k_1} \searrow \ \cdots \ \searrow P_{n-1}^2 \subset P_{n-1}^2 \cup B_{n-1}^{k_{n-1}} \searrow P_n^2 = \{*\},$$

where each $B_i^{k_i}$ is a cell of dimension $k_i \leq 3$ and $P_i^2 \cap B_i^{k_i}$ is a $(k_i - 1)$-cell. Hence the above result can be viewed as the first step in proving (Z) modulo (AC). For example, the 1-collapsibility of the *dunce hat, Bing's house,* the *house with one room*[3] , and the *Abalone*[4] (see Chapter I, §2.2, and Figure 1 below) can be easily proved using *collapsing by adding a cell.* Actually, this method allows one to collapse $P^2 \times I$ to a vertical segment of the form $\{*\} \times I$.

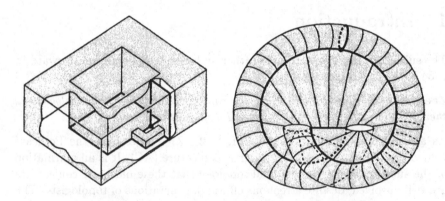

Figure XI.1. House with one room; Abalone

There exist 1-collapsible polyhedra P^2 for which $P^2 \times I$ cannot collapse to $\{*\} \times I$. In §3, we will discuss the examples $K(p, q, r, s)$ of W. B. R. Lickorish which illustrate this phenomenon.

---

[3] The house with one room consists of the surface of a large cube minus a rectangle on roof, plus walls of inner box minus a small door, plus a rectangle minus another door, plus the walls and ceiling of a tunnel connecting the doors.

[4] The *Abalone* is the complex of the trivial group presentation $\langle x, y \mid x, xy^2 x^{-1} y^{-1} \rangle$; it is homeomorphic with the house with one room.

A second general method for collapsing $P^2 \times I$ was proposed by A. Zimmerman [Zi78] (see also [CoMeZi81, CoMeSa85]) and is called *prismatic collapsing*. At first we get rid of the 3-dimensional part of $P^2 \times I$ as follows: for each 2-cell C^2 of P^2 we collapse $C^2 \times I$ to the union of $\partial C^2 \times I$ and a 2-cell $C^* \subset C^2 \times I$ such that the direct product projection maps Int C^* onto Int C^2 homeomorphically. Then we look for a collapse of the resulting 2-dimensional polyhedron. One may say that prismatic collapsing is a very rough method, but exactly this roughness allows one to give an algebraic criterion for prismatic collapsibility of $P^2 \times I$: attaching maps for 2-cells of P^2 have to determine a basis-up-to-conjugation in the free fundamental group of 1-dimensional skeleton of P^2.

Like other conjectures that have significantly influenced the development of mathematics, (Z) is very sharp in the following sense: As a rule, its slightly weakened versions admit positive solutions and its slightly strengthened versions prove to be false. For example, (Z) is true modulo 2-expansions (see Chapter I for the definition of 2-expansion): any contractible 2-dimensional polyhedron P^2 can be expanded by a sequence of 2-expansions to a 2-dimensional polyhedron P_1^2 such that $P_1^2 \times I \searrow \{*\}$ [KrMe83]. One may relax (Z) admitting the repeated multiplication of P^2 by I. In this weakened form, it becomes true: for each contractible P^2 there exist an integer n such that $P^2 \times I^n \searrow \{*\}$, [Di68, Li70]. In fact, $n = 6$ suffices for all P^2 [Co75]. It is surprising that there is such a large gap between the known ($n = 6$) and Zeeman's conjectured ($n = 1$) values of n.

On the other hand, a generalization of (Z) to multidimensional polyhedra is false, since for any $n > 2$ there exist a contractible polyhedron P^n of dimension n such that $P^n \times I$ is not collapsible [Co77]. The proof of non-collapsibility is based on a very specific (one may say *bad*) local structure of P^n. So the idea to investigate (Z) for 2-dimensional polyhedra with a *nice* local structure (such polyhedra are called *special*) seems to be very promising.

The first step in this direction was made by D. Gillman and D. Rolfsen [GiRo83]. They proved that if P^2 is a special spine of a homology 3-ball M^3, then $P^2 \times I$ collapses onto a homeomorphic copy of M^3. It follows that (Z) is true for all special spines of a genuine 3-ball and that (Z) for special spines is equivalent to the Poincaré Conjecture. Later this result was reproved by S. Matveev [Ma87$_1$]. He used another method which proved to be sufficient for clarifying the situation for all special polyhedra. It turned out that (Z) for special polyhedra that cannot be embedded in a 3-manifold is equivalent to the Andrews-Curtis Conjecture [Ma87$_2$]. Concluding this introduction, one can say there exist two possible ways of disproving (Z): via disproving either the Poincaré or Andrews-Curtis Conjectures and via construction of counterexamples with a bad local structure. But we can not exclude the possibility that (Z) is true!

2   Collapsing

Collapsing was defined in Chapter I, §2.2, in several categories. Here we consider collapse of polyhedra, in other words, simplicial collapse with respect to *some* (unspecified) triangulation. It is a geometric and very special type of homotopy equivalence, and is the underlying idea of Whitehead's simple-homotopy theory.

If a polyhedron is *collapsible*, i.e., collapses to a point, then it is contractible. For 1-dimensional polyhedra, the converse is true: each contractible 1-dimensional polyhedron is a tree and hence is collapsible. But it was pointed out in Chapter I that there exist contractible but not collapsible 2-dimensional polyhedra. The famous *dunce hat, Bing's house with two rooms* and slightly less famous *house with one room* (also known as an *abalone* [Ik71]; see Figure 1) serve as examples. Each of them is contractible but not collapsible because of the *can't start*-argument: there is no free edges to start the collapse (see Chapter I, §2.2). For 2-dimensional polyhedra, the *can't start*-argument is a unique obstacle to collapsibility. The following simple criterion is true.

Criterion *A 2-dimensional polyhedron P can be collapsed to a subpolyhedron of dimension ≤ 1 if and only if P contains no 2-dimensional subpolyhedra without free edges.*

In particular, each contractible non-collapsible 2-dimensional polyhedron contains a 2-dimensional subpolyhedron having the *can't start*-property. The criterion makes it very easy to decide whether a given contractible 2-dimensional polyhedron P^2 is collapsible. If it is collapsible, then any attempt succeeds; if not, then it stops on a 2-dimensional subpolyhedron $Q^2 \subset P^2$ which does not depend on the chosen way of collapsing (see Chapter I, §2.2, for the proof).

The situation is much more complicated in dimension 3. The result of collapsing a 3-dimensional polyhedron P^3 depends heavily on a chosen triangulation of P^3 and on a chosen sequence of elementary simplicial collapses. For example, there exist non-collapsible triangulations of a 3-ball [Go68]. Moreover, if K is a collapsible triangulation of a 3-ball, we can not be sure that each sequence of elementary simplicial collapses reduces it to a point; it is possible that we get a 2-dimensional subpolyhedron with the *can't start*-property (for instance, the dunce hat or Bing's house). These are two reasons why the problem of collapsing $P^2 \times I$ is so hard.

Let us recall some simple facts that are useful for altering collapses. First, a collapse can always be replaced by one in which simplices are removed in non-increasing order of dimension; see Chapter I, (14). Second, if P is a collapsible polyhedron, then for *any* given point $x \in P$ one can collapse P onto x (Proof: choose a triangulation K of P such that x is a vertex of K.

Perform on K all collapses of dimension ≥ 2 . We get a tree containing all vertices, and every tree can be collapsed to any of its vertices). Third, if Q is a subpolyhedron of a polyhedron P, and N is a regular neighborhood of Q in P, then a collapse $P \searrow Q$ can be replaced by $P \searrow N \searrow Q$; see [Li70, Co77].

3   Some Special Ways of Collapsing $P^2 \times I$

3.1   Collapsing by adding a cell

A very simple method of collapsing $P^2 \times I$ is called *collapsing by adding a cell*. It works for each polyhedron that becomes collapsible after a single elementary expansion.

Theorem 3.1 ([Di68, Li70, Gi86]) *If Q is a collapsible polyhedron and Q collapses to P by an elementary collapse, then $P \times I \searrow \{*\}$.*

Proof: Since Q collapses onto P by an elementary collapse, we have $Q = P \cup B^n$ and $P \cap B^n = B^{n-1} \subset \partial B^n$, where B^n, B^{n-1} are cells. Note that $P \times I \searrow Q_1$ where $Q_1 = P \times \{0\} \cup (B^{n-1} \times I))$; see Figure 2. Since $Q_1 \approx Q$ and $Q \searrow \{*\}$, we have $P \times I \searrow \{*\}$.

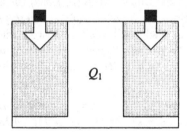

Figure XI.2.

Remark 1 The proof of Theorem 3.1 can be modified to construct a collapse of $P \times I$ to a *vertical* segment of the type $\{*\} \times I \subset P \times I$. To see this, choose a point $x \in P - B^{n-1}$ and collapse $P \times I$ onto $Q_1' = Q_1 \cup (\{x\} \times I)$ instead of Q_1. Since Q_1 is collapsible, it can be collapsed to x. This collapse determines a collapse of Q_1' onto $\{x\} \times I$.

Remark 2 If a contractible polyhedron Q collapses to P by two elementary collapses and $Q - P$ is not connected, then $P \times I \searrow \{*\}$. The proof is similar to the proof of Theorem 3.1; we use *both sides* of I to collapse $P \times I$ onto a homeomorphic copy Q_1 of Q, see Figure 3.

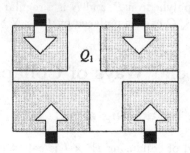

Figure XI.3.

3.2   Non-collapsibility to a vertical segment

The method of collapsing by adding a cell allows one to prove (Z) for a large class of contractible 2-dimensional polyhedra. For example, the *dunce hat* and both of *Bing's houses* belong to this class. But the naive hope the method is all-powerful proved to be false. The point is that there exist many contractible 2-dimensional polyhedra that can not be collapsed to a vertical segment (see Remark 1 to Theorem 3.1). Denote by $K(p,q,r,s)$ the geometrical realization of the group presentation $< a,b \mid a^p b^q, a^r b^s >$, where $ps - rq = \pm 1$. In other words, $K(p,q,r,s)$ is the wedge $K^1 = a \vee b$ of two circles a, b with two 2-cells C_1, C_2 attached along their boundaries by the words $a^p b^q, a^r b^s$, respectively. Each $K(p,q,r,s)$ is a contractible 2-dimensional polyhedron.

Theorem 3.2 ([Li70, Wr71])   *If each of p,q,r,s is not equal to ± 1 , then the polyhedron $K = K(p,q,r,s)$ can not be collapsed to a vertical segment.*

Proof: Suppose $K \times I \searrow \{x\} \times I$ for a point $x \in K$. Then $K \times I$ collapses to a regular neighbourhood $N^3 = N^2 \times I$ of $\{x\} \times I$ in $K \times I$, where $N^2 = N(x, K)$ is a regular neigbourhood of x in K. If $x \in K^1$, we replace x by a point y such that $y \in K - K^1$ and $y \in N^2$; it is clear that $K \times I \searrow N^3 \searrow y \times I$. So we can assume without loss of generality that $x \in \text{Int } C_1$ and $N^2 \subset \text{Int } C_1$. Then $(K \times I) - \text{Int } N^3 \searrow (K \times I) \cap \partial N^3 \approx S^1 \times I$. Since $\pi_1((K \times I) - \text{Int } N^3) \approx < a,b \mid a^r b^s > \ne \mathbb{Z}$, we have a contradiction. $\square$

Actually, the proof is valid for any 2-dimensional polyhedral CW complex K such that $\pi_1(K - \text{Int } C) \ne \mathbb{Z}$ for each 2-cell C of K .

3.3   An *improbable* collapse

It follows from §3.2 that $K(2,3,3,4) \times I$ can not be collapsed to a vertical segment. Nevertheless, in 1973, W. B. R. Lickorish found a very delicate and clever collapse of $K(2,3,3,4) \times I$ to a point [Li73].

Theorem 3.3 $K = K(2,3,3,4)$ *is 1-collapsible.*

Proof: Denote by P the polyhedron $K^1 \times I$ with six 2-cells C_i attached to $K^1 \times I$ along their boundaries, $1 \leq i \leq 6$. The boundary curves l_i of C_i are depicted on Figure 4. The following conditions must be satisfied: 1) l_1 and l_2 trace monotonically the word a^3b^4 beginning with the points $A \in l_1, B \in l_2$; 2) l_3, l_4, l_5 trace monotonically the word a^2b^3 beginning with the points $A \in l_3, B \in l_4, C \in l_5$, respectively. Let H_1, H_2, H_3 be small open 2-discs as shown on Figure 4. We claim that the polyhedron $P_1 = P - (H_1 \cup H_2 \cup H_3) \searrow \{*\}$. The proof is left as an exercise (recall that any attempt to collapse a collapsible 2-dimensional polyhedron does succeed).

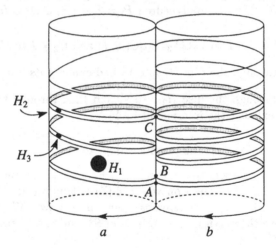

Figure XI.4.

To construct a collapse of $K \times I$, present it as $K^1 \times I$ with two cylinders $C_1 \times I, C_2 \times I$ attached along their lateral surfaces by maps $\Psi_1 = \psi_1 \times Id : \partial C_1 \times I \to K^1 \times I$, $\Psi_2 = \psi_2 \times Id : \partial C_2 \times I \to K^1 \times I$, where $\psi_i : \partial C_i \to K^1, i = 1, 2$, are attaching maps for 2-cells of K. Consider two discs $D_1, D_2 \subset C_2 \times I$ and three discs $D_3, D_4, D_5 \subset C_1 \times I$ as in Figure 5. We require that, for each index $1 \leq i \leq 5$, the disc D_i be attached to the curve $l_i \subset K^1 \times I$, such that the polyhedron $(K_1 \times I) \cup (\cup_{i=1}^{5} D_i)$ coincides with P.

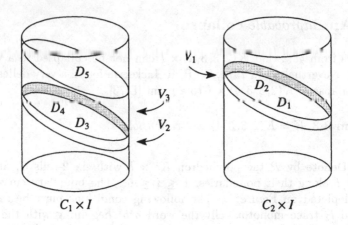

Figure XI.5.

Let V_1 be the 3-ball bounded by $D_1 \cup D_2 \cup (\partial C_2 \times I)$ and let V_2, V_3 be the 3-balls bounded by $D_3 \cup D_4 \cup (\partial C_1 \times I)$ and $D_4 \cup D_5 \cup (\partial C_1 \times I)$, respectively. We collapse $K \times I$ to the polyhedron $P_1 = P - (H_1 \cup H_2 \cup H_3)$ as follows:

1° Collapse $C_1 \times I$ to $V_2 \cup V_3 \cup (\partial C_1 \times I)$ and $C_2 \times I$ to $V_1 \cup (\partial C_2 \times I)$;

2° Pierce through the disc H_1 into V_3 and exhaust its interior;

3° Pierce through the disc H_2 into V_1 and then through the disc H_3 into V_2 and exhaust the interiors of V_1, V_2.

Recalling that $P_1 \searrow \{*\}$, the proof is complete.   □

It is known at the present that if $p = 0, \pm1, \pm2$, then (Z) is true for $K(p,q,r,s)$ [We73, WeWa78]. The question whether (Z) is true for all $K(p,q,r,s)$ with $| p |, | q |, | r |, | s | > 2$ remains open up to now. We leave it to the reader to prove that $K(p,q,r,s)$ embeds in S^3, and is so a spine of B^3, when $ps - qr = \pm1$.

Hint   For a pointed space $(X, \{*\})$ and two maps $f_i : X \to Y_i, i = 1, 2$, define the reduced two-sided mapping cylinder $C(f_1, f_2)$ as

$$(Y_1 \cup (X \times [-1,1]) \cup Y_2)/R,$$

where the equivalence relation R is generated by equalities $(x, -1) = f_1(x)$, $(x, 1) = f_2(x)$, $(*, t) = (*, 0)$ for all $x \in X$ and $t \in [-1, 1]$.

Consider two curves l_1 and l_2 on the torus $T^2 = \partial D^2 \times \partial D^2 \subset \partial(D^2 \times D^2) \approx S^3$ that intersect in a single point $\{*\}$ and represent the classes (p,q) and (r,s) in the torus' homology. Present S^3 as the reduced two-sided mapping cylinder $C(p_1, p_2)$, where p_1 and p_2 are projections of T^2 onto cores $\partial D^2 \times \{0\} \subset \partial D^2 \times D^2$ and $\{0\} \times \partial D^2 \subset D^2 \times \partial D^2$ of the solid tori. Then $K(p, q, r, s) \approx C(p'_1, p'_2)$, where p'_i are the restrictions of p_i onto $l_1 \cup l_2$, $i = 1, 2$.

3.4 Prismatic 1-collapsing

The notion of prismatic 1-collapsing was proposed by A. Zimmermann [Zi78]; see also [CoMeZi81, CoMeSa85]. It is inextricably tied to whether the attaching maps for the 2-cells of CW-complex P determine a basis-up-to-conjugation in the fundamental group of its 1-dimensional skeleton. Following [CoMeSa85], we prefer to work with CW-complexes, even in the PL category. Thus we adopt the notion of a PLCW-complex; see Chapter I, §1.4.

Let P be a 1-dimensional polyhedron P^1 with n 2-cells $C_i, 1 \leq i \leq n$, attached to it along their boundaries. For each i choose a 2-cell $C_i^* \subset C_i \times I$ such that the direct product projection maps $\text{Int } C_i^*$ onto $\text{Int } C_i$ homeomorphically. (Note that this condition is stronger than the one of [CoMeSa85], §(2.6).) We shall refer to the polyhedron $P_v = (P^1 \times I) \cup (\cup_{i=1}^n C_i^*) \subset P \times I$ as a *vertical resolution* of P.

Definition A 2-dimensional PLCW-complex P is called *prismatically 1-collapsible (prismatically collapsible)* if it has a collapsible vertical resolution.

Since $P \times I \searrow P_v$, each prismatically 1-collapsible polyhedron is 1-collapsible.

For simplicity we shall consider in this section only polyhedra that can be obtained from a wedge $P^1 = a_1 \vee \ldots \vee a_n$ of n circles $a_1, \ldots, a_n$ by attaching n 2-cells $C_i, 1 \leq i \leq n$. Particularly, standard complexes of (finite) presentations (see Chapter I, (9)) belong to this class.

The common point of the circles forming the wedge will be denoted by e_0. For each 2-cell C_i choose a basepoint $e_i \in \partial C_i$. We shall suppose that the attaching map $\varphi_i : \partial C_i \to P^1$ satisfies the condition $\varphi_i(e_i) = e_0$. Then each φ_i determines an element $[\varphi_i]$ of the group $\pi_1(P^1, e_0)$ which is isomorphic to the free group $F(a_1, \ldots, a_n)$, see Chapter I, §1.3.

Definition A set of elements $w_1, \ldots, w_n \in F(a_1, \ldots, a_n)$ is called a *basis-up-to-conjugation* if there are elements $g_1, \ldots, g_n$ in $F(a_1, \ldots, a_n)$ such that $\{g_i w_i g_i^{-1} \mid 1 \leq i \leq n\}$ is a basis in $F(a_1, \ldots, a_n)$.

Theorem 3.4 ([CoMeSa85]) *If P^2 is a prismatically 1-collapsible polyhedron with attaching maps φ_i, then the elements $w_i = [\varphi_i]$, $1 \leq i \leq n$, form a basis-up-to-conjugation in $F(a_1, \ldots, a_n)$*

Proof: Let $P_v = (P^1 \times I) \cup (\cup_{i=1}^n C_i^*)$ be a collapsible vertical resolution of P^2. For each $i, 1 \leq i \leq n$, choose a point $x_i \in \text{Int } C_i^*$. Since P_v is collapsible, one can collapse it onto a tree $T \subset P_v$ such that each x_i is an endpoint of T.

Denote by B a regular neighbourhood of $\cup_{i=1}^n x_i$ in P_v. Then B is a union of 2-discs, $B = \cup_{i=1}^n B_i$, and P_v can be collapsed onto $T \cup B$. Let $Q = P_v - \text{Int } B$ and $T_B = (T \cup B) - \text{Int } B$, see Figure 6. Then $Q \searrow P^1 \times I \searrow P^1$ and the collapse $P_v \searrow T \cup B$ induces $Q \searrow T_B$. The deformations $T_B \nearrow Q \searrow P^1$ determine an isomorphism $\alpha : \pi_1(T_B) \to \pi_1(P^1)$. Denote by $v_1, \ldots, v_n$ a basis of $\pi_1(T_B)$ generated by the curves $\partial B_1, ..., \partial B_n \subset T_B$, respectively. It follows easily from the construction of α that for each $i, 1 \leq i \leq n$, the element $\alpha(v_i)$ coincides with w_i up to conjugation. □

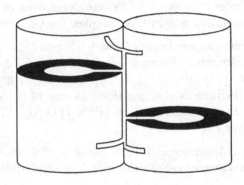

Figure XI.6.

In a sense, the converse of Theorem 3.4 is true. It is a generalization of Theorem 3 of [CoMeSa85].

Theorem 3.5 *If $\{w_1, \ldots, w_n\}$ is a basis-up-to-conjugation of $F(a_1, \ldots, a_n)$ and if P^1 is a wedge of n circles, then there are attaching maps $\varphi_i : (\partial C_i, e_i) \to (P^1, e_0), 1 \leq i \leq n$, such that $[\varphi_i] = w_i$ and the resulting polyhedron $P^2 = P^1 \cup_\varphi (\cup_{i=1}^n C_i)$ is prismatically 1-collapsible.*

For the proof of Theorem 3.5, we need two lemmas.

Lemma 3.6 *Suppose $\beta : \pi_1(P^1, e_0) \to \pi_1(P^1, e_0)$ is an automorphism, where $P^1 = a_1 \vee \ldots \vee a_n$ is the wedge of n circles with the wedge point e_0. Then there is an embedding $h : (P^1 \times I, e_0 \times I) \to (P^1 \times I, e_0 \times I)$ such that h induces the automorphism β of the group $\pi_1(P^1 \times I, e_0 \times \{0\})$, and $P^1 \times I \searrow h(P^1 \times I)$.*

Proof: If the conclusion of Lemma 3.6 is true for two automorphisms, then it is evidently true for their composition. According to the Nielsen theorem, each automorphism of $F(a_1, \ldots, a_n)$ can be presented as a superposition of automorphisms (moves) of two types. Both moves map each generator to itself except the generator a_i which is mapped to a_i^{-1} by the first move, and to $a_i a_j, i \neq j$, by the second move; compare Chapter 1, (27) and (19). The first move can be realized by a homeomorphism h of the form $h_1 \times Id : P^1 \times I \to P^1 \times I$, where the restriction of the homeomorphism $h_1 : P^1 \to P^1$ onto a_j is the identity for $j \neq i$ and reverses the orientation of a_i. It is clear from Figure 7 how one can realize the second move; $P^1 \times I$ collapses onto the shaded area which is the image of $P^1 \times I$ under an embedding $h : P^1 \times I \to P^1 \times I$. $\quad\square$

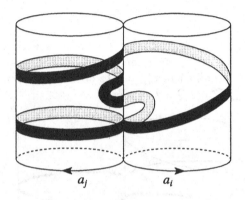

Figure XI.7.

Lemma 3.7 *Let* $< a_1, \ldots, a_n \mid R_1, \ldots, R_n >$ *be a presentation of the trivial group, where* $R_i = S_i a_i S_i^{-1}$ $(1 \leq i \leq n)$ *for some (not necessarily reduced) words* S_i *in generators* $a_1, \ldots, a_n$. *Denote by* r_i *the element of* $F(a_1, \ldots, a_n)$ *corresponding to* R_i. *Then there are attaching maps* $\psi_i : (\partial C_i, e_i) \to (P^1, e_0)$ *such that* $[\psi_i] = r_i$ *and the resulting polyhedron* $Q = P^1 \cup_\psi (\cup_{i=1}^n C_i)$ *is prismatically collapsible.*

Proof: Define words X_i inductively by the rules $X_0 = 1, X_i = S_i X_{i-1}^{-1} X_{i-1}$ for $1 \leq i \leq n$. Then the words R_i and $R_i' = X_i a_i X_i^{-1}$ determine the same element in $F(a_1, \ldots, a_n), 1 \leq i \leq n$. The advantage of new relators R_i' is that a terminal segment of each next conjugator X_i coincides with the previous conjugator X_{i-1}. Choose attaching maps $\psi_i : (\partial C_i, e_i) \to (P^1, e_0)$ to trace monotonically the relators R_i'. To prove that the resulting polyhedron $Q = P^1 \cup_\psi (\cup_{i=1}^n C_i)$ is prismatically 1-collapsible, take 2-cells C_i^* in $C_i \times I$ with $\partial C_i^* \subset \partial C_i \times I$, as drawn in Figure 8.

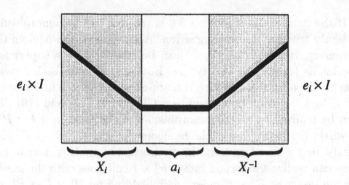

Figure XI.8.

For each i, the image of ∂C_i^* in $P^1 \times I$ is a circle with a *tail* t_i. A little care is needed to have $t_i \subset t_{i+1}$ for all $i, 1 \le i < n - 1$; see Figure 9. Then all tails are contained in t_n, and t_n does not prevent the collapse of $Q_v = P^1 \times I \cup (\cup_{i=1}^n C_i^*) \subset Q \times I$ to a point.   □

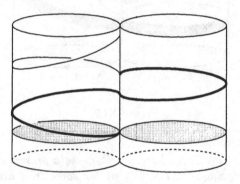

Figure XI.9.

Proof of Theorem 3.5: For $1 \le i \le n$, write $w_i = g_i v_i g_i^{-1}$ where $v_1, \dots, v_n$ is a basis of $F(a_1, \dots, a_n)$. Denote by β the automorphism of $F(a_1, \dots, a_n)$ sending a_i to v_i. If we take $s_i = \beta^{-1}(g_i)$, we have $\beta(s_i a_i s_i^{-1}) = w_i, 1 \le i \le n$. Present each element s_i by a word S_i . According to Lemma 3.7, there are attaching maps $\psi_i : (\partial C_i, e_i) \to (P^1, e_0)$ such that $[\psi_i] = s_i a_i s_i^{-1}$ and the resulting polyhedron $Q = P^1 \cup_\psi (\cup_{i=1}^n C_i)$ has a collapsible vertical resolution $Q_v = (P^1 \times I) \cup (\cup_{i=1}^n C_i^*)$. By Lemma 3.6, construct an embedding

$$h : (P^1 \times I, e_0 \times I) \to (P^1 \times I, e_0 \times I)$$

realizing β. If we re-attach the 2-cells C_i^* of Q_v to $P^1 \times I$ via the embedding h, we obtain a vertical resolution P_v of a polyhedron $P = P^1 \cup_\varphi (\cup_{i=1}^n C_i)$.

Each attaching map $\varphi_i : \partial D^2 \to P^1$ has the form $\varphi_i = ph\psi_i^*$, where $\psi_i^* : \partial D^2 \to P^1 \times I$ is the the attaching map for the 2-cell C_i^* and $p : P^1 \times I \to P^1$ is the projection. Since $\beta(s_i a_i s_i^{-1}) = g_i v_i g_i^{-1} = w_i$, attaching maps φ_i satisfy the conditions $[\varphi_i] = w_i, 1 \leq i \leq n$. By construction, the embedding h can be extended to an embedding $h_v : Q_v \to P_v$. Now we have $P_v \searrow h_v(Q_v) \searrow \{*\}$. $\square$

Note that Theorems 3.4 and 3.5 give a geometrical characterization of a basis-up-to-conjugation in $F(a_1, \ldots, a_n)$: the words $w_1, \ldots, w_n$ form a basis-up-to-conjugation if and only if they represent based homotopy classes in which attaching maps for 2-cells may be chosen so as to yield a prismatically 1-collapsible polyhedron. Simultaneously, Theorem 3.5 allows one to construct a great variety of polyhedra satisfying (Z) : for each basis-up-to-conjugation $w_1, \ldots, w_n$ one can construct an example (we do not assert that the geometrical realization of the presentation $< a_1, \ldots, a_n \mid w_1, \ldots, w_n >$ is prismatically 1-collapsible; compare §7, Problem 1.)

As we have mentioned in the Introduction, prismatic 1-collapsing is a rather rough instrument. There exist many 1-collapsible but not prismatically 1-collapsible polyhedra. For instance, $K(2,3,3,4)$ (see §3.2) is not prismatically 1-collapsible. Since the words $a^2 b^3, a^3 b^4$ do not form a basis-up-to-conjugation, Theorem 3.4 yields a proof. For a more elegant and simple proof one can use the following criterion of A. Zimmermann.

Theorem 3.8   *If* P^2 *is prismatically 1-collapsible, then there is a point* $v \in P^2$ *such that the link of* v *has an articulation point.*

An *articulation point* of a graph G is a point x such that $G - \{x\}$ has more components than G. In fact, Zimmermann's results are more general than Theorem 3.8, at least in the simplicial case; see [Zi78]. Actually, Theorem 3.8 is a consequence of Theorem 3.4 and Whitehead's lemma [Wh36], but a direct geometric proof which is a modification of the one in [Zi78] may be formulated. Note that for a standard complex P^2 of a group presentation, the only candidates for an articulation point in their link are the vertex e_0 and points on free faces of P^2.

Proof of Theorem 3.8:   Let $P_v = (P^1 \times I) \cup (\cup_{i=1}^n C_i^*) \subset P^2 \times I$ be a collapsible vertical resolution of P^2. Define a point $x = (x', t) \in P^1 \times I \subset P_v$ to be *positive* if the segment $\{x'\} \times [t,1] \subset \{x'\} \times I$ has no common points with boundary curves of 2-cells. The point x is called *negative* if the same is true for the segment $\{x'\} \times [0,t]$. Denote by Q the union of all points of P_v that are neither positive nor negative. Since Q contains all 2-cells C_i^*, Q is a 2-dimensional subpolyhedron of P_v. Let T be a triangulation of Q, sufficiently fine for the following arguments. Since P_v is collapsible, there exists a free open 1-dimensional simplex δ of T. Suppose δ is not *vertical*, that is, its projection to $P^1 \times I$ does not consist of a single point. Then δ lies in the boundary curve of a 2-cell C_i^*, and the link in P of the projection onto P of every point $x \in \delta$ is homeomorphic to a closed segment, and hence has an articulation point. Let δ be vertical, that is, have a form $\delta = \{x'\} \times (t_1, t_2) \subset P^1 \times I$, and the collapse from δ enters $P^1 \times I$. Then the link G of δ in $P \times I$ can be identified with the link of x' in P. Since δ is free in T, exactly one vertex v of G belongs to Q. Denote by V_+ and V_- the sets of positive and negative vertices in G. It is easy to see that an edge of G can not have one vertex in V_- and the other in V_+. It follows that if $V_+, V_- \neq \emptyset$, then v is an articulation point of G. If $V_+ = \emptyset$ or $V_- = \emptyset$, then there is an edge in G with both endpoints at v and degree of v is at least 3; hence, v is an articulation point again. The existence of an articulation point in the link of a point in P is proved. $\square$

4   1-Collapsibility Modulo 2-Expansions

Let n-dimensional polyhedra P^n, Q^n have the same simple-homotopy type. It is known that if $n > 2$, then P^n can be transformed to Q^n by a sequence of elementary expansions and collapses of dimension $\leq n + 1$; see [Wa66]. For $n = 2$, however, the question is open up to now; the existence of a 3-dimensional deformation is a conjecture equivalent to the generalized Andrews-Curtis conjecture (AC); see Chapter I, §4.1.

Suppose now that P^2 3-deforms to Q^2. Is it possible to replace the 3-deformation by an expansion $P^2 \nearrow P^2 \times I$ and a collapse $P^2 \times I \searrow Q^2$? It turns out that the answer is positive modulo 2-expansions of P^2.

Theorem 4.1 ([KrMe83])   *If P^2 3-deforms to Q^2, then there exists a 2-dimensional polyhedron P_1^2 such that $P^2 \nearrow P_1^2$ and $P_1^2 \times I \searrow Q^2$.*

Every 3-deformation of P^2 to Q^2 can be replaced by a sequence of elementary expansions and collapses of dimension 3 such that each elementary expansion is followed by an elementary collapse onto a 2-dimensional polyhedron; see [Wr75]. Hence, Theorem 4.1 can be proved by an induction on the number of elementary transient moves. The inductive step is given by the following lemma.

Lemma 4.2   *Let $P^2 \nearrow P^2 \cup B^3$ be an elementary expansion and let $P^2 \cup B^3 \searrow Q^2$ be an elementary collapse. Then for every 2-dimensional* *expansion $Q^2 \nearrow Q_1^2$ there is a 2-dimensional expansion $P^2 \nearrow P_1^2$ such that $P_1^2 \times I$ collapses onto a homeomorphic copy of $Q_1^2 \times I$.*

Proof: We think of Q^2 as being embedded into $P^2 \cup B^3$ such that $Q^2 = (P^2 \cup B^3) \cap Q_1^2$. The boundary ∂B^3 of B^3 consists of two 2-cells

$$B_1^2 = P^2 \cap B^3 \text{ and } B_2^2 = Cl(\partial B^3 - B_1^2).$$

Choose a small 2-cell $A \subset B_2^2$ such that $A \cap Cl(Q_1^2 - Q^2) = \emptyset$. Then there exists a 2-dimensional expansion of P^2 to $P_1^2 = (P^2 \cup \partial B^3) - \text{Int } A$.

Consider the polyhedron

$$Q_1' = ((Q_1^2 - \text{Int } A) \times \{0\}) \cup (\partial A \times I) \cup ((\partial B^3 - \text{Int } A) \times \{1\}) \subset P_1^2 \times I.$$

Since $(\partial A \times I) \cup ((\partial B^3 - \text{Int } A) \times \{1\})$ is a 2-cell, Q_1' is homeomorphic with Q_1^2. One can show that the homeomorphism of Q_1^2 onto Q_1' can be extended to an embedding $i : Q_1^2 \times I \to P_1^2 \times I$ (see Figure 10) such that $P_1^2 \times I \searrow i(Q_1^2 \times I)$.□

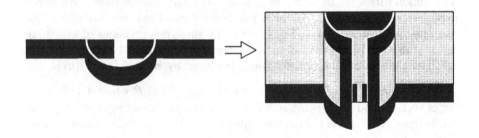

Figure XI.10.

Using Theorem 4.1, one can construct many examples of 2-dimensional polyhedra satisfying (Z) : an appropriate 2-expansion of every polyhedron P^2 3-deformable to a point is 1-collapsible.

5   Zeeman Conjecture for Special Polyhedra

5.1   A quick survey on special polyhedra

As we have mentioned in the Introduction, the generalization of (Z) to multi-dimensional polyhedra is false. The proof is based on a bad local structure of the counterexample. It seems to be natural therefore to investigate (Z) for 2-dimensional polyhedra having a nice local structure. Fake surfaces and special polyhedra fit this setting perfectly.

Definition   A compact 2-dimensional polyhedron is called a *fake surface* if each of its points has a neighbourhood homeomorphic to one of the types pictured on Figure 11.

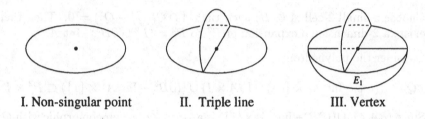

I. Non-singular point   II. Triple line   III. Vertex

Figure XI.11.

The singularities II, III are, in a sense, the only stable ones. All other singularities that a 2-dimensional polyhedron can have are unstable. For example, a fourfold line can be destroyed by small deformations of attaching maps of 2-cells. In R^3, soap films exhibit singularities exactly of types II and III. The concept of fake surfaces was introduced by Hiroshi Ikeda [Ik71].

A fake surface P stratified by subsets $V(P) \subset S(P) \subset P$ where $V(P)$ is the set of vertices and $S(P)$ is the singular (multi)graph consisting of vertices and triple lines. The connected components of $P - S(P)$ are called *2-components* of F.

Definition   A fake surface P is called *a special polyhedron* if the stratification $V(P) \subset S(P) \subset P$ is a cellular one; that is $V(P) \neq \emptyset$, each component of $S(P) - V(P)$ is an open arc and each 2-component of P is an open 2-cell. A spine of a 3-manifold is said to be *special* if it is a special polyhedron.

Sometimes the adjective *standard* rather than *special* is used. We regard these terms as synonyms. It is known that every 2-dimensional polyhedron can be 3-deformed to a special one [Wr75], and that every 3-manifold has a special spine [Ca65]. Both of *Bing's houses* (see the Introduction) are examples of special spines for the 3-ball.

Consider the modifications (moves) of special polyhedra described by Figure VIII.11. The moves T_1 and T_3 alter a small neighbourhood E_1 of a vertex, the move T_2 changes a small neighbourhood of an edge. The part of the polyhedron which does not appear in the figures is understood to be fixed. Note that exactly one of the polyhedra pictured on Figure VIII.11 can not be realized inside a 3-manifold. The following theorems were proved by S. Matveev and independently by R. Piergallini .

Theorem 5.1 *Let P be homeomorphic to a special spine of a 3-manifold M; let Q be a special polyhedron. Then Q also is homeomorphic to a special spine of M if and only if P transforms to Q by a sequence of the moves $T_1^{\pm 1}, T_2^{\pm 1}$.*

Theorem 5.2 *A special polyhedron P 3-deforms to a special polyhedron Q if and only if P transforms to Q by a sequence of the moves $T_i^{\pm 1}, 1 \leq i \leq 3$.*

For the proofs see [Ma87$_1$, Ma87$_2$, Pi88].

One can easily prove that if a special polyhedron has at least two vertices, then every T_1 -move on it can be expressed through moves $T_2^{\pm 1}$. Hence, with a few exceptions, we need only one type of moves in Theorem 5.1 and two types of moves in Theorem 5.2.

5.2 An equivalence of (Z) for special spines and (P)

In 1983, D. Gillman and D. Rolfsen advanced the investigation of (Z) by proving the following theorem.

Theorem 5.3 *If P is a special spine of a homology 3-ball M, then $P \times I$ collapses to a subset homeomorphic with M.*

We shall say that a topological space X is *acyclic* if $H_i(X; \mathbb{Z}) = 0$ for all $i > 0$. Actually the original proof of Theorem 5.3 (see [GiRo83]) is valid for every acyclic spine $P \subset M$ having a fake surface structure. We prefer to propose here another proof. It follows immediately from Lemmas 5.4-5.6 below and gives, by the way, the same embedding of M into $P \times I$.

Let P be a fake surface such that all 2-components of P are orientable. Define an *orientation* of P to be a collection of orientations of its 2-components such that the following condition holds:

- *on every edge of P two of the three induced orientations are opposite to the third.*

The notion of orientability for fake surfaces was introduced in [GiRo91]. We would like to point out that an orientable 3-manifold can have a non-orientable special spine as well as an orientable special polyhedron can serve as a special spine of a non-orientable 3-manifold.

Figure XI.12.

Lemma 5.4 *Every acyclic fake surface P is orientable.*

Proof: Since $S(P)$ is a graph with vertices of order 2 and 4, there exist a 1-cycle $\gamma \in C_1(S(P); \mathbb{Z})$ all of whose coefficients are ± 1. By the acyclity there exist a 2-chain $\beta \in C_2(S(P); \mathbb{Z})$ with $\partial \beta = \gamma$. Consider the union Σ of 2-simplices on which β has even coefficients. One checks that Σ is a closed surface, using the fact that if three integer numbers add to ± 1, then either two or none of them are even. But since P is 2-dimensional and acyclic, we conclude that Σ is actually empty. In particular, β has odd coefficients for all 2-simplices of P. The orientation α is obtained now by reversing the orientations of all 2-simplices with negative coefficients (since if three odd numbers add to ± 1, then they can not have the same sign). $\square$

We shall think of the polyhedron E_1 (a typical neighbourhood of a vertex in a special polyhedron) as comprised of a vertex, four edges and six 2-cells. Although E_1 is not a special polyhedron (it has a boundary), it makes sense to consider orientations of E_1 as collections of orientations of 2-cells such that, for every edge, two of the three induced orientations are opposite to the third.

Exercise   For every two orientations α, β of E_1, there is a unique homeomorphism $h : E_1 \to E_1$, up to isotopy, transforming α to β.

Hint   We present E_1 as a cone over a one-dimensional skeleton of a regular tetrahedron and use the large symmetric group of the tetrahedron to match α with β.

We turn our attention to the notion of branched surface [FlOe84].

Definition   A *branched surface* F is a space locally modelled on the space E_1' shown on Figure 12.

The branched surfaces that we use are mainly embedded in 3-manifolds. It should be pointed out that we have in mind the smooth category now; the upper and lower sheets of E_1' have to be tangent (*pinched*) to the middle square, and the placement of E_1' in R^3 shown on Figure 12 describes the behavior of F inside a 3-manifold up to a diffeomorphism.

Each branched surface is (homeomorphic to) a fake surface, but not every fake surface has a branched surface structure. For example, let L be the simplest spine of the lens space $L_{3,1}$, that is, a disc D^2 with identifications by the free action of the group $\mathbb{Z}_3$ on ∂D^2. Then L can not be realized as a branched surface.

An *orientation* of a branched surface F is an orientation of the underlying fake surface such that the following condition $(*)$ holds:

$(*)$ *for every pair of 2-components adjacent to an edge e of F, the induced orientations of e coincide if and only if the 2-components are pinched together along e.*

The model branched surface E_1' has two orientations. They are induced from two possible orientations of a plane by orthogonal projection of E_1' into the supporting plane of the middle square. Fix an orientation α_0 of E_1' .

Lemma 5.5 *If a special polyhedron P is orientable, then P has an oriented branched surface structure.*

Proof: Let $N(S(P), P)$ be a small regular neighbourhood in P of the singular graph $S(P)$. Decompose it into a union of homeomorphic copies A_i, $1 \leq i \leq n$, of E_1 such that the intersection of each two copies is either empty or consists of Y-shaped graphs (a Y-shaped graph is a wedge of three segments). By assumption we are given an orientation α of P. For each i, $1 \leq i \leq n$, α restricts to an orientation α_i of A_i . We use the above exercise to construct a homeomorphism (called a chart) $\varphi_i : E_1' \to A_i$ sending α_0 to α_i. The charts φ_i determine branched surface structures on all of A_i. Hence for each Y-shaped graph Y we had used for cutting we have two pinchings coming from both *sides* of Y. In view of condition $(*)$ and compatibility of α_i these two pinchings coincide. Therefore, we can extend the branched surface structures on A_i to an oriented branched surface structure on $N(S(P), P)$ and then on P .   □

Lemma 5.6 *Let a special spine P of an orientable 3-manifold M with boundary admit an oriented branched surface structure. Then $P \times I$ collapses to a subset homeomorphic with M .*

Proof: Realize P as a branched surface in M and identify M with a regular neighbourhood of P . Then M can be decomposed into segments with endpoints on ∂M. The behavior of the decomposition (denote the decomposition by ξ) in a neighbourhood N of a vertex is shown on Figure 13. The decomposition space M/ξ can be identified with P. The embedding of M into $P \times I$ that we have to construct will have the form $\varphi = \pi \times p : M \to (M/\xi) \times I$ where $\pi : M \to M/\xi$ is the decomposition projection and $p : M \to I$ maps each segment of the decomposition into a subsegment of I linearly. For a neighbourhood N of every vertex, we define $p_{|N}$ to be the orthogonal projection onto I , see Figure 13. The orientability assumptions allow us to extend the projections to a map $p : M \to I$ such that $\varphi = \pi \times p$ is an embedding. Since the intersection of $\varphi(M)$ with every segment $\{*\} \times I$ is connected, $(M/\xi) \times I$ collapses to $\varphi(M)$.   □

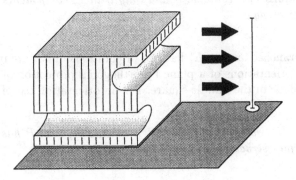

Figure XI.13.

Let us state now two important consequences of Theorem 5.3.

Corollary 5.7 (Z) *is true for all special spines of a 3-ball.*

Corollary 5.8 (Z), *restricted to special spines (of 3-manifolds) , is equivalent to the Poincaré Conjecture* (P).

The proofs of both corollaries are easy, since if P is a special spine of a 3-ball B, then $P \times I \searrow B \searrow \{*\}$. For the implication (Z) $\Rightarrow$ (P); see Chapter I, §4.2.

Remark: Let P be a special spine of a (not necessarily orientable) 3-manifold M. Every open 2-component C^2 of P^2 has the normal line bundle and every open edge e of P has the normal disc bundle. By a *normal orientation* of C^2 or e in M we mean an orientation of the corresponding normal bundle. If C^2 is adjacent to e, then a normal orientation of C^2 induces a well-defined normal orientation of e. Define a *normal orientation* of P to be a collection of normal orientations of its 2-components such that the following holds: for every edge e in P two of the induced normal orientations are opposite to the third. Certainly, every orientable special spine in an orientable 3-manifold is normally orientable. A careful analysis of the proof of Lemma 5.6 shows that actually we need only the existence of a normal orientation. In other words, the conclusion of Lemma 5.6 holds for every normally orientable special spine.

5.3 An equivalence of (Z) for unthickenable special polyhedra and (AC)

This section consists mainly in application of Theorems 5.1 and 5.2 to an investigation of (Z). Let a special polyhedron Q^2 be obtained from a special polyhedron P^2 by one of the moves $T_i^{\pm 1}, 1 \leq i \leq 3$. The question is whether the 1-collapsibility of P^2 implies the 1-collapsibility of Q^2. It turns out that the answer is *Yes* for all $T_i^{\pm 1}$ with one exception: if P^2 is transformed to Q^2 by the move T_3^{-1} and Q^2 is a spine, then we can say nothing. Nevertheless, an unexpected result of S. Matveev (see Theorem 5.15 below) allows one to prove the equivalence stated in the title.

A spacious property for special spines

At first we prove that the 1-collapsibility of a special polyhedron is preserved under applying the moves $T_1^{\pm 1}, T_2^{\pm 1}$.

Theorem 5.9 ([Ma87₁]) *Suppose a special polyhedron Q is obtained from a special polyhedron P by a sequence of the moves $T_1^{\pm 1}, T_2^{\pm 1}$. Then there is an embedding $\psi : P \times I \to Q \times I$ such that $Q \times I \searrow \psi(P \times I)$.*

It follows from Theorems 5.1 and 5.9 that every special spine Q of a 3-manifold M has the following *spacious* property: for every special spine $P \subset M$ a homeomorphic copy of $P \times I$ lies in $Q \times I$ such that $Q \times I \searrow P \times I$.

It is certainly sufficient to prove Theorem 5.9 for the case when Q is obtained from P by one of the moves $T_1^{\pm 1}, T_2^{\pm 1}$. Let us recall the definitions of the moves T_1^{-1}, T_2 in a slightly different form, and introduce two additional moves $T_0, \tilde{T}_1^{-1}$ (see Figure 14). Each move removes a fragment F_i and replaces it by the fragment F_i'. The remaining part of the transformed special polyhedron is understood to be fixed. We define the *boundary* of a fragment to be the union of its closed free edges. Clearly, $\partial F_i = \partial F_i'$ for all i, $0 \le i \le 3$.

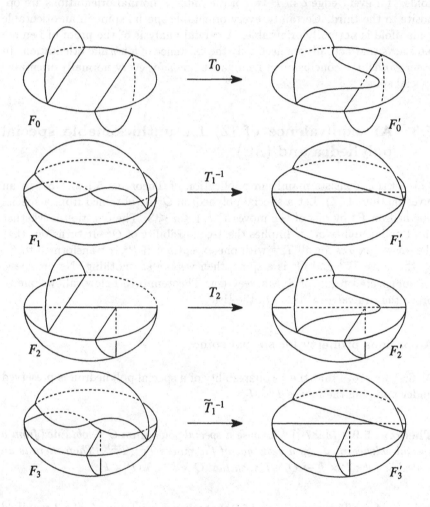

Figure XI.14.

Lemma 5.10 *For every* i, $0 \le i \le 3$, *there exists an embedding* $\psi_i : F_i \times I \to F_i' \times I$ *such that* $\psi_i = Id$ *on a neighbourhood of* $\partial F_i \times I$, *and* $F_i' \times I \searrow \psi_i(F_i \times I)$.

Proof: It is convenient to consider F_i, $0 \le i \le 3$, as the horizontal standard disc D^2 with the additional part $X_i = Cl(F_i - D^2)$ attached to it along

$Y_i = D^2 \cap X_i$. Similarly, present F_i' as $F_i' = D^2 \cup X_i'$ where $X_i' = Cl(F_i' - D^2)$ and $Y_i' = D^2 \cap X_i'$.

Let us modify the identity embedding of the disc $D^2 \times \{0\}$ into $D^2 \times I$ on a smaller round disc $D_0 \times \{0\} \subset \text{Int } (D^2 \times \{0\})$ by taking the center of $D^2 \times \{0\}$ into a point $(x_0, 1/2)$ and using a conical construction to get an embedding $\varphi_1 : D^2 \times \{0\} \to D^2 \times I$. The point x_0 must lie outside D_0; see Figure 15.

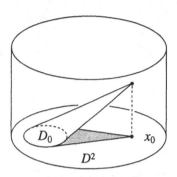

Figure XI.15.

Since $\varphi_1(D^2 \times \{0\}$ lies in $D^2 \times I$ with a collar, one can extend φ_1 to an embedding $\varphi : D^2 \times I \to D^2 \times I$ such that $\varphi = Id$ on a neighbourhood of $\partial D^2 \times \{0\}$. Let us choose the collar and the attaching graphs Y_i, Y_i' in such a fashion that $p\varphi(Y_i \times I) = Y_i'$ where p is the direct product projection. To be more precise, take Y_0, Y_2 and Y_1', Y_3' as shown on Figure 16; the exact placements of Y_0', Y_2' and Y_1, Y_3 can be found from the condition $p\varphi(Y_i) = Y_i'$. Extend the embedding φ to $X_i \times I$ by taking it into $X_i' \times I$. The embedding $\psi_i : F_i \times I \to F_i' \times I$ so obtained is the one desired. The visualization of the collapse is left to the reader.   □

Proof of Theorem 5.9:   Let Q be obtained from P by one of the moves $T_0, T_1^{-1}, T_2, \tilde{T}_1^{-1}$. For a suitable i, $0 \le i \le 3$, define an embedding $\psi : P \times I \to Q \times I$ taking $\psi = Id$ outside $F_i \times I$ and $\psi = \psi_i$ on $F_i \times I$, where $\psi_i : F_i \times I \to F_i' \times I$ is the embedding described in Lemma 5.10. Clearly $Q \times I \searrow \psi(P \times I)$. It remains to note that the move T_1 is a special case of the move T_0 and that the move T_2^{-1} is a composition of T_2 and $\tilde{T}_1^{-1}$.   □

Corollary 5.11 *The property of a special polyhedron to be 1-collapsible is preserved under the moves $T_1^{\pm 1}$ and $T_2^{\pm 1}$.*

Proof: If $P^2 \times I \searrow \{*\}$ and P^2 transforms to Q^2 by a sequence of the moves $T_1^{\pm 1}$, $T_2^{\pm 1}$, then $Q^2 \times I \searrow \psi(P^2 \times I) \searrow \{*\}$.   □

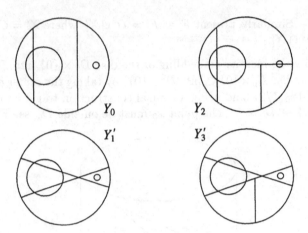

Figure XI.16.

The following corollary generalizes Theorem 5.3.

Corollary 5.12 *If Q is a special spine of a 3-manifold M with boundary, then $Q \times I$ collapses to a subpolyhedron homeomorphic with M.*

Proof: One can easily show that every 3-manifold M has a normally orientable special spine P . Actually the special spine constructed in [GiRo91] is normally orientable. It follows from the remark to Lemma 5.6 that $P \times I \searrow M$. Since $Q \times I \searrow P \times I$ by Theorem 5.9, we have $Q \times I \searrow M$.   □

Remark Let L denote the 2-complex associated with the group presentation $< a \mid a^3 >$. Then L is a fake surface and L is a spine of the punctured lens space $L(3,1)_*$. Nevertheless $L \times I$ can not be collapsed to a subset homeomorphic with $L(3,1)_*$; see [Gi86]. This does not contradict the above; L is not a special spine because it has no vertices.

1-Collapsibility of unthickenable special polyhedra

A special polyhedron P is called *unthickenable* if it can not be embedded in a 3-manifold. We know from Theorem 5.2 that every 3-deformation of one special polyhedron to another can be replaced by a sequence of moves $T_i^{\pm 1}$, $1 \le i \le 3$. The move T_3 transforms every special polyhedron to an unthickenable one. Our aim now is to clarify how T_3 affects the 1-collapsibility. At first we prove a more general theorem than is required for the purpose.

Theorem 5.13 *Let Y be a collapsible subpolyhedron of a polyhedron X. If $(X/Y) \times I \searrow \{*\}$, then $X \times I \searrow \{*\}$.*

Proof: Let $y \in Z = X/Y$ be the point corresponding to Y. Since $Z \times I \searrow \{*\}$, there is a collapsible triangulation K of $Z \times I$ such that $y \times I$ is the underlying space of a subcomplex $J \subset K$. The triangulation K may be chosen so that $S = \mid St(J'', K'') \mid$ has a form $S = N_1 \times I \subset Z \times I$, where N_1 is a regular neighbourhood of y in Z. Let $G = \mathrm{Cl}(Z \times I - S)$. Denote by by N_2 a regular neighbourhood of Y in X. Identify $\partial N_1 = N_1 \cap Cl(Z - N_1)$ with $\partial N_2 = N_2 \cap Cl(X - N_2)$.

For every vertex $v \in J'$ the pairs $(\mid St(v, K'') \mid, \mid St(v, K'') \mid \cap G)$ and $(N_1 \times I_v, \partial N_1 \times I_v)$ are homeomorphic, where $I_v = \mid St(v, J'') \mid$. Hence there exist a homeomorphism $h : (N_1 \times I, \partial N_1 \times I) \to (S, S \cap G)$ such that

$$h(\partial N_1 \times I_v) = \mid St(v, K'') \mid \cap G$$

for every vertex $v \in J'$. It is clear that $Z \times I = G \cup_r (N_1 \times I)$ and that $X \times I$ is homeomorphic to the polyhedron $W = G \cup_r (N_2 \times I)$, where r is the restriction of h onto $\partial N_1 \times I = \partial N_2 \times I$; see Figure 17.

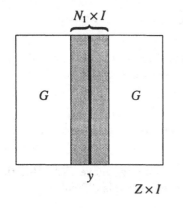

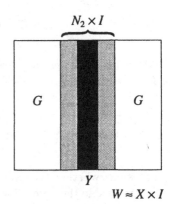

Figure XI.17.

For every subcomplex $L \subset K$ define a subpolyhedron $W(L) \subset W$ by the formula

$$W(L) = (\mid L \mid \cap G) \cup_r (N_2 \times \mid St(L'' \cap J'', J'') \mid).$$

Using the collapsibility of Y, one can easily verify that if $K \searrow^s L$, then $W(K) \searrow W(L)$. To be more precise, each elementary simplicial collapse

$L \searrow L_1 = L - (\sigma \cup \delta)$ can be followed by a collapse of $W(L)$ onto $W(L_1)$. It is evident if the free face δ of σ is not contained in J, and for the case $\delta \in J$; see Figure 18, where the case $\sigma = L$ is shown. It remains to note that $W(K) = W \approx X \times I$ and $W(*) \searrow \{*\}$. □

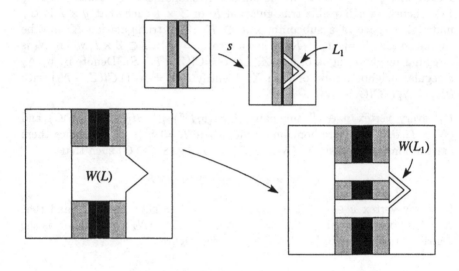

Figure XI.18.

Corollary 5.14 *Let a special polyhedron Q be obtained from a special polyhedron P by the move T_3. If $P \times I \searrow \{*\}$, then $Q \times I \searrow \{*\}$.*

Proof: Let $E_5 \subset Q$ be a fragment appeared under the move T_3. Since $Q/E_5 = P$ and $P \times I \searrow \{*\}$, it follows from Theorem 5.13 that $Q \times I \searrow \{*\}$ □

We do not know whether the same result holds for the move T_3^{-1}. But the following remarkable theorem allows one to overcome this difficulty.

Theorem 5.15 *Let a special polyhedron P 3-deform to an unthickenable special polyhedron Q. Then one can pass from P to Q by a sequence of the moves $T_1^{\pm 1}$, $T_2^{\pm 1}$ and T_3 (that is, without using T_3^{-1}).*

For the proof, see [Ma87₂].

Corollary 5.16 (Z) *is true for all unthickenable special polyhedra that can be 3-deformed to a point.*

Proof: Suppose an unthickenable special polyhedron Q can be 3-deformed to a point. Then *Bing's house B* can be 3-deformed to Q. It follows from Theorem 5.15 that one can pass from B to Q by a sequence of moves $T_1^{\pm 1}$, $T_2^{\pm 1}$ and T_3. By Theorem 5.9 and Corollary 5.14, these moves preserve the 1-collapsibility. Since $B \times I \searrow \{*\}$, we have $Q \times I \searrow \{*\}$. $\qquad\square$

Corollary 5.17 (Z) *restricted to unthickenable special polyhedra is equivalent to* (AC).

Proof: (AC) implies (Z) for unthickenable special polyhedra by Corollary 5.16. The inverse implication follows from [Wr73]; see also Chapter I, Remarks 1,2 to Theorem 3.1.

Combining Corollaries 5.8 and 5.17, we have

Theorem 5.18 ([Ma87$_2$]) (Z) *restricted to special polyhedra is equivalent to the union of* (P) *and* (AC).

Theorem 5.18 may bring some doubt to the common belief (Z) is false.

6   Generalizing (Z) to Higher Dimensions

Let us reformulate (Z) as follows: the direct product of a contractible 2-dimensional polyhedron and the 1-dimensional cube is collapsible. There are two ways for generalizing (Z) to higher dimensions: one may increase the dimension of the ball, and one may increase the dimension of the polyhedron. Let us consider the first possibility.

6.1 n-Collapsing

Definition   A compact polyhedron P is called *n-collapsible* if $P \times I^n \searrow \{*\}$.

Theorem 6.1 ([Di68, Li70]) *Every contractible polyhedron is n-collapsible for some n.*

Proof: According to [Wh39], there is a finite sequence of elementary expansions transforming P to a collapsible polyhedron. The *collapsing by adding a*

cell method (section 3.1) allows one to shorten the sequence at the expense
of multiplying it by I.   □

There arises a natural question concerning an estimate for the number n.
The proof of Theorem 6.1 does not give a uniform upper bound, since the
number of elementary expansions can be arbitrarily large. First results in
this direction were obtained by M. Cohen [Co75].

Theorem 6.2 *Let P be a spine of a PL manifold M with boundary ∂M.
Then $P \times Cone(\partial M)$ collapses to a subset PL homeomorphic with M.*

Proof: One can identify M with the mapping cylinder M_f of a map $f : \partial M \rightarrow$
P. Present M_f and $Cone(\partial M) \times P$ as quotient spaces of $(\partial M \times I) \cup P$ and
$\partial M \times I \times P$, respectively. Then the rule $(x,t) \rightarrow (x,t,f(x))$ for $(x,t) \in$
$\partial M \times I$ determines an embedding $\psi : M_f \rightarrow Cone(\partial M) \times P$. Starting from
$\partial M \times \{0\} \times P$, we collapse $Cone(\partial M) \times P$ onto $\psi(M_f) \approx M$ along line
segments of the form $(x,t,y) \subset Cone(\partial M) \times P$, $0 \leq t \leq 1$. Certainly a little
care is needed to be sure that the collapse is PL.   □

Corollary 6.3 *Every spine P of a n-ball B^n is n-collapsible.*

Proof: $P \times B^n \approx Cone(\partial B^n) \times P \searrow B^n \searrow \{*\}$.   □

Corollary 6.4 *Every compact contractible n-dimensional polyhedron P^n is
m-collapsible, where $m = 6$ for $n = 2$ and $m = 2n$ for $n \geq 3$.*

Proof: Embed P^n in R^m (see, for example, [Fe70]) and consider a regular
neighbourhood N of P^n. The higher dimensional Poincaré Conjecture implies
that N is an m-ball. By Corollary 6.3, P^n is m-collapsible.   □

From Corollary 6.3, we know that every spine of the 3-ball is 3-collapsible.
R. Edwards and D. Gillman have improved this result by one [EdGi83].

Theorem 6.5 *Every spine of the 3-ball is 2-collapsible.*

6.2 Disproving (Z) for higher dimensional polyhedra

The first counterexample to higher dimensional (Z) was exhibited by M. Cohen [Co77]. His paper is based on Rothaus' result concerning the existence of Whitehead torsion values which cannot be realized by two-dimensional additions to a complex (see [Rot77] and Chapter I, §4.4) and a principle of stopping collapses at regular neighbourhoods of subpolyhedra that generalizes the one of the proof of Theorem 3.2 (compare Chapter I, (54)).

Theorem 6.6 *For every $n \geq 3$ there exists a contractible n-dimensional polyhedron P^n such that $P^n \times I$ is not collapsible.*

Now the existence of rather nice 3-dimensional counterexamples is known. Consider the 2-complex P which may be described as the boundary of a regular dodecahedron, with opposite faces identified under a rotation through $\pi/5$ (see Figure VIII.1a). P is a (special!) spine of the punctured Poincaré homology sphere. As I. Bernstein, M. Cohen and R. Connelly have shown, the suspension ΣP is not 1-collapsible [BeCoCo78], although it is clearly contractible. Concerning n-collapsibility, they showed that for every integer $n \geq 1$ there exists a contractible polyhedron of dimension $n + 4$ which is not n-collapsible. The latter result uses J. W. Cannon's and R. D. Edwards' work on non-combinatorial triangulations of spheres.

Recently, J. Bracho and L. Montejano [BrMo91] have constructed $(n + 1)$-dimensional polyhedra A^{n+1} of the (simple-) homotopy type of S^1 such that $A^{n+1} \times I^q$ collapses to S^1 iff $q \geq n$. In contrast to the previous result, the noncollapsibility part of their theorem does not depend on the PL-structure but is related to the Ljusternik-Schnirelman category.

7   Open Problems

1 Does there exist a presentation $< a_1, \ldots, a_n \mid R_1, \ldots, R_n >$ such that the reduced words $R_1, \ldots, R_n$ form a basis in $F(a_1, \ldots, a_n)$ but the 2-complex associated with the presentation is not prismatically collapsible?

2 Let a special polyhedron Q be obtained from a special polyhedron P by applying the move T_5 shown on Figure VIII.11. Does the 1-collapsibility of Q imply the 1-collapsibility of P?

  If *Yes* then it follows from Corollary 5.16 that the Poincaré conjecture is true in the following weakened (and natural!) form: every 3-manifold that 3-deforms to a point is a genuine 3-cell.

3 Does (Z) hold for higher-dimensional polyhedra having a nice local structure? As a candidate for polyhedra having a nice local structure one may take higher-dimensional special polyhedra; see [Ma73].

4 Are all $K(p, q, r, s)$ 1-collapsible (see [Li73])? It is known that every $K(p, q, r, s)$ is a spine of a 3-ball (see §3.3).

5 Denote by Y_n the wedge of n segments. Is it true that for every contractible 2-dimensional polyhedron P^2 there is an integer n such that $P^2 \times Y_n \searrow \{*\}$? This would imply that P is 2-collapsible (compare Theorem 6.5).

6 Two special polyhedra P and Q are said to be *T-equivalent*, if there is a sequence of moves $T_1^{\pm 1}, T_2^{\pm 1}$ transforming P to Q. Define an *imaginary* 3-manifold to be a T-equivalence class of special polyhedra (Motivation: Theorem 5.1). The notion of singular 3-manifolds introduced by F. Quinn [Qu81] (see Chapter VIII, §2) can be considered as a geometrical realization of an imaginary 3-manifold.

Try to develop a theory of imaginary 3-manifolds.

7 Investigate the following generalization of (Z): If K^2 is homotopically equivalent to S^1, then $K^2 \times I \searrow S^1$ (The analogue to Theorem 6.6 for this situation has been proved in [BrMo91], compare §6.2). It is sensible also to replace S^1 by a wedge of circles and/or 2-spheres.

8 (D. Gillman) **Conjecture** If P^2 is a fake surface and $P^2 \times I$ is collapsible, then Punctured Real Projective 3-space does not embed in $P^2 \times I$.

This conjecture fails without the first hypothesis by a theorem of Zhong-mou Li [Li93]. It fails without the second hypothesis by a theorem of [Ma87$_1$]. The conjecture may be strengthened by replacing *collapsible* with *contractible*, or by replacing *Real Projective 3-space* by *any nontrivial lens space*.

Chapter XII

The Andrews-Curtis Conjecture and its Generalizations

Cynthia Hog–Angeloni and Wolfgang Metzler

1 Introduction

For the following survey on the (generalized) Andrews-Curtis problem we assume that the reader is familiar with relevant material from previous chapters. In particular, we refer to Chapter I, § 4.1 for the origin of $(AC^{(')})$ and to Chapters X and XI for relations to $(Z^{(')})$ resp. to the Whitehead asphericity conjecture. Throughout this chapter all complexes will be compact, connected, and all presentations will be finite. Because of Chapter I, Theorem 2.4 we may speak interchangeably of 3-deformations and of Q^{**}-equivalences, according to the specific context.

1.1 Some balanced presentations of the trivial group

A Q^{**}-*trivialization* of a finite presentation $\mathcal{P}$ is a Q^{**}-transformation of $\mathcal{P}$ into a presentation of type $\langle a_1, \ldots, a_g | a_1, \ldots, a_g \rangle$; Q^{*}- and Q-*trivializations* are defined analogously. No Q^{**}-trivialization is known for the following examples, although these balanced presentations can be shown to define the trivial group:

(1)   a) $\langle a, b, c | c^{-1}bc = b^2,\ a^{-1}ca = c^2,\ b^{-1}ab = a^2 \rangle$, see Rapaport [Ra68$_2$];

b) $\langle a,b | ba^2b^{-1} = a^3,\ ab^2a^{-1} = b^3 \rangle$, see Crowell-Fox [CrFo63], p. 41;

c) $\langle a,b | aba = bab,\ a^4 = b^5 \rangle$, see Akbulut Kirby [AkKi85].

They serve as potential counterexamples to disprove (AC).

We add some comments and facts about these examples:

Example a) belongs to a series of presentations with n generators a_i and cyclically indexed defining relations $a_{i+1}^{-1}a_i a_{i+1} = a_i^2$ which, for $n \geq 4$ present nontrivial infinite groups (see B.H. Neumann in Kurosch [Ku53], p. 376). For $n = 2$, the presentation can easily be trivialized by a Q-transformation.

Each of the relators in *Example* b) defines a one-relator, non-hopfian group, (see Lyndon Schupp [LySc77], p. 197). If in b) one relator is replaced by any other such that the resulting group π is perfect, then $\pi = \{1\}$ still holds. Moreover, the exponents 2 and 3 in the remaining relator may be replaced by n and $n + 1$ without loosing the triviality of π, Miller Schupp [MiSc79].

Example c) corresponds to a handle decomposition of the Akbulut-Kirby 4-sphere [AkKi85] which was later shown to be smoothly standard by Gompf [Go91]. The presentations $\langle a,b | aba = bab,\ a^n = b^{n+1} \rangle$ all yield the trivial group. For $n \geq 3$ it is unknown whether they are Q^{**}-trivial, which is the case for $n = 2$, see Gersten [Ge88$_2$].

2 Strategies and Characterizations

2.1 Considerations on Length

The early works on (AC) contribute to the question whether there exists some analogy between Nielsen's or Whitehead's algorithm for free groups (Lyndon Schupp [LySc77], Chapter 1) and the problem of finding a Q^{**}-trivialization for a given balanced presentation of $\pi = \{1\}$. Rapaport [Ra68$_1$] yields a negative answer to certain expectations of Andrews Curtis [AnCu66]. Theorem 6 of her paper contains a (sub-)family of presentations, for which the "obvious" Q-trivializations by chains of elementary transformations pass through intermediate stages, where the sum of the lengths of (cyclically reduced) defining relators is bigger than in the beginning and in the end[1]. In fact, the height of these "mountains" depends exponentially on a parameter of the family, and no competing trivializations seem to be known that do better. But in Rapaport's examples this phenomenon can be bypassed by lumping together

---

[1] Compare the discussion after Theorem 3.4 below.

consecutive changes of one defining relator R_j to a new type of elementary Q-transformation, namely

(2) $R_j \to w R_j^{\pm 1} w^{-1} \cdot S$, where S is a consequence of the remaining relators.

Gersten [Ge88$_2$] has taken up Rapaport's discussion and, with the use of a slightly modified basic terminology which involves the notion of "pieces" of small cancellation theory and, in particular, takes care of (2), was able to exhibit weakly monotonous Q^{**}-trivializations for "all Andrews-Curtis trivializable presentations known to the author". Moreover he proved that such a "direct path" in his sense does not exist for Example (1)a) above.

As a comment in ([Ra68$_2$], p. 149) shows, Rapaport herself was hesitant to allow (2) as an elementary move: in general, length reduction by (2) embodies a decision problem which may already be unsolvable in the restricted case of balanced presentations of $\pi = \{1\}$. At least, when passing to non-balanced presentations and omitting explicit knowledge on π, the danger of "trying to solve" an unsolvable decision problem becomes apparent:

(3) The presentations $\mathcal{P} = \langle a_1, \ldots, a_g | R_1 \ldots, R_h, 1, \ldots, 1 \rangle$
and $\mathcal{Q} = \langle a_1, \ldots, a_g | 1, \ldots, 1, a_1, \ldots, a_g \rangle$

with equal number of defining relators are Q- resp. Q^*- resp. Q^{**}-equivalent iff $\langle a_1, \ldots, a_g | R_1, \ldots, R_h \rangle$ is a presentation of $\pi = \{1\}$.

Example (1)a) can be Q^{**}-transformed into a presentation with 2 generators and 2 relators, (see [Ra68$_2$], p. 141). Such a "reduction" will in general lead to a considerable increase of lengths of the remaining relators. Thus it may be worth while to "simplify" a presentation in the other direction: Passing to a triangulation of its standard complex, a given presentation can be Q^{**}-transformed until

(4) all relators are of length of at most 3.

Furthermore, it can be shown that a 3-deformation between two finite, simplicial 2-complexes gives rise to a chain of *simplicial expansions and collapses of dim. ≤ 3 in which the 3-dimensional moves are transient*, compare [KrMe83] and Chapter I, § 2.3. These immediately translate into a list of generating Q^{**}-moves such that (4) holds throughout the transformation. Keeping in mind the simplicial steps, it is possible to give a purely algebraic argument that this list generates all Q^{**}-equivalences between presentations of type (4) (Exercise).

A dual property may be achieved by shifting complexes and 3-deformations into general position, compare Chapter I, § 3.1 and Chapter XI, § 5: One may state (AC'') in terms of presentations and (new) elementary Q^{**}-transformations such that

(5) each generator (including its inverse) occurs altogether at most three times in the defining relators.

2.2   Algebraic aspects

Let $\mathcal{P} = \langle a_1, \ldots, a_g | R_1, \ldots, R_h \rangle$ and $\mathcal{Q} = \langle a_1, \ldots, a_g | S_1, \ldots, S_h \rangle$ be presentations with equal relator subgroups $N = N(R_j) = N(S_j) \subseteq F(a_i)$; in particular, $\mathcal{P}$ and $\mathcal{Q}$ present the same group $\pi = F(a_i)/N$. In order to obtain criteria whether $\mathcal{P}$ and $\mathcal{Q}$ are Q-equivalent, we take up notions and results from Chapter II:

To $\mathcal{P}$ we associate a (bigger) free group $F(a_i, r_j)$, $i = 1, \ldots, g$, $j = 1, \ldots, h$ and a projection $p : F(a_i, r_j) \to F(a_i)$ given by $a_i \to a_i$, $r_j \to R_j$. Let $E(\mathcal{P})$ be the normal closure of the r_j in $F(a_i, r_j)$. $F(a_i)$ operates on $E(\mathcal{P})$ and on N by conjugation; p induces an $F(a_i)$-equivariant surjection $\partial_\mathcal{P} : E(\mathcal{P}) \to N$. The kernel of $\partial_\mathcal{P}$ is the *group $I(\mathcal{P})$ of identities* of $\mathcal{P}$. It contains the *Peiffer elements* $(r, s) = r \cdot s \cdot r^{-1} \cdot \partial_\mathcal{P}(r)s^{-1}\partial_\mathcal{P}(r)^{-1}$, $r, s \in E(\mathcal{P})$ and their normal[2] closure the *Peiffer group $P(\mathcal{P})$*. The corresponding data for $\mathcal{Q}$ are constructed and denoted in analogy to those for $\mathcal{P}$.

That the relators of $\mathcal{P}$ are consequences of those of $\mathcal{Q}$, can be expressed as follows: There exists an $F(a_i)$-equivariant homomorphism $\varphi : E(\mathcal{P}) \to E(\mathcal{Q})$ leading to a commutative diagram

(6)

$$
\begin{array}{ccc}
E(\mathcal{P}) & \xrightarrow{\varphi} & E(\mathcal{Q}) \\
\partial_\mathcal{P} \searrow & & \swarrow \partial_\mathcal{Q} \\
& N &
\end{array}
$$

We call such a homomorphism φ a *presentation morphism*; the usual wording on composition of morphisms and on isomorphisms can be made here.

Expressing the defining relators of $\mathcal{P}$ in different ways as consequences of the S_j, yields morphisms φ, ψ which differ by identities of $\mathcal{Q}$, and vice versa.

With this terminology we have

---

[2] The group generated by the (r, s) *is* normal (Exercise).

Lemma 2.1 $\mathcal{P} = \langle a_1, \ldots, a_g | R_1, \ldots, R_h \rangle$ and $\mathcal{Q} = \langle a_1, \ldots, a_g | S_1, \ldots, S_h \rangle$ with $N = N(R_j) = N(S_j)$ are Q-equivalent iff there exists a presentation isomorphism $\varphi : E(\mathcal{P}) \to E(\mathcal{Q})$.

Proof: Elementary Q-transformations (conjugation, inversion, multiplication) immediately give rise to presentation isomorphisms; hence a Q-equivalence yields an isomorphism φ. Conversely, if $\varphi : E(\mathcal{P}) \to E(\mathcal{Q})$ is a presentation isomorphism, then the $F(a_i)$-conjugates of the $r_1, \ldots, r_h$ and of the $\varphi^{-1}(s_1), \ldots, \varphi^{-1}(s_h)$ are both (ordinary) bases of $E(\mathcal{P})$. The "Nielsen theorem for free groups with operators" (see the references given in Chapter I. § 4.1) yields that the r_j can then be converted into the $\varphi^{-1}(s_j)$ by a sequence of a) free transformations and b) conjugations by elements of $F(a_i)$. They result in a Q-transformation $\mathcal{P} \to \mathcal{Q}$, as desired. □

A presentation morphism $\varphi : E(\mathcal{P}) \to E(\mathcal{Q})$ yields a continuous map $K_\mathcal{P} \to K_\mathcal{Q}$ which is the identity on the 1-skeleta; furthermore $I(\mathcal{P})$ and $P(\mathcal{P})$ are sent to the corresponding groups associated to $\mathcal{Q}$. These maps give rise to a characterization of Q-equivalence which can be related to geometric arguments of previous chapters:

Theorem 2.2 ([Me79$_1$]) *Two presentations* $\mathcal{P} = \langle a_1, \ldots, a_g | R_1, \ldots, R_h \rangle$ *and* $\mathcal{Q} = \langle a_1, \ldots, a_g | S_1, \ldots, S_h \rangle$ *with* $N = N(R_j) = N(S_j)$ *are Q-equivalent iff there exists a presentation morphism* $\varphi : E(\mathcal{P}) \to E(\mathcal{Q})$ *that induces a homotopy equivalence* $K_\mathcal{P} \to K_\mathcal{Q}$ *and an isomorphism* $P(\mathcal{P}) \to P(\mathcal{Q})$.

Proof: If $\mathcal{P}$ and $\mathcal{Q}$ are Q-equivalent, then, by the preceding lemma, there is an isomorphism $\varphi : E(\mathcal{P}) \to E(\mathcal{Q})$. This induces isomorphisms $I(\mathcal{P}) \to I(\mathcal{Q})$, $P(\mathcal{P}) \to P(\mathcal{Q})$, hence an isomorphism from $\pi_2(K_\mathcal{P}) \approx I(\mathcal{P})/P(\mathcal{P})$ to $\pi_2(K_\mathcal{Q}) \approx I(\mathcal{Q})/P(\mathcal{Q})$ (see Chapter II, Theorem 2.7). As the fundamental groups are also mapped isomorphically onto each other, Whitehead's theorem (Chapter II, Theorem 2.11) yields that φ induces a homotopy equivalence $K_\mathcal{P} \to K_\mathcal{Q}$.

Conversely, if φ induces an isomorphism $P(\mathcal{P}) \to P(\mathcal{Q})$ and an isomorphism $I(\mathcal{P})/P(\mathcal{P}) \to I(\mathcal{Q})/P(\mathcal{Q})$, then φ also induces an isomorphism $I(\mathcal{P}) \to I(\mathcal{Q})$. By the commutativity of (6), φ maps $E(\mathcal{P})/I(\mathcal{P})$ isomorphically to $E(\mathcal{Q})/I(\mathcal{Q})$; together with $I(\mathcal{P}) \xrightarrow{\approx} I(\mathcal{Q})$ we get that $\varphi : E(\mathcal{P}) \to E(\mathcal{Q})$ is an isomorphism itself which, by the preceding lemma, gives rise to a Q-transformation $\mathcal{P} \to \mathcal{Q}$. □

Q^*- and Q^{**}-transformations can be covered by an extended concept, compare [Me79$_1$]. This paper also contains a discussion of special cases (e.g. balanced presentations of the trivial group), and derives tests which result from the above classification (e.g. via Fox calculus). In contrast to Theorem 2.2, which can be interpreted as measuring the "gap" between homotopy type and Q-equivalence, the following considerations focus on an algebraic description of the potential difference between *simple*-homotopy of 2-complexes and Q^{**}-equivalence. Again we assume that $\mathcal{P} = \langle a_1, \ldots, a_g | R_1, \ldots, R_h \rangle$ and $\mathcal{Q} = \langle a_1, \ldots, a_g | S_1, \ldots, S_h \rangle$ fulfill $N(R_j) = N(S_j)$ $(= N)$. If, in addition,

(7) all $R_j S_j^{-1}$ are contained in the commutator subgroup $N^{(1)} = [N, N]$, then $K_{\mathcal{P}}$ and $K_{\mathcal{Q}}$ are simple-homotopy equivalent.

This holds because we then get a chain equivalence of equivariant chain complexes where all chain groups are mapped by identity matrices: a commutator of relators contributes the trivial (equivariant) 2-chain to such a map. □

But there is a converse to this fact:

Theorem 2.3 ([Qu85], [Ho-AnMe90]) *If K^2 and L^2 are compact, connected CW complexes that are simple-homotopy equivalent, then these complexes can be 3-deformed until corresponding presentations $\mathcal{P} = \langle a_1, \ldots, a_g | R_1, \ldots, R_h \rangle$ and $\mathcal{Q} = \langle a_1, \ldots, a_g | S_1, \ldots, S_h \rangle$ are achieved which fulfill $N = N(R_j) = N(S_j)$ and $R_j S_j^{-1} \in N^{(1)}$.*

Quinn [Qu85] relates the statement to 4-dimensional handlebody theory and derives an elementary simple-homotopy move for 2-complexes which helps to interpret a 4-deformation. We sketch the (algebraic)

Proof given in [Ho-AnMe90]: K^2 and L^2 can be 3-deformed until $\mathcal{P}$ and $\mathcal{Q}$ have the same generators, the same number of defining relators, $N(R_j) = N(S_j)$ holds, and the given simple-homotopy equivalence maps all equivariant chain groups (with bases given by the cells) by identity matrices. In particular, the R_j resp. S_j give rise to fundamental systems $\tilde{R}_j$ resp. $\tilde{S}_j$ of 2-cells such that

(8) $\tilde{R}_j$ is mapped to $\tilde{S}_j$ for each j.

This corresponds to a presentation morphism $\varphi : E(\mathcal{P}) \to E(\mathcal{Q})$, such that

(9) the $\varphi(r_j)s_j^{-1}$ are contained in the kernel of the natural map $\theta_{\mathcal{Q}} : E(\mathcal{Q}) \to C_2(\tilde{K}_{\mathcal{Q}})$.

But by Reidemeister [Re49] (compare also Chapter II, Lemma 2.4), $\ker \theta_Q$ is normally generated by commutators and Peiffer identities . As the latter are trivialized under $\partial_Q : E(Q) \to N$, we get that $\partial_Q(\varphi(r_j))s_j^{-1} = R_j S_j^{-1}$ is a product of commutators of N. $\qquad \square$

Because of Theorem 2.3, (7) can be considered as the essential principle in constructing potential counterexamples to (AC'). But potential invariants to disprove (AC') can be pushed up to arbitrarily high commutators (unlike those for distinctions in Chapters III and VII):

Theorem 2.4 ([Ho-AnMe91]) *If* $\mathcal{P} = \langle a_1, \ldots, a_g | R_1, \ldots, R_h \rangle$ *and* $Q = \langle a_1, \ldots, a_g | S_1, \ldots, S_h \rangle$ *fulfill* $N = N(R_j) = N(S_j)$ *and if* $R_j S_j^{-1} \in N^{(1)}$ *is true, then for each* $n \in \mathbb{N}$ *there exists a* Q-*transformation* $\mathcal{P} \to \mathcal{P}' = \langle a_1, \ldots, a_g | R_1', \ldots, R_h' \rangle$ *such that* $R_j' S_j^{-1} \in N^{(n)}$ *holds for all* j. $\qquad \square$

Here $N^{(n)}$ denotes the n-th commutator subgroup which is defined inductively by $N^{(0)} = N$ and $N^{(n)} = [N^{(n-1)}, N^{(n-1)}]$. The (AC)-case of Theorem 2.4 is due to W. Browning [Br76$_2$]. We relate Theorem 2.4 to presentation morphisms and identities:

Let $\mathcal{P} = \langle a_1, \ldots, a_g | R_1, \ldots, R_h \rangle$ and $Q = \langle a_1, \ldots, a_g | S_1, \ldots, S_h \rangle$ fulfill $N = N(R_j) = N(S_j)$ and $R_j S_j^{-1} \in N^{(n)}$. Then there exist presentation morphisms $\varphi : E(\mathcal{P}) \to E(Q)$, $\psi : E(Q) \to E(\mathcal{P})$ with $\psi\varphi(r) \cdot r^{-1} \in I(\mathcal{P}) \cap E^{(n)}(\mathcal{P})$ for all $r \in E(\mathcal{P})$, $\varphi\psi(s) \cdot s^{-1} \in I(Q) \cap E^{(n)}(Q)$ for all $s \in E(Q)$. These deviations from the identity are contained in subgroups which can be rewritten as follows:

By Reidemeister's main step of the proof in [Re49], the property

(10) $I \cap [E, E] = [I, E]$ holds for every presentation,

compare the proof of Lemma 2.4 in Chapter II.But the general argument[3] of [Re49] directly covers the fact

(11) $I \cap [W, W] = [I \cap W, W]$ for every $F(a_i)$-invariant subgroup $W \subseteq E$.

Thus we get

$$I \cap E^{(n)} = [I \cap E^{(n-1)}, E^{(n-1)}] = \left[[I \cap E^{(n-2)}, E^{(n-2)}], [E^{(n-2)}, E^{(n-2)}]\right] = \ldots,$$

---

[3]Reidemeister's proof is given for pre-crossed modules $C \xrightarrow{\partial} G$ which have a section $H : \partial C \to C$. Such a section – which is not assumed to be G-equivariant – exists in the case of a free group $G = F(a_i)$, hence also for the restricted pre-crossed module $W \xrightarrow{\partial} G$.

i.e. *the elements of $I \cap E^{(n)}$ can inductively be expressed as products of $n - th$ commutators of E such that one innermost bracket of each factor is of the type $[u, v]^{\pm 1}$, $u \in I$.* As $[u, v] = uvu^{-1}v^{-1}$ equals the Peiffer element (u, v) for $u \in I$, $[I, E]$ is contained in P. Similarly, the statement in italics implies that $I \cap E^{(n)}$ is contained in P_{2^n}, where P_m is the m-th term of the *Peiffer central series* which has been introduced by H. Baues and D. Conduché, see [BauCo90], Prop. 2.5, and [Bau91]. This Peiffer central series is defined inductively by: "$P_1 = E$, P_m is (normally) generated by all (u, v), $u \in P_i$, $v \in P_j$ such that $i + j = m$" (e.g. $P = P_2$).

Hence φ in particular induces an isomorphism $E(\mathcal{P})/P_{2^n}(\mathcal{P}) \to E(\mathcal{Q})/P_{2^n}(\mathcal{Q})$, the inverse being induced by ψ.

The Theorems 2.2, 2.3, 2.4 and the above discussion reveal that, by killing (higher) commutators or (higher) Peiffer elements, essential information to disprove (AC') may get lost. Instead of passing to such quotients it may be worth while to use projections into perfect or simple groups.

2.3   Aspects concerning singular 3-manifolds

By the classification of 3-deformations of 2-complexes in terms of singular 3-manifolds and their surgery operations, see Chapter VIII, § 2.1, (AC') can be rephrased as the expectation that

(12) singular 3-manifolds $\check{M}_1^3, \check{M}_2^3$ that are simple-homotopy equivalent can be transformed into each other[4] by a finite sequence of Quinn's surgery operations (2,0), (0,2), (4,3).

The special case with $\check{M}_2^3 = D^3$ is (AC). This motivates the search for criteria which guarantee resp. exclude the possibility to remove all singular points from an $\check{M}^3$ by the operations (2,0), (0,2), (4,3). Of course, $\pi_1(\check{M}^3)$ should be a 3-manifold group, and the question may be formulated under the restriction that the (simple-)homotopy type of $\check{M}^3$ can be realized by a 3-manifold without singularities (compare Chapter I, (end of)§ 3.1, § 4.1, (61) and (62)). Singular 3-manifolds also give rise to the question whether manifold duality can be applied resp. modified in order to distinguish presentation classes.

---

[4]up to (p.l.) homeomorphism

3 Q^{**}-Transformations and Presentations of Free Products

3.1   Semisplit presentations and nonsplittable homotopy types

Given a finite presentation of a free product $\pi = G * H$, Grushko's theorem see Stallings [St65$_1$] yields *sorted* generators, i.e. there exists a Q^{**}-transformation (a free transformation of the generators) such that the resulting generators are partitioned into $\{a_i, b_j\}$ where a_i resp. b_j project to generators of G resp. H. Each defining relator thereby has become a word in $F(a_i, b_j)$ which (is trivial or) decomposes uniquely as a product of reduced segments alternately being contained in the factors $F(a_i)$ and $F(b_j)$. (The number of segments is the length with respect to the free product $F(a_i) * F(b_j)$.) It is natural to ask whether there exists a further Q^{**}-transformation which also splits the relators. This holds if the presentation class $\Phi(\mathcal{P})$ contains the spine of a compact, connected 3-manifold. Stallings' proof of Kneser's conjecture (see Hempel [He76], compare Jaco [Ju86]) then gives rise to the desired decomposition.

A modification of this proof for singular 3-manifolds still yields:

Theorem 3.1 ([HoLuMe85]) *Every presentation class of a finitely presentable free product $\pi = G * H$ contains a finite presentation $\langle a_i, b_j | R_k \rangle$ such that the a_i resp. b_j project to generators of G resp. H and each defining relator R_k consists of at most two segments.*

In the case of one segment, R_k is a *split relator* for either G or H; in the other case $R_k^{\pm 1} = S_k(a_i) \cdot T_k(b_j)$ holds with $S_k = 1$ resp. $T_k = 1$ being a relation in G resp. H. Note that, in general, *both* segments are nevertheless not amongst the defining relators. A presentation of $G * H$ as in Theorem 3.1 will be called *semisplit*. We give an algebraic **proof** of this theorem based on the idea of subdividing the 2-cells which correspond to the defining relators of a presentation with sorted generators $a_1, \ldots, a_g$, $b_1, \ldots, b_{g'}$ (as above), see [HoLuMe85]:

For a given defining relator, at least one of the segments must map to 1, say $w(a_i)$. If the number of segments is greater than 2, we prolong the presentation by a new generator $b_{g'+1}$ for the second factor and the relation $b_{g'+1} = w(a_i)$ (which consists of 2 segments). In the given relator we can now replace w by $b_{g'+1}$, yielding a relator with fewer segments. We apply this process until every nontrivial defining relator has a number of segments not exceeding 2. (All steps are Q^{**}-transformations).   □

But as opposed to the 3-manifold (sub-)case, *a presentation class* Φ *of* $\pi = G * H$ *does not always split as a sum* $\Phi_1 + \Phi_2$ *for appropriate presentation classes* Φ_1 *of* G, Φ_2 *of* H. Examples for this phenomenon can be obtained by the following principle, compare Chapter I, § 4.4:

(13) Let $\tilde{G}$ and $\tilde{H}$ be finitely presented groups with elements $x \in \tilde{G}$, $y \in \tilde{H}$ of finite relatively prime order. Define G, H to be the quotients of $\tilde{G}$ resp. $\tilde{H}$ by adding the relations $x = 1$ resp. $y = 1$. Then $\pi = G * H$ can be presented by forming the "disjoint" union of presentations of minimal deficiency for $\tilde{G}, \tilde{H}$, enlarged by the relation $x = y$.

As a concrete example, consider $\mathcal{P} = \langle a_1, a_2, b_1, b_2 | a_1^2 = 1, a_1 a_2 a_1^{-1} = a_2^3$, $b_1^3 = 1, b_1 b_2 b_1^{-1} = b_2^4, a_2^2 = b_2^3 \rangle$. The first and the second relator together imply $a_2^8 = 1$, the third and the forth $b_2^{63} = 1$. Hence a_2^2 and b_2^3 have coprime orders which implies $a_2^2 = b_2^3 = 1$ by the last relator. Now $\mathcal{P}$ turns out to be a semisplit presentation of $(\mathbb{Z}_2 \times \mathbb{Z}_2) * (\mathbb{Z}_3 \times \mathbb{Z}_3)$. But $\chi(K_\mathcal{P}) = 2$; and for presentations $\mathcal{P}_1$ resp. $\mathcal{P}_2$ of $G = \mathbb{Z}_2 \times \mathbb{Z}_2$ resp. $H = \mathbb{Z}_3 \times \mathbb{Z}_3$ we have that $\chi(K_{\mathcal{P}_i})$ is at least 2 which yields $\chi(K_\mathcal{Q}) \geq 2 + 2 - 1 = 3$ for every split presentation $\mathcal{Q}$ of $(\mathbb{Z}_2 \times \mathbb{Z}_2) * (\mathbb{Z}_3 \times \mathbb{Z}_3)$. Hence *the homotopy type of* $K_\mathcal{P}$ *cannot be split as a one-point union*[5] $K_{\mathcal{P}_1} \vee K_{\mathcal{P}_2}$ *which implies that* $\mathcal{P}$ *cannot be* Q^{**}-*transformed to a split presentation* $\mathcal{Q}$. Similar examples, where semisplit presentations "save" Euler characteristic in comparison with split presentations, exist for $(\mathbb{Z}_{m_1} \times \mathbb{Z}_{n_1}) * (\mathbb{Z}_{m_2} \times \mathbb{Z}_{n_2})$, if the greatest common divisors fulfill $(m_1, n_1) \neq 1 \neq (m_2, n_2)$ and $((m_1, n_1), (m_2, n_2)) = 1$, see [HoLuMe85] and [Ho-An88].

This phenomenon can also be rephrased as follows:

The deficiency[6]*of a finitely presentable group* π (*i.e.* $\min(\# \ (relators) - \# \ (generators))$, *taken over all finite presentations of* π) *in general is not additive under the operation of forming the free product of groups.*

See Chapter VII, § 3.7 for a criterion that guarantees additivity.

---

[5]Because of Chapter I, § 2.3, standard complexes can be replaced in the statement by arbitrary finite, connected 2-complexes of the corresponding groups.

[6]Compare Chapter I, footnote 27.

3.2   Simple-homotopy type, 3-deformations and semi-split Q^{**}-transformations

That the (simple-)homotopy theory of 2-complexes with π_1 a free product $G * H$ is not merely a "sum" of those of the factors, is also the underlying expectation of our further research. One may ask, for instance, the (open) questions[7]

(14) whether every finite, connected K^2 with $\pi_1(K^2) = G * H$ that is homotopy equivalent to $K_{\mathcal{P}_1} \vee K_{\mathcal{P}_2}$ can also be split up to simple-homotopy type, i.e. whether $K^2 \nearrow_\searrow K_{\mathcal{Q}_1} \vee K_{\mathcal{Q}_2}$ holds for suitable presentations $\mathcal{P}_1, \mathcal{Q}_1$ resp. $\mathcal{P}_2, \mathcal{Q}_2$ of G resp. H.

and, similarly,

(15) whether a splitting with respect to simple-homotopy type can always be improved to a splitting by 3-deformations.

Counterexamples to (15) would disprove (AC').

As for *sh*-type and 3-deformations, we cite and comment on two results, the proofs of which make use of semisplit presentations of free products (with several factors):

Theorem 3.2 ([Me90]) *Let K_0^2 be a finite, connected CW-complex, let K_i^2, $i = 1, \ldots, n$ be standard complexes of the presentation $\langle a, b | a^2 = [a, b] = b^4 = 1 \rangle$ of $\mathbb{Z}_2 \times \mathbb{Z}_4$, let K^2 be $K_0^2 \vee K_1^2 \vee \ldots \vee K_n^2$, and let $\tau_0 \in \mathrm{Wh}(\pi_1(K_0^2))$ be given. If n (depending on τ_0) is big enough, then there exists a CW complex $L_{\tau_0}^2$ and a homotopy equivalence $f : L_{\tau_0}^2 \to K^2$ such that $\tau(f) = \tau_0 \in \mathrm{Wh}(\pi_1(K^2)) = \mathrm{Wh}(\pi_1(K_0^2))$ holds.* $\quad\square$

In [Me90] it is shown that $L_{\tau_0}^2 \nearrow_\searrow K^2$ does not always hold, thereby distinguishing homotopy type and simple homotopy type in dimension 2, see Chapter VII § 5.6, 5.7, 5.8 for M. Lustig's treatment of this phenomenon. As the homotopy type of $L_{\tau_0}^2$ splits (via K^2), it may well be that some $L_{\tau_0}^2$ don't allow any simple-homotopy splitting (for the given factorization of the fundamental group), which would imply that (14) has a negative answer.

The main "trick" in the proof of Theorem 3.2 is that the relators of the $\mathbb{Z}_2 \times \mathbb{Z}_4$-factors can be Q-transformed to consist of 2 segments in order also to carry information on squares and on commutators of a presentation of $\pi_1(K_0^2)$.

---

[7]Compare § 2.3 above.

By the same idea and by use of Theorem 2.3, it is possible to convert ρh equivalences into 3-deformations, if additional $\mathbb{Z}_2 \times \mathbb{Z}_4$-factors are admitted:

Theorem 3.3 ([Ho-AnMe90]) *Let K_0^2, L_0^2 be finite, connected CW-complexes which are simple-homotopy equivalent. Then by forming the one-point unions $K^2 = K_0^2 \vee K_1^2 \vee \ldots \vee K_n^2$ and $L^2 = L_0^2 \vee K_1^2 \vee \ldots \vee K_n^2$, where the K_i^2, $i = 1, \ldots, n$ are standard complexes of the presentation $\langle a, b | a^2 = [a, b] = b^4 = 1 \rangle$ of $\mathbb{Z}_2 \times \mathbb{Z}_4$, and if n is big enough (depending on K_0^2, L_0^2), K^2 and L^2 fulfill $K^2 \overset{3}{\nearrow\!\!\searrow} L^2$.* □

This theorem covers cases where all complexes which are specified in its formulation live on the level of $\chi_{\min}$. In particular, in contrast to the one-point union with 2-spheres, the assumption $K_0^2 \nearrow\!\!\searrow L_0^2$ cannot generally be weakened to $K_0^2 \simeq L_0^2$: The examples of [Me90] for $L_{\tau_0}^2 \simeq K^2$ which are not simple-homotopy equivalent remain simple-homotopy distinct after stabilization with finitely many $\mathbb{Z}_2 \times \mathbb{Z}_4$-complexes $K_i^2, i = 1, \ldots, n$ as in Theorems 3.2, 3.3. Similarly to Theorem 2.4, Theorem 3.3 prevents from fruitless attempts to obtain (AC')-disproofs: A (potential) invariant which "survives" the above $\mathbb{Z}_2 \times \mathbb{Z}_4$-stabilization is ineffective.

We now turn to our approach of a disproof of (AC'), by summarizing recent and not yet published work in progress:

A Q^{**}-transformation between two finite, semisplit presentations $\mathcal{P}, \mathcal{Q}$ of $\pi = G * H$ may pass through intermediate stages with non-sorted generators; and even if sorted ones are given, the relators may have more than 2 segments. But the property of being semisplit is preserved by the following Q^{**}-transformations:

(16) a) transformations of type $R_k \rightarrow v R_k^{\pm 1} v^{-1}$ or of type $R_k \rightarrow v R_k v \cdot w R_\ell w^{-1}$ resp. $w R_\ell w^{-1} \cdot v R_k v^{-1}$, $k \neq \ell$, if the resulting relator also has at most 2 segments *(elementary semisplit Q-transformations),*

b) free transformations, prolongations (and their inverses) amongst generators which project to one factor (G resp. H), and generalized prolongations (and their inverses) of type $R = a^{-1} \cdot w(b_j)$ resp. $R = b^{-1} \cdot w(a_i)$, such that w is contained in the kernel of $F(b_j) \rightarrow H$ resp. $F(a_i) \rightarrow G$, and a resp. b denotes a new generator for the complementary factor *(semisplit generator transformations).*

Finite compositions of these transformations are called *semisplit Q^{**}-transformations*. Note that (16)a) leads to restrictions for the conjugators v and w: If, for instance, the initial and the resulting relators have exactly 2 segments, $v^{\pm 1}, w^{\pm 1}$ can be assumed to be segments of R_k and R_ℓ. The possibilities correspond to the inversion of a $G * H$-equation $S_k(a_i) = T_k(b_j)$ to become $S_k^{-1} = T_k^{-1}$ or to the multiplication of two such equations. This is a Nielsen-like behaviour of transformations rather than that of a general Q-transformation.

By a geometric argumentation which uses diagrams, it is possible to prove that "bad" intermediate stages (as mentioned above) can be avoided:

Theorem 3.4 (Metzler) *Let $\mathcal{P}, \mathcal{Q}$ be finite, semisplit presentations of $\pi = G * H$ and let $\mathcal{P} \to \mathcal{Q}$ be a Q^{**}-transformation which induces the identity[8] map on π. Then there exists a semisplit Q^{**}-transformation from $\mathcal{P}$ to $\mathcal{Q}$.* □

This theorem is the basis of our current work on (AC'): The Nielsen-like behaviour of semisplit Q-transformations between non-split relators may give rise to a modification of known tests for Nielsen equivalence in order to show that certain Q^{**}-equivalences don't exist at all. In particular, we hope to obtain obstructions against a (total) splitting of 3-deformation types which allow a splitting with respect to sh-type, i.e. to disprove (AC') via a negative answer to (15). But one has to take into consideration that – in contrast to Whitehead's algorithm – *peak-reduction cannot generally be achieved for semisplit Q^{**}-transformations* with respect to the number of relators with 2 segments. This is shown by an example due to A. Sieradski [Si77] which is one case in a Q^{**}-classification of a whole family of presentations. The example also yields that *a factorization of a presentation class of $\pi = G * H$ with freely indecomposable G, H, if possible, is in general not unique:*

Let $\mathcal{P}_1$ resp. $\mathcal{Q}_1$ be presentations of $\mathbb{Z}_5 \times \mathbb{Z}_5 \times \mathbb{Z}_5$ given by $\langle a_1, a_2, a_3 | a_1^5, a_2^5, a_3^5, [a_1, a_2], [a_1, a_3], [a_2, a_3] \rangle$ resp. $\langle a_1, a_2, a_3 | a_1^5, a_2^5, a_3^5, [a_1^2, a_2], [a_1, a_3], [a_2, a_3] \rangle$. $\mathcal{P}_2$ resp. $\mathcal{Q}_2$ are defined as $\mathcal{P}_1$ resp. $\mathcal{Q}_1$, but with generators b_i instead of the a_i.

(17) $\mathcal{P}_1 + \mathcal{P}_2$ can be Q-transformed into $\mathcal{Q}_1 + \mathcal{Q}_2$

---

[8] If the automorphism is different from the identity and if $G \neq \mathbb{Z}$ and $H \neq \mathbb{Z}$ are freely indecomposable, then Theorem 3.4 can be extended by use of the elementary automorphisms of free products which were obtained by Fouxe-Rabinovitch [Fo-Ra40].

by the following chain of semisplit Q-transformations:

$$a_1^5 = 1 \quad , \quad a_2^5 = 1 \quad , \quad [a_1, a_2] = 1 \quad , \quad b_1^5 = 1 \quad , \quad b_2^5 = 1 \quad , \quad [b_1, b_2] = 1$$

$$\downarrow$$

$$a_1^5 = a_2^5 \quad , \quad a_2^5 = [a_1^2, a_2] \quad , \quad '' \quad , \quad b_1^5 = b_2^5 \quad , \quad b_2^5 = [a_1^2, a_2] \quad , \quad ''$$

$$\downarrow (\alpha)$$

$$'' \quad , \quad '' \quad , \quad [a_1^2, a_2]^3 = 1 \quad , \quad '' \quad , \quad '' \quad , \quad ''$$

$$\downarrow$$

$$'' \quad , \quad '' \quad , \quad b_2^{15} = 1 \quad , \quad '' \quad , \quad '' \quad , \quad ''$$

$$\downarrow$$

$$'' \quad , \quad '' \quad , \quad b_2^{15} = [b_1^2, b_2] \quad , \quad '' \quad , \quad '' \quad , \quad ''$$

$$\downarrow (\beta)$$

$$'' \quad , \quad '' \quad , \quad '' \quad , \quad '' \quad , \quad '' \quad , \quad [b_1^2, b_2]^3 = 1$$

$$\downarrow$$

$$'' \quad , \quad '' \quad , \quad '' \quad , \quad '' \quad , \quad '' \quad , \quad b_2^{45} = 1$$

$$\downarrow$$

$$'' \quad , \quad '' \quad , \quad '' \quad , \quad '' \quad , \quad '' \quad , \quad b_2^5 = [a_1^2, a_2]^{10}$$

$$\downarrow (\gamma)$$

$$'' \quad , \quad '' \quad , \quad '' \quad , \quad '' \quad , \quad '' \quad , \quad b_2^5 = 1$$

$$\downarrow$$

$$'' \quad , \quad '' \quad , \quad [b_1^2, b_2] = 1 \quad , \quad b_1^5 = 1 \quad , \quad [a_1^2, a_2] = 1 \quad , \quad ''$$

$$\downarrow$$

$$a_1^5 = 1 \quad , \quad a_2^5 = 1 \quad , \quad [b_1^2, b_2] = 1 \quad , \quad b_1^5 = 1 \quad , \quad [a_1^2, a_2] = 1 \quad , \quad b_2^5 = 1.$$

The stages of the transition are given line by line. We have not listed the relations which contain a_3 and b_3, as they remain unchanged throughout the transformation and are not used to modify the others. All transitions are immediate except for (α), (β), (γ) which are explained (by use of the commutator identity $[x \cdot y, z] = xyzy^{-1}x^{-1}z^{-1} = x[y, z]x^{-1} \cdot [x, z]$) as follows: (α) The first and the second relation (of the line in question) together imply $a_1^5 \leftrightarrows a_2$ and $[a_1^2, a_2] \leftrightarrows a_1$; hence modulo these relations we have $[a_1, a_2] = [a_1^6, a_2] = [a_1^2, a_2]^3$. Similarly, (β) can be derived from the third and forth relation. For (γ), we note that the argument for (α) can be applied again to yield $[a_1^2, a_2] = a_1 \cdot [a_1, a_2]a_1^{-1}[a_1, a_2] = a_1[a_1^2, a_2]^3 a_1^{-1} \cdot [a_1^2, a_2]^3 = [a_1^2, a_2]^6$, whence $[a_1^2, a_2]$ has order 5 (modulo the first two relations).

(Each) $K_{\mathcal{P}_i}$ differs from (each) $K_{\mathcal{Q}_j}$ with respect to homotopy (and homology) type[9], see [Me76], [Si77] and Chapter III, §1. Moreover, $\mathbb{Z}_5 \times \mathbb{Z}_5 \times \mathbb{Z}_5$ is freely indecomposable. This implies the statements in italics which precede (17). □

Concrete counterexamples to (15) we expect by a combination of the principles (7) and (13) above. Already in the case of $H = \{1\}$ (where the factorization of $\pi = G * H$ degenerates and the splitting property (15) trivially holds), such a combination yields presentations for which we don't know whether in general they are Q^{**}-equivalent:

(18) $\mathcal{P} = \langle a_1, \ldots, a_g, b_1, b_2 | R_1(a_i), \ldots, R_h(a_i), S(a_i) \cdot b_1^{-1}, b_1^3 b_2^{-5}, b_1^3(b_1 b_2)^{-2} \rangle$ and $\mathcal{Q} = \langle a_{1_1}, \ldots, a_g | R_1(a_i), \ldots, R_h(a_i), S(a_i) \rangle$ define sh-equivalent standard complexes, if S has finite order coprime to 6 modulo the relators $R_1, \ldots, R_h$.

By the last two relators of $\mathcal{P}$, $\tilde{H}$ is the binary icosahedral group of order 120, in which b_1 has order 6. $S(a_i) = b_1$ then implies $S(a_i) = b_1 = 1$, whence $b_2 = 1$ follows. As $\tilde{H}$ is perfect, the last three relators of $\mathcal{P}$ can be Q-transformed to become $S(a_1) \cdot v^{-1}, b_1 \cdot v_1^{-1}, b_2 v_2^{-1}$, where v and the v_i are contained in $[N, N]$. Hence, (7) above yields that $\mathcal{P}$ and $\langle a_1, \ldots, a_g, b_1, b_2 | R_1, \ldots, R_h, S(a_i), b_1, b_2 \rangle$ define sh-equivalent standard complexes. □

Such examples might even remain inequivalent when passing to the (coarser?) equivalence classes for which Q^{**}-transitions *and* the addition or deletion (if possible) of a balanced presentation of the trivial group are admitted, i.e. they may constitute counterexamples to (AC') which are *specific for the group* which is presented.

Most of the results of this section generalize to free products with amalgamation, HNN-extensions and graphs of groups, compare Hog-Angeloni [Ho-An88]. A degenerate HNN-case analogous to (18) is that

(19) $\mathcal{P} = \langle b_1, b_2, t | t b_1 t^{-1} b_2^{-1}, b_1^3 b_2^{-5}, b_1^3 (b_1 b_2)^{-2} \rangle$ and $\mathcal{Q} = \langle b_1, b_2, t | b_1, b_1^3 b_2^{-5}, b_1^3 (b_1 b_2)^{-2} \rangle$ define sh-equivalent standard complexes

which (so far) we can't Q^{**}-transform into each other [10]. In particular, if we require the common subcomplex $L = K^1 \cup e_2^2 \cup e_3^2$ of $K_{\mathcal{P}}$ and $K_{\mathcal{Q}}$ to be fixed throughout a 3-deformation, where K^1 denotes the 1-skeleton and e_2^2 resp. e_3^2 correspond to the second resp. the third relator, we might (at least) get a counterexample to (rel. AC').

---

[9] This property would get lost, if a_3, b_3 and the relators which didn't enter the above semisplit Q-transformation were omitted from the $\mathcal{P}_i$ resp. $\mathcal{Q}_j$.

[10] $\mathcal{Q}$ immediately is seen to Q^{**}-transform to the presentation $\langle t|1 \rangle$ of $\mathbb{Z}$ with $\chi = \chi_{min} + 1$.

4   Some Further Results

This article and previous chapters cannot possibly cover all publications which are related to the (generalized) Andrews-Curtis problem. Thus we close with a list of some relevant contributions including indications of the specific aspects which are treated:

R. Craggs' papers [Cr79₁] and [Cr79₂] in particular deal with free splitting homomorphisms analogously to the Stallings-Jaco characterization of the 3-dimensional Poincaré conjecture. Similarly, the double coset characterization in Heegaard theory of 3-manifolds (J. Birman [Bi75₁], [Bi75₂]) has an analogue for Q-equivalences; the latter are characterized by double cosets in automorphism groups of free groups, see [Me85].

Craggs' publications [Cr88] and [Cr89] focus on 4-manifold aspects of (AC') (handle decompositions resp. embeddings of 2-complexes); [Cr89] contains an approach to control the linking phenomenon of Chapter I, § 3.2.

Finally, we recommend the paper of T.D. Cochran and J.P. Levine [CoLe91]. It shows that (AC) may enter geometric considerations different from the original motivations which were mentioned in Chapter I, § 4.1: Higher-dimensional homology boundary links are fusions of boundary links iff every Q^{**}-class of balanced presentations of $\pi = \{1\}$ is invertible (compare Chapter I, § 4.4).

Bibliography

[AbHo92]   H. Abels and S. Holz, *Higher generation by subgroups.* preprint, Universität Bielefeld. [Chapter V]

[Ad55]   J. F. Adams, *A new proof of a theorem of W.H.Cockroft.* J. London Math. Soc. **30** (1955), 482-488. [Chapters I, V]

[Ad75]   S. I. Adian, *The Burnside problem and identities in groups.* Nauka, Moscow, 1975). [Chapter V]

[AkKi85]   S. Akbulut, R. Kirby, *A potential smooth counterexample in dimension 4 to the Poincaré conjecture, the Schoenflies conjecture, and the Andrews-Curtis conjecture.* Topology **24** (1985), 375-390. [Chapters I, XII]

[Al19]   J. W. Alexander, *Note on two three-dimensional manifolds with the same group.* Trans. Amer. Math. Soc. **20** (1919), 339-342. [Chapter I]

[Al20]   J. W. Alexander, *Note on Riemann Spaces.* Bull. Amer. Math. Soc. **26** (1920), 370-372. [Chapter VIII]

[Al90]   J. M. Alonso, *Inégalités isopérimétrique et quasiisométries.* C. R. Acad. Sci. Paris, **311**, Série I, 1990, 761–764. [Chapter VI]

[Al92]   J. M. Alonso, *Combings of Groups* in: Algorithms and classification in combinatorial group theory. G. Baumslag and C. F. Miller III eds., Springer Verlag, Math. sci. res. inst. publ., 1992, 165–177. [Chapter VI]

[AlBr92]   J. M. Alonso and M. R. Bridson, *Semihyperbolic groups.* Report No. **13** (1992), Univ. of Stockholm, Sweden, 59. [Chapter VI]

[AnCu65]   J. J. Andrews and M. L. Curtis, *Free groups and handlebodies.* Proc. Amer. Math. Soc. **16** (1965), 192-195. [Chapters I, XII]

[AnCu66]   J. J. Andrews and M. L. Curtis, *Extended Nielsen operations in free groups.* Amer. Math. Monthly **73** (1966), 21-88. [Chapters I, V, X, XII]

[BaBoPr]   Y.-G. Baik, W. A. Bogley and S. J. Pride, *Asphericity of positive length four relative presentations.* in preparation. [Chapters V, X]

[BaHoPr92] Y.-G. Baik, J. Howie and S. J. Pride, *The identity problem for graph products of groups.* J. Algebra, to appear. [Chapter IV, V, X]

381

382   BIBLIOGRAPHY

[BaPr]   Y.-G. Baik and S. J. Pride, *Generators of the second homotopy module of presentations arising from group constructions.* preprint, University of Glasgow. [Chapters V, VII, X]

[Ba91]   P. Bandieri, *Platonic decompositions of 3-manifolds.* Ital. B., **5** (1991), n.7, 745-756. [Chapter VIII]

[BaPeVa84] W. Barth, C. Peters, A. Van de Ven. *Compact complex surfaces.* Springer-Verlag, Berlin, 1984. [Chapter IX]

[Ba64]   H. Bass, *Projective modules over free groups are free.* J. Algebra 1 (1964), 367-373. [Chapter III]

[Ba68]   H. Bass, *Algebraic K-Theory.* W.A. Benjamin, New York, 1968. [Chapter IX]

[Ba73]   H. Bass, *Unitary algebraic K-theory in Algebraic K-Theory III: Hermitian K-theory and Geometric Applications.* in: Springer LNM **343**, 57-265, 1973. [Chapter IX]

[Ba88]   S. Bauer, *The homotopy type of of a 4-manifold with finite fundamental group.* in: Algebraic Topology and Transformation Groups, Springer LNM **1361**, 1-6, 1988. [Chapter IX]

[Bau91]   H. J. Baues, *Combinatorial homotopy and 4-dimensional complexes.* de Gruyter Expositions in Mathematics 2, De Gruyter, New York (1991). [Chapter XII]

[BauCo90]   H. J. Baues, D. Conduché, *The central series for Peiffer commutators in groups with operators.* J. Algebra **133** (1990), 1-34. [Chapter XII]

[Be80₁]   W. H. Beckmann, *Completely Aspherical 2-complexes* (Thesis, Cornell University, 1980). [Chapter V]

[Be80₂]   W. H. Beckmann, *A certain class of non-aspherical 2-complexes.* J. Pure Appl. Algebra **16** (1980), 243-244. [Chapter V]

[Be75]   H. Behr, *Präsentationen von Chevalleygruppen über* **Z**. Math. Z. **141** (1975), 235-241. [Chapters V, X]

[Be74]   G. Bergman, *Modules over coproducts of rings.* Trans. Amer. Math. Soc. **200**, (1974), 1-33. [Chapter III]

[BeCoCo78] I. Berstein, M. Cohen, R. Connelly, *Contractible, non-collapsible products with cubes.* Topology **17** (1978), 183-187. [Chapter XI]

[Bi75₁]   J. S. Birman, *Poincaré's conjecture and the homeotopy group of a closed orientable 2-manifold.* J. Austr. Math. Soc. **17** (1975), 214-221. [Chapter XII]

[Bi75₂]   J. S. Birman, *On the equivalence of Heegaard splittings of closed, orientable 3-manifolds.* Ann. Math. Study **84** (1975), 137-164. [Chapter XII]

[Bi75₃] J. S. Birman, *Braids, Links, and Mapping Class Groups*. Ann. of Math. Studies No. **82**, Princeton University Press (1975). [Chapter VIII]

[Bo91] W. A. Bogley, *An embedding for π_2 of a subcomplex of a finite contractible two-complex*. Glasgow Math. J. **33** (1991), 365-371. [Chapters V, X]

[Bo92] W. A. Bogley, *Unions of Cockcroft two-complexes*. Proc. Edinburgh Math. Soc, to appear. [Chapter IV, V]

[BoGu92] W. A. Bogley and M. A. Gutiérrez, *Mayer-Vietoris sequences in homotopy of 2-complexes and in homology of groups*. J. Pure Appl. Algebra **77** (1992), 39-65. [Chapter IV, V, X]

[BoPr92] W. A. Bogley and S. J. Pride, *Aspherical relative presentations*. Proc. Edinburgh Math. Soc. **35** (1992), 1-39. [Chapter IV, V, X]

[BoZi85] M. Boileau and H. Zieschang, *Nombre de ponts et générateurs méridiens des entrelacs de Montesinos*. Comment. Math. Helvetici **60** (1985), 270-279. [Chapter VII]

[Bo55] W. W. Boone, *Certain simple unsolvable problems in the theory of groups. I,II,III,IV*. Nederl. Akad. WetensCh. Proc. Ser. A. **57** (1954), 231-237, 492-497; **58** (1955), 252-256, 571-577. [Chapter VI]

[BrMo91] J. Bracho, L. Montejano, *The scorpions: examples in stable non-collapsibility and in geometric category*. Topology **30** (1991), n. 4, 541-550. [Chapter XI]

[BrDy81] J. Brandenburg and M. N. Dyer, On J. H.C. Whitehead's aspherical question I, Comment. Math. Helv. **56** (1981), 431-446. [Chapter IV, V, X]

[BrDySt83] J. Brandenburg, M. N. Dyer and R. Strebel, On J. H.C. Whitehead's aspherical question II. in: *Low Dimensional Topology* (S. Lomonaco, editor), Contemp. Math. **20** (1983), 65-78. [Chapter IV, V, X]

[Bri92] M. Bridson, *On the geometry of normal forms in discrete groups*. preprint, Princeton University, 1992. [Chapter VI]

[BrS80] S. D. Brodskiĭ, *Equations over groups and groups with one defining relator (Russian)*. Uspehi Mat. Nauk. **35** (1980), 183. [Chapters V, X]

[Br72] E. H. Brown Jr., *Generalizations of the Kervaire Invariant*. Annals of Math. **95** (1972), 368-383. [Chapter IX]

[Br69] E. M. Brown, *The Hauptvermutung for 3-complexes*. Trans. Amer. Math. Soc. **144** (1969), 173-196. [Chapter I]

[Br82] K. Brown, *Cohomology of Groups*, Graduate texts in Mathematics, vol. 87, Springer-Verlag, New York, 1982. [Chapters II, III, IV, VII, IX, X]

[BrGe84] K. Brown and R. Geoheagan, *An infinite dimensional torsion-free FP_∞-group*. Invent. Math. **77** (1984), 367-381. [Chapter IV]

[BrCo74] M. Brown, M. M. Cohen, *A proof that simple-homotopy equivalent poly-hedra are stably homeomorphic.* Michigan Math. J. 21 (1974), 181-191. [Chapter I]

[Br80] R. Brown, *On the second relative homotopy group of an adjunction space: An exposition of a theorem of J. H.C. Whitehead.* J. London Math. Soc. (2) **22** (1980), 146-152. [Chapters V, X]

[Br84] R. Brown, *Coproducts of crossed P-modules: Applications to second homotopy groups and to the homology of groups.* Topology **23** (1984), 337-345. [Chapter IV, V, X]

[Br92] R. A. Brown, *Generalized group presentations and formal deformations of CW-complexes.* Trans. Amer. Math. Soc., to appear. [Chapters I, XII]

[BrHi78] R. Brown and P. J. Higgins, *On the connection between the second relative homotopy groups of some related spaces.* Proc. London Math. Soc. (3), **36** (1978), 193-212. [Chapter IV, V, X]

[BrHu82] R. Brown and J. Huebschmann, *Identities among relations.* in: *Low-Dimensional Topology* (R. Brown and T. L. Thickstun, editors), London Math. Soc. Lecture Note Series **48** (1982), 153-202. [Chapter IV, V, X]

[Br76₁] W. J. Browning, *A relative Nielsen theorem.* Cornell University, Ithaca N.Y. (1976), manuscript. [Chapters I, XII]

[Br76₂] W. J. Browning, *The effect of Curtis-Andrews moves on Jacobian matrices of perfect groups.* Cornell University, Ithaca N.Y. (1976), manuscript. [Chapter XII]

[Br78] W. J. Browning, *Homotopy types of certain finite CW-complexes with finite fundamental group.* Ph.D. Thesis, Cornell University 1978. [Chapters III, IX]

[Br79₁] W. J. Browning, *Pointed lattices over finite groups.* ETH pre-print (unpublished), February 1979. [Chapter III]

[Br79₂] W. J. Browning, *Truncated projective resolutions over a finite group.* ETH pre-print (unpublished), April 1979. [Chapter III]

[Br79₃] W. Browning, *Finite CW-complexes of cohomological dimension 2 with finite abelian π_1.* ETH preprint, (unpublished), May 1979. [Chapter III]

[BuZi85] G. Burde and H. Zieschang, *Knots.* de Gruyter Series in Mathematics 5, De Gruyter, New York (1985). [Chapters V, VII, X]

[CaSh71] S. E. Cappell and J. L. Shaneson. *On four-dimensional surgery and applications.* Comment. Math. Helv. **46** (1971), 500-528. [Chapter IX]

[CaEi56] H. Cartan and S. Eilenberg, *Homological Algebra*, Princeton University Press, Princeton, N. J. 1956. [Chapters II, IV]

[Ca65] B. G. Casler, *An embedding theorem for connected 3-manifolds with boundary.* Proc. Amer. Math. Soc. **16** (1965), 559-566. [Chapters I, XI]

[CaSp92]   A. Cavicchioli and F.Spaggiari, *The classification of 3-manifolds with spines related to Fibonacci Groups, Groups and classifying spaces.* Lecture Notes in Mathematics, Springer-Verlag, Berlin-New York, to appear. [Chapter VIII]

[Ce68]   J. Cerf, *Sur les difféomorphismes de la sphère de dimension trois* ($\Gamma_4 = 0$). Springer Lecture Notes in Math. **53** (1968). [Chapter I]

[Ch74]   T.A. Chapman, *The topological invariance of Whitehead torsion.* Amer. J. of Math. **96** (1974), 488-497. [Chapter I]

[Ch67]   D. R. J. Chillingworth, *Collapsing three-dimensional convex polyhedra.* Proc.Camb. Phil. Soc. **63** (1967), 353-357. [Chapter I]

[Ch80]   D. R. J. Chillingworth, *Correction: Collapsing three-dimensional convex polyhedra.* Math. Proc.Cam. Phil. Soc. **88** (1980), 307-310. [Chapter I]

[ChCoHu81]   I. Chiswell, D. J. Collins and J. Huebschmann, *Aspherical group presentations.* Math. Z. **178** (1981), 1-36. [Chapters V, X]

[Chu41]   A. Church, *The calculi of lambda conversion.* Princeton University Press, Princeton (1941). [Chapter VI]

[CoLe91]   T.D.Cochran, J.P. Levine, *Homology boundary links and the Andrews-Curtis conjecture.* Topology 30 (1991), 31-239. [Chapter XII]

[Co51]   W. H. Cockcroft, *Note on a theorem by J. H.C. Whitehead.* Quart. J. Math. Oxford (2) **2** (1951), 159-160. [Chapters V, X]

[Co54]   W. H. Cockcroft, *On two-dimensional aspherical complexes.* Proc. London Math. Soc. (3) **4** (1954), 375-384. [Chapter IV, V, X]

[CoSw61]   W. H. Cockcroft and R. G. Swan, *On the homotopy type of certain two-dimensional complexes.* Proc. London Math. Soc. (3) **11** (1961), 193-202. [Chapters II, V, X]

[Co89]   D. E. Cohen, *Combinatorial group theory: a topological approach.* Cambridge University Press, Cambridge, 1989. [Chapter VI]

[Co91]   D. E. Cohen, *Isodiametric and isoperimetric inequalities for group presentations.* Intern. J. Alg.Comp. 1, 315-320, 1991. [Chapter VI]

[Co78]   J. M. Cohen, *Aspherical 2-complexes.* J. Pure Appl. Algebra **12** (1978), 101-110. [Chapters V, X]

[Co69]   M. M. Cohen, *A general theory of relative regular neighbourhoods.* Trans. Amer. Math. Soc. **136** (1969), 189-229. [Chapter I]

[Co73]   M. M. Cohen, *A Course in Simple-Homotopy Theory.* GTM 10, Springer-Verlag, New York · Heidelberg · Berlin (1973). [Chapters I, VII, XII]

[Co75]   M. M. Cohen, *Dimension estimates in collapsing $X \times I^q$.* Topology **14** (1975), 253-256. [Chapter XI]

[Co77]   M. M. Cohen, *Whitehead torsion, group extensions and Zeeman's conjecture in high dimensions.* Topology **16** (1977), 79-88. [Chapters I, IX, XI, XII]

[CoMeSa85]   M. M. Cohen, W. Metzler, K.Sauermann, *Collapses of K × I and group presentations.* Amer. Math. Soc.Contemp. Math. **44** (1985), 3-33. [Chapters I, XI]

[CoMeZi81]   M. M. Cohen, W. Metzler, A. Zimmermann, *What does a basis of F(a,b) look like?.* Math. Ann. **257** (1981), 435-445. [Chapters I, XI]

[CoHu82]   D. J. Collins and J. Huebschmann, *Spherical diagrams and identities among relations.* Math. Ann. **261** (1982), 155-183. [Chapters V, X]

[CoPe85]   D. J. Collins and J. Perraud, *Cohomology and finite subgroups of small cancellation quotients of free products.* Math. Proc.Camb. Phil. Soc. **97** (1985), 243-259. [Chapters V, X]

[CoFl64]   P. E. Conner and E. E. Floyd, *Differentiable periodic maps.* Springer-Verlag, Berlin (1964). [Chapter IX]

[CoTh88]   D. Cooper and W. P.Thurston, *Triangulating 3-manifolds using 5 vertex link types.* Topology **27** (1988), 23-25. [Chapter VIII]

[CoDePa90]   M. Coornaert, T. Delzant, and A. Papadopoulos, *Notes sur les groupes hyperboliques de Gromov.* Lecture Notes in Mathematics 1441, Springer Verlag, Berlin–Heidelberg–New York (1990). [Chapter VI]

[Cr79₁]   R. Craggs, *Free Heegaard diagrams and extended Nielsen transformations, I.* Michigan Math. J. **26** (1979), 161-186. [Chapters I, XII]

[Cr79₂]   R. Craggs, *Free Heegaard diagrams and extended Nielsen transformations, II.* Ill. J. Math. 23 (1979), 101-127. [Chapters I, XII]

[Cr88]   R. Craggs, *On the algebra of handle operations in 4-manifolds.* Topology and its appl. 30 (1988), 237-252. [Chapters I, XII]

[Cr89]   R. Craggs, *Freely reducing group readings for 2-complexes in 4-manifolds.* Topology 28 (1989), 247-271. [Chapters I, XII]

[CrHo87]   R. Craggs, J. Howie, *On group presentations, coproducts and inverses.* Ann. Math. Study **111** (1987), 213-220. [Chapters I, XII]

[CrFo63]   R. H. Crowell, R. H. Fox, *Introduction to knot theory.* Ginn and Co. (1963). [Chapter I]

[CuRe62]   C. W. Curtis and I. Reiner. *Representation theory of finite groups.* John Wiley & Sons, New York, 1962. [Chapter IX]

[De12]   M. Dehn, *Über unendliche diskontinuierliche Gruppen.* Math. Ann. **71**, 116–144, 1912. [Chapter VI]

[DeMe88]   G. Denk, W. Metzler, *Nielsen reduction in free groups with operators.* Fund. Math. **129** (1988), 181-197. [Chapters I, XII]

[Di68] P. Dierker, *Note on collapsing $K \times I$ where K is a contractible polyhedron* Proc.A.M.S. **19** (1968), 452-428 [Chapter XI]

[Do83] S. K. Donaldson, *An application of gauge theory to the topology of 4-manifolds.* J. Diff. Geom. 18, 269-316, (1983). [Chapter IX]

[Do90] S. K. Donaldson, *Polynomial invariants for smooth 4-manifolds.* Topology 29, 257-316, (1990). [Chapter IX]

[DuElGi92] A. Duncan, G. Ellis, N. Gilbert, *A Meyer-Vietoris sequence in group homology and the decomposition of relation modules.* preprint, Heriot-Watt University and University College Galway. [Chapter IV, V]

[DuHo1] A. J. Duncan and J. Howie, *One relator products with high-powered relators.* preprint, Heriot-Watt University. [Chapters V, X]

[DuHo92_1] A. J. Duncan and J. Howie, *Weinbaum's conjecture on unique subwords of non-periodic words.* Trans. Amer. Math. Soc. **115** (1992), 947-954. [Chapter V]

[DuHo92_2] A. J. Duncan and J. Howie, *One relator products with high-powered relators.* preprint, Heriot-Watt University. [Chapter V]

[Du76] M. J. Dunwoody, *The homotopy type of a two-dimensional complex.* Bull. London Math. Soc. **8** (1976), 282-285. [Chapters I, II]

[Du80] M. J. Dunwoody, *Answer to a conjecture of J. M.Cohen.* J. Pure Appl. Algebra **16** (1980), 249. [Chapters V, X]

[Dy76] M. N. Dyer, *Homotopy classification of (π, m)-complexes.* J. of Pure and Appl. Alg. 7 (1976), 249-282. [Chapter III]

[Dy79] M. N. Dyer, *Trees of homotopy types of (π, m)-complexes.* London Math. Soc. Lecture Note Series 36 (1979), 251-254. [Chapter III]

[Dy81] M. N. Dyer, *Simple homotopy types for (G,m)-complexes.* Proc. Amer. Math. Soc. 111-115, (1981). [Chapter IX]

[Dy85] M. N. Dyer, *Invariants for Distinguishing between Stably Isomorphic Modules.* J. Pure Appl. Algebra **37** (1985), 117-153. [Chapters III, VII]

[Dy86] M. N. Dyer, *A topological interpretation for the bias invariant.* Proc. Amer. Mth. Soc. 89 (1986), 519-523. [Chapter III]

[Dy87_1] M. N. Dyer, *Subcomplexes of two-complexes and projective crossed modules.* in: *Combinatorial Group Theory and Toplogy* (S. M. Gersten and J. Stallings, editors) Ann. of Math. Studies **111** (1987), 255-264. [Chapter IV]

[Dy87_2] M. N. Dyer, *Localization of group rings and applications to 2-complexes.* Comm. Math. Helvetici, **62** (1987), 1-17. [Chapter IV]

[Dy91_1] M. N. Dyer, *Cockcroft 2-complexes.* preprint, University of Oregon. [Chapter IV, X]

[Dy91₂]   M. N. Dyer, *Aspherical 2-complexes and the homology of groups*. preprint,
               University of Oregon. [Chapter IV]

[Dy93]   M. N. Dyer, *Groups with no infinite perfect subgroups and aspherical 2-
               complexes*. Comment. Math. Helv., **68** (1993), 333-339. [Chapter X]

[DyHa9?]   M. N. Dyer and J. Harlander, *A note on Cockcroft complexes*. Glasgow
               Math. J., to appear. [Chapter V]

[DySi73]   M. N. Dyer and A. J. Sieradski., *Trees of homotopy types of two-
               dimensional CW-complexes*. Comm. Math. Helv. **48** (1973), 31-44. [Chap-
               ters II, III]

[Ed88]   M. Edjvet, *On a certain class of group presentations*. Math. Proc.Comb.
               Phil. Soc. **105** (1988), 25-35. [Chapters V, X]

[Ed91]   M. Edjvet, *On the asphericity of one-relator relative presentations*.
               preprint, University of Nottingham. [Chapter V]

[EdHo91]   M. Edjvet and J. Howie, *The solution of length four equations over groups*.
               Trans. Amer. Math. Soc. **326** (1991), 345-369. [Chapters V, X]

[EdEw90]   A. Edmonds and J. Ewing, *Topological realization of equivariant inter-
               section forms*. preprint, (1990). [Chapter IX]

[EdGi83]   R. Edwards, D. Gillman, *Any spine of the cube is 2-collapsible*. Can. J.
               Math. **35** (1983), 43-48. [Chapter XI]

[ElPo86]   G. Ellis and T. Porter, *Free and projective crossed modules and the second
               homology group of a group*. J. Pure and Appl. Algebra, **40** (1986), 27-31.
               [Chapter IV]

[Ep92]   D. B. H. Epstein, J. W. Cannon, D. F. Holt, S.V.F. Levy, M.S. Paterson,
               W.P. Thurston, *Word processing in groups*. Jones and Bartlett, Boston–
               London, 1992. [Chapter VI]

[EvMa77]   B. Evans and J. Maxwell, *Quaternion actions of S^3*. Amer. J. Math. **101**
               (5) (1979), 1123-1130. [Chapter VIII]

[Fe70]   R. Fenn, *Embedding polyhedra*. Proc. London Math. Soc. 2 (1970), 316-318
               [Chapter XI]

[Fe83]   R. Fenn, *Techniques in Geometric Topology*. London Mathematical Society
               Lecture Note Selries **57** (Cambridge University Press, Cambridge, 1983).
               [Chapters III, V, X]

[FlOe84]   W. Floyd, U. Oertel, *Incompressible surfaces via branched surfaces*. Topol-
               ogy **23** (1984), 117-125. [Chapter XI]

[Fo-Ra40]   D. I. Fouxe-Rabinovitch, *Über die Automorphismengruppe der freien Pro-
               dukte, I*. Math. Sb. **8** (1940), 265-276. [Chapters I, XII]

[Fo53]   R. H. Fox, *Free differential calculus I*. Ann. of Math. (1953), 547-560.
               [Chapters II, III]

[Fr35]   W. Franz, *Über die Torsion einer Überdeckung.* J. reine angew. Math. 173
              (1935), 245-254. [Chapter I]

[Fr76]   B.M. Freed, *Embedding contractible 2-complexes in E^4.* Proc. Amer. Math.
              Soc. **54** (1976), 423-430. [Chapter I]

[Fr82]   M.H. Freedman, *The topology of 4-dimensional manifolds.* J. Diff. Geom.
              17 (1982), 357-453. [Chapter I]

[Fr84]   M. H. Freedman, *The disk theorem for four-dimensional manifolds.* in
              Proc. Int Conf. Warsaw, 647-663, (1984). [Chapter IX]

[FrQu90]   M. H. Freedman and F.Quinn, *Topology of Four-Manifolds.* Princ. Math.
              Series **39**, Princeton University Press, (1990). [Chapter IX]

[Ga68]   T. Ganea, *Homology et extensiones centrales de groupes.* C. R. Acad. Sci.
              Paris, **266** (1968), 556-558. [Chapter IV]

[Ge83]   S. M. Gersten, *Conservative groups, indicability, and a conjecture of Howie*
              J. Pure Appl. Algebra **29** (1983), 59-74. [Chapter IV, V, X]

[Ge86]   S. M. Gersten, *Products of conjugacy classes in a free group: A counterex-*
              ample. Math. Z. **192** (1986), 167-181. [Chapter V]

[Ge87$_1$]   S. M. Gersten, *Reducible diagrams and equations over groups.* in: *Essays*
              in Group Theory (S. M. Gersten, editor), MSRI Publications **8** (1987),
              15-73. [Chapters V, VI]

[Ge87$_2$]   S. M. Gersten, *Branched coverings of 2-complexes and diagrammatic re-*
              ducibility. Trans. Amer. Math. Soc. **303** (1987), 689-706. [Chapter V]

[Ge88$_1$]   S. M. Gersten, *The isoperimetric inequality and the word problem.* unpub-
              lished, 1988. [Chapter VI]

[Ge88$_2$]   S. M. Gersten, *On Rapaport's example in presentations of the trivial group.*
              University of Utah, Salt Lake City (1988), preprint. [Chapter XII]

[Ge90]   S. M. Gersten, *Isoperimetric and isodiametric functions of finite presen-*
              tations. preprint, 1990. [Chapter VI]

[Ge91$_1$]   S. M. Gersten, *Dehn functions and l_1-norms of finite presentations.* Pro-
              ceedings of the workshop on algorithmic problems, C. F. Miller III and
              G. Baumslag editors, Springer Verlag, 1991. [Chapter VI]

[Ge91$_2$]   S. M. Gersten, *The double exponential theorem for isodiametric and*
              isoperimetric functions. Intern. J. Alg.Comp. 1, 321-327, 1991. [Chap-
              ter VI]

[Ge92]   S. M. Gersten, *Bounded cocycles and combings of groups.* Intern. J. of
              Algebra and Comp. **2** (3) (1992), 307-326. [Chapter VI]

[GeRo62]   M. Gerstenhaber and O. S. Rothaus, *The solution of sets of equations in*
              groups. Proc. Nat. Acad. Sci. USA **68** (1962), 1531-1533. [Chapters V, X]

[Gh90]   E. Ghys, *Les groupes hyperboliques* Séminaire Bourbaki, 42ème année, n.
                72 (1990). [Chapter VI]

[GhHa90]   E. Ghys and P. de la Harpe (eds.), *Sur les groupes hyperboliques d'après
                Mikhael Gromov*. Birkhäuser, Boston (1990). [Chapter VI]

[GhHa91]   E. Ghys and P. de la Harpe, *Infinite groups as geometric objects*. in: *Er-
                godic theory, symbolic dynamics and hyperbolic spaces* T. Bedford, M.
                Keane, C. Series, eds., 299-314, Oxford University Press, Oxford (1991).
                [Chapter VI]

[Gi92]   N. D. Gilbert, *Central extensions of groups and an embedding question of
                J. H.C. Whitehead*. ArCh. Math. **58** (1992), 114-120. [Chapter IV, V, X]

[Gi93]   N. D. Gilbert, *Identities between sets of relations*. J. Pure Appl. Algebra
                83 (1993), 263-276. [Chapter V]

[GiHi89]   N. D. Gilbert and P. J. Higgins, *The non-abelian tensor product of groups
                and related constructions*. Glasgow Math. J. **31** (1989), 17-29. [Chapter V]

[GiHo$_1$]   N. D. Gilbert and J. Howie, *Threshold subgroups for Cockcroft 2-
                complexes*. preprint, Heriot-Watt University. [Chapters V, X]

[GiHo$_2$]   N. D. Gilbert and J. Howie, *Cockcroft properties of graphs of 2-complexes*.
                Proc. Roy. Soc. Edinburgh. (to appear). [Chapters V, X]

[GiHo92]   N. D. Gilbert and J. Howie, *Threshold subgroups for Cockcroft 2-
                complexes*. preprint, Heriot-Watt University. [Chapter IV]

[Gi86]   D. Gillman, *Bing's house and the Zeeman Conjecture*. Topology and its
                Appl. **24** (1986), 147-151. [Chapter XI]

[GiRo83]   D. Gillman, D. Rolfsen, *The Zeeman conjecture for standard spines is
                equivalent to the Poincaré conjecture*. Topology **22**(1983), n.3, 315-323.
                [Chapter XI]

[GiRo91]   D. Gillman, D. Rolfsen, *Three-manifolds embed in small 3-complexes*. In-
                ternational J. Math. **3** (1991), n.2, 179-183. [Chapter XI]

[Go84]   R. E. Gompf, *Stable diffeomorphisms of compact 4-manifolds*. Top. and
                its appl. **18**, 115-120, (1984). [Chapter IX]

[Go91]   R. E. Gompf, *Killing the Akbulut-Kirby 4-sphere, with relevance to the
                Andrews-Curtis and Schoenflies problems.*, Topology **30** (1991), 97-115.
                [Chapters I, XII]

[Go68]   R. E. Goodrick, *Non simplicially collapsible triangulations of I^n*.
                Proc.Camb. Phil. Soc. **64** (1968), 31-36. [Chapter I]

[GoSh86]   F. Gonzalez-Acuña and H. Short, *Knot surgery and primeness*. Math.
                Proc.Camb. Phil. Soc. **99** (1986), 89-102. [Chapters V, X]

[GrHa81]   M. J. Greenberg and J. R. Harper, *Algebraic Topology: A First Course*,
                Benjamin/Cummings, Mendlo Park (1981).

[Gr87]   M. Gromov, *Hyperbolic groups* in: *Essays in group theory.* (S. Gersten, ed.), Springer Verlag, Math. sci. res. inst. publ., 1987, 75–263. [Chapter VI]

[Gr79]   K. W. Gruenberg, *Free abelianized extensions of finite groups.* Lond. Math. Soc. Lecture Notes Series **36** (1979), 71-104. [Chapter I]

[Gr91]   K. W. Gruenberg , *Homotopy classes of truncated projective resolutions- a new look at Browning's work.* Queen Mary College, London, (1991), preprint. [Chapter III]

[Gu87]   N. Gupta, *Free Group Rings*, Contemp. Math. **66** (Amer. Math. Soc., 1987). [Chapters V, X]

[GuLa91]   M. Gutierrez, M. P. Latiolais, *Partial homotopy type of finite two-complexes.* Math. Zeit. **207** (1991), 359-378. [Chapter III]

[GuLa93]   M. Gutierrez, M. P. Latiolais, *Two-complexes with Fundamental group a semi-direct product of cyclics.* Boletin de la Sociedad Matematica Mexicana (to appear). [Chapter III]

[GuRa81]   M. A. Gutiérrez and J. G. Ratcliffe, *On the second homotopy group.* Quart. J. Math. Oxford (2) **32** (1981), 45-55. [Chapter IV, V, X]

[Ha49]   M. Hall, Jr., *Subgroups of finite index in free groups.* Canadian Math. J. **1** (1949), 187-190. [Chapters V, X]

[HaKr88]   I. Hambleton and M. Kreck, *On the classification of topological 4-manifolds with finite fundamental group.* Math. Ann. **280**, 85-104, (1988). [Chapter IX]

[HaKr90]   I. Hambleton and M. Kreck, *Smooth structures on algebraic surfaces with finite fundamental group.* Invent. Math. **102** (1990), 109-114. [Chapter IX]

[HaKr92$_1$] I. Hambleton and M. Kreck, *Cancellation of lattices and finite two-complexes.* to appear J. f. reine u. angew. Math. [Chapters IX, III]

[HaKr92$_2$] I. Hambleton and M. Kreck, *Cancellation of hyperbolic forms and topological four-manifolds.* to appear J. f. reine u. angew. Math. [Chapter IX]

[HaKr92$_3$] I. Hambleton and M. Kreck, *Cancellation, elliptic surfaces and the topology of certain four-manifolds.* to appear J. f. reine u. angew. Math. [Chapter IX]

[HaKr88]   I. Hambleton and M. Kreck, *On the classification of topological 4-manifolds with finite fundamental group.* Math. Ann. **280**, 85-104, (1988). [Chapter IX]

[HaKrTe92] I. Hambleton, M. Kreck and P. Teichner, *Four-manifolds with fundamental group of order 2.* preprint, (1992). [Chapter IX]

[Ha91]   J. Harlander, *Minimal Cockcroft subgroups.* (to appear in Glasgow J. Math.). [Chapter IV]

[Ha92]   J. Harlander, *Solvable groups with cyclic relation module*. University of
              Oregon, Eugene OR. (1992), preprint, to appear in: J. Pure Appl. Algebra.
              [Chapter I]

[Ha81]   A. E. Hatcher, *Hyperbolic structures of arithmetic type on some link com-
              plements*. J. London Math. Soc. (2) **27** (1981), 345-355. [Chapter VIII]

[Ha89]   A. E. Hatcher, *Notes on basic 3-manifold topology*. Cornell University,
              Ithaca N.Y. (1989), preprint. [Chapter I]

[He76]   J. Hempel, *3-Manifolds*. Annals of Math. Studies **86**, Princeton University
              Press, Princeton, New Jersey, 1976. [Chapter VIII]

[He90]   J. Hempel, *The lattice of branched covers over the figure-eight knot*. Topol-
              ogy Appl. **34** (1990), 183-201. [Chapter VIII]

[Hi40]   G. Higman, *The units of group rings*. Proc. London Math. Soc. (2) **46**
              (1940), 231-248. [Chapters V, X]

[Hi74]   H. M. Hilden, *Every closed, orientable 3-manifold is a 3-fold branched cov-
              ering space of S^3*. Bull. Amer. Math. Soc. **80** (1974), 1243-1244. [Chap-
              ter VIII]

[Hi76]   H. M. Hilden, *Three-fold branched coverings of S^3*. Amer. J. Math. **98**
              (1976), 989-997. [Chapter VIII]

[HiLoMo83]   H.M. Hilden, M.T. Lozano, and J.M. Montesinos, *The Whitehead link,
              the Borromean rings, and the knot 9_{46} are universal*. Collect. Math. **34**
              (1983), 19-28. [Chapter VIII]

[HiLoMo85$_1$] H. M. Hilden, M. T. Lozano, and J. M. Montesinos, *On knots that are
              universal*. Topology **24** (1985), 499-504. [Chapter VIII]

[HiLoMo85$_2$] H. M. Hilden, M. T. Lozano, and J. M. Montesinos, *On the universal group
              of the Borromean rings*. [Chapter VIII]

[HiSt71]   P. Hilton and U. Stammbach, *A course in homological algebra*. Springer-
              Verlag, Berlin-Heidelberg-New York (1971). [Chapter IV]

[Ho83]   C. Hog, *Pseudoflächen und singuläre 3-Mannigfaltigkeiten*. Staatsexamen-
              sarbeit, Frankfurt 1983. [Chapter VIII]

[Ho-An88]   C. Hog-Angeloni, *Beiträge zum (einfachen) Homotopietyp zweidimension-
              aler Komplexe zu freien Produkten und anderen gruppentheoretischen Kon-
              struktionen*. Thesis, Frankfurt/Main (1988). [Chapters I, VII, XII]

[Ho-An90$_1$]   C. Hog-Angeloni, *A short topological proof of Cohn's theorem*. Springer
              Lecture Notes in Math. **1440** (1990), 90-95. [Chapter III]

[Ho-An90$_2$]   C. Hog-Angeloni, *On the homotopy type of 2-complexes with a free product
              of cyclic groups as fundamental group*. Springer Lecture Notes in Math.
              1440 (1990), 96-108. [Chapter III]

[Ho-An92]   C. Hog-Angeloni, University of Frankfurt, Germany (1992), manuscript. [Chapter VIII]

[Ho-AnLaMe90] C. Hog-Angeloni, P. Latiolais, W. Metzler, *Bias ideals and obstructions to simple-homotopy equivalence.* Springer Lecture Notes in Math. **1440** (1990), 109-121. [Chapters III, VII]

[HoLuMe85]   C. Hog, M. Lustig, W. Metzler, *Presentations classes, 3-manifolds and free products.* Springer Lecture Notes in Math. **1167** (1985), 154-167. [Chapters I, XII]

[Ho-AnMe90]   C. Hog-Angeloni, W. Metzler, *Stablilization by free products giving rise to Andrews-Curtis equivalences.* Note di Matematica 10, Suppl. n. 2 (1990), 305-314. [Chapters III, XII]

[Ho-AnMe91]   C. Hog-Angeloni, W. Metzler, *Andrews-Curtis-Operationen und höhere Kommutatoren der Relatorengruppe.* J. Pure Appl. Algebra **75** (1991), 37-45. [Chapter XII]

[Ho31]   H. Hopf, *Über die Abbildungen der dreidimensionalen Sphäre auf die Kugelfläche.* Math. Ann. **104** (1931), 637-665. [Chapter I]

[Ho41]   H. Hopf, *Fundamentalgruppe and zweite Bettische Gruppe.* Comment. Math. Helv. **14** (1941), 257-309. [Chapter II]

[Ho79]   J. Howie, *Aspherical and acyclic 2-complexes.* J. London Math. Soc. (2) **20** (1979), 549-558. [Chapters V, X]

[Ho81$_1$]   J. Howie, *On the fundamental group of an almost-acyclic 2-complex.* Proc. Edinburgh Math. Soc. **24** (1981), 119-122. [Chapters V, X]

[Ho81$_2$]   J. Howie, *On pairs of 2-complexes and systems of equations over groups.* J. reine angew. Math. **324** (1981), 165-174. [Chapters V, X]

[Ho82]   J. Howie, *On locally indicable groups.* Math. Z. **180** (1982), 445-461. [Chapters V, X]

[Ho83$_1$]   J. Howie, *The solution of length three equations over groups.* Proc. Edinburgh Math. Soc. **26** (1983), 89-96. [Chapters V, X]

[Ho83$_2$]   J. Howie, *Some remarks on a problem of J. H. C. Whitehead.* Topology **22** (1983), 475-485. [Chapters V, X]

[Ho84]   J. Howie, *Cohomology of one-relator products of locally indicable group.* J. London Math. Soc. (2) **30** (1984), 419-430. [Chapters V, X]

[Ho85]   J. Howie, *On the asphericity of ribbon disc complements.* Trans. Amer. Math. Soc. **289** (1985), 281-302. [Chapters V, X]

[Ho87]   J. Howie, *How to generalize one-relator group theory.* in: *Combinatorial Group Theory and Topology*, S. M. Gersten and J. R. Stallings, eds., Annals of Mathematics Studies **111** (Princeton University Press, 1987). [Chapters V, X]

[Ho89]   J. Howie, *The quotient of a free product of groups by a single high-powered relator. I. Pictures. Fifth and higher powers.* Proc. London Math. Soc. (3) **59** (1989), 507-540. [Chapters V, X]

[Ho90]   J. Howie, *The quotient of a free product of groups by a single high-powered relator. II. Fourth powers.* Proc. London Math. Soc. (3) **61** (1990), 33-62. [Chapters V, X]

[HoSc79]   J. Howie and H. R. Schneebeli, *Groups of finite quasi-projective dimension.* Comment. Math. Helvetici **54** (1979), 615-628. [Chapters V, X]

[HoSc83]   J. Howie and H. R. Schneebeli, *Homological and topological properties of locally indicable groups.* Manuscripta Math. **44** (1983), 71-93. [Chapters V, X]

[Hu59]   S. T. Hu, *Homotopy theory.* Academic Press (1959). [Chapter I]

[Hu90]   G. Huck, *Embeddings of acyclic 2-complexes in S^4 with contractible complement.* Springer Lecture Notes in Math. **1440** (1990), 122-129. [Chapter I]

[HuRo92]   G. Huck and S. Rosebrock, *Ein verallgemeinerter Gewichtstest mit Anwendungen auf Baumpräsentationen,* Math. Z. **211** (1992), 351-367. [Chapters V, VI]

[HuRo93]   G. Huck and S. Rosebrock, *A bicombing that implies a sub-exponential Isoperimetric Inequality.* to appear in the Edinborough Proceedings, 1993. [Chapter VI]

[Hu69]   J. F. P. Hudson, *Piecewise linear topology.* W.A. Benjamin (1969). [Chapter I]

[Hu79]   J. Huebschmann, *Cohomology theory of aspherical groups and of small cancellation groups* J. Pure Appl. Algebra **14** (1979), 137-143. [Chapters V, X]

[Hu81]   J. Huebschmann, *Aspherical 2-complexes and an unsettled problem of J. H.C. Whitehead.* Math. Ann. **258** (1981), 17-37. [Chapters V, X]

[Hu35]   W. Hurewicz, Kon. WetensCh. Amsterdam, **38** (1935), 112-9; 521-8; **39** (1936), 117-125, 215-24. [Chapter II]

[Ig79₁]   K. Igusa, *The generalized Grassmann invariant.* Brandeis University, Waltham(Mass) (1979), preprint. [Chapter V]

[Ig79₂]   K. Igusa, *The Borel regulator on pictures.* Brandeis University, Waltham(Mass) (1979), preprint. [Chapter V]

[Ik71]   H. Ikeda, *Acyclic fake surfaces.* Topology **10** (1971), 9-36. [Chapter XI]

[Ik71]   H. Ikeda, *Acyclic fake surfaces.* Topology 10 (1971), 9-36. [Chapter I]

[Iv92]   S. V. Ivanov, *On the Burnside problem on periodic groups.* Bull. Amer. Math. Soc. **27** (1992), 257-260. [Chapter V]

[Ja69]   W. Jaco, *3-manifolds with fundamental group a free product*. Bull. Amer. Math. Soc. **75** (1969), 972-977. [Chapters I, XII]

[JaOr84]   W. Jaco and U. Oertel, *An algorithm to decide if a 3-manifold is a Haken manifold*. Topology **23** No. 2 (1984), 195-205. [Chapter VIII]

[Jo80]   D. L. Johnson, *Topics in the theory of group presentations*. Lond. Math. Soc. Lect. Notes **42** (1980), Cambridge University Press. [Chapter VI]

[Ju86]   A. Juhasz, *Small cancellation theory with a weakened small cancellation hypothesis. 1. The basic theory*. Israel J. Math. **55** (1) (1986), 65–93. [Chapter VI]

[Ju87]   A. Juhasz, *Small cancellation theory with a weakened small cancellation hypothesis. 2. The word problem. 3. The conjugacy problem*. Israel Journ. of Math. **58** (1987), 19–53. [Chapter VI]

[Ju89]   A. Juhasz, *Small cancellation theory with a unified small cancellation condition*. Journ. of London Math. Soc. (2) **40** (1989) 57-80. [Chapter VI]

[vK32]   E. R. van Kampen, *Komplexe in euklidischen Räumen*. Abh. Math. Sem. Univ. Hamburg **9** (1932), 72-78, correction: 152-153. [Chapter I]

[vK33]   E. R. van Kampen, *On some lemmas in the theory of groups*. Amer. J. Math. **55** (1933), 268-273. [Chapters V, X]

[Ka72]   I. Kaplansky, *Fields and Rings*. (University of Chicago Press (1972). [Chapters V, X]

[KaZi92]   R. Kaufmann and H. Zieschang, *On the rank of NEC groups*in: *Discrete groups and Geometry*. Eds. W.J. Harvey and C. Maclachlan, LMS Lecture Note Series 173, Cambridge University Press (1992), 137-147. [Chapter VII]

[KiPi93]   C. Kilgour and S. J. Pride, *Cockcroft presentations*. University of Glasgow, preprint, (1993). [Chapter X]

[Ki78]   R. C. Kirby, *Problems in low dimensional manifold theory*. Amer. Math. Soc. Proc. of Symposia in Pure Math. **32** (1978), 273-312. [Chapter I]

[Ki89]   R. C. Kirby, *The topology of 4-manifolds*. Springer Lecture Notes in Math. **1374** (1989). [Chapter I]

[KiSi77]   R. C. Kirby and L. Siebenmann, *Foundational essays on topological manifolds, smoothings and triangulations*. Princeton University Press, Annals of Math. Studies No. **88** (1977). [Chapter IX]

[KiTa89]   R. C. Kirby and L. R. Taylor, *Pin structures on low-dimensional manifolds*. in: *Geometry of Low- Dimensional Manifolds: 2*. edited by S.K.Donaldson and C.B.Thomas, London Math.Soc. Lecture Note Series 151, Cambridge University Press 1989. [Chapter IX]

[Kl92]   A. A. Klyachko, *A funny property of the sphere and equations over groups*. preprint, Moscow State University (1992). [Chapters V, X]

[Kn29]   H. Kneser, *Geschlossene Flächen in dreidimensionale Mannigfaltigkeiten.* Jahresber. DeutsCh. Math.-Verein. **38** (1929), 248-260. [Chapter VIII]

[Kr79]   M. Kreck, *Isotopy classes of diffeomorphisms of* $(k - 1)$-*connected almost parallelizable manifolds.* in: Algebraic Topology, Aarhus 1978, Springer LNM **763** (1979), 643-661. [Chapter IX]

[Kr84$_1$] M. Kreck, *Smooth structures on closed 4-manifolds up to connected sum with* $S^2 \times S^2$. preprint, (1984). [Chapter IX]

[Kr84$_2$] M. Kreck, *Some closed 4-manifolds with exotic differentiable structure.* Springer Lecture Notes in Math. **1051** (1984), 246-262. [Chapter I]

[Kr85]   M. Kreck, *Surgery and Duality.* To appear as a book in the Vieweg-Verlag, Wiesbaden. As a preprint of the Johannes-Gutenberg-Universität Mainz 1985 available under the title: *An Extension of Results of Browder, Novikov and Wall about Surgery on Compact Manifolds.* [Chapter IX]

[KrSc84]   M. Kreck and J. A. Schafer, *Stable and unstable classification of manifolds: some examples.* Comment. Math. Helv. **59** (1984), 12-38. [Chapter IX]

[KrMe83]   R. Kreher, W. Metzler, *Simpliziale Transformationen von Polyedern und die Zeeman-Vermutung.* Topology **22** (1983), 19-26. [Chapters I, XI]

[Kr85]   S. Krstić, *Systems of equations over locally p-indicable groups.* Invent. Math. **81** (1985), 373-378. [Chapters V, X]

[Ku53]   A. G. Kurosch, *Gruppentheorie.* Akademie-Verlag (1953). [ÄCh. XII]

[La86]   M. P. Latiolais, *The simple homotopy type of finite 2-dimensional CW-complexes with finite abelian* π_1, Trans. of the A.M.S. **293** (2) (1986), 655-661. [Chapter III]

[La91]   M. P. Latiolais, *When homology equivalence implies homotopy equivalence for 2-complexes.* J. Pure & A Alg. **76** (1991), 155-165. [Chapters III, IX]

[LaU86]   U. Lattwin, *Spaltung der Homotopieklassen und Präsentierungen zu freien Produkten von Gruppen.* Ph.D. Thesis, Dortmund (1986). [Chapter III]

[LeWi90]   R. Lee and D. M. Wilczyński, *Locally flat 2-spheres in simply connected 4-manifolds.* Comment. Math. Helv. **65** (1990), 388-412. [Chapter IX]

[Le62]   F. Levin, *Solutions of equations over groups.* Bull. Amer. Math. Soc. **68** (1962), 603-604. [Chapters V, X]

[Le90]   F. Levin, lecture Bochum 1990, unpublished. [Chapter VII]

[Le77]   D. W. Lewis, *Forms over real algebras and the multi-signature of a manifold.*, Adv. Math. **23** (1977), 272-284. [Chapter IX]

[Li93]   Z. Li, *Every 3-manifold embeds in* $Y \times Y \times I$ *where* Y *is a wedge of three segments.* Proc. Amer. Math. Soc., to appear.

[Li62]   W. B. R. Lickorish, *A representation of orientable combinatorial 3-manifolds.* Ann. of Math. (2) **76** (1962), 531-540. [Chapter VIII]

[Li70]   W. B. R. Lickorish, *On collapsing $X^2 \times I$. Topology of Manifolds*, ed. by J. C. Cantrell and C. H. Edwards, 157-160, Markham Publishing Co., Chicago (1970). [Chapter XI]

[Li73]   W. B. R. Lickorish, *An improbable collapse*. Topology **12** (1973), 5-8. [Chapter XI]

[Lu91₁]   M. Lustig, *Nielsen Equivalence and Simple-Homotopy Type*. Proc. London Math. Soc. **62** (1991), 537-562. [Chapters I, VII, XII]

[Lu91₂]   M. Lustig, *On the Rank, the Deficiency and the Homological Dimension of Groups: The Computation of a Lower Bound via Fox Ideals*. in: *Topology and Combinatorial Group Theory*, Ed. P. Latiolais, Lecture Notes in Math. **1440** (1991), Springer Verlag, 164-174. [Chapter VII]

[Lu93]   M. Lustig, *Infinitely many pairwise homotopy inequivalent 2-complexes K_i with fixed $\pi_1(K_i)$ and $\chi(K_i)$*. University of Bochum, Germany (1993), preprint. [Chapter III]

[LuMo91]   M. Lustig, Y. Moriah, *Nielsen Equivalence in Fuchsian Groups and Seifert Fibered Spaces*. Topology **30** (1991), 191-204. [Chapter VII]

[LuMo92]   M. Lustig, and Y. Moriah, *On the complexity of the Heegaard structure of hyperbolic 3-manifolds*. preprint (1992). [Chapter VII]

[LuMo93₁] M. Lustig and Y. Moriah, *Generalized Montesinos Knots, Tunnels and $\mathcal{N}$-Torsion*. Math. Ann. **295** (1993), 167-189. [Chapter VII]

[LuMo93₂] M. Lustig and Y. Moriah, *Generating Systems for Groups and Reidemeister-Whitehead Torsion*. J. of Algebra **157** (1993), 170-198. [Chapter VII]

[LuPr92]   M. Lustig and V. Preusser, *Relator identities in groups and stably free modules*. University of Bochum, Germany (1992), preprint. [Chapter III]

[Ly50]   R .C. Lyndon, *Cohomology theory of groups with a single defining relation*. Ann. of Math. (2) **52** (1950), 650-655. [Chapters V, X]

[Ly66]   R. C. Lyndon, *On Dehn's algorithm*. Math. Ann. **166** (1966), 208-228. [Chapters V, VI, X]

[LySc77]   R. C. Lyndon and P. E. Schupp, *Combinatorial Group Theory*. Springer-Verlag (1977). [Chapters I, XII]

[Ly90]   I. G. Lysionok, *On some algorithmic properties of hyperbolic groups*. Math. USSR Izvestiya 35 (1) (1990), 145-163. [Chapter VI]

[Ly92]   I. G. Lysionok, *The infinity of Burnside groups of exponent 2^k for $k \geq 13$*. Uspekhi mat. nauk, **47** (1992), 201-212. [Chapter V]

[Ly93]   I. G. Lysionok, *Infinite Burnside groups of even exponent*. preprint, Steklov Institute, Moscow (1993). [Chapter V]

[LyPr]   I. G. Lysionok and S. J. Pride, *The structure of the second homotopy module of presentations of split extensions of groups.* in preparation. [Chapter V]

[MacL68]   S. MacLane, *Homology,* Academic Press, 1968. [Chapters II, IV]

[MaWh50]   S. MacLane and J. H. C. Whitehead, *On the 3-type of a complex.* Proc. Nat. Acad. Sci. **36** (1950), 41-48. [Chapter II]

[Ma30]   W. Magnus, *Über diskontinuierliche Gruppen mit einer definierenden Relation. (Der Freiheitssatz).* J. Reine Angew. Math. **163** (1930), 141-165. [Chapter V]

[Ma54]   A. A. Markov, *Theory of algorithms,* Trudy Mat. Inst. Steklov **42** (1955), Moscow–Leningrad. [Chapter VI]

[Ma71]   B. Maskit, *On Poincaré's Theorem for Fundamental Polygons.* Adv. in Math. **7** (1971), 219-230. [Chapter VIII]

[Ma67]   W. S. Massey, *Algebraic topology: An introduction.* Harcourt, Brace & World (1967). [Chapter I]

[Ma80]   W. S. Massey, *Singular Homology Theory.* Springer-Verlag, New York, 1980. [Chapter II]

[Ma73]   S. V. Matveev, *Special spines of piecewise linear manifolds.* Math. Sb. **92** (1973), n.134, 282-293. (Russian; English transl. in Math.USSR Sb. **21** (1973), 279-291. [Chapter XI]

[Ma87₁]   S. V. Matveev, *Transformations of special spines and the Zeeman Conjecture.* Izv. AN SSSR **51** (1987), n.5, 1104-1115. (Russian; English transl. in Math. USSR Izvestiya **31** (1987), 423-434. [Chapter XI]

[Ma87₂]   S. V. Matveev, *Zeeman Conjecture for unthickenable special polyhedra is equivalent to the Andrews-Curtis Conjecture.* Sibirskii Matematicheskii Zhurnal **28** (1987), n. 6 , 66-80. (Russian). [Chapters VIII, XI]

[Ma61]   B. Mazur, *A note on some contractible 4-manifolds.* Ann. of Math. **73** (1961), 221-228. [Chapter I]

[Me93]   H. Meinert, *A quantitative approach to bicombings.* preprint, Frankfurt/M (1993). [Chapter VI]

[Me83]   W. W. Menasco, *Polyhedra representation of link complements.* Contemporary Math. **20** (1983), 305-325. [Chapter VIII]

[Me67]   W. Metzler, *Beispiele zu Unterteilungsfragen bei CW-und Simplizialcomplexen.* ArCh. Math. **18** (1967), 513-519. [Chapter I]

[Me76]   W. Metzler, *Über den Homotopietyp zweidimensionaler CW-Komplexe und Elementartransformationen bei Darstellungen von Gruppen durch Erzeugende und definierende Relationen.* J. reine angew. Math. **285** (1976), 7-23. [Chapters I, XII]

[Me79₁]   W. Metzler, *Äquivalenzklassen von Gruppenbeschreibungen, Identitäten und einfacher Homotopietyp in niederen Dimensionen.* Lond. Math. Soc. Lecture Note Series **36** (1979), 291-326. [Chapters I, XII]

[Me79₂]   W. Metzler, *Two-dimensional complexes with torsion values not realizable by self-equivalences.* Lond. Math. Soc. Lecture Note Series **36** (1979), 327-337. [Chapters I, XII]

[Me85]   W. Metzler, *On the Andrews-Curtis conjecture and related problems.* Amer. Math. Soc.Contemp. Math. **44** (1985), 35-50. [Chapters I, VIII, XI, XII]

[Me90]   W. Metzler, *Die Unterscheidung von Homotopietyp und einfachem Homotopietyp bei zweidimensionalen Komplexen.* J. reine angew. Math. **403** (1990), 201-219. [Chapters I, XII]

[Mi92]   C. F. Miller III, *Decision problems for groups – survey and reflections* in: *Algorithms and classification in combinatorial group theory*, G. Baumslag and C.F. Miller III eds., Springer Verlag, Math. sci. res. inst. publ., 1992, 1, 3–59. [Chapter VI]

[MiSc79]   C. F. Miller and P. E. Schupp, *Letter to M.M.Cohen.* (Oct. 1979). [Chapter XII]

[Mi62]   J. Milnor, *A unique factorisation theorem for 3-manifolds.* Amer. J. Math. **84** (1962), 1-7. [Chapter VIII]

[Mi57]   J. Milnor, *Groups which act on Sⁿ without fixed points.* Amer. J. Math. **79** (1957), 623-630. [Chapter VIII]

[Mi71]   J. Milnor, *Introduction to Algebraic K-Theory.* Ann. of Math. Study 72, Princeton University Press (1971). [Chapter VII]

[MiSt74]   J. W. Milnor and J. D. Stasheff, *Characteristic classes.* Princeton University Press, Annals of Math. Studies No. **76** (1974). [Chapter IX]

[MiPrRe86] W. J. R. Mitchell, J. Przytycki and D. Repovš, *On spines of knot spaces.* Magdalen College, Cambridge U.K. et al. (1986), preprint. [Chapter I]

[Mo51]   E. Moise, *Affine structures in 3-manifolds. V. The triangulation theorem and Hauptvermutung.* Ann. Of Math. (2) **56** (1952), 96-114CH8]

[Mo77]   E. E. Moise, *Geometric topology of dimensions 2 and 3.* Springer-Verlag (1977). [Chapter I]

[Mo74]   J. M. Montesinos, *A representation of closed, orientable 3-manifolds as 3-fold branched coverings of S³.* Bull. Amer. Math. Soc. **80** (1974), 531-540. [Chapter VIII]

[Mo69]   M. S. Montgomery, *Left and right inverses in group algebras.* Bull. Amer. Math. Soc. **75** (1969), 539-540. [Chapters V, X]

[Mo90]   L. Mooney, *Massey Products in Groups*, (Thesis, University of Oregon, 1990). [Chapters V, X]

[MoSh93]   Y. Moriah and V. Shpilrain, *Non-tame automorphisms of extensions of periodic groups*, to appear in Israel J. Math. [Chapter VII]

[Mu60]   J. R. Munkres, *Obstructions to smoothing of piecewise differentiable homeomorphisms*. Ann. of Math. **72** (1960), 521-554. [Chapter I]

[Ne43]   B. H. Neumann, *Adjunction of elements to groups*. J. London Math. Soc. **18** (1943), 4-11. [Chapters V, X]

[Ne49]   B. H. Neumann, *On ordered division rings*. Trans. Amer. Math. Soc. **66** (1949), 202-252. [Chapters V, X]

[Ne68]   L. Neuwirth, *An algorithm for the construction of 3-manifolds from 2-complexes*. Proc.Camb. Phil. Soc. **64** (1968), 603-613. [Chapters I, VIII]

[Ne73]   J.P. Neuzil, *Embedding the dunce hat in S^4*. Topology **12** (1973), 411-415. [Chapter I]

[Ni19]   J. Nielsen, *Über die Isomorphismen unendlicher Gruppen ohne Relation*. Math. Ann. **79** (1919), 269-272. [Chapters I, XII]

[No55]   P. S. Novikov, *On the algorithmic unsolvability of the word problem in group theory*. Trudy Mat. Inst. Steklov **44** (1955), 143 (Russian). [Chapter VI]

[Ol91]   A. Y. Ol'shanskii, *The geometry of defining relations in groups*. Kluwer Academic Publishers (1991). [Chapters V, X]

[Ol65]   P. Olum, *Self-equivalences of pseudo-projective planes*. Topology **4** (1965), 109-127. [Chapter II]

[Os78]   R. P. Osborne, *The simplest closed 3-manifolds*. Pac. J. Math., **74** (1978), 481-495. [Chapter VIII]

[OsSt74]   R. P. Osborne and R.S. Stevens, *Group presentations corresponding to spines of 3-manifolds I*. American J. Math, **96** No. 3, (1977), 454-471. [Chapters I, VIII]

[OsSt77]   R. P. Osborne and R. S. Stevens, *Group presentations corresponding to spines of 3-manifolds II*. Trans. Amer. Math. Soc., **234** (1977), 213-243; III, Trans. Amer. Math. Soc. **234** (1977), 245-251. [Chapter VIII]

[Pa57]   C. D. Papakyriakopoulos, *On Dehn's lemma and the asphericity of knots*. Ann. of Math. **66** (1957), 1-26. [Chapters V, X]

[Pa63]   C. D. Papakyriakopoulos, *Attaching 2-dimensional cells to a complex*. Ann. Math. **78** (1963), 205-222. [Chapters II, IV, V, X]

[Pe49]   R. Peiffer, *Über Identitäten zwischen Relationen*. Math. Ann. **121** (1949), 76-99. [Chapters II, IV, V, X]

[Pe86]   N. Perron. *Pseudo-isotopies et isotopies en dimension quartre dans la categorie topologique*. Topology 25, 381-397, (1986). [Chapter IX]

[Pi88]   R. Piergallini, *Standard moves for standard polyhedra and spines*. Supplemento ai Rendiconti del Circolo Matematico di Palermo Serie 11 (1988), n. 18 , 391-414. [Chapter XI]

[Po04]   H. Poincaré, *Cinquieme complément à l'analysis situs*. Rend.Circ. Mat. Palermo, 18 (1904), 45-110. [Chapter VIII]

[Pr87]   S. J. Pride, *Groups with presentations in which each defining relator involves exactly two generators*. J. London Math. Soc. (2) 36 (1987), 245-256. [Chapters V, X]

[Pr88]   S. J. Pride, *Star-complexes, and the dependence problems for hyperbolic complexes*. Glasgow Math. J. 30 (1988), 155-170. [Chapters V, VI, X]

[Pr89]   S. J. Pride, *Involutary presentations, with applications to Coxeter groups, NEC-groups and groups of Kanevskiǐ*. J. Algebra 120 (1989), 200-223. [Chapters V, X]

[Pr91]   S. J. Pride, *Identities among relations of group presentations* in: *Group theory from a geometrical viewpoint, Trieste 1990* (E. Ghys, A. Haefliger, A Verjovsky, editors), World Scientific Publishing (1991), 687-717. [Chapter IV, V]

[Pr92$_1$]   S. J. Pride, *The (co)homology of groups given by presentations in which each defining realtor involves at most two types of generators*. J. Austral. Math. Soc. (Series A) 52 (1992), 205-218. [Chapter V]

[Pr92$_2$]   S. J. Pride, *Examples of presentations which are minimally Cockcroft in several different ways*. preprint, University of Glasgow. [Chapter IV, V]

[PrSt89]   S. J. Pride and R. Stöhr, *Relation Modules of groups with presentations in which each relator involves exactly two types of generators*. J. London Math. Soc. (2) 38 (1988), 99-111. [Chapter V]

[PrSt90]   S. J. Pride and R. Stöhr, *The (co)-homology of aspherical Coxeter groups*. J. London Math. Soc. (2) 42 (1990), 49-63. [Chapter V]

[Pu58]   D. Puppe, *Homotopiemengen und Ihre Induzieten Abbildungen I*. Math. Zeit. 69 (1958), 299-344. [Chapter III]

[Qu81]   F. Quinn, *Presentations and 2-complexes, fake surfaces and singular 3-manifolds*. Virginia Polytechnic Institute, Blackburg Va. (1981), preprint. [Chapters I, VIII, XI, XII]

[Qu83]   F. Quinn, *The stable topology of 4-manifolds*. Top. and its appl. 15 (1983), 71-77. [Chapter IX]

[Qu85]   F. Quinn, *Handlebodies and 2-complexes*. Springer Lecture Notes in Math. 1167 (1985), 245-259. [Chapters I, XII]

[Ra68$_1$]   E. S. Rapaport, *Remarks on groups of order 1*. Amer. Math. Monthly 75 (1968), 714-720. [Chapters I, XII]

[Ra68₂]   E. S. Rapaport, *Groups of order 1, some properties of presentations.* Acta
               Math. **121** (1968), 127-150. [Chapters I, XII]

[Ra80]   J. G. Ratcliffe, *Free and projective crossed modules.* J. London Math. Soc.
               22 (1980), 66-74. [Ch. IV, V]

[Ra83]   J. G. Ratcliffe, *Finiteness conditions for groups.* J. Pure Appl. Algebra **27**
               (1983), 173-185. [Chapter X]

[Re32]   K. Reidemeister, *Einführung in die kombinatorische Topologie.* Vieweg
               (1932). [Chapter I]

[Re33]   K. Reidemeister, *Zur dreidimensionalen Topologie.* Abh. Math. Sem. Univ.
               Hamburg **9** (1933), 189-194. [Chapter I]

[Re34]   K. Reidemeister, *Homotopiegruppen von Komplexen.* Abh. Math. Sem.
               Univ. Hamburg **10** (1934), 211-215. [Chapters II, V, X]

[Re35]   K. Reidemeister, *Homotopieringe and Linsenraüme,* Abh. Math. Sem.
               Univ. Hamburg **11** (1935), 102-109. [Chapters I, II]

[Re36]   K. Reidemeister, *Kommutative Fundamentalgruppen.* Monatsh. Math.
               Phys. **43** (1936), 20-28. [Chapter I]

[Re49]   K. Reidemeister, *Über Identitäten von Relationen.* Abh. Math. Sem Univ.
               Hamburg, **16** (1949), 114-118. [Chapters II, IV, XII]

[Re50]   K. Reidemeister, *Complexes and homotopy chains.* Bull. Amer. Math. Soc.
               56 (1950), 297-307. [Chapters V, X]

[Res61]   Y. G. Reshetnyak, *On a special kind of mapping of a cone onto a polyhedral
               disk.* Math. Sbornik, V. 53 (95), 39-52 1961, engl. translation: J. Stallings,
               UC Berkeley [Chapter VI]

[Ri82]   E. Rips, *Generalized small cancellation theory and applications. I, The
               word problem.* Isreal J. Math. **41** (1982), 1-146. [Chapters V, X]

[Rol76]   D. Rolfsen, *Knots and Links* (Publish or Perish, 1976). [Chapter V]

[Ro90]   S. Rosebrock, *A reduced spherical diagram into a ribbon-disk complement
               and related examples.* in: *Topology and Combinatorial Group Theory,* M.
               P. Latiolais, ed., Lecture Notes in Math. **1440** (Springer, 1990), 175-185.
               [Chapters V, X]

[Ro91]   S. Rosebrock, *On the realization of Wirtinger presentations as knot groups,*
               (Preprint, J. W. Goethe Universität-Frankfurt, 1991). [Chapters V, VI, X]

[Rot76]   O. S. Rothaus, *On the non-triviality of some group extensions given by gen-
               erators and relations.* Bull. Amer. Math. Soc. **54** (1976), 284-286. [Chap-
               ters I, V, X, 12]

[Rot77]   O. S. Rothaus, *On the non-triviality of some group extensions given
               by generators and relations.* Ann. Math. **106** (1977), 599-612. [Chap-
               ters I, V, X, XII]

[Ro73]   J. J. Rotman, *The theory of groups*. Allyn and Bacon, Boston (1973).
              [Chapter VI]

[Ro79]   C. P. Rourke, *Presentations of the trivial groups*. in: Topology of Low
              Dimensional Manifolds (R. Fenn, editor), Lecture Notes in Mathematics
              722 (Springer, 1979), 134-143. [Chapter V]

[RoSa72]   C. P. Rourke and B. J. Sanderson, *Introduction to piecewise-linear topol-
              ogy*. Springer Verlag (1972). [Chapter I]

[Sche73]   B. Schellenberg, *On the self-equivalences of a space with non-cyclic fun-
              damental group*. Math. Ann. **205** (1973), 333-344. [Chapter II]

[Schu64]   H. Schubert, *Topology*. Teubner (1964), English edition: Allyn and Bacon
              (1968). [Chapter I]

[Se33]   H. Seifert, *Topologie dreidimensionaler gefaserter Raum*, Acta Math. **60**
              (1933), 147-238, Translation by W. Heil, memo. notes, Florida State Uni-
              versity (1976). [Chapter VII]

[SeTh45]   H. Seifert and W. Threlfall, *Lehrbuch der Topologie*. Chelsea Publishing
              Company (1945). [Chapter VIII]

[Se73]   J. P. Serre, *A course in arithmetic*. Springer-Verlag, Berlin (1973). [Chap-
              ter IX]

[Se80]   J. P. Serre, *Trees*. Springer-Verlag, Berlin-Heidelberg-New York (1980).
              [Chapter IV, V]

[Sh74]   I. R. Shafarevic, *Basic algebraic geometry*. Springer-Verlag, Berlin (1974).
              [Chapter IX]

[Sh81]   H. Short, *Topological Methods in Group Theory: The Adjunction Problem*.
              (Thesis, University of Warwick, 1981). [Chapters V, X]

[Sh90]   H. Short, *Groups and combings*. preprint, ENS Lyon 1990. [Chapter VI]

[Sh91]   H. Short, ed, *Notes on word hyperbolic groups*. in: *Group theory from
              a geometrical viewpoint*, E. Ghys, A. Haefliger, A. Verjovsky, eds., 3-63,
              World Scientific Publ., Singapore (1991). [Chapter VI]

[Si76]   A. J. Sieradski, Combinatorial isomorphisms and combinatorial homotopy
              equivalences, J. Pure Appl. Alg. **7** (1976), 59-95. [Chapter II]

[Si77]   A. J. Sieradski, *A semigroup of simple homotopy types*. Math. Z. 153
              (1977), 135-148. [Chapters II, III, VII, XII]

[Si80]   A. J. Sieradski, *Framed links for Peiffer identities*. Math. Z. **175** (1980),
              125-137. [Chapters V, X]

[Si81]   A. J. Sieradski, *A coloring test for asphericity*. Quart. J. Math. Oxford
              (2)**34** (1983), 97-106. [Chapters V, X]

[Si84]   A. J. Sieradski, *A combinatorial interpretation of the third homology of a
              group*. J. Pure Appl. Algebra, **33** (1984), 81-96. [Chapter IV]

[Si86]   A. J. Sieradski, *Combinatorial squashings, 3-manifolds, and the third homology of groups.* Invent. Math. **84** (1986), 121-139. [Chapter VIII]

[Si92]   A. J. Sieradski, *An introduction to topology and homotopy.* PWS-Kent Publishing Company, Boston (1992). [Chapters I, II]

[SiDy79]   A. J. Sieradski and M. L. Dyer. *Distinguishing arithmetic for certain stably isomorphic modules.* J. Pure & Appl. Alg. **15** (1979), 199-217. [Chapter IX]

[Sil81]   J. R. Silvester, *Introduction of Algebraic K-Theory* Chapman and Hall Ltd. (1981). [Chapter III]

[Si33]   J. Singer, *Three dimensional manifolds and the Heegaard diagrams.* Trans. Amer. Math. Soc. 35 (1933), 88-111. [Chapter I]

[So73]   C. Soulé, *Groupes operant sur un complexe simplicial avec domaine fondamental.* C.R. Acad. Sci. Series A **276** (1973), 607-609. [Chapter V]

[St59]   J.R. Stallings, *Grushko's theorem II: Kneser's conjecture.* Notices of Amer. Math. Soc. No. 559–165, 531–532 (1959) [Chapter VIII]

[St62]   J.R. Stallings, *On the recursiveness of sets of presentations of 3-manifold groups.* Fund. Math. **51** (1962), 191-194. [Chapter I]

[St65$_1$]   J.R. Stallings, *A topological proof of Grushko's theorem on free products.* Math. Z. **90** (1965), 1-8. [Chapters I, XII]

[St65$_2$]   J.R. Stallings, *Whitehead Torsion of Free Products.* Ann. of Math. **82** (1965), 354-363 [Chapter VII]

[St73]   U. Stammbach, *Homology in group theory.* Springer-Verlag, New York (1973). [Chapter IV]

[St82]   P. Stefan, *On Peiffer transformations, link diagrams, and a question of J. H.C. Whitehead.* in: *Low-Dimensional Topology* (R. Brown and T. L. Thickstun, editors), London Math. Soc. Lecture Note Series **48** (1982), 203-213. [Chapters V, X]

[St75]   R. S. Stevens, *Classification of 3-manifolds with certain spines.* Trans. Amer. Math. Soc. **205** (1975), 151-166. [Chapter VIII]

[St80]   J. Stillwell, *Classical topology and combinatorial group theory.* Springer Verlag (1980). [Chapters I, VIII]

[Sti82]   J. Stillwell, *The word problem and the isomorphism problem for groups.* Bull. Amer. Math. Soc. (new series) **6** (1982), 33–56. [Chapter VI]

[St68]   B. Stong, *Notes on cobordism theory.* Princeton University Press (1968). [Chapter IX]

[St74]   R. Strebel, *Homological methods applied to the derived series of groups.* Comment. Math. Helv. **49** (1974), 302-322. [Chapter IV, V, X]

[Sw65]   R. G. Swan, *Minimal resolutions for finite groups.* Topology 4 (1965), 193-208 [Chapter VII]

[Sw70]   R. G. Swan, *K-Theory of Finite Groups and Orders*. Springer Lecture Notes in Math **149** (1970). [ChapterIII]

[Sw73]   G. A. Swarup, *On embedded spheres in 3-manifolds*. Math. Ann. **203** (1973), 89-102. [Chapter VIII]

[Sw74]   G. A. Swarup, *On a theorem of C. B. Thomas*. J. London Math. Soc. (2) **8** (1974), 13-21. [Chapter VIII]

[Ta91]   M. Takahashi, *Framed-link representations of 3-manifolds*. Tsukuba J. Math., **15** (1991), 79-83. [Chapter VIII]

[Th54]   R. Thom, *Quelques propriétés des varietés différentiables*. Comment. Math. Helv. **28** (1954), 17-86. [Chapter IX]

[Th67]   C. B. Thomas, *The oriented homotopy type of compact 3-manifolds*. Proc. London Math. Soc., **19** (1967), 31-44. [Chapter VIII]

[Th78]   C. B. Thomas, *Free actions by finite groups on S^3*. Proc. Sympos. Pure Math. No. 32, Part I (American Mathathematical Society, Providence, R.I.) (1978), 125-130. [Chapter VIII]

[Th79]   C. B. Thomas, *On 3-manifolds with finite solvable fundamental group*. Invent. Math. **52** (1979), 187-197. [Chapter VIII]

[Ti08]   H. Tietze, *Über die topologischen Invarianten mehrdimensionaler Mannigfaltigkeiten*. Monatsh. Math. Phys. **19** (1908), 1-118. [Chapters I, II]

[Tu88]   V. G. Turaev, *Towards the topological classification of geometric 3-manifolds*. Topology and geometry—Rohlin Seminar, Lecture Notes in Math. **1346** (1988), 291-323. [Chapter VIII]

[Tur37]   A. M. Turing, *On computable numbers with an application to the Entscheidungsproblem*. Proc. Lond. Math. Soc. (2) 42, 230–265, 1936, A correction, Proc. Lond. Math. Soc. (2) **43** (1937), 544–546. [Chapter VI]

[Wa80]   J. Wagoner, *A picture description of the boundary map in algebraic K-theory*. Lect. Notes in Math. **966** (1980), 362-389. [Chapter V]

[Wa68]   F. Waldhausen, *On irreducible 3-manifolds which are sufficiently large*. Ann. of Math. **87** (1968), 56-88. [Chapter VIII]

[Wa78]   F. Waldhausen, *Algebraic K-theory of generalized free products I, II*. Ann. of Math. **108** (1978), 135-256. [Chapter VII]

[Wa65]   C. T. C. Wall, *Finiteness conditions for CW-complexes*. Ann. of Math. **81** (1965), 56-69. [Chapter III]

[Wa66]   C. T. C. Wall, *Formal deformations*. Proc. London Math. Soc. **16** (1966), 342-354. [Chapters I, XI, XII]

[Wa70]   C. T. C. Wall, *Surgery on Compact Manifolds*. Academic Press, New York (1970). [Chapter IX]

[Wa76]   C. T. C. Wall, *Classification of hermitian forms VI. Group rings*. Annals of Math. **103** (1975), 1-80. [Chapter IX]

[Wa79]   C. T. C. Wall, *List of problems*. Lond. Math. Soc. Lecture Note Series **36** (1979), 369-394. [Chapters I, XII]

[Wa80]   C. T. C. Wall, *Relatively 1-dimensional complexes*. Math. Z. **172** (1980), 77-79. [Chapters I, XII]

[Wa60]   A. H. Wallace, *Modifications and cobounding manifolds*. Canad. J. Math. **12** (1960), 503-528. [Chapter VIII]

[We67]   C. Weber, *Plongements de polyèdra dans le domaine métastable*. Comm. Math. Helv. **42** (1967), 1-27. [Chapter I]

[WeSe33]   C. Weber and W. Seifert, *Die beiden Dodekäderräume*. Math. Z. **37** (1933), 237-253. [Chapter VIII]

[We73]   D. E. Webster, *Collapsing $K \times I$*. Proc.Cambridge Phil. Soc. **74** (1973), 39 - 42. [Chapter XI]

[WeWa78]   D. E. Webster, L. W. Wajda, *On Zeeman's Conjecture. The collapsing of $K(2, q, r, s) \times I$*. Manuscript, 1978. [Chapter XI]

[We71]   J. E. West, *Mapping cylinders of Hilber cube factors*. General Top. and its Appl. **1** (1971), 111-125. [Chapter I]

[Wh36]   J. H. C. Whitehead, *On certain sets of elements in a free group*. Proc. London Math.Soc. (2) **41** (1936), 48-56. [Chapter XI]

[Wh39]   J. H. C. Whitehead, *Simplicial spaces, nuclei and m-groups*. Proc. London Math. Soc. **45** (1939), 243-327. [Chapters I, Chs. II, X]

[Wh41$_1$]   J. H. C. Whitehead, *On adding relations to homotopy groups*. Ann. of Math. **42** (1941), 409-428. [Chapters I, IV, X]

[Wh41$_2$]   J. H. C. Whitehead, *On incidence matrices, nuclei and homotopy types*. Ann. of Math. **42** (1941), 1197-1239. [Chapters I, VIII, XII]

[Wh46]   J. H. C. Whitehead, *Note on a previous paper entitled "On adding relations to homotopy groups."* Ann. Math. **47** (1946), 806-810. [Chapters V, X]

[Wh48]   J. H. C. Whitehead, *On the homotopy type of ANR's*. Bull. Amer. Math. Soc. **54** (1948), 1133-1145. [Chapter II]

[Wh49$_1$]   J. H. C. Whitehead, *Combinatorial homotopy I*. Bull. Amer. Math. Soc. **55** (1949), 213-245. [Chapters I, II]

[Wh49$_2$]   J. H. C. Whitehead, *Combinatorial homotopy II*. Bull. Amer. Math. Soc. **55** (1949), 453-496. [Chapters II, X]

[Wh50]   J. H. C. Whitehead, *Simple homotopy types*. Amer. J. Math. **72** (1950), 1-57. [Chapters I, XII]

[Wi90]   D. M. Wilczyńzski, *On the topological rigidity of pseudo-free group actions on 4-manifolds I.* preprint (1990). [Chapter IX]

[Wo91]   A. R. Wolf, *Inherited asphericity, links and identities among relations.* J. Pure Appl. Alg. **71** (1991), 99-107. [Chapters V, X]

[Wr71]   P. Wright, *Collapsing $K \times I$ to a vertical segments.* Proc.Camb. Phil. Soc. **69** (1971), 71-74. [Chapter XI]

[Wr73]   P. Wright, *Formal 3-deformations of 2-polyhedra.* Proc. Amer. Math. Soc. **37** (1973), 305-308. [Chapters I, XI, XII]

[Wr75]   P. Wright, *Group presentations and formal deformations.* Trans. Amer. Math. Soc. **208** (1975), 161-169. [Chapters I, XI, XII]

[Wr77]   P. Wright, *Covering 2-dimensional polyhedra by 3-manifold spines.* Topology **16** (1977), 435-439. [Chapters I, XII]

[Yo76]   S. F. Young, *Contractible 2-complexes.* Christ's College, University of Cambridge U.K., preprint (1976). [Chapters I, XII]

[Ze63-66]   E. C. Zeeman, *Seminar on combinatorial topology.* I.H.E.S.-Notes, Paris and: University of Warwick, Coventry U.K. (1963-1966). [Chapter I]

[Ze64$_1$]   E. C. Zeeman, *On the dunce hat.* Topology **2** (1964), 341-358. [Chapters I, XI]

[Ze64$_2$]   E. C. Zeeman, *Relative simplicial approximations.* Proc.Camb. Phil. Soc. **60** (1964), 39-42. [Chapter II]

[Zi81]   H. Zieschang, *Finite groups of mapping classes of surfaces.* Lecture Notes in Mathematics, **875**, Springer-Verlag, Berlin-New York, 1981. [Chapter VIII]

[Zi88]   H. Zieschang, *On Heegaard diagrams of 3-manifolds.* Asterisque No. **163/164** (1988), 247-280. [Chapters VII, VIII]

[Zi78]   A. Zimmermann, *Eine spezielle Klasse kollabierbarer Komplexe $K^2 \times I$.* Thesis, Frankfurt/Main (1978). [Chapter XI]

Index

α-homotopy, 62
$\Sigma_d(K)$, 101
1-collapsibility modulo 2-expansions, 348
1-collapsible, 335
2-bridge knot, 243
2-expansions, 348
3–deformation, 376
3-deformation type, 22, 28, 122
$C(\tilde{K})$, equivariant chain complex, 80
$C(\mathcal{P})$-chain complex, 84
$\mathcal{CM}$, $\mathcal{CM}_G$, $\mathcal{PCM}$, $\mathcal{PCM}_G$, 127

Abalone, 336
absolutely Cockcroft, 155
action of the cohomology group, 95
acyclic complex, 322, 326
algebraic 2-type, 92
algebraic normal 1-type, 287
algorithm, 190
Andrews-Curtis conjecture, 21, 45, 47, 319
articulation point, 347
aspherical, 48, 90, 167, 273, 309
aspherical presentation, 148
asphericity, combinatorial (CA), 167, 174
asphericity, diagrammatic (DA), 171, 332
attaching map, 2, 11, 13

balanced presentation, 8, 259
barycentric subdivision, 52
based map, 62
based space, 62
basic Peiffer element, 67
basis-up-to-conjugation, 343
bias, 97, 104, 246

bicombing, 199
Bing's house, 19, 336
bordism group, 288
bored manifold, 280
boundary of a cell, 2
boundary point, 274
bounded combing, 199
braid, 240
branched covering, 252
branched surface, 352
branched surface structure, 353
Browning obstruction, 113
Burau representation, 240
Burnside group, 182

"can't start"-argument, 18, 20, 338
category, $\mathcal{CM}_G$, $\mathcal{PCM}$, $\mathcal{PCM}_G$, 127
category, crossed modules $\mathcal{CM}$, 127
cell complex, 1
cellular chain complex, 76
cellular homeomorphism, 2
cellular homology group, 77
centralizer, 332
chain complex $C(\mathcal{P})$, 84
chain complex, cellular, 76
change of ring procedure, 80
characteristic map, 11, 25
Chern manifold, 300
Cockcroft, 148, 154, 312, 315
Cockcroft, absolutely, 155
Cockcroft threshold, 155
cohomological dimension, 329
cohomology of a group, 89
collapse, 12, 31
collapsible, 17, 20, 43, 280, 338, 361
collapsing by adding a cell, 339
combinatorial approximation, 57
combinatorial complex, 9, 10, 57

combinatorial equivalence, 5, 6
combinatorial map, 57
combinatorial map (strong sense), 203
combinatorially aspherical (CA), 167, 174
combinatorially reducible (CR), 170
combing, 199
comparison theorem, 88
complement of a picture, 161
conjugacy problem, 189
conservative group, 322
conservative radical, 323
contractible, 8, 17, 47
Coxeter group, 227
crossed commutator, 129
crossed extension, 132
crossed module, 66, 125
crossed module homomorphism, 66, 127
crossed module, coproduct, 140
crossed module, direct sum, 143
crossed module, free, 71, 127
crossed module, projective, 132
curtain system, 61
curvature, 158, 188
CW-complex, 2
cycle, 203
cycletest, 206

decidable, 190
deficiency (=directed deficiency), 28, 152, 227, 374
defining relation, 7, 60
deformation, 12, 45
Dehn algorithm, 172, 191
Dehn function, 195
diagrammatically aspherical (DA), 171, 332
diagrammatically reducible (DR), 171
dipole, 169
direct sum formula, 232
direct sum map, 143
directed deficiency, see: deficiency
dodecahedral space, hyperbolic, 255
dodecahedral space, spherical, 255
dunce hat, 18, 40, 258, 336

efficient, 228, 229
efficient group, 152
efficient normal factorization, 317
efficient presentation, 152
Eichler's condition, 111
elementary automorphism, 283
elementary collapse, 12
elementary expansion, 12
equations over groups, 171, 321, 334
equivariant bicombing, 199
equivariant cellular chain complex, 80
equivariant map, 80
Euler characteristic, 8, 28, 222
expanded presentation, 73, 87
expansion, 12

fake surface, 10, 350
fibered cell, 256
Fox ideal, 149, 220, 222
Fox-Reidemeister derivative, 85
framed link, 61, 68, 330
framing curve, 41
free crossed module, 71, 127
free differential calculus, 85
free face, 12, 23
free pre-crossed module, 128
free product, 232, 235, 248
Fuchsian group, 237
fundamental class ζ_M, 268
fundamental group, 7, 40
fundamental sequence for K, 63
fundamental sequence for $\mathcal{P}$, 68

(G, d)-complex, 99, 149
G-crossed module, 66, 126
G-morphism, 127
generalized Andrews-Curtis conjecture, 45
generalized dunce hat, 43
generalized prolongation, 22
generalized Zeeman conjecture, 48
generator of a presentation, 7, 60
geodesic path, 197
geometric 3–manifold, 273
geometric normal 1-type, 287
geometrically split, 332

graph of groups, 179
group of identities, 67, 127, 368
group presentation, 7, 60, 262
group, perfect, 136
group, Rosset, 149
group, superperfect, 134, 136

Haken manifold, 273
handle decomposition, 257
handlebody, 30, 32, 41
Hauptvermutung, see: combinatorial
        equivalence
Heegaard splitting, 238, 252, 262
Heegaard-diagram, 32
highway system, 59
homological dimension, 229
homology 3-ball, 337
homology equivalence, 105
homology of a group, 89
homomorphism, crossed module, 127
homotopy action, 62
homotopy equivalence test, 246
homotopy module, 157
homotopy type, 28, 122, 375
Hopf's formula, 91
house with one room, 336
Hurewicz's Theorems, 77
hyperbolic dodecahedral space, 255

identifiable cells, 256
identity, 67
identity property, 139
identity property, left, 148
identity property, right, 148
incidence number, 76
inner point, 274
interval presentation, 209
intrinsic skeleta, 10
isodiametric function, 197
isoperimetric function, 195
isoperimetric inequality, 196

k-invariant, 92
knot, 320
knot space, 35

labeled oriented graph, 319, 333

L-Cockcroft complex, 148, 149
left identity property, 148
lens space, 35
L-identity property, 148
link graph (= Whitehead graph), 33,
        170, 203
localization, 111, 122
locally finite group, 321
locally indicable, 333, 334
locally indicable group, 322
locally nonperfect, 326, 334
locally residually finite, 321, 334
lower central series, 317, 319

Mac Lane-Whitehead theory, 91
manifold of dimension five, 43
manifold of dimension four, 43, 281
manifold of dimension three, 31, 40
map bias, 103, 104
maximal tower lifting, 327
Mazur-manifold, 42, 44
minimal subgroup, 154
model of a presentation, 61; see: stan-
        dard complex
Montesinos knot, 243
morphism, crossed module, 66
moves of special polyhedra, 274, 350

n-collapsible polyhedron, 361
n-generating, 179
Nielsen equivalence, 233, 234
Non-Euclidean crystallographic group,
        226
normal 1-smoothing, 288
normal 1-type, see: geometric normal
        1-type
normal form in a group, 191
normal orientation, 355
$\mathcal{N}$-torsion, 230, 231, 234, 246
null-H_2 threshold, 156
null-homotopy, 70, 330

one-relator product, 185
ordinary cycle, 210
orientation of a fake surface, 351

oriented branched surface structure, 353

paired identity, 261
partial homotopy equivalence, 118
patterns of contact, 253
Peiffer central series, 372
Peiffer element, 67, 127, 128, 368, 372
Peiffer group, 67, 127, 368
Peiffer identity, 371
pentagulation, 253
perfect group, 136, 324, 334
perfect radical, 325
picture, 158, 160
piecewise Euclidean, 212
pin-cushion, 258
PLCW-complex, 8, 14, 15, 24, 343
Poincaré conjecture, 45, 47
pre-crossed module, 66, 127, 371
pre-crossed module, free, 128
presentation, 7, 60, 262
presentation class, see: Q**-class
presentation morphism, 368, 369
presentation, efficient, 152
prime 3–manifold, 272
prime factorization, 272
prismatically (1-) collapsible, 343
product cell, 57, 256
projective crossed module, 132, 134
prolongation, 21
proper power in G, 312
pseudo-projective plane, 85

Q**-class, 8, 21, 22, 26, 29, 39, 372
Q**-transformation, 21
Q**-trivialization, 365
Q*-transformation, 21
Q-equivalence, 369
Q-transformation, 21
quadratic A-module, 291
quasi isometric, 195

railroad-system (RR-system) , 263, 266
rank, 225
rank test, 226
rank, $(A, \mathbb{Z})$–free, 282

rank, $(A, \mathbb{Z})$-hyperbolic, 291
recurrent arc, 267
recursive, 190
recursive presentation, 190
reducible 2-complex, 313
regular neighbourhood, 30, 43
Reidemeister complex, see: combinatorial complex
Reidemeister-Fox derivative, 85
relation module, 50
relative barycentric subdivision, 53
relative generalized Andrews-Curtis conjecture, 46
relator of a presentation, 7, 60
residually nilpotent, 319, 333
resolution of module, 88
ribbon disc, 320, 333
right identity property, 148
Rosset group, 149
RR-system, 263, 266

second homotopy module, 157
Seifert fibre space, 239
semisplit presentation, 373
semisplit Q**-transformation, 377
sides of a hyperspace, 55
sign of an ordered simplex, 55
simple-homotopy, 246, 248, 249
simple-homotopy equivalence test, 248
simple-homotopy extension, 14
simple-homotopy type, 12, 16, 17, 28, 122, 370, 376
simplicial approximation, 52
simplicial approximation theorem, 53
simplicial curtain, 56
simplicial linkage, 54
singular 3-manifold, 39, 274, 372
singular point, 274
skeleton, 1
space form, 272
spanning tree, 7
special polyhedron, 10, 20, 36, 38, 274, 350
special spine, 38, 274
spherical dodecahedral space, 255

spherical modification, 82
spine, 12, 17, 47, 253
split extension, 181
squashable complex, 256
squashing, 258, 260
squashing map, 256
stable range, 281
stably diffeomorphic, 287
stably homeomorphic, 287
staggered 2-complex, 314
Stallings' Splitting Theorem, 270
standard complex, 9, 61
standard polyhedron, see: special poly-
       hedron
star covering, 52
star graph, 170, 203
subdivision, 5, 8
superperfect group, 134, 136, 324
surgery, 252
syllable-decomposition, 264
syllable-reduced, 267
symmetrized presentation, 191

T-equivalent special polyhedra, 364
Tarski monster, 182
taut identity, 259
test for homotopy equivalence, 246
test for simple-homotopy equivalence,
       248
thickenable, 358
thickening, 31, 257, 286
Tietze transformation, 28
torsion of a map or a matrix, 112
tower, 315, 326
transient move, 14, 23, 367
transvection, 291
triangulation, 252
triple torus, 254
trivial presentation, 29
twisted presentation, 108
type FL, 329

unimodular element, 283
universal link, 252
unthickenable, 358

vacuum sealing, 257
van Kampen diagram, 193
vertex star, 52
vertical resolution, 343

w_2-type, 295
weak identity property, 152
weak topology, 2, 3
weight function, 203
weight of a subgroup, 149
weight test, 170, 205
Whitehead asphericity conjecture, 48,
       309
Whitehead graph (= link graph), 33,
       170, 203
Whitehead group, 16, 49, 230
Whitehead's Theorem, 74
Wirtinger presentation, 209
word hyperbolic, 197
word metric, 193
word problem, 189

$\mathbb{Z}$-covering, 327
Zeeman's conjecture, 47, 335